THERMOPHOTOVOLTAIC GENERATION OF ELECTRICITY

THERMOPHOTOVOLTAIC GENERATION OF ELECTRICITY

Fourth NREL Conference

Denver, Colorado   October 1998

EDITORS
Timothy J. Coutts
John P. Benner
Carole S. Allman
National Renewable Energy Laboratory

American Institute of Physics

AIP CONFERENCE PROCEEDINGS 460

Woodbury, New York

Editors:

Timothy J. Coutts
National Renewable Energy Laboratory
1617 Cole Boulevard
Golden, CO 80401-3393
U.S.A.

Email: tim_coutts@nrel.gov

John P. Benner
National Renewable Energy Laboratory
1617 Cole Boulevard
Golden, CO 80401-3393
U.S.A.

Email: john_benner@nrel.gov

Carole S. Allman
National Renewable Energy Laboratory
1617 Cole Boulevard
Golden, CO 80401-3393
U.S.A.

Email: carole_allman@nrel.gov

The Article on pp. 30–35 was authored by U. S. Government employees and is not covered by the below mentioned copyright.

Authorization to photocopy items for internal or personal use, beyond the free copying permitted under the 1978 U.S. Copyright Law (see statement below), is granted by the American Institute of Physics for users registered with the Copyright Clearance Center (CCC) Transactional Reporting Service, provided that the base fee of $15.00 per copy is paid directly to CCC, 222 Rosewood Drive, Danvers, MA 01923. For those organizations that have been granted a photocopy license by CCC, a separate system of payment has been arranged. The fee code for users of the Transactional Reporting Service is: 1-56396-828-2/ 99 /$15.00.

© 1999 American Institute of Physics

Individual readers of this volume and nonprofit libraries, acting for them, are permitted to make fair use of the material in it, such as copying an article for use in teaching or research. Permission is granted to quote from this volume in scientific work with the customary acknowledgment of the source. To reprint a figure, table, or other excerpt requires the consent of one of the original authors and notification to AIP. Republication or systematic or multiple reproduction of any material in this volume is permitted only under license from AIP. Address inquiries to Office of Rights and Permissions, 500 Sunnyside Boulevard, Woodbury, NY 11797-2999; phone: 516-576-2268; fax: 516-576-2499; e-mail: rights@aip.org.

L.C. Catalog Card No. 99–60390
ISBN 1-56396-828-2
ISSN 0094-243X
DOE CONF- 981055

Printed in the United States of America

CONTENTS

Preface xi

OVERVIEWS

Session A 3
    T. J. Coutts
Session 1: Fundamental Aspects 4
    J. P. Benner
Session 2: Monolithic Interconnected Modules (MIMS) 6
    D. R. Riley
Session 3: Radiators and Optical Control 9
    J. A. Mazer
Session 4: GaSb-Related Materials and Devices 10
    G. W. Charache
Session 5: Characterization 12
    J. Luther
Session 6: Systems Design and Experience 14
    D. Krommenhoek

SESSION A: PROGRAMMATIC

Thermophotovoltaic Potential Applications for Civilian and Industrial
Use in Japan 17
    H. Yamaguchi and M. Yamaguchi
Army Thermophotovoltaic Efforts 30
    J. S. Kruger, G. Guazzoni, and S. J. Nawrocki

SESSION 1: FUNDAMENTAL ASPECTS

The Optical Constants of n- and p-Doped $In_{0.66}Ga_{0.34}As$ on InP (001)
Including the Burstein–Moss Shift: Experiment and Modeling 39
    T. Holden, F. H. Pollak, J. L. Freeouf, G. W. Charache, and J. E. Raynolds
Enhanced Electro-Magnetic Energy Transfer Between a Hot and Cold
Body at Close Spacing due to Evanescent Fields 49
    J. E. Raynolds
Theoretical Maximum Efficiencies for Thermophotovoltaic Devices 58
    G. D. Cody
A Numerical Semiconductor Device Model for TPV Cells 68
    J. L. Gray and A. M. El-Husseini
Theoretical Determinations of Semiconductor Band Structures
and Optical Properties, and Applications to TPV Devices 74
    C. B. Geller, T. S. Blazeck, W. Wolf, and W. Mannstadt

Optimisation of InGaAsP Quantum Well Cells for Hybrid
Solar-Thermophotovoltaic Applications 83
    C. Rohr, J. P. Connolly, K. W. J. Barnham, I. Ballard, P. R. Griffin,
    J. Nelson, C. Button, and J. Clark

A TPV System Model and an Analysis of TPV System Efficiency 93
    J. L. Gray and A. M. El-Husseini

Efficiency and Power Density Potential of Thermophotovoltaic
Systems Using Low Bandgap Photovoltaic Cells 103
    A. Heinzel, J. Luther, G. Stollwerck, and M. Zenker

SESSION 2: MIMS

Square Cones for TPV; Experiments and Computer Simulations 115
    L. Broman, K. Jarefors, E. Lindberg, and J. Marks

High-Performance, Lattice-Mismatched InGaAs/InP Monolithic
Interconnected Modules (MIMs) 121
    N. S. Fatemi, D. M. Wilt, R. W. Hoffman, Jr., M. A. Stan, V. G. Weizer,
    P. P. Jenkins, O. S. Khan, C. S. Murray, D. Scheiman, and D. Brinker

High-Performance, 0.6-eV, $Ga_{0.32}In_{0.68}As/InAs_{0.32}P_{0.68}$ Thermophotovoltaic
Converters and Monolithically Interconnected Modules 132
    M. W. Wanlass, J. J. Carapella, A. Duda, K. Emery, L. Gedvilas,
    T. Moriarty, S. Ward, J. D. Webb, and X. Wu

Growth and Properties of InGaAs/FeAl/InAlAs/InP Heterostructures
for Buried Reflector/Interconnect Applications in InGaAs
Thermophotovoltaic Devices 142
    S. A. Ringel, R. N. Sacks, L. Qin, M. B. Clevenger, and C. S. Murray

n/p/n Tunnel Junction InGaAs Monolithic Interconnected Module (MIM) 152
    D. M. Wilt, C. S. Murray, N. S. Fatemi, and V. Weizer

Monte Carlo Analysis of a Monolithic Interconnected Module
with a Back Surface Reflector 161
    C. T. Ballinger, G. W. Charache, and C. S. Murray

SESSION 3: RADIATORS AND OPTICAL CONTROL

High Temperature Optical Properties of Thermophotovoltaic
Emitter Components 177
    D. E. Pierce and G. Guazzoni

Microstructured Tungsten Surfaces as Selective Emitters 191
    A. Heinzel, V. Boerner, A. Gombert, V. Wittwer, and J. Luther

Thermal Spray Approach for TPV Emitters 197
    C. J. Crowley, N. A. Elkouh, and P. J. Magari

Temperature Gradient Effects in an Erbium Aluminum Garnet
Selective Emitter 214
    B. S. Good, D. L. Chubb, and A. M. Pal

SESSION 4: GaSb-RELATED MATERIALS AND DEVICES

Bulk Crystal Growth of Antimonide Based III-V Compounds for TPV Applications 227
    P. S. Dutta, A. G. Ostrogorsky, and R. J. Gutmann

Pseudo-Closed Box Diffusion of Zn into InGaAsSb and AlGaSb for TPV Devices 237
    A. W. Bett, B. Y. Ber, M. G. Mauk, J. T. South, and O. V. Sulima

Interfacial Recombination in In(Al)GaAsSb/GaSb Thermophotovoltaic Cells 247
    V. B. Khalfin, D. Z. Garbuzov, H. Lee, G. C. Taylor, N. Morris, R. U. Martinelli, and J. C. Connolly

Extending the Cutoff Wavelength of Lattice-Matched GaInAsSb/GaSb Thermophotovoltaic Devices 256
    C. A. Wang, H. K. Choi, D. C. Oakley, and G. W. Charache

SESSION 5: CHARACTERIZATION

FTIR and FT-PL Spectroscopic Analysis of TPV Materials and Devices 269
    J. D. Webb, L. M. Gedvilas, M. R. Olson, X. Wu, A. Duda, M. W. Wanlass, and K. M. Jones

Recombination Lifetime of $In_xGa_{1-x}As$ Alloys Used in Thermophotovoltaic Converters 282
    R. K. Ahrenkiel, R. Ellingson, S. Johnston, J. Webb, J. Carapella, and M. W. Wanlass

RF Photoreflectance Characterization of Binary and Quasi-Binary Substrates and Antimonide-Based TPV Devices 289
    S. Saroop, J. M. Borrego, R. J. Gutmann, P. S. Dutta, A. G. Ostrogorsky, and G. W. Charache

Thermophotovoltaic Cell Temperature Measurement Issues 301
    T. Moriarty and K. Emery

A Single TPV Cell Power Density and Efficiency Measurement Technique 312
    L. Fraas, M. Groeneveld, G. Magendanz, and P. Custard

Measurement Techniques for Single Junction Thermophotovoltaic Cells 317
    L. R. Danielson, J. R. Parrington, G. W. Charache, G. J. Nichols, and D. M. Depoy

Optical Properties of Thin Semiconductor Device Structures with Reflective Back-Surface Layers 327
    M. B. Clevenger, C. S. Murray, S. A. Ringel, R. N. Sacks, L. Qin, G. W. Charache, and D. M. Depoy

Lessons Learned on Closed Cavity TPV System Efficiency Measurements 335
    C. K. Gethers, C. T. Ballinger, and D. M. Depoy

SESSION 6: SYSTEMS DESIGN AND EXPERIENCE

Design of a Thermophotovoltaic Battery Substitute 351
    E. F. Doyle, F. E. Becker, K. C. Shukla, and L. M. Fraas

Component Development for 500 Watt Diesel Fueled Portable Thermophotovoltaic (TPV) Power Supply 362
    C. L. DeBellis, M. V. Scotto, L. M. Fraas, J. Samaras, R. C. Watson, and S. W. Scoles

Development Status on a TPV Cylinder for Combined Heat and Electric Power for the Home 371
    L. Fraas, J. Samaras, H.-X. Huang, M. Seal, and E. West

Portable TPV Generator Based on Metallic Emitter and 1.5-Amp GaSb Cells 384
    V. C. Rumyantsev, V. P. Khvostikov, S. V. Sorokina, V. I. Vasil'ev, and V. M. Andreev

Operating Experience of a Portable Thermophotovoltaic Power Supply 394
    F. E. Becker, E. F. Doyle, and K. C. Shukla

Interfacing a Small Thermophotovoltaic Generator to the Grid 403
    W. Durisch, B. Grob, J.-C. Mayor, J.-C. Panitz, and A. Rosselet

POSTER SESSIONS
CHARACTERIZATION

Performance Status of 0.55eV InGaAs Thermophotovoltaic Cells 417
    S. Wojtczuk, P. Colter, G. Charache, and D. DePoy

Fabrication and Electrical Characterization of 0.55eV N-on-P InGaAs TPV Devices 427
    W. Nishikawa, D. Joslin, D. Krut, J. Eldredge, A. Narayanan, M. Takahashi, M. Haddad, M. M. Al-Jassim, and N. H. Karam

Effect of Elevated Temperatures on the Performance of an InP Cell Illuminated by a Selective Emitter 438
    Z. Chen and H. W. Brandhorst, Jr.

Integrated Development and Testing of Multi-Kilowatt TPV Generator Systems 446
    E. M. West and W. R. Connelly

FUNDAMENTAL ASPECTS

Theoretical Prediction of the Plasma Frequency and Moss–Burstein Shifts for Degenerately Doped InAs, $In_{1-x}Ga_xAs$ and $InP_{1-y}As_y$ 457
    J. E. Raynolds, C. B. Geller, G. W. Charache, T. Holden, F. H. Pollak, W. Maanstadt, R. Asahi, and A. J. Freeman

RADIATORS AND OPTICAL CONTROL

Emittance Theory for Cylindrical Fiber Selective Emitter 463
    D. L. Chubb

Fibrous Selective Emitter Structures from Sol-Gel Process.................. 472
    K. C. Chen

SYSTEMS

Commercial GaSb Cell and Circuit Development for the Midnight Sun® TPV Stove 480
    L. Fraas, R. Ballantyne, S. Hui, S.-Z. Ye, S. Gregory, J. Keyes, J. Avery, D. Lamson, and B. Daniels

Use of a Thermophotovoltaic Generator in a Hybrid Electric Vehicle 488
    O. Morrison, M. Seal, E. M. West, and W. Connelly

Cylindrical TPV Array Characterization 497
    W. R. Connelly and E. M. West

High Power Density AEM Combustion for TPV Applications 505
    A. S. Kushch and S. M. Skinner

MIMs

A Study of Contacts and Back-Surface Reflectors for 0.6-eV $Ga_{0.32}In_{0.68}As/InAs_{0.32}P_{0.68}$ Thermophotovoltaic Monolithically Interconnected Modules... 517
    X. Wu, A. Duda, J. J. Carapella, J. S. Ward, J. D. Webb, and M. W. Wanlass

GaSb-RELATED MATERIALS AND DEVICES

p-GaSb/n-GaAs Heterojunctions for Thermophotovoltaic Cells Grown by MOVPE 525
    L. Zheng, G. M. Sweileh, S. K. Haywood, C. G. Scott, M. Lakrimi, N. J. Mason, and P. J. Walker

Phase Stability in $Ga_xIn_{(1-x)}As_ySb_{(1-y)}/GaSb$ Heterostructures............... 535
    Y-C. Chen, V. Bucklen, M. Freeman, R. P. Cardines, Jr., G. Nichols, P. Sanders, G. Charache, and K. Rajan

$Al_xGa_{1-x}Sb$ Window Layers for InGaAsSb/GaSb Thermophotovoltaic Cells 545
    J. T. South, Z. A. Shellenbarger, M. G. Mauk, J. A. Cox, P. E. Sims, R. A. Mueller, and J. D. Meakin

Author Index... 555

PREFACE

This was the fourth of the NREL conferences devoted to thermophotovoltaic generation of electricity. Once again, it was held in Colorado, this time at the Adams Mark Hotel in Denver, during the period October 11-14, 1998. As the Chairs were planning the conference, they once again examined their motivations for organizing it and came to the conclusion that these remained very much the same as they were at the time of the first conference in 1994: these being to provide an opportunity for members of this small community to interact with one another: to familiarize themselves with the latest developments in the field: to archive the state of development of the technology at this time: and, hopefully, to attract interest from federal funding organizations in a technology that appears to have much to offer both the military and non-military sectors of society.

Attendance at this Conference has remained almost constant at approximately 120 but, this time, the number of submitted abstracts increased to approximately 60, this representing a 50% increase over previous years. The reasons for this are not entirely clear, but the implication is that the community, which has changed in its membership since TPV1, is perhaps working 50% harder! In any event, to accommodate the additional submissions, it was necessary to include a poster session of approximately 20 posters at TPV4. This had the benefit of easing the schedule during the day and providing a further opportunity for attendees of the Conference to interact socially. This was the first time a poster session had been organized and the organizers are encouraged by the response, both of attendees and of poster presenters, nearly all of whom provided a written paper for publication in this volume. There were also several tabletop exhibits during the Poster Session with contributions from JX Crystals, Inc. and the American Institute of Physics.

Professor Michael Seal and his colleagues from the Vehicle Research Institute of the Western Washington University, added substantially to the Conference by bringing with them their TPV/Hybrid car that remained on display outside the hotel for the duration of the meeting. Professor Seal and his colleagues were kind enough to give attendees and others rides in the car to experience the sensation of almost sports-car acceleration. In addition, Professor Seal was interviewed by Denver's Channel 9 News which, hopefully, brought additional publicity and visibility to the topic.

Like many conferences, we have a website giving a great deal of information about the Conference to help prospective attendees. The site address is <www.tpv.org>. This is a general TPV site with links to all previous TPV conferences, as well as to much other information on TPV. Instructions to Authors, the actual program with hot links to abstracts are all available, both before and after the Conference. The Proceedings, of course, are once again published by the American Institute of Physics, which has published the previous three conferences.

At TPV3, we asked the Session Chairs to provide a written review of their sessions in which they highlight significant advances made. The collection of summaries was published in the Proceedings of TPV3 and proved to be very popular. We have, therefore, repeated that exercise in this volume.

In recent weeks, the Co-chairs have visited approximately nine organizations with interests in TPV in order to assess the needs and potential directions of possible future TPV programs. Much of the information that we learned was extremely thought provoking, and all of it will undoubtedly be of great help to us in planning any new program which may or may not emerge. We shall be writing an internal report discussing our findings, which will become available and will be posted on the TPV website for future reference.

Although the Session Chairs have written their own views of their sessions, and we are in agreement with most of these, one or two major points seemed to emerge from the conference. Firstly, there is certainly a general impression amongst those from industry, that the current semiconductor converters are adequate for first-generation products. Secondly, the most important technological issue now is thermal management and the need to develop better approaches for recirculation of sub-bandgap photons. Thirdly, it is clear that the performance of selective radiators (emitters) has greatly improved and this approach is rivalling the broad-band radiator systems. Finally, there is a growing need for standardization and characterization of devices and systems. A TPV system is far more complicated than a PV system and there are many more adjustable parameters. Because of this, standardization will inevitably be more complicated. There is a danger that the conditions under which the performance of a system is characterized do not correspond to those for which the system was designed. Given that any characterization facility will inevitably be limited in scope, perhaps there is a need to address computational issues related to characterization under non-standard conditions.

Timothy J. Coutts
John P. Benner
Carole S. Allman

OVERVIEWS

The following seven short contributions were prepared by the Chairs of the Sessions. The General Chairmen asked these individuals to introduce and summarize their sessions and to provide short written versions of their thoughts. The individual sessions are listed below. It is hoped that this summary will provide the casual reader with an overview of the entire conference without needing to delve deeply into the main body of the proceedings.

Session Number	Session Chair	Session Title
A	Timothy J. Coutts	Programmatic
1	John P. Benner	Fundamental Aspects
2	David R. Riley	Monolithic Interconnected Modules (MIMS)
3	Jeffrey A. Mazer	Radiators and Optical Control
4	Greg W. Charache	GaSb-Related Materials and Devices
5	Joachim Luther	Characterization
6	Daniel Krommenhoek	Systems Design and Experience

Session A

In the opening session, after Dr. Timothy Coutts of NREL had introduced the Conference with the above remarks, a presentation was made by Dr. Jeffrey Mazer, who is a PV Program Manager in the Deptartment of Energy's Office of Energy Efficiency and Renewable Energy. Dr. Mazer talked about possible interests of the Department in TPV and the prospects of future support. Their specific interest is in the use of TPV in domestic stand-alone gas furnaces. Electricity is required to operate gas furnaces and heating systems are therefore at risk of any power outages. These are not uncommon during ice-storms etc. and the consequences are unpleasant. If the electrical power for the pumps and blowers etc. could be provided by TPV panels, then the furnaces could be regarded as stand-alone. Dr. Mazer also said that the DoE has a longer-term interest in the recovery of high-temperature industrial waste heat. There are many excellent examples of situations in which this could be valuable and the potential for the recovery of high-grade energy is considerable. Mazer said that a new initiative had originally been planned for FY2000 although this had subsequently been cancelled. In addition, an SBIR program was intended to be launched during FY99, which could be expected to be of benefit to start-up companies.

Dr. Stanley Bull, who is Associate Director of Science and Technology at NREL, then discussed the potential importance of TPV to NREL. The topic is entirely compatible with NREL's mission and it could become very important to energy-related issues nationally. TPV has grown at NREL in a manner that is similar to other programmatic developments in that we have always insisted any work for others program must be closely related to NREL's mission. The NREL Directors have a small amount of discretionary funds and funding was provided to the co-chairs of this conference to assist their efforts to ascertain industrial perspectives of R&D needs. Dr. Bull also commented on the increasing level of interest from overseas and commented on the fact that attendees came not only from the USA but also from Sweden, Japan (which is due to start its own TPV program in FY2000), Switzerland, Germany, the UK, Russia, and Canada.

Session 1: Fundamental Aspects
Chair: John P. Benner, National Renewable Energy Laboratory

The results and predictions presented during this session provided a comprehensive view of the ideal and practical performance potential of thermophotovoltaic (TPV) systems and TPV cells, as well as exhibiting recent insights and novel component concepts that may cause us to reassess the current practical limits. This prompts a comparison to similar analyses from the First NREL Conference on Thermophotovoltaic Generation of Electricity. At that time, the potential of engineering new cells with response tuned to system needs had rekindled interest in TPV. However, with participants primarily from the cell and selective emitter fields, few present possessed the experience with the system to appreciate the magnitude of losses expected from the other components. An extended presentation by one of the founders of the field, David C. White, gave the attendees a somewhat discouraging though realistic systems assessment. He first noted that if the system performance is analyzed from the contributions of each of the six major components, and if efficiency of each were the engineer's favorite first guess of 90%, then the system would deliver a little more than 50%. Including some thermodynamics, temperature limits of materials, and a more careful look at the radiant transfer system he reached a practical value of about 13%. Parametric variations surrounding selection of emitter temperature and cell bandgap were not included since he assumed the use of silicon cells and practical limits for the operating temperature of the emitter. He included the caveat that his calculations were based on too little experimental data to be taken as a goal.

During Session 1 of this Fourth TPV Conference, speakers addressed many of the issues not considered by White – tradeoffs between power density and conversion efficiency, burner efficiency versus emitter temperature, optical generation rates relative to angular dependence of incident light, and influence on the bandwidth of the radiant transfer system on converter performance. Discussion of the interdependence of the PV cell efficiency and output power density provides an interesting contrast with White's paper. However, White's assumption of efficiencies near the thermodynamic limit for the cell while projecting low radiant transfer efficiencies proved to be completely consistent with the parametric analysis performed by Gray. This showed that as the bandwidth is narrowed the cell efficiency approaches ideal, but the power output falls toward zero. And, as we were cautioned four years ago, the importance of optimizing each component was confirmed in the paper analyzing the potential of low bandgap based systems. Here the authors concluded that the realistic range of efficiencies is comparable to White's estimates, adding that power densities of about 1 W/cm^2 should be achievable. More importantly, the growing base of experimental data suggests that perhaps these values can now be taken as reasonable technology goals.

No fundamental session would be complete without opening avenues through which the present goals might be exceeded. Our session did not disappoint us.

Coverage of the materials properties, novel device designs, and new approaches for energy transfer stimulated ideas for capturing some improvements. But George Cody's reminder that in solar PV, silicon cells are now above predictions of the diode models, provided the incentive to keep some attention above the near term issues.

Session 2: Monolithic Interconnected Modules (MIMS)
Session Chair: David R. Riley, Bettis Atomic Power Laboratory

A concerted effort to develop Monolithic Interconnected Modules (MIMs) has been realized during the past three years. The key driver for this development is that MIMs provide the process engineer an opportunity to develop a single semiconductor device with tuned material, electrical and optical properties that can be matched to any specific operating environment. While significant advances continue to be gleaned, difficult challenges, primarily associated with spectral utilization of fully processed devices, still need to be overcome.

MIMs provide exceptional flexibility regarding the development of large area system applications. The power output of MIMs can be tuned to any geometric or temperature boundary imposed on a system by matching the area of each cell to the incident photon flux. This is important because most systems have non-uniform temperature profiles associated with the emission from the coupled radiator. For example, if the MIMs are coupled to a cylindrical radiator source, not only are there temperature profiles to contend with, but non-uniform irradiance on the surface of the devices, due to proximity of the devices to the radiator, also must be considered. By mapping the photon flux, either by radiometry measurements of the candidate radiator or by modeling of the emission properties, the cell areas on the MIMs can be varied to "current-match" and operate in series. Each cell is optimized to achieve the maximum electrical efficiency by minimizing the cell area (i.e., increasing the number of cells per unit area) and, thus, reducing losses associated with joule heating. However, the design engineer must balance decreased cell area and reduced current density with active area (i.e., minimizing the number and size of interconnects and grid fingers) to maximize power density. This trade sets the boundary conditions for the electrical network design of the MIM.

Five presentations were given on MIM technology at this conference. Two of them focused on the accomplishments related to the development of low bandgap (i.e., < 0.74 eV) MIMs. The advances in this area were realized by improving the material and optical performance of the as-grown MIM structures. Accordingly, regarding materials development, the MIM development programs have utilized InP as the substrate of choice because it can be prepared as semi-insulating, single crystalline and epi-ready. Additionally, ternary InGaAs materials can be grown onto this substrate, either as lattice-matched with a bandgap of 0.74 eV, or as lattice-mismatched with bandgaps less than 0.74 eV. For low radiator temperature applications, the latter growth regime is important to increase the amount of useful light available for conversion.

Two presentations were given, one by the National Renewable Energy Laboratory and one by Essential Research Inc. / NASA Lewis Research Center, on efforts related to the growth of high quality 0.6 eV MIM structures. Double heterostructure, InGaAs MIMs were grown with InPAs buffer layers. These buffer layers accommodate the lattice-mismatch between the InP substrate and the 0.6 eV diode layers. The most important results presented were: 1) high quality, 0.6 eV double-heterostructure diodes were grown by organo-metallic vapor phase epitaxy (OMVPE); 2) the InPAs buffer layers successfully pinned the misfit dislocations to the graded layers; 3) the number of threading dislocations in the active diode region was minimal ($< 10^{-5}$ cm^2) with no measurable influence on electrical performance; 4) lifetimes of several microseconds were measured in these materials; 5) active area absolute internal quantum efficiencies approached unity even near the bandedge; 6) fill factors of prototypical structures were greater than 70%; and 7) weighted reflectance of the long wavelength useless radiation was greater than 90%.

Efficiency predictions based on these structures' measured reflectance and quantum efficiency, and on measured fill factor of test diodes, were calculated to be near 20% at 1000°C. The successful growth and characterization of these structures over the past several years represent a significant advancement towards the realization of useful MIMs for lower temperature TPV applications.

Two additional speakers presented novel attempts to improve the optical and electrical performance of MIMs by incorporating either lattice-matched metal layers or tunnel-junctions below the diode layers. Ohio State University successfully grew, via molecular beam epitaxy (MBE), lattice-matched FeAl layers on InAlAs/InP substrates (the InAlAs served to prevent iron from diffusing into the substrate). It was expected that the FeAl layers would improve electrical conductivity by serving as the lateral conduction layer (LCL) in the MIM. It would also simultaneously reflect the non-absorbed light back through the device without requiring it to transmit through a semiconductor-based LCL, which can parasitically absorb radiation. Fully strained FeAl layers were grown on InP based substrates in single crystalline form. Furthermore, InGaAs active layers were then successfully grown onto the FeAl layers. Although the concept was successfully demonstrated, these structures suffered from parasitic absorption of incident radiation at wavelengths between about 1-5 microns.

A new n/p/n tunnel junction MIM was described by NASA Lewis to replace the p/n diode with a n/p diode while maintaining a n-doped lateral conduction layer in the MIM architecture. Current state-of-the-art MIMs that use p-doped emitters are hindered by free carrier absorption and high sheet resistance in this layer. The use of a tunnel junction allows the elimination of thick (> 0.1 _m), heavily p-doped layers in the MIM architecture. In addition, the n/p diode may result in decreased dark current (reverse saturation current) and improved bandedge spectral response because of the longer minority carrier diffusion length associated with a p-doped base. A new design was presented to incorporate a p/n tunnel junction between the base and the LCL. Initial structures were grown with 500Å tunnel junctions. These devices displayed minimal voltage drop across the tunnel junction (< 3 mV/junction under 1200K-blackbody illumination). Further developments of these structures are planned in the near term and may lead to the highest performance MIM devices that can be developed.

The final presentation focused on the modeling of MIM optical performance parameters, specifically light trapping due to structural features integral to a MIM. The "Racer X" Monte Carlo photonics code was used to model the photon path of incident light rays through a MIM. The presentation discussed the effects of light trapping due to etch trenches on the front surface, absorption in the reactive metal on the back side of the grid fingers, and the losses associated with non-specular back surface reflectors. The key result was that the etch trenches caused substantial losses associated with light trapping due to refraction of light in the trenches and subsequent absorption of useless light. This light trapping was substantially increased by the addition of highly reflective grid fingers into the trenches. Efforts to design new MIMs that do not suffer from parasitic light trapping are being developed.

MIMs have successfully been developed with device performance equivalent to that of single junction TPV devices. However, the advantages that MIMs have in the area of system matching are significant and represent the key driving force for the continued development of these devices. The fact that MIMs have built in spectral control (all the incident light is transmitted into the device, with only the useful light being absorbed in an ideal MIM device), make it attractive for high power applications. Further, with the significant improvements gleaned in material quality of low bandgap devices, diode efficiencies of nearly 20% are now being realized. These advances, and advantages, suggest that MIMs definitely rival other TPV device technologies, and likely will be utilized for both low power and high power applications.

Session 3: Radiators and Optical Control
Session Chair: Jeffrey A. Mazer, U.S. Department of Energy

This session dealt with the emittance characteristics and fabrication of radiators (emitters). Because thermophotovoltaic systems employ high-concentration power densities, the issues of radiator reliability and effective radiation coupling are highly important. The four papers presented in this session all addressed some aspect of these issues. The paper entitled "High Temperature Optical Properties of Thermophotovoltaic Emitter Components" reported measurements on quartz, sapphire, and translucent-polycrystalline sapphire to determine infrared properties as a function of temperature for these materials.

Metal surfaces have promise as selective emitters because of their thermal stability at high temperature and low emissivity in the far infrared. The paper entitled "Microstructured Tungsten Surfaces as Selective Emitters" investigated the enhancement of short wavelength emissivity produced by microstructuring of such surfaces. Experimental tungsten surfaces were fabricated with the use of holographically defined photoresist masks. Measured angle-dependent emittance at 1500 K was compared with predictions from a rigorous grating theory analysis of the structure.

The paper entitled "Thermal Spray Approach for TPV Emitters" investigated the emittance characteristics of rare-earth oxides and cobalt-doped spinel that were deposited by spraying thin coatings on alumina and silicon carbide substrates. Results were reported for a test system operating at 1600 K.

Large temperature gradients across the thickness of a thin-film selctive emitter are known to degrade performance. The fourth paper, entitled "Temperature Gradient Effects in an Erbium-Doped YAG Selective Emitter" reported computational results of a conduction and radiative model for heat transfer across the thickness of an erbium-doped YAG emitter.

The four papers presented in this session indicate that the problem of radiator optimization is receiving a considerable degree of attention through sophisticated development of both measurement and theory.

Session 4: GaSb-Related Materials and Devices
Chair: Greg W. Charache, Lockheed Martin, Corp.

Low bandgap ($E_g < 0.72$ eV) antimonide-based devices continued to make dramatic progress since the last TPV conference. This year, papers were presented on four InGaAsSb growth techniques: bulk crystal, liquid phase epitaxy (LPE), organometallic vapor phase epitaxy (OMVPE) and molecular beam epitaxy (MBE). The epitaxially-grown antimonide-based devices with bandgaps of 0.55 eV or less are currently the highest performing devices for TPV applications, with demonstrated minority carrier diffusion lengths of 15-20 microns. Further development of the bulk crystal work may yield high-quality cost-effective low-bandgap materials that reduce the high temperature material requirements of current radiators and recuperators.

LPE growth of single layers of n-type 0.53 eV InGaAsSb on GaSb was processed into TPV diodes using zinc diffusion from a pseudo-closed box (PCB). This method has been used previously to form high-quality GaSb diodes. The surface properties of these diodes were further optimized using anodic oxidation. Fabricated diodes with open circuit voltages of ~300 mV at 1 A/cm^2 were presented. These diodes are approaching the material quality already demonstrated using OMVPE/MBE and confirm the importance of attaining a low surface recombination velocity.

The bulk crystal growth of Sb-based III-V compounds was highlighted with the demonstration of single-phase crack-free quaternary alloys, which were termed "quasi-binary". This novel material was formed by synthesizing $(GaSb)_x(InAs)_{1-x}$ using a vertical bridgeman furnace with a submerged baffle. GaSb-rich material with bandgaps ranging from 0.72 eV (100% GaSb) to 0.55 eV (90% GaSb, 10% InAs) have been demonstrated. Further work will involve quantifying the minority carrier properties of these crystals.

OMVPE-grown InGaAsSb has achieved bandgaps as low as 0.48 eV. These metastable alloys are showing signs of decreasing material quality with lower bandgaps, however even the 0.48 eV material demonstrates internal quantum efficiencies of > 85%. Initial microstructural and optical characterization of these alloys has shown signs of lateral composition modulation and alloy clustering. Further detailed studies are required to correlate the microstructure with device properties in a effort to optimize growth conditions.

The paper presented on the MBE growth of InGaAsSb quantified the effects of surface recombination on thick emitter structures. Modeling results indicate that further reduction of surface recombination velocity could improve the short circuit current by 10% and the open circuit voltage up to 50 mV on 0.55 eV devices.

Finally, a summary of the reported dark currents versus bandgap for TPV devices was shown [Fig. 1]. The dark current represents one component of the overall device efficiency and serves as a convenient quantifiable number to compare various technologies. It should be noted that the best dark currents achieved for all technologies compare well with reported InGaAs detector data and are within two orders of magnitude of theoretical limits. This confirms the successful transition of optoelectronic communications technology to power production. In addition, for bandgaps less than 0.55 eV, the antimonide-based materials are now demonstrating routinely lower dark currents than InGaAs-based devices. In fact the low bandgap antimonides are achieving comparable dark currents to the best Ge-devices available today.

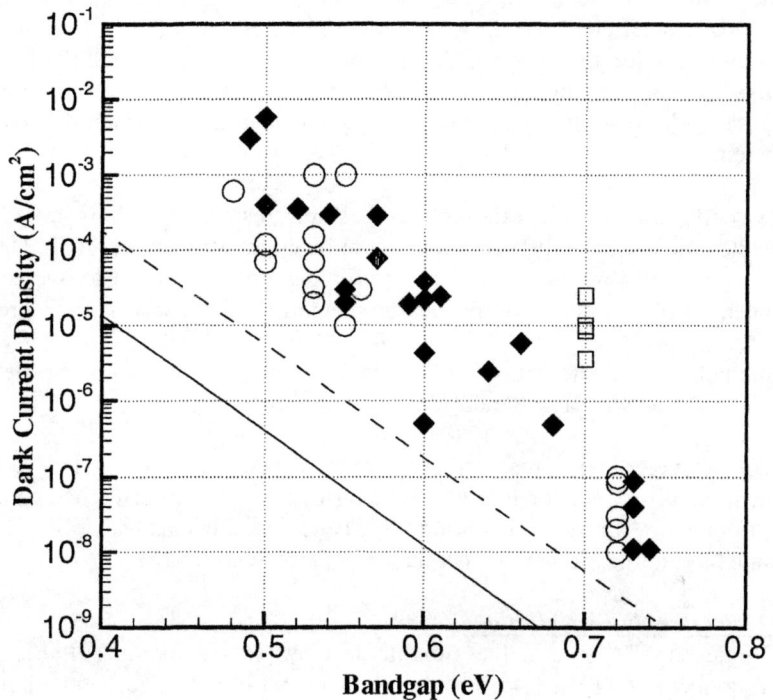

Figure 1 - TPV voltaic dark current density versus bandgap: Germanium (□), InGaAs (♦), InGaAsSb and GaSb (○), Non-photon recycled radiative limit (---), Photon recycled radiative limit (___).

Session 5: Characterization
Chair: Joachim Luther, Fraunhofer Institute for Solar Energy Materials

The session was dedicated to the electrical and thermal characterization of TPV cells and materials. The content was complementary to the presentations on radiation emitters, optical materials and filters in other sessions of the conference.

TPV cells characterization
For the accurate measurement of TPV cells, it is essential to control the temperature of the cell. In a paper by Moriarty et al., it was shown, that three methods are efficient: (i) fast V method: Voc is recorded immediately after opening a shutter, this original Voc is then maintained by cooling the sample (ii) analysis of dark IV curves to obtain additional information of the cell's behavior (iii) flash illumination. The high intensity flash simulator turned out to be most efficient.

TPV cells can have a relatively low shunt resistance. This may lead to considerable errors in the determination of quantum efficiencies. R.L. Danielson et al. presented a measurement system utilizing a probe resistor to overcome this problem. A robot system for the measurement of IV characteristics was presented.

A method to evaluate MOVPE TPV cells (including a back surface field) in a production line was presented by A. Narayanan et al.

M.B. Clevenger et al. presented ray tracing analysis of very thin TPV cells with optical back reflector or buried reflecting metal layer. Interference effects can deteriorate the IR reflectivity of ultra thin cells. After his analysis MIM structures should have a thickness of greater than 30 micrometer.

TPV systems characterization
L. Fraas et al. presented a flexible test system which is suitable for the measurement of TPV cells, radiation emitters and filters. The standard emitter is a SiC radiator. The necessity of measurement standards was addressed.

The analysis of a closed cavity TPV system was presented by C.K. Gethers et al. The test system has improved considerably during the last years.

Material characterization
It was demonstrated by J.D. Webb et al. that IR - Fourier transform spectroscopy represents a very useful method to investigate into TPV cell materials as well as filters. Defect levels and band gaps may easily be measured.

New lifetime data for InGaAs materials were presented by R. Ahrenkiel et al. Time resolved femto second upconversion spectroscopy made it possible to measure Auger and radiative recombination coefficients at highest doping levels.

RF photoreflectance spectroscopy was used by S. Saroop et al. to measure lifetimes in atimonide-based materials. The surface recombination velocity of MOVPE grown InGaAsSb was determined for different cap materials. SRH recombination dominated the lifetime in the case of low doped GaSb substrates.

Conclusion

The session demonstrated clearly the progress made in the detailed characterisation of TPV cells and materials. On the other hand it was obvious that some standards for figures of merit and for the detailed characterisation of TPV cells are needed. The same is necessary for other system components as well as for complete systems. For a detailed cell characterization the EQE, IR reflectance and IV curves at different radiation levels and temperatures should be measured. There is a definite need for simple standard test procedures being available in all TPV R&D laboratories.

Session 6: Systems Design and Experience
Chair: Daniel Krommenhoek, Quantum Group, Inc.

Six integrated systems that have been built and tested were evaluated in this session. TPV demonstration sizes ranged from 5 to 990 watts. Three of these initial demonstrations were for portable power, with the others for a battery substitute, combined heat and power and finally for grid connection feasibility. All demonstration units have spectral control on the emitter and/or in front of the cells. The cells were either GaSb or Si in these demonstration units. The emitter cell combination was matched. Thus fibrous or ceramic composite ytterbia was used with Si cells, and SiC or textured tungsten or Fe-Cr-Al-Co-C alloy was used with GaSb.

In general, the Si cells do not have an optical filter since these cells are used with ytterbia that has a narrow emission band closely matched to silicon cells. For the GaSb cell modules, optical filters are utilized in front of the GaSb cells since the photon source was broad and resembled a black body spectrum even though the spectrum was tailored in two of the designs. The optical filters reflect lower energy photons back to the heat source. Filters tested were both dielectrics and tandem dielectrics with plasma filters. Benefits of improved filters were evidenced in the system modules. However, for this new technology, there is still a need for high efficiency spectral control.

The photon losses are significant for the region between the emitter and cavity. Most concepts have air in the space between this region, and one had a vacuum and the other had water to absorb and transfer radiation prior to the photons reaching the cells. However, the main photon losses are probably due to low spectral utilization caused by electrical grids, electrical cell connections, incomplete reflection back from the optical coatings (including filters) on cells, photon absorption in non TPV cell areas and stray photon leakage at ends and sides of the emitter cell assembly. One concept tested shingled cells to eliminate cell electrical connector losses, and another used water filters to protect cells and use this waste heat for alternate purposes. Since many photons escape and do not reach active electrical TPV cell regions, stray photons represent a significant loss for integrated TPV systems as seen in the test data.

All of these demonstrations except one were based on gaseous hydrocarbon fuels; this one concept used thermally vaporized liquid diesel fuel.

In terms of overall system efficiency improvements can be realized by concentrating on the following three areas. These are: (1) effectively taking the energy from the fuel to the emitter photons to improve the combustion to emitter energy conversion; (2) narrowing the wavelength of photons that reach the active TPV cell by selective emitters and cell filter technology development to improve the spectral control; and (3) reducing the stray photons between the emitter and cells by designing the photon collection geometry to reduce parasitic photon losses.

SESSION A: PROGRAMMATIC

Thermophotovoltaic Potential Applications for Civilian and Industrial Use in Japan

Hiromi Yamaguchi
Ishikawajima-Harima Heavy Industries Co., Ltd., 3-1-15 Toyosu, Koto-ku, Tokyo, 135-8732, Japan

Masafumi Yamaguchi
Toyota Technological Institute, 2-12-1 Hisakata, Tempaku, Nagoya, 468-8511, Japan

Abstract. Investigative research on potential market for TPV power sources in Japan has been focused on how TPV can contribute to energy conservation and environmental protection and harmony. The application needs for TPV were surveyed in comparison with conventional engine or turbine generators and developing power generation technologies such as fuel cells or chemical batteries, etc. The investigation on the performance of commercial generators shows that regarding system efficiency, TPV can compete with conventional generators in the output power class of tens of kW. According to the sales for small scale generators in Japan, most of the generators below 10kW class are utilized mainly for construction, communication, leisure, and that 10-100kW class generators are for cogeneration in small buildings. Waste heat recovery in dispersed furnaces is another potential application of compact TPV cells. Exhaust heat from small scale incinerators and industrial furnaces is undesirable to be recovered into electricity due to excessive heat loss of the smaller steam turbine generators. Solar powered TPV is also of our concern as a natural energy use.

From the viewpoint of applicability for TPV, portable generators, cogeneration systems, and solar power plants were selected for our system consideration. Intermediate report on the feasibility study concerning such TPV systems is given as well as the review of the current status of competing power generation technologies in Japan.

INTRODUCTION

Motivation

In recent years, the spread of solar power generation and portable electrical equipment powered by solar cells has made people more familiar with photovoltaic cells. Thermophotovoltaics (TPV) attracts our attention since the combination of simple elements, a radiation heat source and photovoltaic cells can easily produce electricity. TPV generation is also fascinating to be suitable for various kinds or scales of heat sources, which indicates that a variety of applications for TPV are expected to appear on the market. There has been published some literature regarding market potential for TPV (1-3) in the US up to date. On the other hand, Japanese energy policy states that, by yr. 2030, more than 10 percent of domestic energy demand should be supplied with substitutive energy such as natural renewable energy, waste heat recovery in garbage furnaces, cogeneration of heat and electricity, fuel cell power generation, and so on, which is estimated to reduce the CO_2 emission to the 1990 yr.

level. It suggests that compact power supplies like TPV generators can be one of the choices as dispersed systems which can satisfy the electricity demand on-site.

Research Organization

This investigative research has been conducted in the committee group in cooperation with academic, governmental, and industrial sector members (see Appendix). The committee is organized by the Engineering Advancement Association of Japan (ENAA), which was established in 1978 with support by the Ministry of International Trade and Industry (MITI). The research program started in the middle of 1997 and are scheduled to continue until the end of March in 1999.

Research Direction

In this research, we directed our efforts to study how TPV can satisfy potential needs, and which applications of TPV can contribute to primary energy saving and environmental protection and harmony. Therefore, attention was paid to the point that a TPV use should not increase the total primary energy consumption. First, we review the status of competing technologies for TPV : conventional engine or turbine generators and various developing generation technologies such as fuel cells, chemical batteries, thermoelectrics, etc. Second, some systems are proposed for potential TPV use in Japan, from the viewpoint of applicability and contribution to energy and environmental issues. Finally, we discuss the requirements for TPV performance in some of applications.

COMPETING TECHNOLOGY STATUS

The current status of potential competing power supplies for TPV in Japan will be reviewed for both conventional and developing technologies in this section. Since most of recent literature on TPV deal with systems with the electric power output of less than tens of kW, the systems for our study were focused on small scale or dispersed power supplies with this power range.

Internal combustion engine and turbine generators

Figure 1 shows the system efficiency of existing generation systems with the output power ranging from 1kW to 1MW, which include a gasoline engine, a gas engine, a diesel engine, and a steam turbine generation system. System efficiency exceeds 30% in the range of above 10kW output, except of waste heat recovery in incinerators. However, below 10kW the efficiency decreases drastically to 10-20% due to increase of heat loss from the generator. Generators of 10-1,000kW output power are mainly used for cogeneration, with the backup function at emergency, in office buildings, hotels, hospitals, restaurants, etc., while less than 10kW class of generators are demanded for construction, communication, agriculture, leisure, and outdoor shops, etc. The annual sales for such portable generators made in Japan in 1997 are shown in

Fig.2. Japanese portable generators are also exported to numerous countries, most of which are in East South Asia and North America.

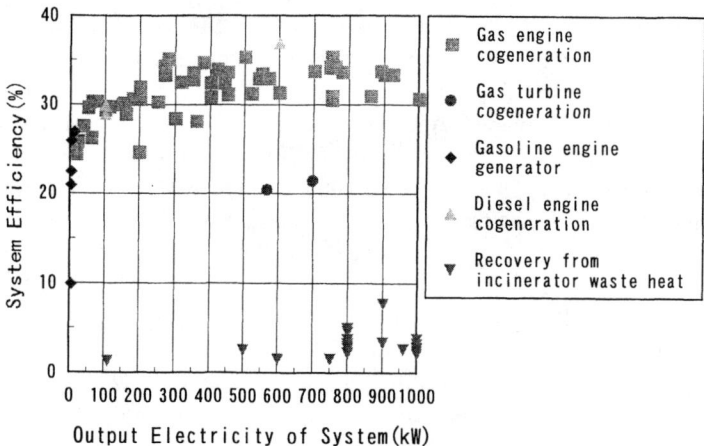

FIGURE 1. System efficiency of existing generation systems with output below 1MW in Japan.
(The data is provided by Junichi Ochiai of Ishikawajima-Harima Heavy Industries.)

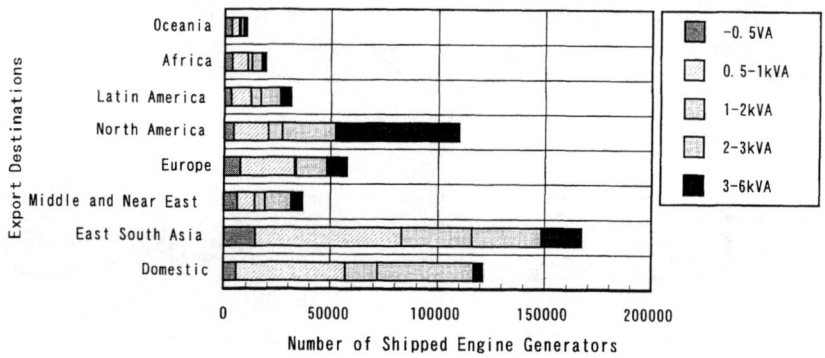

FIGURE 2. Annual sales of Japanese portable generators with output below 6kW in 1997.

Fuel cells

One of the national plans for advanced energy development is to increase the amount of the power supply by fuel cells to 200MW until yr. 2000 and to 2,600MW until yr. 2010. Some types of fuel cells are commercially available from a few Japanese electric companies. Toshiba Corporation provides 200kW class of phosphoric acid fuel cells (PAFC) for domestic and overseas use, the annual sales of which were 40 systems in 1996. Those systems were installed for cogeneration. Portable PAFCs for emergency use are provided on trial by Sanyo Electric Corporation. They do also provide portable polymer electrolyte fuel cells (PEFC). It appears necessary to reduce the system size and cost in order to expand a market.

Chemical batteries

High performance chemical batteries have been developed to be in practical use with the rapidly increasing demand for portable electric appliances such as cellulars, computers, and video cameras. Among a variety of batteries, nickel-hydride and lithium batteries have a big market with 102 billion and 209 billion yen, respectively, for the 1997 annual sales. These kinds of batteries demonstrate a high power density. Typical values are 80Wh/kg for a NiH battery and 125Wh/kg for a Li battery, which are approximately double that of a conventional lead cell. Since 1992, batteries of larger capacity have been developed for excessive power storage and for electric vehicles in the national project. The project specifications of a module for an electric vehicle are 3kWh in capacity, 150kWh/kg in power density, and 1,000 cycles in charge and discharge cyclic tolerance. The power density obtained up to date almost satisfies the requirement, but some problems are left in reliability and safety.

Thermoelectrics

Thermoelectric power generation is not utilized in a large scale at present time. However, its potential for waste heat recovery is extensively studied in both Japanese government and industry, because a thermoelectric device can convert even below a few hundred degrees of heat to electricity. Recovery of low grade heat from industrial plants and incinerators is the target for development of thermoelectric devices and units. There have been several examples of field tests of thermoelectric generation systems. Thermoelectric material with higher conversion efficiency is required to emerge to improve cost performance of a total system.

Solar power generation

In Japan, solar power generation systems have been developed in the Sunshine National Project since 1974. As one of the fruitful results, 3kW solar photovoltaic power systems for residence have been equipped on the roofs of over ten thousand houses. It was reported that such a solar power system made residents save electricity by 48%. The cost for a system is still high, three million yen, one third of which was subsided by the government. The government plans to increase solar power generation to 400MW by 2000 yr. and to 4,600MW by 2010. It is estimated that the mass production effect reduces the cost of silicon cell panels to be less than ¥200/W.

The development of solar thermal power generation also started in the national project. Two types of 1MW pilot plants, using a central receiver and a distributed collector, respectively, were built and succeeded in operation for the first time in the world (6,7). However, the development was stopped in 1987, because of the poor economical feasibility of the technology. Duration of sunshine at the plant site is 2,200hr per year and solar irradiance is $0.75kW/m^2$, while in California desert 3,500hr and $0.9kW/m^2$, respectively. Solar radiation in Japan is not sufficient to drive steam turbine generators efficiently throughout the year.

APPLICATION SURVEY

There have been numerous articles on TPV application potential in various fields, which cover both terrestrial and space use, or both military and civilian use : that is portable generators, space power supplies, cogeneration systems, hybrid vehicles, combined cycle power plants, solar power plants, grid independent equipment, etc.

Since it was suggested that increasing CO_2 emission causes global warming, reduction of fossil fuel consumption tends to be promoted in several developed countries. Much effort on energy saving and fossil fuel substitution has also been made by the Japanese government and industry. In such a severe circumstance, it is desirable that TPV use should not increase the consumption of primary energy resources. Accordingly it is important to decrease the present energy consumption by using more efficient generators. One possibility is that a conventional generator with lower conversion efficiency could be substituted with a TPV generator. Another is that heat which remains unused in nature or industry could be effectively applied to TPV.

As mentioned in the previous section, it is in the power class of lower than tens of kW that TPV could compete with conventional mechanical generators. Most of current low power generators are of engine driven type. Since a TPV generator consists essentially of no moving parts, it is possibly superior to an engine generator in quietness and high reliability at starting up and for a long operation. The demand for portable generators is analyzed from their sales. As shown in Fig.3, demand for construction and communication has the majority in the total amount. Generators for leisure are demanded by 4% and they could be replaced with TPV generators if such quiet power supplies are provided at a lower price.

Recovery from waste heat in the civilian and industrial fields is a significantly contemporary issue in Japan. Sixty percent of primary energy supply is wasted into heat, some of which originates in exhaust from small scale incinerators and industrial furnaces. Small scale exhaust heat is undesirable to be recovered into electricity due to excessive heat loss of the smaller steam turbine generators, accordingly the recovery from small scale heat is another potential application of compact solid state converters such as TPV. The problem for TPV application to waste heat is that the temperature of the exhaust gas from those furnaces is not high enough to efficiently operate TPV emitters under development. However, it is possible to use such exhaust heat recharged to premium heat by adding combustion fuel. It is one of the candidates for TPV use.

As described above, solar thermal power generation in Japan faced to economical problems due to insufficient sunshine. It should be true for solar TPV. But considering the contribution to global energy issues, it doesn't matter if solar TPV is feasible only for domestic use. It is important for us to develop solar TPV systems assigned to foreign sites under fully sunshine as an international cooperative work.

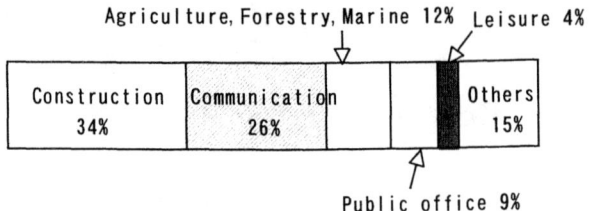

FIGURE 3. Portable generator demand in Japan 1994.

SYSTEM CONSIDERATIONS

From the viewpoint of applicability for TPV, portable generators, cogeneration systems, and solar power plants were selected for our system feasibility study. The intermediate results will be presented below for a portable generator and a cogeneration system. Since a solar power plant using TPV is still under study, it is not mentioned here.

Portable generators

Essentially portable generators are provided for many and unspecified consumers. Therefore portable generators should be designed to satisfy the major needs of numerous consumers. The first requirement of consumers is output. Other requirements are considered as follows:

- Weight-to-power ratio
- Volume-to-power ratio
- Price-to-power ratio
- Continuous operation time
- Fuel availability
- Noise

The specifications of typical existing portable generators are listed in published articles (4,5,8). We calculated system efficiency and other specific performance values based on the specifications of each generator to make quantitative comparison between generators.

The first three items shown above depend on system efficiency. Figure 4 represents system efficiency of generators in the power class of less than 5kW. In the range of 1-5kW, gasoline or diesel engine generators demonstrate around 20% efficiency, while below 1kW, system efficiency decreases gradually to less than 10%. In the power class of below 1kW, TPV generators can be sufficiently competitive to engine type of generators at present status. Higher efficiency ensures a longer continuous operation at the same amount of fuel. On the other hand, fuel cells show much higher efficiency at such a low power level. However, fuel cells require to prepare hydrogen or methanol fuel which is unpopular at present.

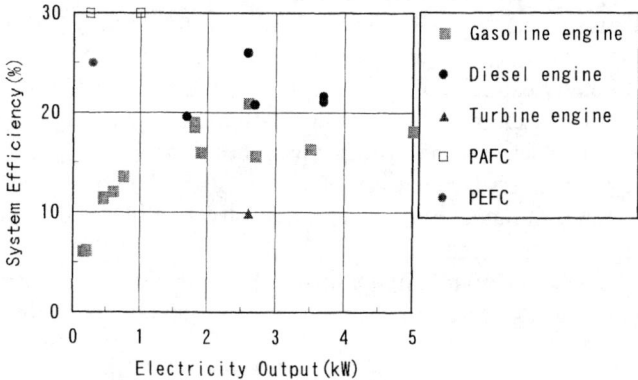

FIGURE 4. System efficiency of existing portable generators with output below 5kW.

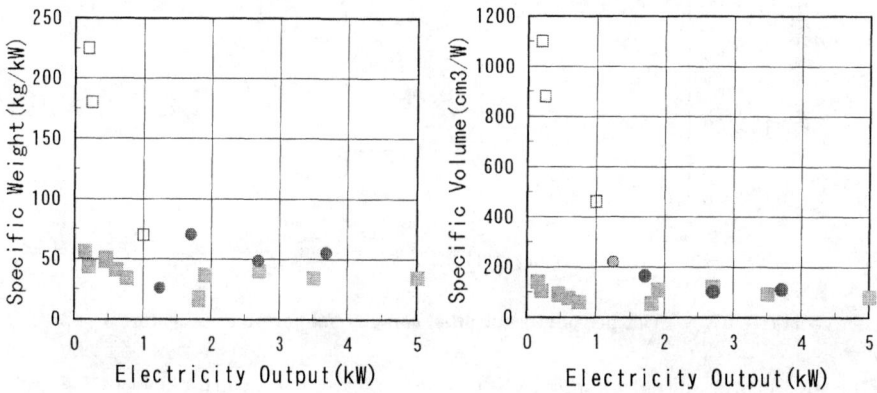

FIGURE 5. Specific weight of portable generators: fuel weight excluded (for legend cf Fig.4).

FIGURE 6. Specific volume of portable generators (for legend cf Fig.4).

Important indications of portability are weight and volume. One of the promising features for TPV is high power density resulted from high radiation density. Specific weight and volume of existing portable generators are shown in Fig.5 and Fig.6. Both values for mechanical engine generators slightly decrease with increase of their output power. The projected weight and volume of TPV for competition with engine generators should be estimated to be approximately below 30kg/kW and below 100cm^3/W. Fuel cells are significantly heavier and more bulky.

Besides generators, batteries are also competing technology for portable TPV. Let us compare performance of recent batteries with that of gasoline engine generators. As an example is taken a commercial gasoline engine capable of 4h operation at 5kW AC. It possesses 180kg in weight including fuel and 410,000cm^3 in total volume. Typical NiH battery shows specific power to weight and to volume as 80Wh/kg and 300Wh/liter, respectively. These values correspond to 260kg in weight and 73,000cm^3 in volume for a battery capacity of 5kW for 4hr, assumed that DC to AC conversion efficiency is 90%. This example manifests that batteries are generally heavier but more compact than engine generators in the output power of a few kW.

As for noise, consumers generally prefer to silent generators, and the more for leisure or maintenance of outdoor equipment. Recently, silent type of engine generators are available with the sonic emission level of below 60dB, but fuel cells are still more silent as 40dB. As there are no moving parts in a TPV system except for pumps or fans for fuel supply and gas exhaust, it suggests that TPV should realize the similar sonic level.

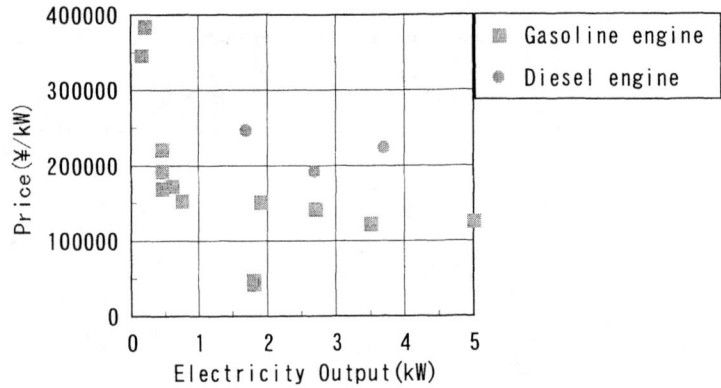

FIGURE 7. Price per power output of commercial portable generators.

Price per power output changes with system efficiency. A critical point seems to exist around 1kW power output. The scale demerit of price clearly appears in the power range of below 1kW, which is due to increasing heat loss and difficulty in mechanical building. If TPV portable generators can be provided at the unit price of around ¥100,000/kW, they should be economically feasible at least in the power range of this study, below 5kW.

Cogeneration systems

The smallest generators for currently existing cogeneration systems in Japan are of 10kW class. A micro cogeneration system for a residence is considered to be useful for energy conservation. Because heat is provided by a power generation system of the residence, it can be utilized on-site to make water or air hot, while most of heat in a central thermal power station remains wasted. Residential cogeneration requires about

1kW power generator. In this power range, TPV, gas engine generators and fuel cells are possible options. However, TPV should be preferred because the absence of moving parts makes TPV systems resident-friendly as featured to be silent and highly reliable. Before TPV is justified for residential systems, primary energy saving and cost performance should be estimated in comparison with a conventional energy supply system, which supplies all electricity from central power stations and all fuel from central gas tanks. In this estimation, the TPV cogeneration system is assumed to be operated under the following conditions:

(1) 1kW TPV generator is driven by combustion of town gas.
(2) TPV efficiency of 10% or 20%.
(3) Electricity is essentially supplied from a TPV generator which is operated according to electricity demand, and the shortage supply is compensated with electricity from a central power station.
(4) Heat for hot water and heating is supplied in the form of hot water stored in a thermal storage bath which is heated by exhaust gas from a TPV generator. The rate of thermal storage loss is 20%.
(5) A TPV generator is operated until the total amount of stored heat reaches the daily demand.

The dependence of daily consumption of electricity and heat for a residence upon season was investigated with the governmental support. The Institute of Research and Innovation made a report on the feasibility of residential cogeneration using a gas engine generator. This report shows that the average demand for electricity or heat in a residence strongly depends upon seasons. Therefore, in the case of a 1kW TPV cogeneration system of limited output, it is necessary to compensate the electricity shortage with conventional energy supply. Energy consumption and running cost of TPV systems are calculated on assuming the energy demand patterns in Fig.8. The results are described in comparison between a TPV system and a conventional energy supply system.

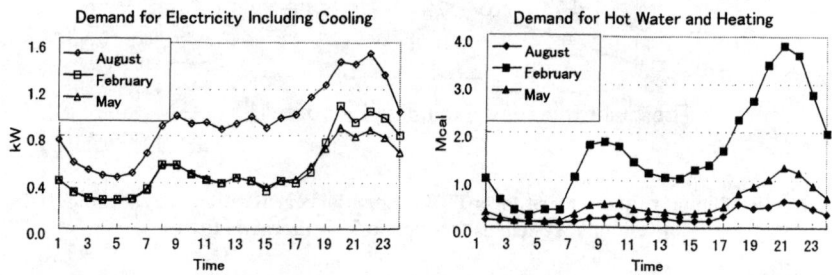

FIGURE 8. Daily energy consumption of a residence in Japan.

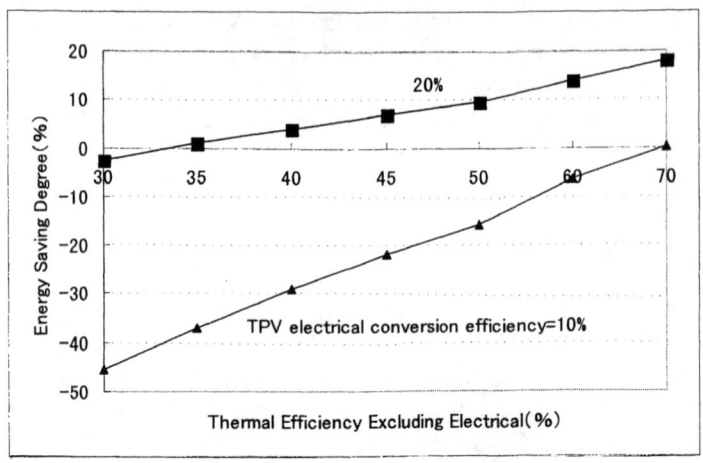

FIGURE 9. Ratio of primary energy consumption for a TPV system to for a conventional system as function of TPV thermal efficiency.

Figure 9 represents the ratio of primary energy consumption of a TPV system to that of a conventional system. Thermal efficiency shown here indicates efficiency just for usable heat. The degree of energy saving increases with thermal efficiency of the system. In the case of 20% electrical conversion efficiency, a TPV system exceeds a conventional one in primary energy saving at more than 35% thermal efficiency, while, in the case of 10% electrical conversion efficiency, it exceeds at more than 70% thermal efficiency.

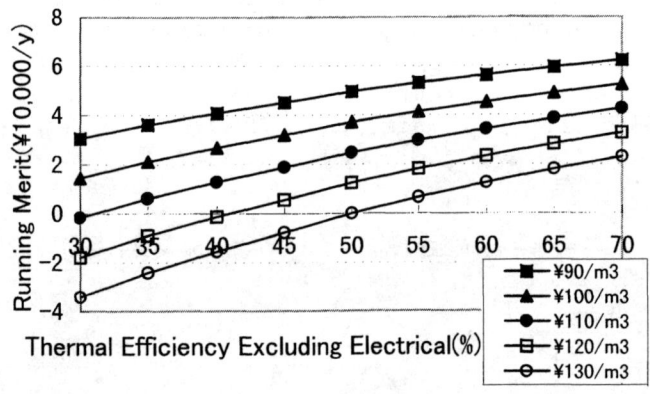

FIGURE 10. Annual running merit for a TPV residential cogeneration system compared with a conventional system as function of TPV thermal efficiency.

Subsequently the cost of a TPV system is estimated on the assumptions that TPV electrical efficiency is 20% and electric charges is ¥23/kWh. The running merit was calculated as function of thermal efficiency with gas charges as parameter (see Fig.10).

The running merit indicates the amount of the annual running cost of a TPV system substituted from that of a conventional system. Let us take one example. If annual running merit is projected to be ¥40,000/year, a system is priced ¥200,000 to pay 5 years. In the case of ¥90/m^3 gas charges, such a TPV system is required to realize thermal efficiency of 40% for electrical efficiency of 20%.

Improvement of System Efficiency

Most of TPV prototype models fabricated to date are driven by a combustion heat source. It has been demonstrated that TPV efficiency is less than that of internal combustion engine generators or other small power supplies. The question is how to improve TPV efficiency. Main factor to lower its system efficiency is heat loss in combustion flue gas. Many of researchers propose (9-11) that a thermal recuperator is most effective to increase system efficiency. Essentially, a recuperator with infinite length of a flow duct can recover almost all the heat from exhaust gas. But, a big recuperator causes a system to be bulky and expensive.

One of the possible solutions is presented here. We propose a regenerative burner system to be applied for recuperation in a TPV system. The concept of a regenerative burner was advocated over a hundred years ago to be applied for recuperation in a furnace. The system possesses two burners accompanied with each recuperator duct made of porous material, which efficiently stores and releases heat. During a certain period, one burner is fired and exhaust gas passes through a porous duct accompanied to unfired another burner. During a successive period, combustion air comes in and is preheated through the porous duct, and consequently another burner is fired efficiently. This cycle is repeated to provide alternative recuperation, which leads to so high efficiency as, demonstrated in practical furnaces, approximately 85% for heat recovery. Recently, a Japanese company developed compact porous recuperators consisting of ceramic honeycomb. It enables us to make the system for heat recovery much more compact. However, it should be taken into account that the control of alternative flow and fire needs a driving power.

SUMMARY AND FUTURE WORK

Potential needs for TPV technology and generators in Japan were investigated focused on contribution to energy conservation and environmental harmony. From the viewpoint of applicability for TPV, portable generators, cogeneration systems, and solar power plants were selected for our feasibility study. The requirements for some of TPV systems were stated here. TPV generators can compete in system efficiency with mechanical engine generators in the power range of below 1kW at present status. However, it can be said that higher system efficiency drives the market of TPV with tens of kW class. In comparison with conventional generation systems, the project values concerning efficiency, dimensions and cost were estimated for small power TPV systems such as portable generators or residential cogeneration systems. The amount of CO_2 and NOx gas emission should be evaluated to compare the effect of TPV and conventional systems upon global environment as well.

This work is a preliminary study ahead of intensive research on TPV devices and systems. On the next step, it is necessary to design TPV systems satisfying the above requirements and to verify experimentally the economical and technical feasibility of the systems.

ACKNOWLEDGEMENTS

This paper is based on the work funded by the Engineering Advancement Association of Japan (ENAA). The authors gratefully acknowledge all the committee members and Dr. Takashi Amano of IMURA Material R&D Corporation for technical information and discussion, especially Dr. Hiroyuki Iwai and Dr. Keiji Takimoto of Osaka Gas Corporation for providing feasibility study report on residential cogeneration systems.

REFERENCES

1. Ralph, E.L. and FitzGerald, M.C., "Systems/Marketing Challenges for TPV," AIP Conference Proceedings 321 - The First NREL Conference on Thermophotovoltaic Generation of Electricity, AIP Press., 315-321 (1994).
2. Rose, M.F., "Competing Technologies for Thermophotovoltaics," AIP Conference Proceedings 358 - The Second NREL Conference on Thermophotovoltaic Generation of Electricity, AIP Press., 213-220 (1995).
3. Ostrowski, L. J., Pernisz, U. C. and Frass, L. M., "Thermophotovoltaic Energy Conversion: Technology and Market Potential," AIP Conference Proceedings 358 - The Second NREL Conference on Thermophotovoltaic Generation of Electricity, AIP Press., 251-260 (1995).
4. *NIKKEI MECHANICAL* 1997.2.17, no.500 (1997).
5. Shindo, K., et. al, "Small Power Systems Using Fuel Cells," *Sanyo Technical Review* **29**, Nov. 17-25 (1997).
6. Tanaka, T., *Journal of JSES* **2**, 9-15 (1991).
7. Tanaka, T., *Journal of JSES* **4**, 52-58 (1991).
8. http://www.honda.co.jp/power/industry/line_up/.
9. Becker, F. E., Doyle E. F. and Shukla, K., "Development of a Portable Thermophotovoltaic Power Generator," AIP Conference Proceedings 401 - The Third NREL Conference on Thermophotovoltaic Generation of Electricity, AIP Press., 329-339 (1997).
10. Guazzoni, G. and McAlonan, M., "Multifuel TPV Generator," AIP Conference Proceedings 401 - The Third NREL Conference on Thermophotovoltaic Generation of Electricity, AIP Press., 341-354 (1997).
11. DeBellis, C. L., Scotto, M. V., Scoles, S. W. and Frass, L., "Conceptual Design of 500W Portable Thermophotovoltaic Power Supply Using JP-8 Fuel," AIP Conference Proceedings 401 - The Third NREL Conference on Thermophotovoltaic Generation of Electricity, AIP Press., 355-367 (1997).

APPENDIX

The research organization and the committee members are shown below:

Engineering Advancement Association of Japan (ENAA)

TPV Research Committee

Chairperson: Masafumi Yamaguchi, Toyota Technological Institute
Committee members: Yasuhiro Hayakawa, Shizuoka University
Hiroo Yugami, Tohoku University
Kunihisa Eguchi, National Aerospace Laboratory
Koichi Sakuta, Electrotechnical Laboratory
Toshihisa Masuda, Electrotechnical Laboratory
Toshio Mimaki, Central Research Institute of Electric Power Industry
Hiroyuki Iwai, Osaka Gas Corporation
Hiroshi Kurita, Japan Energy Corporation
Shigeki Tokita, TYK Corporation
Hideto Ikeda, Ishikawajima-Harima Heavy Industries Corporation
Joji Shinohara, Ishikawajima-Harima Heavy Industries Corporation
Secretariat: Hiromi Yamaguchi, Ishikawajima-Harima Heavy Industries Corporation

Army Thermophotovoltaic Efforts

John S. Kruger*, Guido Guazzoni**, and Selma J. Nawrocki***

*U.S. Army Research Office, P.O. Box 12211, Research Triangle Park 27709-2211,
**U.S. Army CECOM C2D, Ft. Monmouth, NJ 27703, and
***U.S. Army CECOM C2D, Ft. Belvoir, VA 22060-5817

Abstract. A presentation and description of the several efforts in Thermophotovoltaic (TPV) Energy Conversion for power generation supported/monitored by the Army is provided with their more recent technical status and results. The efforts are related to small business (SBIR, STTR) contracts, academic research grants (MURI), and contracts awarded as the result of specialized solicitations. This paper covers a number of Army potential uses of the TPV power generation and is an attempt to give a more cohesive and integrated picture of the various military interests in TPV. With the exception of low power (<10 W) units, all Army potential uses of TPV power sources will demand operation with logistically available fuels.

BACKGROUND

Throughout this century, the Department of the Defense (DoD) demand for power has continued to increase. Today, electrical power requirements for military missions range from small compact, lightweight portable power sources required to operate small electronic devices and recharge batteries (50-W to 3-kW output) to large megawatt power plants required to operate air bases, seaports and base camps. Military Services of the 21st Century (today through the year 2010) and the Army After Next (AAN) (from 2010 and out) will require electrical power of even greater levels to carry out mission requirements for a wide range of user needs. As increasingly advanced weapon systems, with large power consumption requirements, enter the tactical battlefield, the need for more reliable, highly mobile and logistically supportable mobile electric power sources will also increase. Further, the DoD military forces of the future will be based primarily in CONUS (continental U.S.). Thus current and future battlefield scenarios will require the ability to rapidly deploy troops, and transport tactical and support power assets to ensure that military mission objectives can be met. Deployability here demands that systems must be smaller, lighter and more reliable. The inability to deploy rapidly is an internal threat that will negatively impact the success of future missions. An evaluation of the industrial base has validated that commercial power systems do not meet these requirements, as they are unique to the military.

The U.S. Army Research Office (ARO) and the Defense Advanced Research Projects Agency (DARPA) have sponsored multi-million 6.1 basic research efforts dollar (at least $16 M total to date) in state-of-the-art (SOA) TPV technology for tactical battlefield applications. Several Small Business Innovative Research (SBIR), Small Business Technology Transfer (STTR), and Broad Agency Announcement (BAA) programs have been initiated which address the R&D challenges associated with the development and demonstration of SOA TPV power systems for use on the tactical battlefield. The work is being done in support of commercialization and to further major Army initiatives to lower power costs while extending mission time and reducing the weight burden of the tactical soldier. Each effort is devised to ensure that the resulting TPV system design/prototype has been formatted for a wide range of user needs. The Armyís Communication-Electronics

Command (CECOM) and the National Aeronautics and Space Administration (NASA) act as the agents for these contracts.

This paper focuses on the Armyís current technical efforts to advance/adapt the SOA TPV power technologies for DoD use on the tactical battlefields of the AAN ñ battlefields which continue to become highly automated and technically sophisticated.

INTRODUCTION

The battlefields of the AAN will require quiet, reduced IR-signature, highly reliable power systems capable of providing 50 to 500 Watts of continuous power, of starting and operating on JP-8 / DF-2 fuels, and of being portable (less than 60 pounds). To date, there are no DoD or commercial (U.S. or foreign) logistic-fuel burning, manportable power systems in this power range available to meet the operational/performance requirements of the future battlefield.

To meet these power requirements, the DoD has tasked CECOM Research, Development and Engineering Center (RDEC) to establish, coordinate, and execute a Technology-Based Portable Power Program, which ensures that all Army and DoD power requirements are met. The CECOM Power Program is structured to meet the most demanding power and energy requirements, while balancing the need for high technical performance and operational readiness with the constraints we must live under today, that is, dwindling resources and lack of skilled manpower. Among the power technologies being investigated is TPV. The Army research and development (R&D) community is working together to advance and to adapt TPV technology to achieve DoD and Army goals for future power systems.

The Army goal is to enhance electrical power generation, storage, and conditioning capabilities required to support multiple tactical applications such as tactical operations centers (TOCs); communications and weapon systems; sensors, target acquisition, combat service support and battery chargers to decrease the logistics burden associated with batteries. Technology development support thrusts are aimed at lowering acquisition cost, reducing operations and support (O&S) costs, executing Army advanced technology demonstrations (ATDs), and furthering modernization. Of the SOA power technologies evaluated, TPV has many advantages over existing power technologies. It has potential to overcome the limitations of size, weight, and power found in existing commercial power sources in the 50 ñ 500-Watt range.

TPV power sources are refuelable sources of power that can be made to operate on multiple logistically available fuels and alternative fuels. TPV systems, in competitive package volumes, will be quieter and have lower vibration characteristics than engine-driven power sources of similar ratings. At the lowest end of this range, TPV might be more efficient than the very small engines appropriate for advanced, compact, lightweight power systems. At the lower ratings, the delivered energy cost is expected to be much lower than what is obtainable from either primary or rechargeable batteries. For example, the BA-5590 LiSO2 battery has a minimum cost of $350/kW-hr. Higher energy density replacements will cost even more. Equipment operating lifetimes should far exceed those of rechargeable batteries whose life is measured in 10s or 100s of recharge cycles. At the higher end of the power output range, the low noise and vibration, along with the multi-fuel capability, are the principal advantages of TPV power technology.

Adversely, TPV power systems will not be absolutely noiseless like a battery. They will have a hot exhaust gas flow and will present some IR signature. TPV systems may have some orientation restrictions. Such characteristics may preclude TPV from being integrated into very compact electronic devices (although there have been direct battery replacements, such as one reported at this Conference). They may be used as external power sources or battery chargers.

TPV challenges in this power range include an anticipated substantial initial cost disadvantage (estimated to be 2 to 5 times higher) and uncertain ruggedness compared to military requirements. This latter characteristic could lead to substantially higher life-cycle costs than engine-driven systems of similar ratings. IR signature will require significant suppression measures. These are the challenges that DARPA, ARO, CECOM, and the rest of the Army R&D community are working to overcome.

ARMY TPV APPLICATIONS

With the deployment (introduction) of new SOA communications and weapons systems for the 21st Century, the need for small, lightweight, logistic-fuel-burning power sources is imperative. Much of this future equipment will depend on rechargeable batteries as the principal power source. Continuous exchange of primary (non-rechargeable) batteries is very costly and becoming a serious supply problem. Several primary batteries are being replaced with a single rechargeable battery of the same form and fit. In order to maintain battlefield readiness, these new batteries need a reliable recharger.

And so CECOM is developing power source technology to provide man-portable, multi-fuel, and silent energy source for forward-area battery charging. To effect this, military personnel must be provided with lightweight battery chargers capable of operating on logistically available fuel. The forward-area/frontline soldier will accept the replacement of the throwaway battery with a rechargeable one if the recharging can be done on site, with a self-powered, tactical battery charger that uses the same JP-8 or DF-2 fuels as the military vehicles and larger engine-driven generator sets.

A TPV power source, operating on DF-2 or JP-8, can be engineered as a portable battery charger with different power levels (from a few Watts to a few hundred Watts) to support several power applications:

A single-battery charger (for batteries of the size and energy content of a BB-5590, the rechargeable version). Such a configuration will allow a soldier to recharge one battery at a time, possibly without having to remove the battery from the equipment it powers. This capability will allow the dismounted soldier who carries multiple batteries for communications and night vision to periodically recharge the batteries without having to exchange them for new ones.

A 100 ñ 200-W recharger, which can be mounted on a tactical vehicle and directly use the vehicleís fuel supply. Such a configuration would provide forward-area multiple battery recharging (6 ñ 10 batteries at the same time) in close proximity to the soldierís equipment.

A 500-W Power System/Battery Recharger, which would be capable of starting and operating on logistically available fuels. Such a system should be environmentally compatible and lightweight, and provide high energy density for long duration missions. Use of such a configuration would extend mission time and reduce the load carried by the soldiers. The resulting system would provide not only the forward-area-recharging capability in close proximity to the supported equipment, but also meet the expanded power needs for surveillance and communications.

TPV RESEARCH

The Army is sponsoring/monitoring research to support the future deployment of TPV equipment on the battlefield. The ARO is supporting work under the DoD Multidisciplinary University Research Initiative (MURI), and under Department of the Army STTR Programs.

DoD MURI Investigation

The MURI Center at Western Washington University (WWU), supported by its small business partner, JX Crystals, Inc. (JXC), is carrying out a five-year program (at $900 K per year) to produce a 3- to 5-kW generator of electric power, while studying the underlying issues affecting TPV technology. The stated goals are to initially use natural gas, propane, or butane for a lightweight system producing some 250 W per kg, and aiming for overall efficiencies above 25%.

This work involves developing a basic understanding of TPV technology, processes, components and subsystems. The agenda includes studying cells (Si, GaSb, and InGaAs, multi-fuel operation (gaseous and liquid), waste heat recuperation, military and commercial (dual-use) possibilities, and the scalability of power levels. An important component of this and every other MURI grant is the training of scientist and engineers through direct support of graduate students.

Over the course of this research, WWU has dramatically increased the power output to nearly 1 kW, while significantly reducing the overall weight by paying close attention to the interplay of components and subsystems. A prototype TPV automobile based in part on this sponsored work was shown at this Conference.

Army Small Business TPV Efforts

As part of the Army sponsorship of R&D, the STTR program encourages small companies to collaborate with research partners (such as a university, non-profit organization, or national laboratory). STTR contracts involve a 12-month proof-of-principal (POP) phase (for about $100 K), followed by a two-year demonstration phase (at up to $250 K per year). STTR seeks to transfer technology from the research institution to the small business for demonstration and eventual commercialization. ARO currently has two TPV STTR investigations underway, which are discussed below, for TPV power sources in the 100- to 150-W range.

Likewise, the SBIR program seeks novel research to meet Army needs. Typically an SBIR effort consists of a first POP phase of 6 months costing about $70 K, and a possible second phase for two years at about $350 K per year. CECOM at Ft. Monmouth has sponsored two recent SBIR tasks (one finished after phase I) which are presented later in this paper. These efforts concentrate on battery chargers.

Quantum Group, Inc. STTR Work

Quantum Group, Inc. (QGI), supported by the University of Delaware, is preparing TPV demonstration to recharge several batteries simultaneously. During the first phase, QGI decided on an ytterbia mantle and Si photocell configuration, and analyzed optical reflections, thermal isolation, and air-cooling.

During the ongoing second phase they face some technical hurdles, namely: ruggedness of components and the overall system; operation independent of orientation; and system integration and performance optimization. The demonstration unit is expected to be ready shortly.

JX Crystals, Inc. STTR Contract

JXC has again working with WWU to work on a TPV generator of at least 150 W. During the first POP phase the team developed a better understanding of thermal design and discovered a special infrared (IR) emitter material to closely march the spectral characteristics of their GaSb photocells.

During phase II, JXC and WWU are concentrating on power generation with gaseous or liquid fuels (to meet the Armyí goal of multi-fuel capability), controlling the spectrum with their proprietary emitter and dielectric filter combination, and increasing system efficiency. Demonstrations units are expended in 1999.

Essential Research SBIR Effort

During the phase I work the Essential Research produced an interesting detector embodiment, namely an InGaAs/InP Monolithic Integrated Module (MIM), developed a gold Back Surface Reflector (BSR) to reject out-of-band radiation, and demonstrated a SiC emitter and propane burner, with the help of Teledyne Brown Energy Systems. At the end of phase I, CECOM down selected to a single SBIR contract. So the Essential work ended there.

JX Crystals, Inc. SBIR Investigation

This effort produced a number accomplishments during phase I. JXC built a 20-W demonstration unit, which operates on propane-butane. The unit has an internal rechargeable battery for start-up. Despite only 1% overall efficiency, the demonstrator can nevertheless recharge a single BB-5590 battery.
The phase II goals seek liquid-fuel operation, scale up to the 200-W level, and a ruggedized emitter for military operation.

Army In-house Work

One of the authors (Guazzoni) has been carrying out an investigation with a collaborator from William Patterson University, who is using the CECOM facilities. This effort focuses on characterizations of substrates and filters at temperatures higher than room temperature. This work is presented in another paper at this Conference.
With the support of additional funds made available by Congress, CECOM is planning to award at least one 12-month contract for a TPV flashlight/low-power electric generator combination. Proposals have already been received and evaluated.

CURRENT SYSTEMS DEVELOPMENT

The essential components of TPV systems can be made using available technology. All of the basic components of TPV systems have been shown in laboratory experiments and prototypes, or in small power generation demonstration systems. TPV systems have not been developed for Army field use or troop testing. The development risk is still considered to be moderate-to-high since practically-sized systems that go beyond the laboratory have not been designed, constructed and tested. The greatest need is for system work, along with concurrent continued component development and improvement. Component risk mitigation or performance improvement issues include: emitter spectral range and emitter efficiency; emitter ruggedness (particularly when hot); photocell conversion efficiency; filter or reflector effectiveness; materials performance at high temperature; compact recuperator/heat exchanger design and construction for efficiency enhancement; liquid fuel flow and combustion control; and electronic-power-conversion systems improvement.
In 1997, DARPA sponsored a BAA solicitation that resulted in the award of two contracts for the development of a 500-Watt, diesel-fuel-burning TPV system. The work is technically monitored by CECOM at Ft. Belvoir.

The two contractors (EDTEK, Inc. and McDermott Technologies, Inc.) will be demonstrating breadboard prototypes in the December 1998/January 1999 timeframe that are capable of starting and operating on DF-2. The objective of this work is to design, fabricate and test a 500-W portable, diesel-fueled TPV power system with overall efficiency of at least 8%. Both systems are projected to be 8 ñ 12 % efficient.

EDTEK Development Investigation

This work, supported by Brookhaven National Lab, completed a prototype design of a 500-W system, made up of a Power Converter Assembly (PCA) and a Burner/Emitter/Recuperator (BER) Assembly. In the course of this effort EDTEK developed a 3-dimensional thermo-optical model. With the fabrication and preliminary testing nearing completion of a fully integrated laboratory prototype, the system demonstration is expected with a few months.
The predicted system efficiency is over 15%, with an air-cooled PCA, starting and operating on DF-2, within the 15-pound weight.

McDermott Technologies Development Contract

With the help of JXC and VSE, Inc., McDermott completed design for the 500-W unit. The PCA consists of 380 GaSb photocells in a cylindrical array of 20 circuits, providing a gross output of 600 Watts when the emitter temperature is at 1700 K. The BER consists of a SiC emitter coupled with a high-intensity diesel burner and annular recuperator. The BER efficiency is about 65% and the emitter temperature 1600 K.
A demonstration of the fully integrated laboratory prototype is expected in January 1999. The predicted system efficiency should be in the range of 7-9% for the air-cooled PCA, with starting and operation on DF-2. The overall weight should be right at 15 pounds.

CONCLUSIONS

Based on current SOA, TPV appears to be a promising technology for satisfying both commercial and military power requirements in the low power range (50 ñ 500 W). We have noted that the Army is supporting TPV R&D on several converging avenues. University research is studying the fundamentals and systems interactions. Small businesses are developing medium-size units for demonstrations as power supplies and battery chargers. And collaborations among DARPA, CECOM, NASA are investigating 500-W TPV power sources. In all but the very smallest devices, liquid fuel (logistically available DF-2 or JP-8) capability should be achievable.

SESSION 1: FUNDAMENTAL ASPECTS

The Optical Constants of n- and p-Doped $In_{0.66}Ga_{0.34}As$ on InP (001) Including The Burstein-Moss Shift: Experiment and Modeling

Todd Holden[a], Fred H. Pollak[a], J. L. Freeouf[b], G.W. Charache[c], and J.E. Raynolds[c]

(a) Physics Department and NY State Center for Advanced Technology in Ultrafast Photonic Materials and Applications, Brooklyn College of CUNY, Brooklyn, NY 11210
(b) Interface Studies, Inc., Katonah, NY 10536
(c) Lockheed-Martin Corp., Schenectady, NY 12301

Abstract. The complex optical constants (real and imaginary components of the dielectric function and index of refraction) and absorption coefficient in the range 0.3–5.5 eV have been evaluated at 300K using spectral ellipsometry for a series of n-(5.7×10^{17} cm^{-3} $< n <$ 5.5×10^{19} cm^{-3}) and p-(6.5×10^{17} cm^{-3} $< p <$ 5.0×10^{19} cm^{-3}) doped relaxed $In_{0.66}Ga_{0.34}As$ grown by metalorganic chemical vapor deposition on InP (001). We have observed the fundamental absorption edge, spin-orbit split E_1-R_1, $(E_1+\Delta_1)$-R_1 doublet and E_2 feature. The data has been fit using a comprehensive model based on the electronic energy-band structure near critical points plus relevant excitonic and band-to-band Coulomb enhancement (BBCE) effects. The intrinsic band gap of 0.612 ± 0.01 eV corresponds to an In composition of $66 \pm 1\%$. The Burstein-Moss (BM) shift was accounted for using a Fermi level filling factor in addition to the excitonic and BBCE terms. While for the p-type samples the BM shift exhibited only parabolic effects, the n-type samples had pronounced non-parabolicity at the highest doping level, in agreement with a bandstructure calculation. By accounting for the BBCE term we have obtained the binding energy, R_1, of the 2D exciton associated with E_1-R_1, $(E_1+\Delta_1)$-R_1 features. Except for Holden *et al.* [Phys. Rev. B <u>56</u>, 4037 (197)] this quantity has not been obtained from any previous ellipsometric or other optical studies.

INTRODUCTION

Thermophotovoltaic (TPV) generation of electricity is attracting attention due to a number of factors including advances in materials and designs, as well as a widening appreciation of the large number of applications that can be addressed using TPV-based generators. The attractions include the wide range of fuel sources and the potentially high power density outputs. One main approach to TPV generation is broad band radiation coupled with converters with bandgaps in the range 0.4-0.7 eV (1).

Using spectral ellipsometry in the range 0.3-5.45 eV we have evaluated the complex optical constants (real and imaginary components of the dielectric function and index of refraction) and absorption coefficient at room temperature for a series of n-(5.7×10^{17} cm^{-3} $< n <$ 5.5×10^{19} cm^{-3}) and p-(6.5×10^{17} cm^{-3} $< p <$ 5.0×10^{19} cm^{-3}) doped relaxed $In_{0.66}Ga_{0.34}As$ grown by metal organic chemical vapor deposition (MOCVD) on InP (001). We have also observed the fundamental absorption edge, spin-orbit split E_1-R_1, $(E_1+\Delta_1)$-R_1 doublet [transitions along the <111> directions of

CP460, *Thermophotovoltaic Generation of Electricity: Fourth NREL Conference*
edited by T. J. Coutts, J. P. Benner, and C. S. Allman
© 1999 The American Institute of Physics 1-56396-828-2/99/$15.00

the Brillouin zone (BZ)], and E_2 feature. The experimental data has been fit using a comprehensive model based on the electronic energy-band structure near critical points plus relevant excitonic and band-to-band Coulomb enhancement (BBCE) effects. The BBCE terms at both the fundamental gap as well as the E_1-R_1, $(E_1+\Delta_1)$-R_1 doublet have been neglected in the past analysis of ellipsometric studies except for the recent work of Holden et al. (2). The intrinsic band gap of 0.612 ± 0.01 eV corresponds to an In composition of $66\pm1\%$. The Burstein-Moss (BM) shift of the fundamental absorption edge was accounted for using a Fermi level filling factor in addition to the excitonic and BBCE terms. While for the p-type samples the BM shift exhibited only parabolic effects, the n-type samples had pronounced non-parabolicity at the highest doping level. These observations for the n-type material are in agreement with a recent bandstructure calculation (3). By accounting for the BBCE term we obtained the binding energy, R_1 ($\sim$ 90 meV), of the 2D exciton associated with E_1-R_1, $(E_1+\Delta_1)$-R_1 features. Except for Ref. 2 this quantity has not been obtained from any previous ellipsometric or other optical studies.

EXPERIMENTAL DETAILS

Samples used in this study were a series of n- and p-doped relaxed $In_{0.66}Ga_{0.34}As$ grown by MOCVD on InP (001) substrates. The epitaxial layers were $\sim$ 1 micron thick. The carrier concentrations were determined by room temperature Hall measurements. The characteristics of the p-and n-type samples are listed in Table I. The optical data in the range 0.75-5.45 eV [ultraviolet (UV)/visible (VIS)/near infrared (NIR)] was taken using an Instruments SA variable angle ellipsometer while for the interval 0.3-1.0 eV [mid infrared (MIR)/far infrared (FIR)] a SENTECH variable angle instrument was employed. Thus there was some overlap between the two intervals. The latter uses a Fourier Transform Infrared Reflectometer as a light source.

To remove the surface oxide an etching procedure was performed on the UV/VIS/NIR ellipsometer with prealigned samples mounted vertically on a vacuum chuck in a windowless cell that maintained the surfaces in a dry nitrogen atmosphere (2). The details of the etching procedure are given in Ref. 2, except in this study we the etch was a saturated KOH/methanol solution instead of HF.

The UV/VIS/NIR measurements were done with a 70° incidence angle, and the MIR/FIR measurements were performed with 60° and 70° incidence angles.

EXPERIMENTAL RESULTS

Shown by the solid lines in Figs. 1 and 2 are the experimental values of the imaginary component of the complex dielectric function, ϵ_2, for p-type samples A and E (displaced by five units for clarity) and for n-type samples F and K (displaced by five units for clarity), respectively, in the range 0.3-5.45 eV at 300K. The real part of the complex dielectric function was also determined, but for clarity is not shown.

The solid lines in Figs. 3 and 4 are the experimental values of ϵ_2 for n-type samples A and E and p-type material F and K, respectively, in the region of the fundamental absorption edge. In Figs. 2 and 4 note the small finite value of ϵ_2 for sample K below ~ 0.5 eV duse to free carrier absorption.

For undoped InGaAs the fundamental absorption edge, E_0, corresponds to a three dimensional (3D) M_0 critical point (CP) at $\vec{k} = 0$ from the highest spin-orbit split valence band (Γ_8^v) to the lowest conduction band (Γ_6^c) at the Γ point, i.e., Γ_8^v-Γ_6^c. The spin-orbit split component is $E_0+\Delta_0$ (Γ_7^v-Γ_6^c). The effect of the impurity related free electrons/holes is to cause a Fermi level filling effect and related blue shift of the absorption edge, i.e., BM shift. The corresponding 2D M_0 CPs

TABLE I. Doping levels and fitting parameters for the fundamental absorption edge. Γ_0 was 60 meV for all samples.

Sample	Doping (10^{17} cm^{-3})	$E_{abs}^{(c)}$ (eV)	"kT" (meV)
A	6.5[a]	0.612[d]	60
B	9.4[a]	0.622	57
C	55[a]	0.658	52
D	87[a]	0.693	52
E	500[a]	0.739	44
F	5.7[b]	0.636	49
G	13[b]	0.660	55
H	28[b]	0.772	53
I	77[b]	0.833	47
J	106[b]	0.955	42
K	550[b]	1.270	50

(a) p-type. (b) n-type. (c) error bars of ± 0.01 eV. (d) $E_{abs} = E_0$.

are labelled E_1 [$L_{4,5}^v(L_3^v)$-$L_6^c(L_1^c)$] and $E_1 + \Delta_1$ [$L_6^v(L_3^v)$-$L_6^c(L_1^c)$]. The E_2 feature is due to transitions along [110] (Σ) or near the X point of the BZ, at an energy of about 4.5 eV. We have also included the influence of the Γ_8^v-L_6^c indirect transition (2).

ϵ_2 in the Region of the Fundamental Absorption Edge

The presence of the exciton at the fundamental gap affects the band-to-band component of the absorption; and, hence a band-to-band Coulomb enhanced (BBCE) function is used (2). Thus even if the exciton is not resolved, the Coulomb interaction still affects the band-to-band lineshape. In addition, for doped samples the Fermi-level filling factor and related BM shift have to be taken into account by:

$$\epsilon_2(E) = \frac{A}{[1+e^{(E_{abs}-E)/"kT"}]E^2}\left[\frac{R_0\Gamma_0}{(E_0-R_0-E)^2+\Gamma_0^2}\right.$$
$$\left.+\int_{-\infty}^{\infty}\frac{\Theta(E'-E_0)}{1-e^{-2\pi z_1(E')}}\frac{\Gamma_0 dE'}{(E-E')^2+\Gamma_0^2}\right] \quad (1)$$

where R_0 is the Rydberg (exciton binding energy), E_{abs} is the absorption edge, Γ_0 is the broadening parameter for both the exciton and BBCE terms (these two parameters have been taken as equal), $z_1(E) = [R_0/(E-E_0)]^{1/2}$, and $\Theta(E)$ is the unit step function. In Eq. (1) the quantity $A \propto (R_0)^{1/2}\mu_0^{3/2}|M_0|^2$ where μ_0 is the reduced interband effective mass at E_0, and M_0 is the matrix element of the momentum between $\Gamma_8^v-\Gamma_6^c$. For non-parabolic bands μ_0 and M_0 are energy dependent.

The term $[1+e^{(E_{abs}-E)/"kT"}]^{-1}$ in Eq. (1) is the Fermi level filling factor. For undoped (or lightly p-doped) samples $E_{abs} = E_0$. The fact that "kT" is in quotation marks will be explained later. In Eq. (1) the first term in the square brackets is a Lorentzian associated with the discrete exciton, while the second term (integral) is the BBCE factor.

ϵ_2 for the Higher Lying Transitions

Following Ref. 2 for the spin-orbit split $E_0+\Delta_0$ transition we used a function similar to Eq. (1), but without the Fermi-level filling term and with $A \rightarrow B$, $E_0 \rightarrow E_0+\Delta_0$, $R_0 \rightarrow R_{so}$ and $\Gamma_0 \rightarrow \Gamma_{so}$. The constant B is proportional to μ_{so} and M_{so} (matrix element between $\Gamma_7^v-\Gamma_6^c$).

The $E_1, E_1+\Delta_1$ CPs are of the 2D M_0 type and, hence, for E_1 we can write:

$$\epsilon_2(E) = \frac{C_1}{E^2}\left[\frac{4R_1\Gamma_1}{(E_1-R_1-E)^2+\Gamma_1^2} + \int_{-\infty}^{\infty}\frac{\Theta(E'-E_1)}{1-e^{-2\pi z_2(E')}}\frac{\Gamma_1 dE'}{(E-E')^2+\Gamma_1^2}\right] \quad (2)$$

where C_1 is a constant, E_1 is the energy of the gap, R_1 is the 2D Rydberg energy, Γ_1 is the broadening parameter for both the exciton and band-to-band transition, and $z_2(E) = [R_1/4(E-E_1)]^{1/2}$.

The $E_1+\Delta_1$ CP also has been described by excitonic and BBCE components, with parameters similar to Eq, (2) with $C_1 \rightarrow C_2$ and $E_1 \rightarrow E_1+\Delta_1$, etc.

In practice only a single broadening parameter, Γ_1, was used for the excitonic as well as the BBCE components for both the E_1 and $E_1+\Delta_1$ transitions, and the same 2D Rydberg (R_1) was used for both features.

The expression for the $\Gamma_8^v-L_6^c$ indirect transition is given by Eq. (4) of Ref.

2. The fitting parameters are the constant D_{ind}, the energy E_{ind}, the broadening parameter Γ_{ind}, and cutoff energy E_c.

For the E_2 feature, the formula for ϵ_2 is given by Eq. (5) of Ref. 2. The fitting parameters are the constant F_2, the energy E_2, and the dimensionless broadening parameter γ.

DISCUSSION OF RESULTS

Shown by the dashed lines in Figs. 3 and 4 are the least-square fits to Eq. (1), plus background terms due to the contribution of higher lying transitions, i.e., $E_0 + \Delta_0$, $\Gamma_8^v - L_6^c$ indirect transition, E_1-R_1, $(E_1+\Delta_1)$-R_1, etc. The obtained values of E_{abs} ($= E_0$) for sample A and E_{abs} (B-M shift) for samples E, F, and K are denoted by arrows in these figures. E_{abs} values for all samples are listed in Table I. The value of $E_0 = 0.612 \pm 0.01$ eV corresponds to an In composition of $(66 \pm 1)\%$ (4). In all cases we used $\Gamma_0 = 60$ meV.

Also listed in Table I are the deduced values for "kT". In order to take into account inhomogeneous broadening, Eq. (1) (with "kT" set equal to its room temperature value of 26 meV) should be convoluted with a Gaussian. However, this is very difficult to do analytically, so to "simulate" inhomogeneous broadening we have used "kT" as an adjustable parameter.

In Fig. 5 the solid squares and triangles are the values of the absorption edge (from Table I) for the *n*-and *p*-type samples, respectively, as a function of (dopant concentration)$^{2/3}$. Representative error bars are shown. For the case of parabolic bands we should have:

$$E_F - E_0 = (\hbar^2/2m_{c(h)}^*)[3\pi^2 n(p)]^{2/3} - E_{BGR} \qquad (3)$$

where $m_{c(h)}^*$ is the effective mass of the electron/hole, $n(p)$ are the electron/hole carrier densities, and E_{BGR} are any band gap reduction effects.

Note that in Fig. 5 for the first five *n*-type samples the relation is linear, as one would expect for parabolic band filling. The solid line is a least-squares fit to a linear function (neglecting the highest doped sample) yielding an intercept (E_0) of 0.599 ± 0.010 eV and a slope of 7.3×10^{-14} eV-cm^2. The deviation of the highest doped sample is due to non-parabolic effects. For the *p*-type samples a least-squares fit to a linear function yields $E_0 = 0.605 \pm 0.010$ eV and a slope of 1.12×10^{-14} eV-cm^2. Note that there is no evidence for nonparabolic effects. Within experimental error the values of E_0 for the *n*-and *p*-type samples in Fig. 5 are the same, and are in good agreement with the value in Table I.

These non-parabolic effects have been accounted for based on electronic band structure calculation using the Full Potential Linearized Augmented Plane Wave method which accounts for non-local exchange and spin orbit effects (3). Good agreement is found between this experiment and theory for the *n*-type samples.

Recently Charache et al. (5) reported on the optical (absorption) and electrical properties of doped n-In$_x$Ga$_{1-x}$As (x = 0.67 and 0.53). The former samples are the same as the n-type samples used in this study. Their obtained value of 0.6 eV for E$_0$ is very close to our value. However, they found deviations from the parabolic band calculation at a doping density of about 7x10^{18} cm^{-3}, while we find that this approximation is still valid for densities as high as ~ 1.0x10^{19} cm^{-3} (see Fig. 5).

Note in Figs 2 and 4 that for sample K, the most heavily doped n-type material, on the low energy side there is a small rise in ϵ_2. This is due to the free carrier plasma absorption observed for these samples (5). From Drude theory:

$$\epsilon_1 + i\epsilon_2 = \epsilon_\infty - \frac{E_p^2}{E^2 + iE\gamma_{fc}}, \quad (4)$$

$$E_p^2 = \frac{4\pi nq^2}{\epsilon_0 m_c^*}$$

where ϵ_∞ is the high energy dielectric function and ϵ_0 is the permittivity of free space. For sample K the quantity E$_p$ ≈ 0.22 eV (5) and so according to Eq. (4) a rise in ϵ_2 is expected below the fundamental absorption edge.

The dashed line in Fig. 1 is a least-squares fit over the entire spectral range for sample A using the above expressions for the fundamental absorption edge [Eq. (1)]; the CPs E$_0$+Δ_0, E$_1$ [Eq. (2)], E$_1$+Δ_1, and E$_2$; as well as the indirect transition Γ_8^v-L_6^c. Since we have not resolved the E$_0$+Δ_0 feature, the value for this gap for x = 0.66 was taken from Ref. 4. For E$_{ind}$ we employed a number using E$_0$ (0.612 eV)

TABLE II. Materials parameters for the fit of ϵ_2 for sample A

Parameter	In$_{0.66}$Ga$_{0.34}$As
E$_0$ (eV)	0.612±0.010
A (eV^2)	0.056±0.003
R$_0$/R$_{so}$ (meV)	3.5
Γ_0 (meV)	60±12
E$_0$+Δ_0 (eV)	0.95 [a]
B (eV^2)	0.046±0.01
Γ_{so} (meV)	100
E$_1$-R$_1$ (eV)	2.56±0.01
C$_1$ (eV^2)	11.4±1
R$_1$ (meV)	92±15
Γ_1 (meV)	200±10
(E$_1$+Δ_1)-R$_1$ (eV)	2.82±0.01
C$_2$ (eV^2)	7.5±0.5
E$_{ind}$ (eV)	1.42 [b]
F$_{ind}$	3±0.5
Γ_{ind} (meV)	400±100
E$_c$ (eV)	4
E$_2$ (eV)	4.51±0.02
D$_2$	3.1±0.1
γ	0.21±0.02

(a) Ref. 3 for x = 0.66. (b) From E$_0$+(Γ_6^c-L_6^c) (Ref. 5) plus the $\Gamma_6^c - L_6^c$ separation (0.81 eV), the latter being obtained from x-ray absorption measurements from core levels to the conduction bands (6). Because of the large number of adjustable parameters in the fit, values of variables such as E_1-R_1, $(E_1+\Delta_1)$-R_1, Γ_1 were initialized by a procedure described in Ref. 2. Listed in Table II are the values for the various parameters used in the fit. Indicated by arrows in Fig. 1 are the obtained values of E_1-R_1, $(E_1+\Delta_1)$-R_1, and E_2.

Our value of the 2D exciton binding energy R_1 ($\sim$ 90 meV) is in good agreement with the general considerations of Ref. 7, based on effective mass and $\vec{k} \cdot \vec{p}$ theories.

The features in the optical spectra of diamond-and zincblende-type semiconductors corresponding to transitions along the equivalent $<111>$ directions of the BZ are primarily excitonic in nature due to the large exciton binding energy, R_1, ($\approx$ 100-300 meV) of the 2D excitons associated with the E_1, $E_1+\Delta_1$ CPs (1,6). Therefore, the optical structure associated with transitions along $<111>$ do not actually correspond to the E_1 and $E_1+\Delta_1$ CPs, but to the excitonic transitions $E_1 - R_1$ and $(E_1+\Delta_1) - R_1$, respectively. Since previous ellipsomteric studies have not included the BBCE term at the E_1 (and E_0) CPs, they have not evaluated R_1 (8).

Pickering et al. (9) have reported an ellipsometric study of strained and relaxed $In_xGa_{1-x}As$ for $0 \leq x \leq 0.53$ as well as InAs in the region of the $E_1 - R_1$ and $(E_1+\Delta_1) - R_1$ features. For $x = 0.66$ the energies of these peaks should be 2.53 eV and 2.79 eV, respectively, which is in good agreement with our data (see Table II).

SUMMARY

Using spectral ellipsometry in the range 0.3-5.45 eV we have evaluated the optical constants of a series of n-and p-doped relaxed $In_{0.66}Ga_{0.34}As$ samples of interest for TPV applications. The experimental data has been fit using a comprehensive model based on the electronic energy-band structure near critical points plus relevant excitonic and BBCE effects. The intrinsic band gap (E_0) of 0.612 ± 0.01 eV corresponds to $x = 0.66 \pm 0.01$. The BM shift was taken into account using a Fermi level filling factor in addition to the excitonic and BBCE terms at the fundamental absorption edge. While for the p-type samples the BM shift exhibited only parabolic effects, the n-type samples had pronounced nonparabolicity at the highest doping level. These results are in agreement with a recent bandstructure calculation. By taking into account the BBCE term we obtained the binding energy R_1 ($\sim$ 90 meV) of the 2D exciton associated with E_1-R_1, $(E_1+\Delta_1)$-R_1 features. Except for Ref. 2 this quantity has not been obtained from any previous ellipsometric or other optical studies.

ACKNOWLEDGEMENTS

The authors TH and FHP acknowledge support from Semiconductor

Characterization Instruments, Inc., NSF grant #DMR-9414209, and the NY State Science and Technology Foundation through its Centers for Advanced Technology program.

REFERENCES

1. G.W. Charache et al., J. Electron. Mat. **27**, 1038 (1998).
2. T. Holden, P. Ram, F.H. Pollak, J.L. Freeouf, B.X. Yang, and M.C. Tamargo, Phys. Rev. B **56**, 4037 (1997).
3. J.E. Raynolds, C.B. Geller, G.W. Charache, T, Holden, and F.H. Pollak, this conference.
4. S.H. Pan, H. Shen, Z. Hang, F.H. Pollak, W. Zhuang, Q. Xu, A.P. Roth, R. Masut, C. LeCelle, and D. Morris, Phys. Rev. B **38**, 3375 (1988).
5. G.W. Charache, D.M. DePoy, J.L. Egley, R.J. Dziendziel, P.R. Sharps, M.L. Timmons, R.E. Fahey, K. Zhang, and J.M. Borrego, AIP Conf. Proc. **401**, 215 (1997).
6. K. Miyano, F.H. Pollak, and G. Charache, private communication.
7. Y. Petroff and M. Balkanski, Phys. Rev. B **3**, 3299 (1971).
8. M.L. Cohen and J. R. Chelikowsky in *Electronic Structure and Optical Properties of Semiconductors, Second Edition* (Springer-Verlag, Berlin, 1988); S. Adachi, Phys. Rev. B **41**, 1003 (1990); S. Adachi and T. Taguchi, Phys. Rev. B **43**, 9569 (1991); S. Ozaki, S. Adachi, M. Dato and K. Ohtsuka, J. Appl. Phys. **79**, 439 (1996); D.E. Aspnes and A.A. Studna, Phys. Rev. B **27**, 985 (1983).
9. C. Pickering, R.T. Carline, M.T. Emeny, N.S. Garawal, and L.K. Howard, Appl. Phys. Lett. **60**, 2412 (1992).

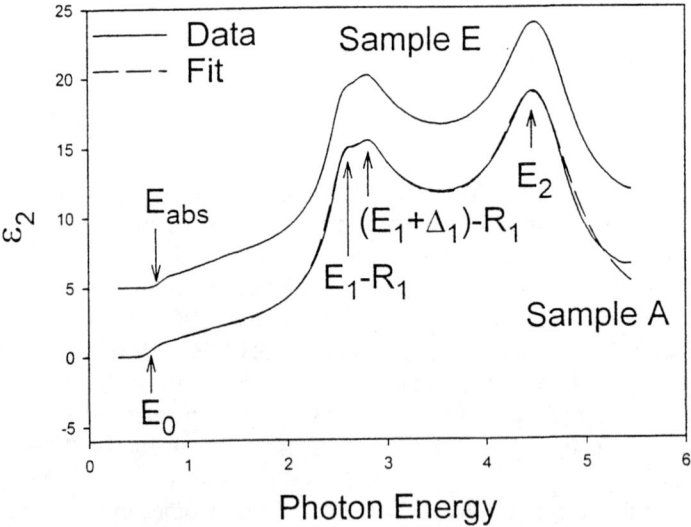

FIGURE 1 Experimental values (solid lines) of ϵ_2 for *p*-type samples A and E. Dashed line is the fit for sample A.

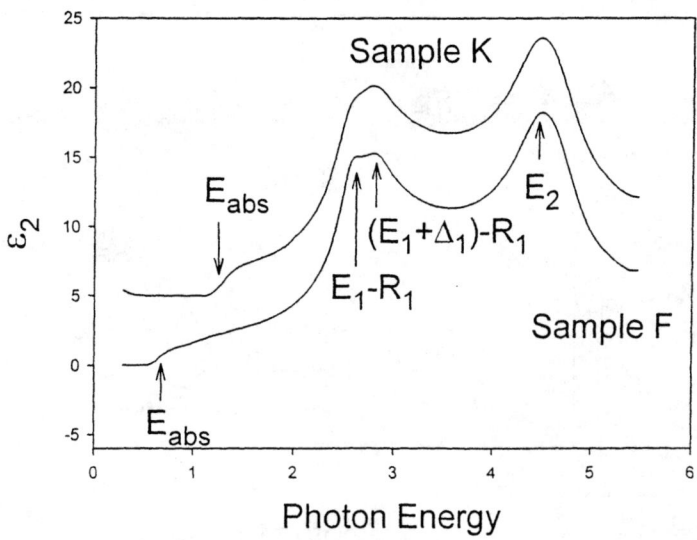

FIGURE 2 Experimental values (solid lines) of ϵ_2 for *n*-type samples F and K.

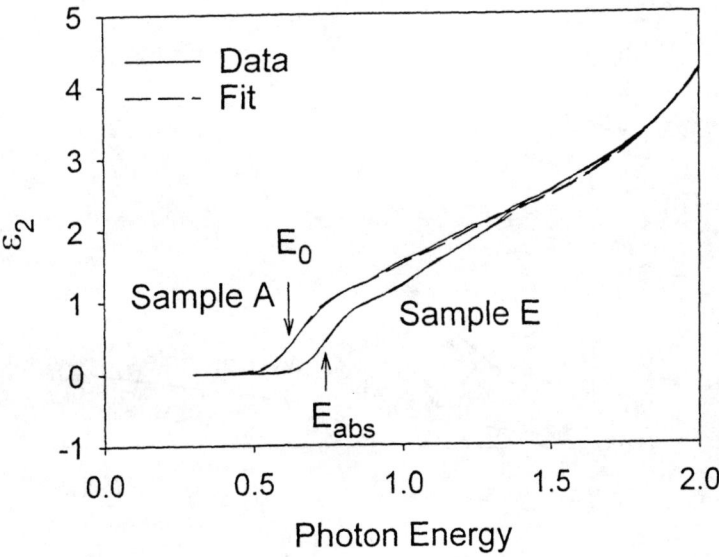

FIGURE 3 Experimental values (solid lines) of ϵ_2 in the region of the fundamental absorption edge for *p*-type samples A and E. Dashed lines are the fits.

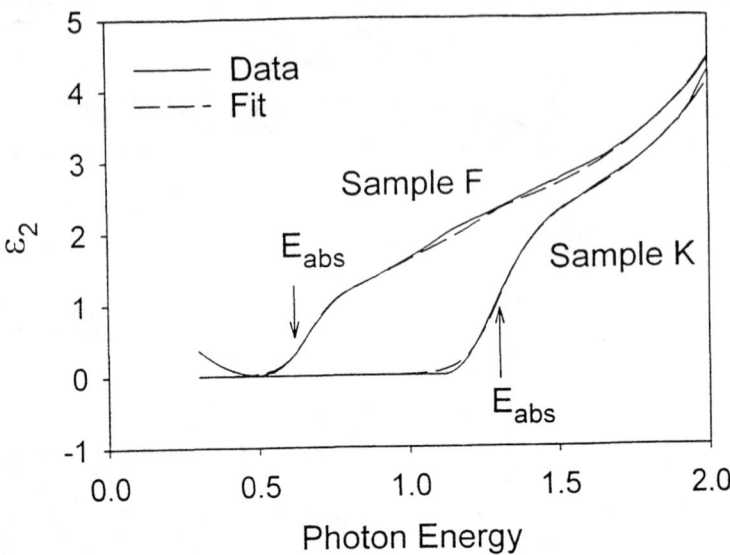

FIGURE 4 Experimental values (solid lines) of ϵ_2 in the region of the fundamental absorption edge for *n*-type samples F and K. Dashed lines are the fits.

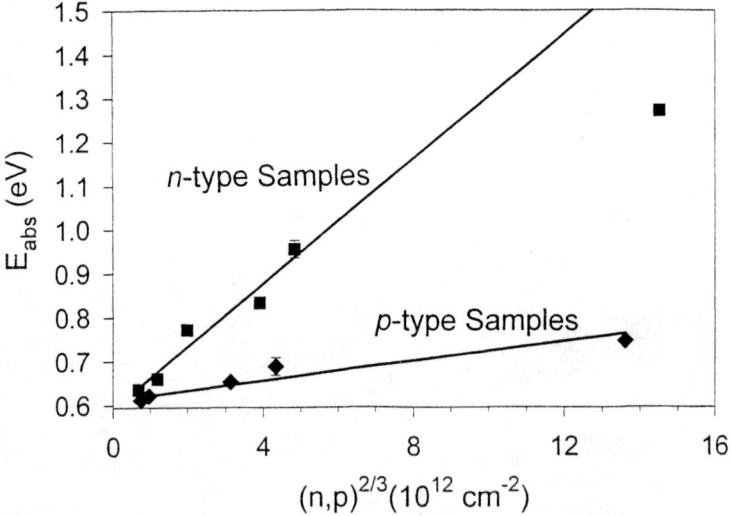

FIGURE 5 Solid squares and triangles are E_{abs} for *n*-and *p*-type samples, respectively, as a function of (dopant concentration)$^{2/3}$. Solid lines are fits to a linear function.

Enhanced electro-magnetic energy transfer between a hot and cold body at close spacing due to evanescent fields

J. E. Raynolds

Lockheed Martin Corp., P. O. Box 1072, Schenectady, NY 12301-1072

ABSTRACT

Theoretical studies have demonstrated that the energy transfer between a hot an cold body at close spacing (on the order of the radiation wavelength) can greatly exceed the limit for black body radiation (i.e. Power = σT^4).[1-9] This effect, due to the coupling of evanescent fields, presents an attractive option for thermo-photovoltaic (TPV) applications (assuming the considerable technical challenges can be overcome).[10-12] The magnitude of the enhanced energy transfer depends on the optical properties of the hot and cold bodies as characterized by the dielectric functions of the respective materials. The present study considers five different situations as specified by the materials choices for the hot/cold sides: metal/metal, metal/insulator, metal/semiconductor, insulator/insulator, and semiconductor/semiconductor. For each situation, the dielectric functions are specified by typical models. An increase in energy transfer (relative to the black body law) is found for all situations considered, for separations less than one micron, assuming a temperature difference of 1000C. The metal/metal situation has the highest increase vs. separation while the semiconductor/semiconductor has the lowest. Factor-of-ten increases are obtained at roughly 0.1 microns for the metal/metal and roughly 0.02 microns for the metal/semiconductor.[13] These studies are helping to increase the understanding of the close-spaced effect in the context of a radiator/TPV context.

INTRODUCTION

The enhancement at close spacing can be described within the framework of statistical mechanics and classical electromagnetism. Any object at non-zero temperature radiates energy, in the form of fluctuating electromagnetic fields, which are produced by the thermal motion (acceleration) of electric charges within the material. There is a component of the electromagnetic field near the surface of a radiating body which has a fluctuating time dependence, but does not propagate. This so-called "evanescent" field decays exponentially with a characteristic distance on the order of the wavelength. At sufficiently close spacings, the evanescent fields produced in one body interact directly with the electric charges in the other body; and, thus, act as sources of electromagnetic waves. These new waves are transmitted or absorbed depending on the properties of the second body. In a sense, the radiation "tunnels" across the gap and represents a net energy transfer.

MATHEMATICAL THEORY

Maxwell's equations of electro-magnetism serve as the starting point for the theoretical treatment. These are:

$$\nabla \cdot \vec{D}(\vec{r}, t) = 4\pi\rho(\vec{r}, t), \tag{EQ 1}$$

$$\nabla \cdot \vec{B}(\vec{r}, t) = 0, \tag{EQ 2}$$

$$\nabla \times \vec{H}(\vec{r}, t) = \frac{1}{c}\frac{\partial}{\partial t}\vec{D}(\vec{r}, t) + \frac{4\pi}{c}\vec{J}(\vec{r}, t), \tag{EQ 3}$$

$$\nabla \times \vec{E}(\vec{r}, t) = -\frac{1}{c}\frac{\partial}{\partial t}\vec{B}(\vec{r}, t), \tag{EQ 4}$$

where $\vec{E}(\vec{r}, t)$, $\vec{D}(\vec{r}, t)$, $\vec{H}(\vec{r}, t)$, and $\vec{B}(\vec{r}, t)$, are the electric field, the displacement field, the magnetic field, and the magnetic induction, respectively, at point $\vec{r}$ and time t and c is the speed of light. The quantities $\rho(\vec{r}, t)$, and $\vec{J}(\vec{r}, t)$ represent the net charge and current densities at point $\vec{r}$ and time t.[14]

By taking the Fourier transform with respect to time of Eqs. 3 and 4 we obtain:

$$\nabla \times \vec{H}(\vec{r}, \omega) = -i\frac{\omega}{c}\varepsilon(\omega)\vec{E}(\vec{r}, \omega) - i\frac{\omega}{c}\vec{g}(\vec{r}, \omega), \tag{EQ 5}$$

and,

$$\nabla \times \vec{E}(\vec{r}, \omega) = i\left(\frac{\omega}{c}\right)\vec{B}(\vec{r}, \omega),$$ (EQ 6)

where $i = \sqrt{-1}$ and the current density resulting from the fluctuating charges has been taken (for convenience) as $\vec{J}(\vec{r}, \omega) \equiv -i\omega\vec{g}(\vec{r}, \omega)/4\pi$. Detailed discussions of the fluctuating fields are presented in the book by Rytov.[8]

Once the fields have been obtained (through Fourier analysis), the energy transfer (energy per unit area, per unit time) is given by the time average of the Poynting vector,

$$P = \frac{c}{4\pi}\langle \vec{E} \times \vec{H} \rangle,$$ (EQ 7)

where the angled brackets denote time averaging. The Poynting vector contains products of the $\vec{g}$ fields which give non-vanishing contributions upon time averaging.

The Poynting vector can be evaluated because products of the $\vec{g}$ fields, when time averaged, are related to the dielectric function. The *fluctuation dissipation theorem* gives

$$\langle g_l(\vec{r}_1, \omega)g^*_m(\vec{r}_2, \omega')\rangle = A(\omega, T)\varepsilon''(\omega)\delta_{lm}\delta(\vec{r}_1 - \vec{r}_2)\delta(\omega - \omega'),$$ (EQ 8)

where $\varepsilon''(\omega)$ is the imaginary part of the dielectric function and δ_{lm} gives zero unless the two components of the $\vec{g}$ field are the same ($l = m$).[7] The Dirac δ-functions $\delta(\vec{r}_1 - \vec{r}_2)$ and $\delta(\omega - \omega')$, contribute to the integrals in Eq. 7 only when $\vec{r}_1 = \vec{r}_2$ and $\omega = \omega'$. The frequency and temperature dependent function $A(\omega, T)$ is given by

$$A(\omega, T) = 2\hbar \coth\frac{\hbar\omega}{2k_B T},$$ (EQ 9)

where $\hbar$ is Planck's constant, T is the temperature, and k_B is Boltzmann's constant.

The most general result for the energy transfer between two semi-infinite media with plane parallel surfaces separated by vacuum of thickness l, including magnetic effects, is given in Ref. 5 as:

$$P = \frac{\hbar}{\pi^2}\int_0^\infty \left(\frac{1}{\exp(\hbar\omega/k_B T_1) - 1} - \frac{1}{\exp(\hbar\omega/k_B T_2) - 1}\right)M(\omega)\omega d\omega$$

(EQ 10)

where,

$$M(\omega) = -\int_0^\infty \left\{ \frac{1}{|\Delta_\varepsilon|^2}\left(\frac{p_1}{\varepsilon_1} - \frac{p^*_1}{\varepsilon^*_1}\right)\left(\frac{p_2}{\varepsilon_2} - \frac{p^*_2}{\varepsilon^*_2}\right) \right.$$

(EQ 11)

$$\left. + \frac{1}{|\Delta_\mu|^2}\left(\frac{p_1}{\mu_1} - \frac{p^*_1}{\mu^*_1}\right)\left(\frac{p_2}{\mu_2} - \frac{p^*_2}{\mu^*_2}\right) \right\} |p|^2 x \, dx$$

In Eq. 11, the following definitions hold:

(EQ 12)
$$p = \sqrt{x^2 - k^2}, \quad p_\upsilon = \sqrt{x^2 - k^2 \varepsilon_\upsilon \mu_\upsilon}, \quad k = \omega/c, \quad (\upsilon = 1,2),$$

and,

$$\Delta_\varepsilon = \left(\frac{p_1}{\varepsilon_1} + p\right)\left(\frac{p_2}{\varepsilon_2} + p\right)e^{pl} - \left(\frac{p_1}{\varepsilon_1} - p\right)\left(\frac{p_2}{\varepsilon_2} - p\right)e^{-pl},$$

(EQ 13)

where l is the separation (gap) between the two semi-infinite materials. The expression for Δ_μ is obtained from Δ_ε upon making the substitution $\varepsilon_\upsilon \leftrightarrow \mu_\upsilon$. For p_υ the root must be chosen such as to give $Re(p_\upsilon) > 0$. In the above equations, the frequency dependence of ε_υ and μ_υ has been suppressed for brevity. The quantities T_1 and T_2 in Eq. 10 are the temperatures of the hot and cold body, respectively. The k defined in Eq. 12 is the wave number of light in vacuum. Thus the quantity p corresponds to propagation in vacuum, while p_υ corresponds to propagation in medium υ.

Consider the dependence on the separation l as exhibited in Eq. 13. The variable p for $x < k$, is purely imaginary. Therefore, the exponentials in Eq. 13 are sinusoidal for all values of l. This gives rise to a contribution to Eq. 7, which survives at large distances. This contribution is the familiar black body radiation. For $x > k$, the variable p is a positive real number making Eq. 13 an exponentially increasing function of the gap l. Therefore, in addition to the black body radiation, Eq. 10 contains a contribution which decreases exponentially with separation. There is thus an additional contribution to the energy transfer at close spacing. This enhancement can be significant as will be demonstrated.

MODEL DIELECTRIC FUNCTIONS

The magnitude of the close-spaced enhancement depends on the properties of the materials used on the hot and cold sides as specified by the frequency dependent dielectric function $\varepsilon(\omega)$. A commonly used model based on classical physics is given by:

$$\varepsilon(\omega) = 1 + \frac{4\pi n e^2}{m} \sum_j \left(\frac{f_j}{\omega_j^2 - i\omega\gamma_j - \omega^2} \right), \qquad \text{(EQ 14)}$$

where n is the number of polarizable molecules per unit volume, each containing a fraction f_j of electrons having resonant frequency ω_j and damping constant γ_j. The electron mass is m and its charge is e.

The dielectric function of Eq. 14 also applies to the special case in which free electrons are present, as in metals. There is no restoring force for free electrons; and, therefore, we must set $\omega_0 \rightarrow 0$. The following form is obtained for the dielectric function:

$$\varepsilon(\omega) = \varepsilon_\infty + \frac{4\pi\sigma\gamma i}{\omega(\gamma - i\omega)}. \qquad \text{(EQ 15)}$$

for frequencies large compared with those associated with the the bound resonances. Equation 15 is the Drude model.[15] The static conductivity is defined as $\sigma = n f_0 e^2 / m \gamma_0$. The conductivity σ and damping constant γ each have units of frequency (the dielectric function is dimensionless).

EVALUATION FOR MODEL MATERIALS

Equation 10 (the energy transfer between a hot and a cold body) was evaluated numerically for various model materials using the analytic forms for the dielectric function described in the previous section.[16] For convenience, we define dimensionless quantities l', ω', and P' as: $l = l'\hbar c / k_B T$, $\omega = \omega' k_B T / \hbar$, and $P = P' P_{bb}$; where $P_{bb} = \pi^2 k_B^2 T^4 / 60 \hbar^3 c^2$ is the transfer rate of black body radiation.

Equation 15 was used to describe metallic materials (with $\varepsilon_\infty = 1$). The two quantities σ and γ were treated as adjustable parameters and were varied to maximize the power transfer (at $l' = 1.0$). Both of these variables have units of frequency and are therefore transformed to dimensionless quantities analogously to ω (i.e.,substitute $\omega \rightarrow \sigma$, and $\omega \rightarrow \gamma$, respectively in the above definitions of the dimensionless frequency). For a good conductor such as Cu or Ag the conductivity and damping constant are given roughly as $\sigma \sim 10^{17} \text{sec}^{-1}$, and $\gamma \sim 10^{13} \text{sec}^{-1}$, respectively. At $T = 1000°C$ the corresponding dimensionless quantities are $\sigma' = 600$, and $\gamma' = 0.06$. Both real and imaginary parts tend to unity for high frequencies. The imaginary part is always positive (as is the case for all dielectric functions), and diverges at low frequencies, and the real part tends to a constant at zero frequency.

For insulators, a single resonant frequency (and damping constant) was considered in Eq. 14. The dielectric function for an insulator is thus written as

$$\varepsilon(\omega) = 1 + \frac{\omega_p^2}{(\omega_0^2 - i\omega\gamma_0 - \omega^2)}, \qquad (EQ\ 16)$$

where the plasma frequency has been defined as $\omega_p = \sqrt{4\pi n z e^2/m}$, and we have taken $f_0/z \to 1$ (all of the valence electrons participating). The dielectric function for an insulator thus has three adjustable parameters: ω_0, γ_0, and ω_p. These variables all have dimensions of frequency and are transformed to dimensionless quantities as discussed previously. Typically, the plasma frequency is of the order $10^{16} sec^{-1}$ which is as large or larger than the frequencies of visible light. The dimensionless plasma frequency is therefore $\omega_p' = 60$. There is a further constraint placed on the resonant frequency and the damping constant which takes the form $\omega_0 > \gamma_0/2$. This constraint follows from the fact that the response of the system must obey causality (i.e., cause precedes effect).

Another situation of interest is one in which the cold side is a semiconductor. To describe a semiconductor, we start with a published model which gives a good fit to the measured dielectric function for GaAs.[17]

A model for a smaller band gap material (appropriate to infrared radiation) was obtained by translating the position of the absorption edge and rescaling the energy axis.

RESULTS

The results obtained by numerically evaluating Eq. 10 for the energy transfer vs. separation are plotted in Fig. 1 for various combinations of materials on the hot and cold sides.[18] The hot side temperature was chosen to be $T = 1273K$ ($1000°C$). A significant enhancement of energy transfer is obtained for all material combinations for separations smaller than one micron. For large separations the energy transfer is somewhat less than that for a black body due to the fact that the materials have emissivity less than unity.

The largest enhancement was found for the situation in which the dielectric function for both the hot and cold side is of the Drude form (Eq. 15). The enhancement (at $l' = 1.0$)) was a monotonically increasing function of the damping parameter which reached its maximum asymptotically for roughly $\gamma' > 50$. The large enhancement (several orders of magnitude) depicted in Fig. 1 corresponds to $\sigma' \sim 1.7$ and $\gamma' \to \infty$. These values were obtained by maximizing the energy transfer at $l' = 1.0$ and are the values were used in all subsequent calculations involving metals.

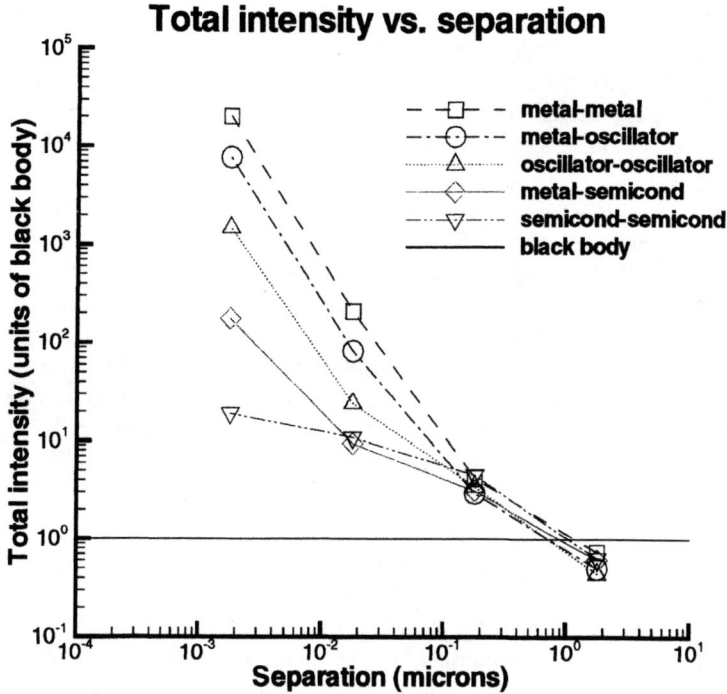

FIGURE 1. Total intensity of energy transfer between hot and cold bodies vs. separation for various combinations of materials. The units of separation were obtained by choosing a hot side temperature of $T = 1273K$ ($1000°C$). The cold side was taken to be at $T = 0K$ (although choosing room temperature makes a negligible difference in the result). The intensity is measured in units of P_{bb} (black body), which for $T = 1273K$ has the value $P_{bb} = 1.5 \times 10^5 W/m^2$. A significant enhancement is found for all material combinations for separations smaller than one micron.

At $l' = 1.0$ the largest energy transfer (still less than P_{bb}) was obtained for intermediate values of σ' (on the order of unity), but was small for large and small values of σ'. This is consistent with the fact that a good metal (large conductivity) has low emissivity. Likewise, in the limit $\sigma' \to 0$ we obtain vacuum, and there is thus no radiation.

A large enhanced energy transfer was also found for the situation in which the hot side was a metal characterized by Eq. 15 and the cold side was an insulator characterized by Eq. 16. The parameters used for the insulator were $\omega_p' = 13.0$, $\omega_0' = 4.0$, and $\gamma_0' = 1.0$. No attempt was made to find the absolute maximum of the energy transfer for this situation (an optimization in the space of five variables). Indeed, there is no guarantee that the dielectric functions obtained by such a procedure would be physically realizable.

The results for the metal hot side/insulator cold side clearly demonstrate that the enhanced energy transfer is a result of the fact that both materials exhibit absorption (i.e., have dielectric functions with non-zero imaginary parts). It is essentially irrelevant that one is a metal and the other is an insulator. This is consistent with the fact that the source of the energy (as well as the energy transfer) is due to the presence of imaginary part of the dielectric function as indicated by Eq. 8.

The energy transfer for the situations not involving metals is less than that with metals. Undoubtedly this is due to less overlap of the imaginary parts of the dielectric functions. Interestingly, however, the energy transfer with the least enhancement at small separation (i.e., between two semiconductors) is equal to or greater than all the others at the two largest separations considered. This emphasizes the fact that although the source of the energy is the imaginary part of the dielectric function, simple conclusions based on the form of the dielectric function cannot always be drawn. Of course the situation with two semiconductors may not be physically realizable since it assumes that the melting temperature is greater than $T = 1000°C$.

CONCLUSIONS

The enhanced energy transfer between a hot and cold body has been evaluated numerically for several model materials for a temperature difference of $1000°C$. The dielectric response of the hot and cold materials dictates the enhancement. Model dielectric functions for a metal, semiconductor, and insulator were considered. The results indicate that significant enhancement occurs only for separations less than one micron.

The largest enhancement occurs for materials which can be described by a metallic (Drude model) dielectric function. The enhancement can be as large as several orders of magnitude (it actually diverges but wasn't proven here[5]). The large enhancement occurs also for insulating materials as long as there is absorption (i.e., non-zero imaginary part of the dielectric function). The smallest enhancement was found between two semiconductors. It was argued that this result occurs because this situation exhibits the smallest overlap between the imaginary parts of the dielectric functions.

REFERENCES

1. Cravalho, C. E. G., Tien, C. L., Caren, R. P., *J. Heat Transfer Trans. ASME, Ser. C*

89, 351 (1967).

2. Polder D., and Van Hove, M., *Phys. Rev. B* **4**, 3303 (1971).

3. Caren, R. P., *J. Heat Transfer, Trans. ASME Ser. C* **94**, 289 (1972).

4. *Ibid*, p. 295.

5. Levin, M. L., Polevoi, V. G., and Rytov, S. M., *Sov. Phys. JETP* **52**, 1054 (1980).

6. Loomis J. J., and Maris, H. J., *Phys. Rev. B* **50**, 18517 (1994).

7. Landau, L. D., and Lifshitz, E. M., *Course of Theoretical Physics, Vol. 5 (Statistical Physics, part 1)*, New York: Pergamon, 1988, pp. 384-393.

8. Rytov, S. M., *Theory of Electric Fluctuations and Thermal Radiation* (Air Force Cambridge Research Center, Bedford, Mass., 1959), AFCRC-TR-59-162.

9. Landau, L. D., and Lifshitz, E. M., *Course of Theoretical Physics, Vol. 8 (Electrodynamics of continuous media)*, New York: Pergamon, 1960, pp. 367-368.

10. R. S. DiMatteo, *Enhanced Semiconductor Carrier Generation Via Microscale Radiative Transfer; MPC - An Electric Power Finance Instrument Policy; Interrelated Innovations in Emerging Energy Technologies*, S. M. thesis, Department of Electrical Engineering and Computer Science, Massachusetts Institute of Technology, (1996).

11. U.S. and PCT Patent Pending, Serial #08-902,817, Robert DiMatteo, Draper Laboratory.

12. M. D. Whale, *A Fluctuational Electrodynamic Analysis of Microscale Radiative Transfer and the Design of Microscale Thermophotovoltaic Devices*, Ph. D. dissertation, Department of Mechanical Engineering, Massachusetts Institute of Technology, (1997).

13. Ref. 12 (above) reports larger enhancements (roughly factor of 5 enhancement for the metal-semiconductor situation at a spacing of $1 \mu m$).

14. Jackson, J. D., *Classical Electrodynamics, Second Edition*, New York: John Wiley & Sons, 1975, p. 227.

15. Ashcroft, N. W., and Mermin, N. D., *Solid State Physics*, Philadelphia: Saunders College, 1976, pp. 1-27.

16. Actually, for computational convenience, the equivalent forms given by Eq.'s 21, and 22, in Ref. 6 were used.

17. Adachi, S., *GaAs and Related Materials: Bulk Semiconducting and Superlattice Properties*, Singapore: World Scientific, 1994, pp. 356-366.

18. The energy transfer which is plotted in Fig. 1 represents the total energy transfer. For TPV applications, the more useful quantity is the usable portion of the energy transfer due to absorption above the band gap.

Theoretical Maximum Efficiencies for Thermophotovoltaic Devices

George D. Cody

Exxon Corporate Research, Clinton Township, Route 22 East, Annandale N. J. 08801,
gdcody@erenj.com

Abstract. Using the thermodynamic model of Shockley and Queisser we compute an upper theoretical limit for the efficiency of thermophotovoltaic (TPV) devices as a function of optical band gap, in the ideal case where radiative recombination is dominant and where the source is taken as a black body. We obtain for TPV diodes with black body source temperatures T_s= 1200K-2500K: optimum bandgaps, E_g in the range 0.2-0.5eV, maximum conversion efficiencies between 30 and 35%, and maximum output powers between 5 and 80 Watts/cm^2. We compare this fundamental calculation of the maximum power output and efficiency with recent projections for the same quantities for TPV devices derived from semi-empirical device modeling of a TPV diode under black body radiation. In contrast to the fundamental calculation, these models give for T_s=1200K-2500K, optimum bandgaps E_g in the range 0.4-0.7eV and maximum output powers and maximum conversion efficiencies of the order of one half of that obtained in the fundamental calculation.

INTRODUCTION

Thermophotovoltaic (TPV) devices are semiconductor diodes that convert photons from a black body radiating source at temperatures below ≈2500C into electricity. Qualitatively they are the black body analog to the semiconductor diodes that convert photons from the sun into electricity in solar photovoltaic (SPV) devices. Quantitatively TPV is significantly different from SPV: (1) TPV devices operate at power densities of the order of 100-1000 times greater than SPV; (2) TPV diodes exhibit intrinsically higher current than SPV diodes; (3) for TPV the solid angle of the source is of the order of 1-4π sr, compared to 7x10^{-5} sr for SPV; (4) the power spectrum of the sun is comparable to a black body source with T_s ≈ 6000K compared to TPV where T_s< 2000C; (5) in TPV the emissivity of the source can be tailored to match the optical band gap of the semiconducting diode with significant improvements in the efficiency of converting the power radiated by the radiating source into electric power produced by the TPV device.

The high power densities reported for TPV devices make them, today, attractive possibilities for co-generation of electricity in remote areas or for remote and portable power. However, critical to any application where TPV must compete with high efficiency power generators such as fuel cells is the *fundamental maximum efficiency* of a TPV diode defined as the ratio of the maximum output power density out of the TPV device at temperature T_c, to the total power density radiated on the TPV device by the black body source at temperature T_s. From SPV experience, efficiencies of the order of 60-80% of this limiting value may be anticipated, but not guaranteed, after a significant and well directed R&D effort[1]. In this note we extend a fundamental thermodynamic model for the maximum efficiency of SPV photo-diodes due to Shockley and Queisser[2] to TPV diodes. We then compare the results of this model to recent semi-empirical models for the maximum efficiency, defined as above, of semiconducting TPV diodes[3] with a direct optical band gap, E_g, as well as recent experimental data. In this paper, we do not discuss improvements in ***overall system efficiency*** obtained by matching the optical emission of the radiant source to the band gap of the semiconductor diode. It is readily shown that improvements of the order of 30% over the fundamental thermodynamic efficiencies reported in this paper would be obtained by the use of selective emission technology as indicated in Figure A-1 of the Appendix to this paper.

DETAILED BALANCE LIMIT OF EFFICIENCY FOR SOLAR PHOTOVOLTAIC - SPV

More than 35 years ago Shockley and Queisser[2] calculated a "theoretically justified upper limit" of the efficiency of a p-n junction solar cell. They noted that: "This limit is a consequence of the nature of atomic processes required by the basic laws of physics, particularly the principle of detailed balance" and compared it to "theoretical limits [derived from] empirical values for the constants describing the characteristics of the solar cell" published in the mid 50's by Prince[4].

As noted by Shockley and Queisser in 1961, "the two limits are not extremely different, the detailed balance limit being at most higher by about 50% in the range of energy gaps considered [and that their] article is concerned with a matter of principle rather than practical values [however the] difference is much more significant [in] estimating potential for improvement. In fact, the detailed balance limit may lie more than twice as far above the achieved values as does the semi-empirical limit, thus suggesting much greater possible improvement."

Figure 1

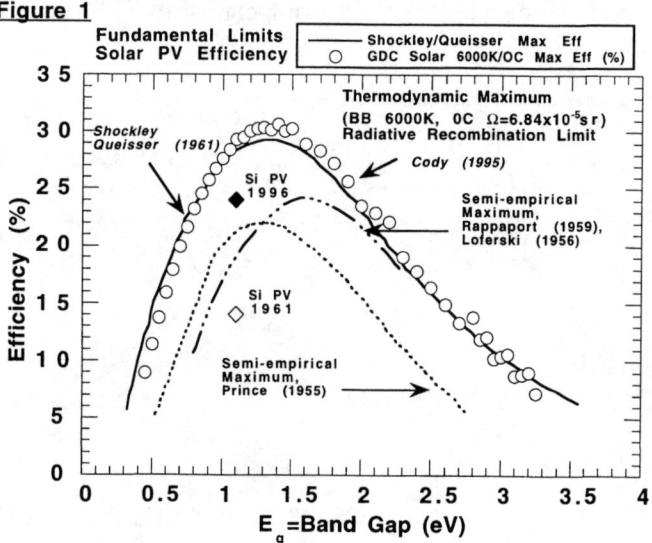

In Figure 1 we reproduce the curves from the first figure of the Shockley and Queisser paper, and add to it the "semi-empirical diode" limit due to Rappaport and Loferski[4, 5] as well as our own calculation of the fundamental limit. This calculation, sketched in the Appendix, is described in more detail elsewhere[1, 6], but is essentially identical to that of Shockley and Queisser. From Fig. 1, we note that after 35 years of R&D, the most recent data for silicon solar SPV cells is about ***15-20% above*** the early semi-empirical limits but still ***20-25% below*** the theoretical maximum limits - in agreement with Shockley and Queisser's observation, that while ***semi-empirical fundamental limits*** may be critical in directing short range R&D, they may be misleading in ***projecting the future*** course of technology.

As early noted by Al Rose[7], one contributor to the difference between the fundamental and semi-empirical limits for the efficiency is the calculation /estimation of the output voltage. The fundamental limit on the voltage is the difference in Fermi levels, or chemical potentials of photoexcited electrons and holes, as set by detailed balance for the photon population in the semiconductor. For real diodes, however, the voltage is set by a variety of recombination mechanisms as well as resistive mechanisms and photon loss mechanisms that are diode, material, junction and surface specific. As noted by Shockley and Queisser, "How closely any existing material can approach the desirable limit of unity for f_e =1 [radiative recombination as fraction of all other recombination mechanisms] is not known. [Indeed, in 1960] existing silicon solar cells fail to fit the current-voltage characteristics predicted on the basis of any of the existing recombination models." The fundamental dependence of output voltage, V_{out}, on the semiconductor bandgap, E_g, in the radiation recombination limit[1] is well known for SPV diodes, specifically $V_{out}(E_g) \approx (E_g - 0.3)$. The fundamental dependence of output voltage on band gap for TPV diodes is significantly different, specifically $V_{out}(E_g) \approx 0.7 E_g$ (Fig. 6). The reader will appreciate the difference between these two expressions for band gaps of the order of 0.5eV or less, and will note the danger in simply extrapolating to TPV, diode models derived from, and perhaps dependent on, SPV experience.

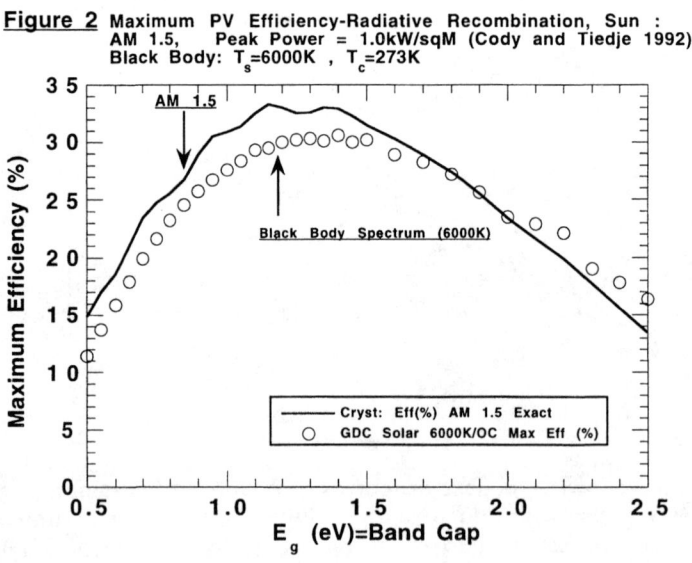

Figure 2 Maximum PV Efficiency-Radiative Recombination, Sun : AM 1.5, Peak Power = 1.0kW/sqM (Cody and Tiedje 1992); Black Body: T_s=6000K, T_c=273K

Figure 1 exhibits the agreement between the calculation of Shockley and Queisser of 35 years ago, and our own recalculation where the incident radiation of the sun is modeled by a black body source at 6000K subtending a solid angle of $\Omega=4.8 \times 10^{-5}$ sr at the earth, and the cell was taken at 273K. These parameters produce a peak power level above the atmosphere of 1.6 kW/sqM, 20% higher than the usual value for the peak power of 1.3kW/sqM (AM0) and 60% higher than that for the surface of the earth (AM1-1.5). For completeness, in Fig. 2 we compare the maximum efficiencies for the black body model ($\approx$ AM0) with the measured AM 1.5 solar spectrum of Matsen et al[8], and note the remarkable agreement between the two spectra. Clearly a black body at 6000K **with power density scaled to 1 kW/sqM** is an excellent approximation to a realistic solar spectrum at the surface of the earth for SPV applications[1].

DETAILED BALANCE LIMIT OF EFFICIENCY FOR THERMOPHOTOVOLTAIC POWER -TPV

Thermophotovoltaic Power or TPV appears to have been introduced by Wedlock[9] in 1963 and was extensively studied in the 80's[10, 11, 12]. There was a revival of interest in the early 90's as evidenced by three NREL/TPV conferences in 1994, 1995[3], and most recently, TPV-3, in May 1997[13]. The author first became aware of TPV in 1995 through articles in the press and on public radio describing the introduction of new TPV products, and was surprised to find no references to the maximum theoretical efficiency that could be obtained for such interesting co-generation power systems.

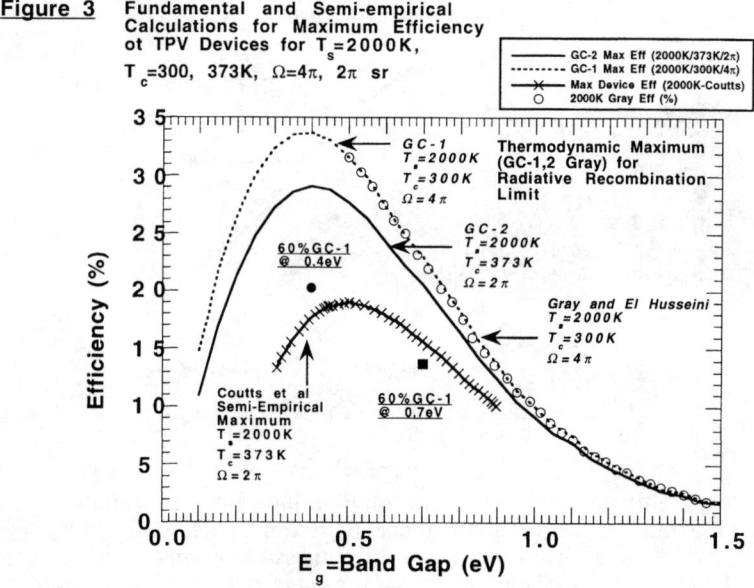

Figure 3 Fundamental and Semi-empirical Calculations for Maximum Efficiency of TPV Devices for $T_s=2000K$, $T_c=300, 373K$, $\Omega=4\pi, 2\pi$ sr

After making the calculations based on the radiation recombination limit model of Shockley and Queisser and the approach of Cody and Tiedje[1] (see Appendix A), it was a pleasant surprise to learn that Gray and El-Husseini, at the Second NREL

TPV conference in 1995[14], had made a similar calculation, and one which agreed with our own for $E_g \geq 0.5\text{eV}$ where the calculations overlapped. Recent articles by Coutts et al[3, 15] reviewing progress in TPV, supply a *semi-empirical diode model* for the maximum performance of TPV devices for $T_S = 1200\text{-}2000\text{K}$ which will serve as a "reality benchmark" for comparison with the fundamental thermodynamic calculation of the maximum efficiency for TPV devices.

In Figure 3 we show two fundamental limits for TPV efficiency obtained by the detailed balance approach of Cody and Tiedje[1]. Curve GC-1 is based on a black body source at $T_S = 2000\text{K}$, a solid angle $\Omega = 4\pi$ sr , and a cell temperature of $T_c = 300\text{K}$. It is in excellent agreement with the calculations of Gray and El Hussieni, who use the same parameters. GC-1 has a peak at $E_g \approx 0.4$ eV. As shown in Figure 4, for $2500\text{K} \geq T_s \geq 1200\text{K}$, the optimum band gap is in the range $0.5 \text{ eV} > E_g > 0.2$ eV.

Curve GC-2 in Figure 3 is also obtained for $T_s = 2000$ K, but with perhaps more realistic parameters for T_c and Ω, namely $T_c = 373$ K and $\Omega = 2\pi$ sr. We note from Fig. 3 that this change leads to a small *reduction* in the maximum efficiency from 34 to 29%, and a slight *increase* in the optimum band gap from 0.38 to 0.40 eV.

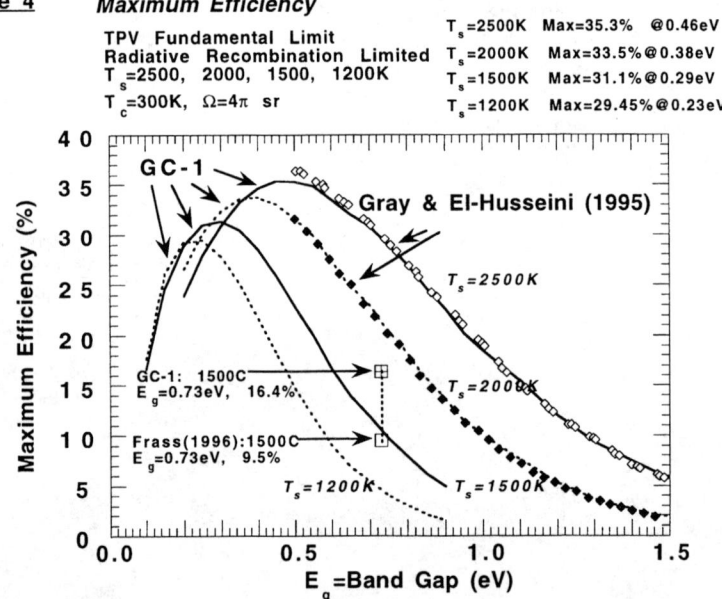

Figure 4 *Maximum Efficiency*
TPV Fundamental Limit
Radiative Recombination Limited
$T_s = 2500, 2000, 1500, 1200\text{K}$
$T_c = 300\text{K}, \Omega = 4\pi$ sr

$T_s = 2500\text{K}$ Max=35.3% @ 0.46eV
$T_s = 2000\text{K}$ Max=33.5% @ 0.38eV
$T_s = 1500\text{K}$ Max=31.1% @ 0.29eV
$T_s = 1200\text{K}$ Max=29.45% @ 0.23eV

More significant differences are found in the comparison shown in Fig. 3 between GC-1 and the semi-empirical calculation of Coutts et al[3] which gives an efficiency (obtained from the optical brightness theorem, that the image can never be brighter than the source, by scaling their calculated maximum output power flux to the input black body flux) about one half that of the fundamental maximum, and an upward shift of the optimum band gap of ≈ 0.1 eV. The scaling of the magnitude of the efficiency by about 2, and the upward shift of about 0.1

eV in the optimum gap, for their model relative to the Shockley Queisser model, holds over the range, T_s = 1200-2000K (Figure 5).

Current estimates for the maximum efficiency are again obtained by taking the output power per unit area for TPV cells and scaling to the black body radiation at the source temperature. As indicated in Figure 4, experimental data of Fraas et al[16] for the output power ($\approx$5.3W/cm^2) of gallium antimonide (GaSb) TPV cells with $E_g\approx$0.73eV and T_s =1500C, divided by the total black body radiation *at 1500C* show that $\approx$60% of GC-1 *at 1500C* is achievable today.

It is of interest to note from Fig. 3 that the maximum efficiency for a TPV device at 2000K at 60% of GC-1 for $E_g = 0.7\ eV$, would be 10% *below* the semi-empirical diode limit of Coutts et al[3] at 0.7eV. However a TPV device at 2000K at 60% of GC-1 for $E_g = 0.4\ eV$ would be 15% *above* the semi-empirical diode limit of Coutts et al. [3] at E_g = 0.4 eV.

Figures 4, 5, and 6 exhibit the maximum efficiencies, output power densities and output voltages derived from the Shockley Queisser detailed balance model for Ts = 2500, 2000, 1500, and 1200K, T_c = 300 K, and Ω=4π sr. We note from Fig. 4, that for temperatures 1200 $\leq$ Ts $\leq$ 2500 K the maximum efficiency is between 30 and 35% and the optimum band gap falls in the range, 0.2<E_g<0.5 eV. In contrast the semi-empirical model of Coutts et al.[3] projects, as shown in Fig. 5, an optimum band gap range in the range 0.4-0.7 eV with a maximum efficiency between 15 and 22%.

Figure 5 Maximum Power Density for Direct Band Gap Semiconductor of Band Gap E_g as a Function of Black Body Temperature T_s:

(1) Semi-Empirical Diode Model: TJC= Coutts et al (1996)
(2) Detailed Balance Model of Shockley and Queisser: GC-1(1995)

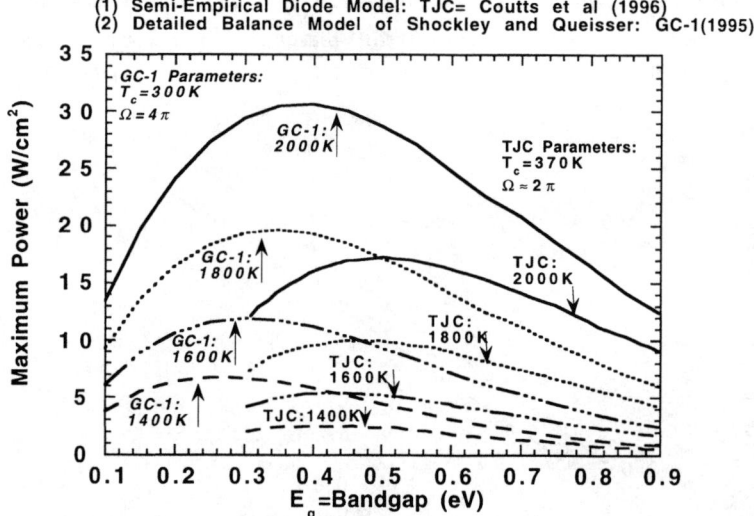

Figure 5 exhibits the significant differences between the derived magnitude of the maximum power, and its dependence on band gap between the fundamental model, GC-1, and the calculated maximum power from the diode model of Coutts et al.(TJC). As noted earlier, the large differences in Figure 5, can not arise from the small differences in T_c (373 - 300K) and $\Omega(2\pi-4\pi)$ for the two models.

The Shockley Queisser model is based on a fundamental model for the recombination of photo-generated holes and electrons, radiative recombination. It then poses, and answers, the question as to what is the maximum efficiency that a p/n diode, dominated by this fundamental recombination mechanism, can achieve with a black body generated photon flux. There are other ***fundamental mechanisms*** for recombination, and one obvious example is Auger recombination[17]. It can be shown that Auger recombination has a relatively small effect on the maximum efficiency of SPV silicon junctions[6] but a similar calculation has yet to be made for low band gap TPV diodes. Furthermore, since it does not appear that the Auger mechanism was the significant limitation in the semi-empirical diode model of Coutts et al[3], we will explore other sources of the difference between the projections of the diode model and the Shockley Queisser model.

The semi-empirical model extrapolates from what is known about diodes made from known semiconductors, and estimates what might be achieved given reasonable improvements in physical parameters. The fundamental model calculates the "best that can be done". One possible source of the difference between the predictions of the two models is that the data base for the diode model may derive from SPV experience, and as shown in Fig. 6, the output voltage for high efficiency TPV diodes, exhibits a significantly different functional dependence on band gap from that obtained for high efficiency SPV diodes.

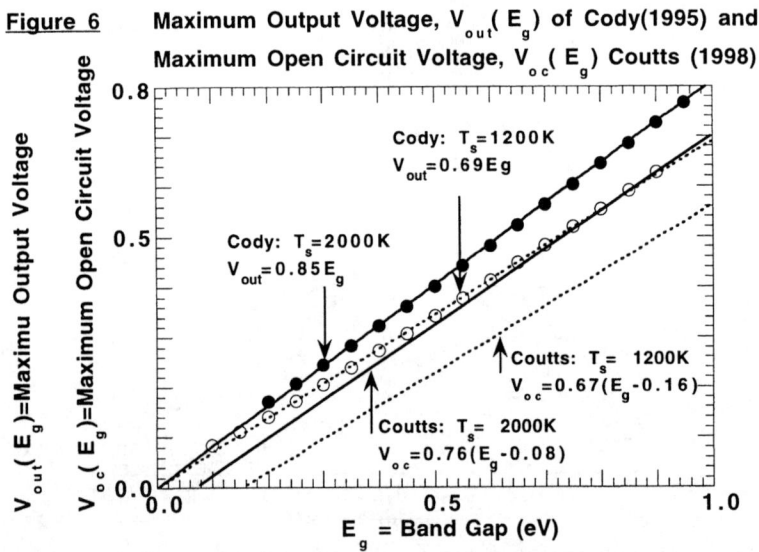

Figure 6   Maximum Output Voltage, $V_{out}(E_g)$ of Cody(1995) and Maximum Open Circuit Voltage, $V_{oc}(E_g)$ Coutts (1998)

We note from Figure 6, that the output voltage for TPV diodes derived from the Shockley Queisser model is a linear function, ***through the origin***, of the optical band gap of the diode, $V_{out} = A\, E_g$ where $A \approx 0.7\text{-}0.9$. Thus the increase in current of the TPV diode obtained by utilizing lower band gap diodes is not off-set by a sharp drop in output voltage, and there is a shift in the maximum of efficiency toward lower band gaps, as exhibited in Figure 5.

This result of the Shockley Queisser model for TPV diodes for large solid angle black body sources with $T_s < 2500K$, is in striking contrast to the predictions of the same model for SPV diodes excited by the sun a black body, low solid angle source at 6000K. Under these circumstances, in the radiative recombination limit, the maximum output voltage is not linear through the origin, but as noted earlier given by, $V_{out} \approx (E_g - B)$ where the intercept, $B \approx 0.3$-0.4 eV[1]. In Fig. 6 we also show the open circuit voltage $V_{oc}(E_g)$ for the TPV diode model of Coutts[15] and note that as in the SPV case, $V_{oc}(E_g)$ is not linear through the origin, but has an intercept between $E_g \approx 0.1$-0.2 eV.

Finally we note in Figs. 4 and 5, the high maximum output power densities and efficiencies for TPV (T_s = 1400-2500K: Output power = 5-80 Watts/cm^2, Efficiency = 30-35%) that make TPV such an attractive possibility for remote co-generation and possibly even vehicular power if 80% of the efficiencies and power densities can be achieved, and if optical enhancement can still further improve overall TPV system efficiencies by another 30%. However, further progress in extending the market for TPV will require equally significant improvements in the cost and performance of electrical energy storage, including high performance batteries as well as novel storage devices such as the high efficiency "electromechanical" battery introduced by Richard Post and collaborators[18].

CONCLUSIONS

The present calculation suggests that the device challenge of low band gap materials ($E_g \leq 0.5$ eV) may have a return in higher efficiency for values of $T_s \leq 2000K$ even without optical enhancement. This view contrasts strongly with the current apparent focus of TPV R&D exhibited at the NREL TPV conferences on semiconductor materials with $E_g \geq 0.5$ eV, as well as with the conclusions drawn from current semi-empirical diode models[15].

In the high risk enterprise of R&D, it may be helpful to derive and explore the implications of rigorous thermodynamic models for the maximum thermodynamic efficiency of TPV diodes as a function of the optical band gap, solid angle of the source, and device and source temperature. Despite the fact that such calculations do not identify the R&D path to the maximum performance, they do place current results in the context of the fundamental limits, and may identify research and development opportunities that are being overlooked in the current R&D focus - for example determining the relative magnitude of Auger and radiative recombination rates for direct band gap semiconductors with band gaps in the range 0.3-0.5 eV.

APPENDIX: FUNDAMENTAL THERMODYNAMIC LIMIT ON TPV EFFICIENCIES

We consider a direct band gap semiconductor with an absorption threshold defined by the band gap, E_g. The semiconductor at temperature, T_c, absorbs photons from a black body source at a temperature T_s which subtends a solid angle, Ω, at the semiconductor. We assume all incident photons to be absorbed, and that the excited electrons and holes fall in energy by rapid thermalization to the band edge. We consider a semiconductor such that radiative recombination is the dominant mechanism for loss of the excited carriers. Clearly these assumptions are major, and will require a thin semiconductor with anti-reflection coatings and optical enhancement of the absorption coefficient[19, 20]. As discussed in detail by Tiedje et al[6] the above assumptions are reasonable even for an indirect semiconductor such as silicon. In any event they do permit us to easily define the maximum efficiency that a semiconductor device can possess and explore the optimum band gap, E_g, for variable temperature T_s, and T_c and solid angle Ω.

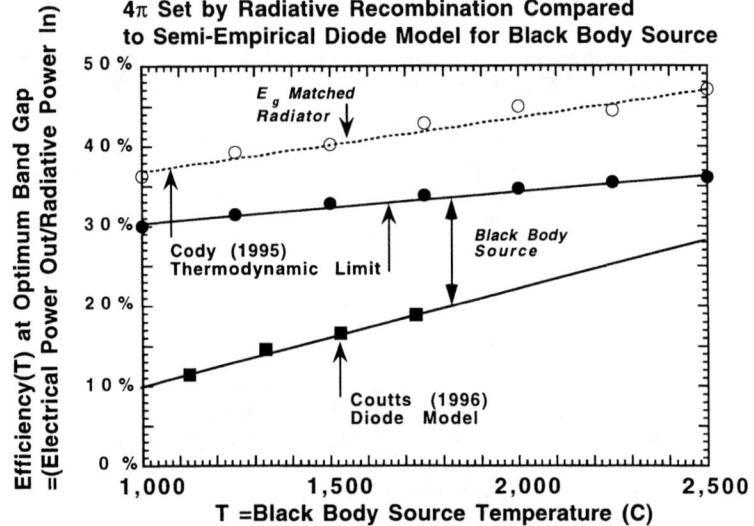

Figure A-1 Efficiency at Optimum Band Gap for Black Body Source and E_g Matched Emitter at Solid Angle 4π Set by Radiative Recombination Compared to Semi-Empirical Diode Model for Black Body Source

Following the simple model of Cody and Tiedje[1] we can show that to first order in $(kT_c)/[E_g(1-T_c/T_s)]$ the output voltage, $V_o(E_g, T_s, T_c)$ of the semiconductor at maximum power is given by

$$V_o(E_g,T_s,T_c)=E_g(1-T_c/T_s)-kT_c\{\ln[(4\pi/\Omega)(T_c/T_s)]+\ln[(kT_c/E_g)(1/(1-T_c/T_s))]\} \quad (A1)$$

If $M(\lambda, T_s)$ is the spectral power emitted by the source in a band width $\Delta\lambda(\mu M)$ at a wavelength $\lambda(\mu M)$:

$$M(\lambda, T_s) = c_1(\Delta\lambda/\lambda^5)(1/[e^{(c_2/\lambda T_s)}-1]) \; [W/M^2] \quad (A2)$$

where $c_1= 3.742 \times 10^{-8} \; [(\mu M)^4 W/M^2(\mu M)^4]$ and $c_2 = 1.439 \times 10^4 \; [K\mu M]$.

We convert $M(\lambda, T_s)$ to the spectral current emitted by the source, $I_o(\lambda, T_s)$ and obtain again to first order in $(kT)/[E_g(1-T_c/T_s)]$ and neglecting small additive terms within the logarithm and where λ is given in μM

$$I_o(\lambda, T_s, T_c) = M(\lambda, T_s) (0.806\lambda) (1-[(kT_c)/(E_g(1-T_c/T_s))]) \text{ [amp/M}^2] \quad (A3)$$

We then obtain for the maximum power out of the TPV device

$$P(E_g, T_s, T_c) = \int_{E_g}^{\infty} V_o(E_g, T_s, T_c) I_o(\lambda, T_s, T_c) d\lambda \quad (A4)$$

If we divide $P(E_g, T_s, T_c)$ by the power emitted by a black body at temperature T_s we obtain the fundamental maximum TPV Efficiency, Eff_{TPV}.

If we divide $P(E_g, T_s, T_c)$ by *the above band gap power* emitted by a black body at temperature T_s we obtain a fundamental limit for the maximum system efficiency obtained by matching the optical emission of the radiant source to the band gap of the semiconductor diode. The differences between these definitions is shown in Figure A-1

REFERENCES

1. G. D. Cody, T. Tiedje, in *Energy and the Environment*, B. Abeles, A. Jacobson, P. Sheng, Eds., World Scientific, Singapore, New Jersey, London, 1992, pp. 147-217.
2. W. Shockley, H. J. Queisser, *Journal of Applied Physics*, **32**, (1961) 510-519.
3. T. J. Coutts, M. W. Wanlass, J. S. Ward, S. Johnson, A review of Recent Advances in Thermophotovoltaics, Twenty Fifth IEEE Photovoltaic Specialists Conference, Washington, DC (IEEE, 1996).
4. J. L. Loferski, *Journal of Applied Physics*, **27**, (1955) 777-784.
5. P. Rappaport, *RCA Review*, **XX**, (1959) 373-397.
6. T. Tiedje, E. Yablonovitch, G. D. Cody, B. G. Brooks, *IEEE Transactions on Electron Devices*, **ED-31**, (1984) 711-716.
7. A. Rose, *Journal of Applied Physics*, **31**, (1960) 1640-1641.
8. R. Matsen, R. Bird, K. Emery, *SERI/TR-612-964* (U. S. Department of Energy, 1981).
9. B. D. Wedlock, *Proceedings of the IEEE*, , (1963) 694-698.
10. R. M. Swanson, Recent Developments in Thermophotovoltaic Conversion, International Electron Devices Meeting (1980).
11. L. D. Woolf, *Solar Cells*, **19**, (1986-1987) 19-38.
12. A. Caruso, G. Piro, *Solar Cells*, , (1986-1987) 123-130.
13. T. J. Coutts, C. S. Allman, J. P. Benner, Eds., *Thermophotovoltaic Generation of Electricity*, the Third NREL Conference, Vol. 401, American Institute of Physics, Woodbury, New York, 1997, pp. 537.
14. J. L. Gray, A. El-Husseini, in *Second NREL Conference on TPV Generation of Electricity* J. P. Bennere, T. J. Coutts, D. S. Ginley, Eds. (AIP Conference Proceedings, Colorado Springs, Colorado, 1995), vol. 358, 3-15.
15. T. J. Coutts, "Principles of, and Progress in, Thermophotvoltaic Generation of Electricity", Fig. 13.11, preprint (1998)
16. L. Fraas, et al., Hydrocarbon Fired Thermophotovoltaic Generator Prototypes Using Low Bandgap Gallium Antimonide Cells, Twenty Fifth IEEE Photovoltaic Specialists Conference, Washington, D.C. (IEEE, 1996).
17. J. I. Pankove, *Optical Processes in Semiconductors*, J. Nick Holonyak, Ed., Solid State Physical Electronics Series, Prentice-Hall, Englewood Cliffs, N.J., 1971, p. 422.
18. R. F. Post, T. K. Fowler, S. F. Post, *Proceedings of the IEEE*, **81**, (1993) 462-474.
19. E. Yablonovitch, G. D. Cody, *IEEE Transactions on Electron Devices*, **ED-29**, (1982) 300-305.
20. E. Yablonovitch, *J. Opt. Soc. Am.*, **72**, (1982) 899-907.

A Numerical Semiconductor Device Model for TPV Cells

Jeffery L. Gray and Ali M. El-Husseini

School of Electrical and Computer Engineering
Purdue University, West Lafayette, IN 47907

Abstract. Detailed numerical modeling has provided valuable insights into the operation and performance limitations of solar cells for both space and terrestrial applications. Detailed numerical modeling of TPV cells should provide similar benefits. In this paper, a detailed numerical model appropriate for TPV cells is presented.

INTRODUCTION

Detailed numerical modeling of TPV cells presents some unique challenges. One of these challenges is the computation of the optical generation rate in the TPV cell. In PV applications, it is reasonable to assume that light impinges on the solar cell at a single angle of incidence. This is not a good assumption for TPV cells. Because of the close proximity of the emitter to the cells, light is incident with a range of incident angles.

In addition, the band structure of narrow band gap materials needed for efficient TPV systems is inherently non-parabolic. This requires a modification to the equations that model electronic transport.

DETAILED NUMERICAL MODELING

The model used here is ADEPT, A Device Emulation Program and Toolbox (1). ADEPT has been modified to handle illumination from diffuse emitters and the non-parabolic band structure of narrow band gap semiconductors.

Detailed numerical modeling of TPV cells requires the solution of three coupled nonlinear partial differential equations

$$\nabla \bullet \varepsilon \vec{E} = q(p - n + N) \tag{1}$$

$$\nabla \bullet J_n = q(R - G) \tag{2}$$

$$\nabla \bullet J_p = q(G - R) \tag{3}$$

with

$$J_n = -q\mu_n n(V + V_n) + kT\mu_n \nabla n \quad (4)$$

$$J_p = -q\mu_p p(V - V_p) - kT\mu_p \nabla p \quad (5)$$

Of concern here is the optical generation rate, G, and the band parameters, V_n and V_p. The derivation of $G(x)$ is presented elsewhere in these proceedings (2). How non-parabolicity is modeled will be discussed here.

Non-Parabolicity

The parabolic band approximation assumes the band gap energy is much greater than kT. For semiconductors with small band gaps, such as those suitable for TPV applications, this is no longer true. A better approximation (3,4) leads to

$$n = N_C F(\varepsilon, \eta) \quad (6)$$

where

$$F(\varepsilon, \eta) = \frac{2}{\pi} \int_0^\infty \frac{\sqrt{y(1 + y/\varepsilon)}\left(1 + \frac{2y}{\varepsilon}\right)}{e^{y-\eta} + 1} dy \quad (7)$$

with $\varepsilon = E_g/kT$ being the reduced band gap and $\eta = \dfrac{E_F - E_C}{kT}$ the reduced Fermi energy. It is easy to show that, under appropriate assumptions, Equation (7) reduces to Boltzmann or Fermi-Dirac (degenerate) statistics. This is illustrated in Figure 1 which shows a plot of the normalized pn product as a function of reduced Fermi energy for several values of the reduced bad gap energy.

SIMULATIONS

Figure 2 shows an I-V characteristic for a GaSb TPV cell (5), which has a band gap of 0.72 eV. As can be seen, the effects of the non-parabolicity are negligible. However, when a 0.55 eV band gap InGaAs TPV cell (6) is simulated (Figure 3), the effect is noticeable.

TPV cells are often characterized under conditions in which the light is all incident at an angle normal to the cell's surface. Since this is not the situation under normal operating conditions in which, assuming a blackbody TPV emitter, there is no preferred

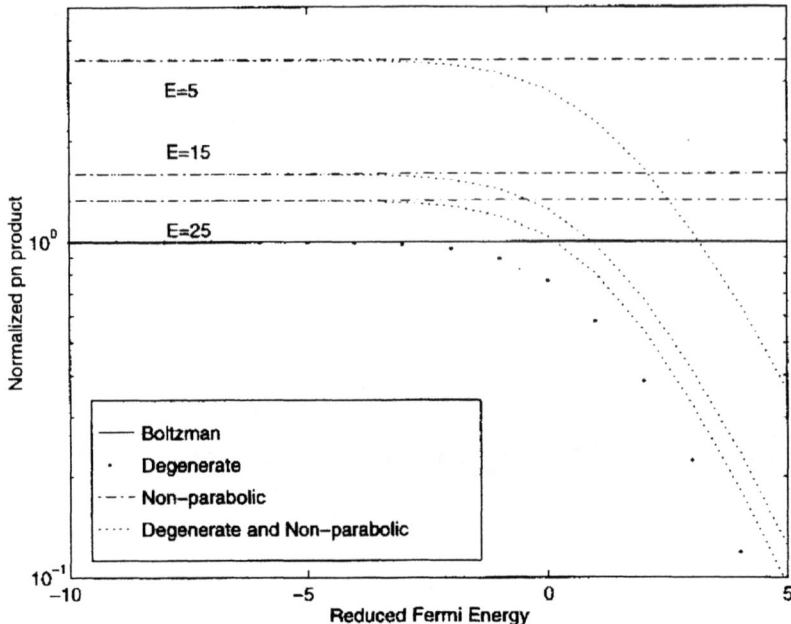

FIGURE 1. Normalized pn product as a function of reduced Fermi energy for various values of the reduced band gap energy.

angle of incidence, extrapolation to operating conditions must be done carefully. When the multiple angles of incidence and reflections are accounted for (2), the external spectral response is markedly affected, as seen in Figures 4 and 5. This influences the predicted short circuit current and the expected cell output power will be less than might be expected.

SUMMARY

A detailed numerical device model has been augmented to handle non-parabolicity in narrow band gap semiconductors and the multiple angles of incidence from blackbody TPV sources. It was demonstrated that these model details significantly affect predicted TPV cell performance.

REFERENCES

1. Gray, J. L., "ADEPT: A General Purpose Numerical Device Simulator for Modeling Solar Cells in One-, Two-, and Three-Dimensions," *Conference Record of the Twenty Second IEEE Photovoltaic Specialists Conference*, pp. 436-439, 1991.
2. Gray, J. L. and El-Husseini, A. M., "A TPV System Model and an Analysis of TPV System Efficiency," presented at the Fourth NREL Thermophotovoltaic Generation of Electricity Conference, Denver, CO, October 1998.
3. Kane, E. O., *J. Phys. Chem. Solids* **1**, 249-256 (1957).
4. Ehrenreich, H., *J. Phys. Chem. Solids* **2**, 131-148 (1957).
5. Fraas, L. M., Avery, J. E., Gruenbaum, P. E., and Sundaram, V. S, "Fundamental Characterization Studies of GaSb Solar Cells, *Conference Record of the Twenty Second IEEE Photovoltaic Specialists Conference*, pp.80-84, 1991.
6. Wojtczuk S., Colter P., Charache G., and Campbell B., "Production Data on 0.55 eV InGaAs Thermophotovoltaic Cells," *Conference Record of the Twenty Fifth IEEE Photovoltaic Specialists Conference*, pp.77-80, 1996.

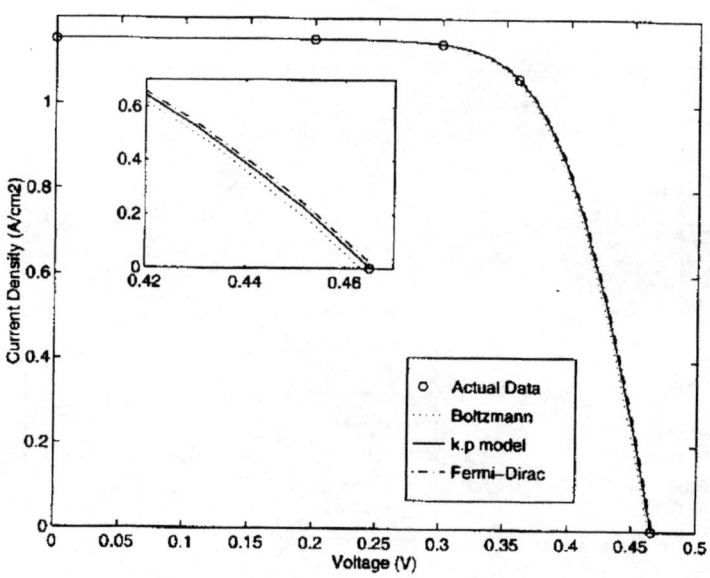

FIGURE 2. Simulated I-V of a GaSb solar cell.

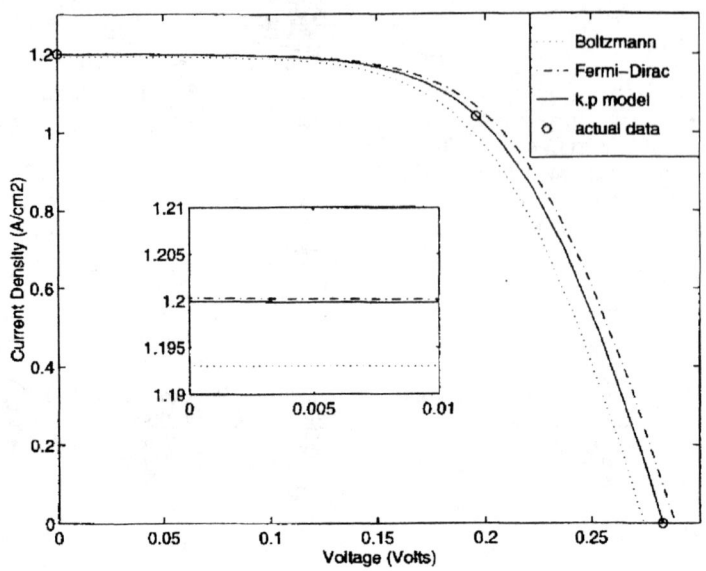

FIGURE 3. Simulated I-V of an InGaAs solar cell.

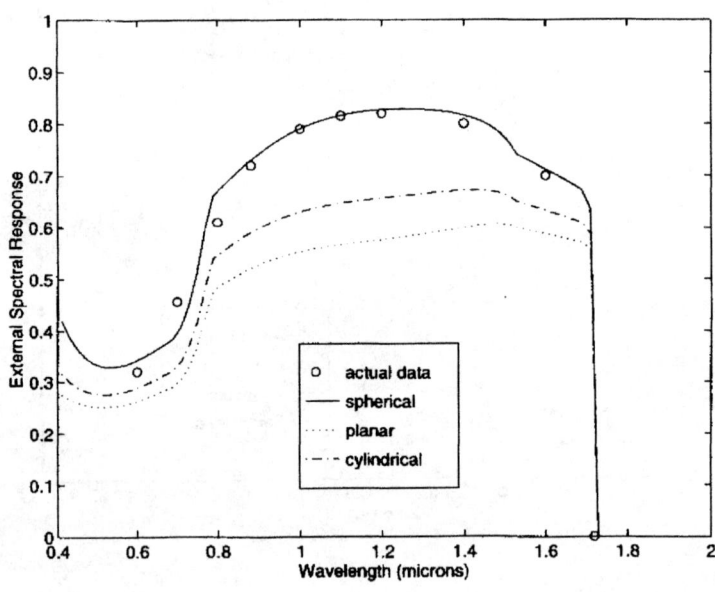

FIGURE 4. Simulated external spectral response of a GaSb solar cell.

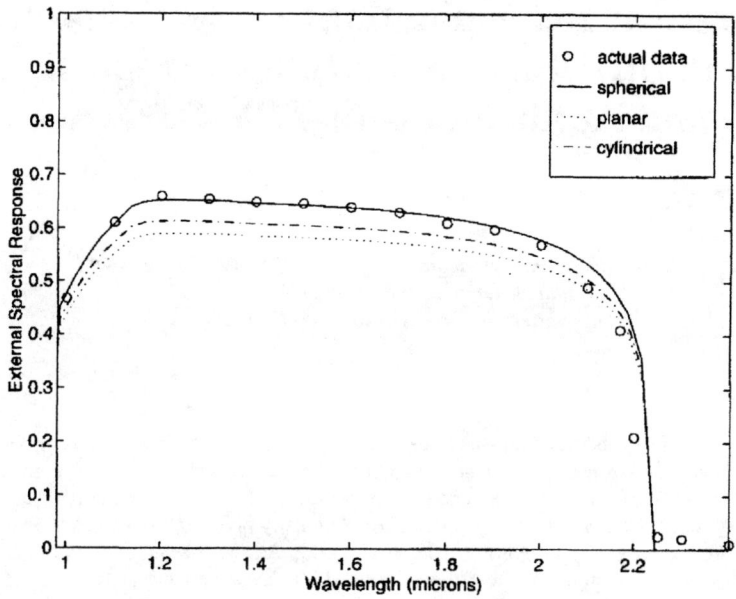

FIGURE 5. Simulated spectral response of an InGaAs solar cell.

Theoretical Determinations of Semiconductor Band Structures and Optical Properties, and Applications to TPV Devices

C. B. Geller[1], T. S. Blazeck[1], W. Wolf[2], and W. Mannstadt[3]

[1]*Bettis Atomic Power Laboratory, West Mifflin, PA 15122-0079*
[2]*c/o Molecular Simulations Inc., San Diego, CA 92121-3752*
[3]*Northwestern University, Evanston, IL 60208-3112*

Abstract. Electron band structures have been calculated from first principles using the Full Potential Linearized Augmented Plane Wave (FLAPW) computational code (1), including nonlocal screened exchange (sX-LDA) and spin-orbit effects (2,3) for representative elemental, binary III-V and II-VI semiconductors, and configurationally optimized $In_xGa_{1-x}As$. Predicted band gaps for narrow gap semiconductors (Ge, InAs, $In_xGa_{1-x}As$ and InSb) are all within 20% of experimental values, compared with errors of over 100% (negative band gaps) obtained with previous calculations based on the Local Density Approximation. Effects of $In_xGa_{1-x}As$ conduction band dispersion on TPV device quantum efficiency are illustrated.

INTRODUCTION

The utility of first-principles quantum computational codes for predicting semiconductor optical properties has been limited by both the relative inaccuracy of the standard Local Density Approximation (LDA) computational technique for predicting excited states, and by the difficulties in modeling dilutely doped materials prototypical of semiconductor devices beyond the rigid band approximation. The inclusion of screened nonlocal exchange within a Generalized Kohn-Sham theoretical framework (4) effects a large, systematic improvement relative to the LDA in the predicted values of fundamental band gaps and conduction band critical points of elemental column IV and binary III-V and II-VI semiconductors. This improvement is achieved without introducing empirical parameters. To account for the effects of band filling/emptying (Moss-Burstein Shift) and electronic band gap narrowing in doped semiconductors, an interpolation methodology is presented based on rigorous computational results for pure and heavily doped materials.

THEORY

As shown in Figure 1, the "sX-LDA" method corrects even LDA starting solutions with the wrong band topology to reasonable band gap values. The computational results designated as "New Theory" in Figure 1 were obtained using the Full Potential Linearized Augmented Plane Wave - Bettis version (FLAPW-B) computational code (2), with screened nonlocal exchange. Additional results on a representative sampling of twelve binary II-VI compounds of the zincblende and wurtzite structures showed an average difference between measured and calculated spin-orbit-corrected sX-LDA band gaps of only 5.5%. FLAPW-B is an all-electron code with a fully relativistic core state treatment and scaler relativistic and perturbative spin-orbit effects included for valence states. Screened nonlocal exchange effects are calculated in a "second variation" procedure in which the LDA solution is obtained first, a contracted basis set of LDA wave functions is formed, and a second variation is initiated with the new screened exchange functional. Computational results for diamond, InAs and GaAs are in qualitative agreement with earlier sX-LDA results obtained with a pseudopotential code by Seidl et. al (4). FLAPW-B currently is practical for self-consistent field calculations on unit cells of up to approximately sixteen atoms on a high-performance workstation, and for perturbative screened exchange calculations on somewhat larger cells.

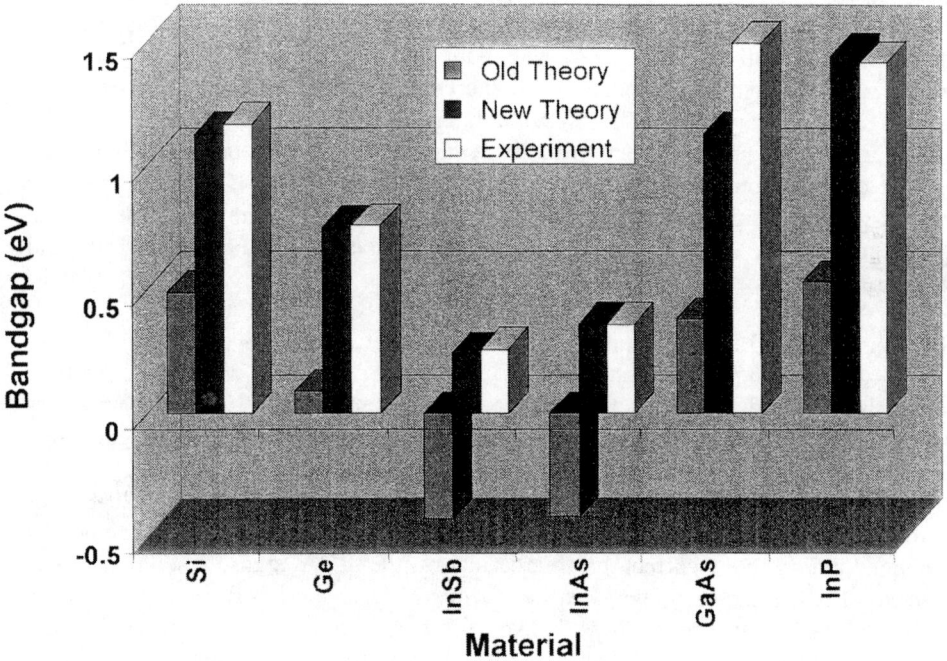

FIGURE 1. Comparison of LDA, sX-LDA and Experimental Band Gaps for Seven Elemental and Binary Semiconductors

The idea behind the dilute limit interpolation scheme is to model a unit cell large enough such that wave functions based on impurity sites have no appreciable overlap. Thus, the impurities can already be considered "isolated" relative to their effect on the band structure, even though their concentration may be orders of magnitude higher than prototypical. It then follows that the energy density of states at the band edge, which (ignoring excitons) results entirely from the presence of the impurity, can be scaled linearly to the impurity concentration of practical interest. This fact allows one to predict optical absorption spectra for dilutely doped materials between the zero concentration and high concentration limits.

InAs, GaAs AND $In_{0.75}Ga_{0.25}As$ RESULTS

$In_{0.75}Ga_{0.25}As$ was selected for study because it has a composition convenient for computation and a band gap of approximately 0.52 eV that is in the range of interest for TPV devices. Eight indium-gallium orderings on the Column III sublattice that are representable with a unit cell of sixteen atoms or fewer were geometrically optimized at the LDA level using FLAPW, holding the lattice parameter fixed at the experimental value. Two configurations whose predicted band gaps most closely approximate the experimental value were calculated to have total energies approximately 7 kcal/mol or more lower than those of the other six. Both low-energy configurations are distinguished from the less energetically favored configurations by the fact that gallium atoms have all indium next nearest neighbors. This observation suggests a strong tendency toward high local symmetry and short-range order in the $In_xGa_{1-x}As$ system, at least in the vicinity of x=0.75. Since high short-range order is difficult to achieve except at certain local stoichiometries, this tendency may in turn promote a tendency toward compositional segregation on longer length scales. Of the two preferred indium-gallium configurations, the one with the highest symmetry, designated "Configuration A," is shown in Figure 2. Configuration A was used for all band structure and optical property calculations.

Self-consistent sX-LDA band structures with spin-orbit effects are compared for InAs and $In_{0.75}Ga_{0.25}As$ in Figures 3 and 4. The $In_{0.75}Ga_{0.25}As$ Brillouin zone, valid for both materials, is used for both plots to facilitate comparison. Lattice parameters appropriate for a temperature of 300 K were assumed. The predicted band gap value for InAs is 0.42 eV, approximately 20% too high relative to experiment. The predicted LDA band gap for InAs is actually *negative*, whereas the residual 20% error in the sX-LDA result is attributed to the approximate nature of the assumed screening function and/or the incorrect band ordering of the LDA starting solution. The predicted Γ - L separation energy in InAs (equivalent to the Γ - R separation energy in the plot) is 1.2 eV, which differs dramatically from published literature values near 0.73 eV (5), but agrees much better with the recent, more accurate measurement by Pollack of 1.1 eV. The band structure of $In_{0.75}Ga_{0.25}As$ has a predicted sX-LDA band gap of 0.50 eV that is in excellent agreement with a value of 0.52 eV interpolated

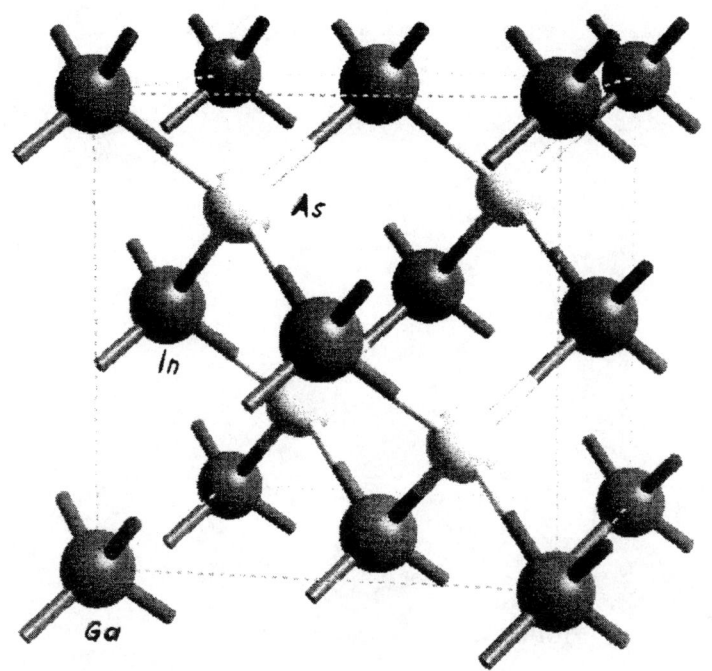

FIGURE 2. In-As Configuration A

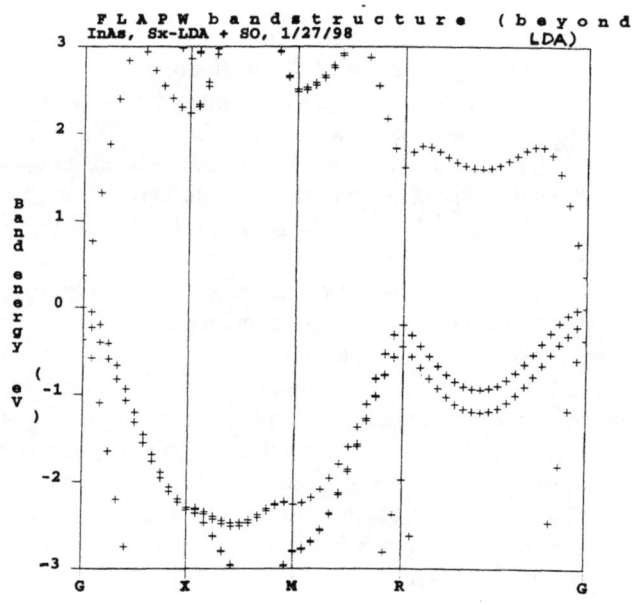

FIGURE 3. InAs Band Structure

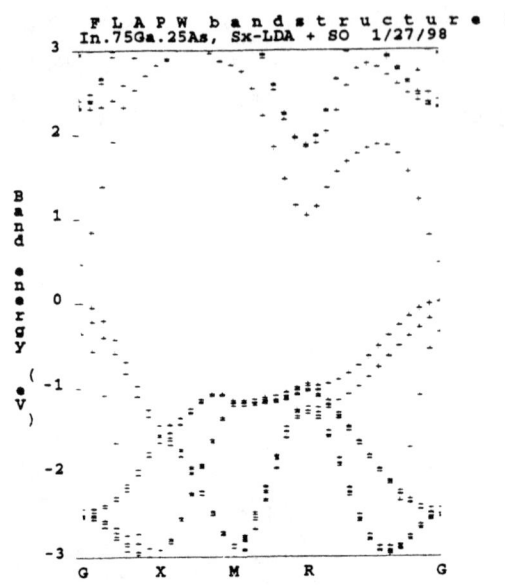

FIGURE 4. $In_{0.75}Ga_{0.25}As$ Band Structure

from experimental measurements on nearby compositions. (The convergence of band gap predictions is estimated to be ±0.03 eV.)

The largest difference between the InAs and the $In_{0.75}Ga_{0.25}As$ band structures is evident in the [111] (i.e., "Γ - R") direction of the Brillouin zone. The predicted Γ - R separation energy in InAs drops from 1.2 eV, compared with a predicted Γ - R separation energy (same direction in k-space, but a different Brillouin zone) of 0.52 eV in $In_{0.75}Ga_{0.25}As$. A character analysis of the conduction band states at Γ and R, taking into account the angular dependencies of the relevant wave functions within the muffin-tin spheres, reveals that while the L point in InAs and the R point in $In_{0.75}Ga_{0.25}As$ both have mixed indium and gallium character, the character of the conduction band states at Γ changes from pure indium s to pure gallium s as x varies from 1.0 to 0.75. The dramatically reduced Γ - R separation energy (more precisely, the separation energy between the first and second critical points in the conduction band) in $In_{0.75}Ga_{0.25}As$ that results from this character shift has important effects for both interband optical absorption and for free carrier absorption. The predicted energies for the first three critical points in the conduction band of InAs are all within 12% of Pollack's measured values. This fact, combined with the excellent band gap prediction for $In_{0.75}Ga_{0.25}As$, provides confidence that the higher critical points in the $In_{0.75}Ga_{0.25}As$ conduction band also are being modeled faithfully with the sX-LDA method as implemented in FLAPW-B.

Predicted and measured absorption coefficients for GaAs are shown in Figure 5, demonstrating excellent agreement, especially at photon energies below 4 eV. Only direct interband transitions are accounted for. Absorption events other than interband transitions do not produce free carriers and should therefore be accounted for separately, whereas indirect interband transitions are of second order. The GaAs results provide confidence in the near-band-gap absorption coefficient predictions for InGaAs.

The spin-orbit split FLAPW-B sX-LDA absorption coefficient curve is shown in Figure 6 for Configuration A of $In_{0.75}Ga_{0.25}As$. Plotted on the same axes is an absorption coefficient curve interpolated from experimental data for InAs and GaAs that had been used in the absence of accurate data for $In_{0.75}Ga_{0.25}As$. The first-principles calculation and the binary compound data interpolation differ by about a factor of two at the band edge where the greatest input intensity exists for a TPV cell illuminated by a 1750 °F (1227 K) blackbody. The magnitude of the effect of this difference on an internal quantum efficiency (QE) prediction is illustrated in Figure 7. Using the STEBS-2D hydrodynamic equation solver (6), internal QE's are predicted for a hypothetical 0.52-eV TPV cell (p-on-n with an n^+ embedded conduction layer and back surface reflector) using transport and recombination parameters similar to those that gave a good fit for a 0.6-eV ERI cell. The only differences in the inputs for the two calculated curves are the assumed absorption coefficient and refractive index.

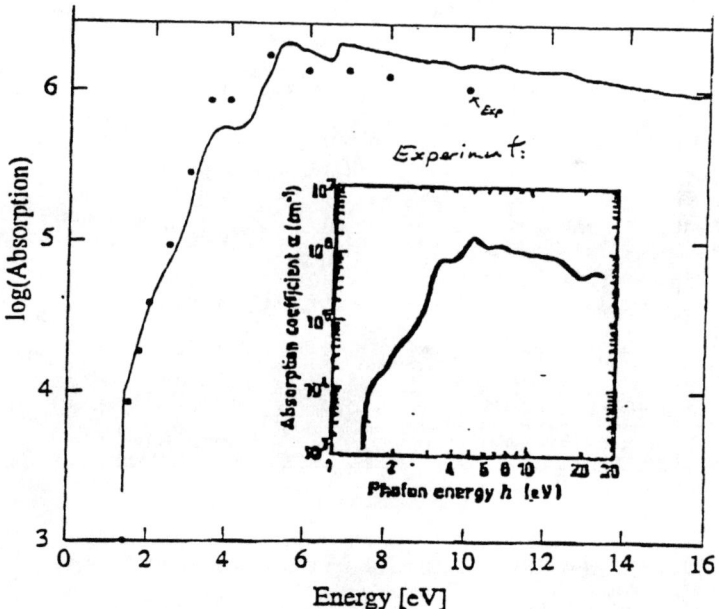

FIGURE 5. FLAPW-B and Measured Absorption Coefficients for GaAs

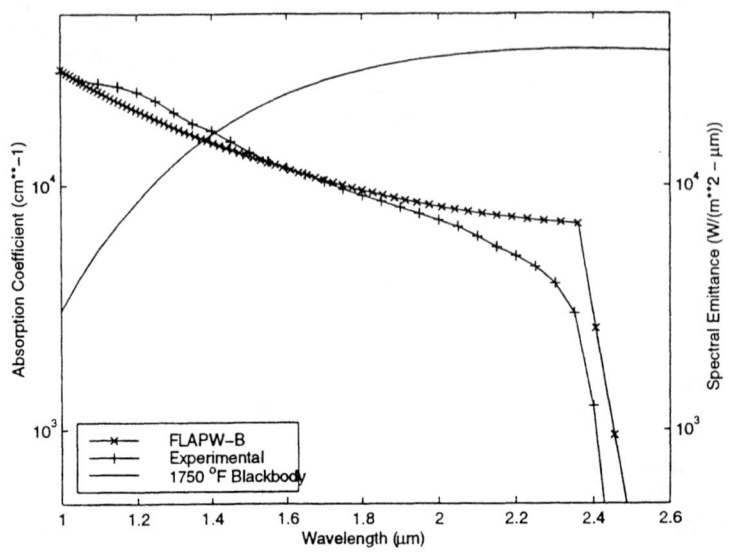

FIGURE 6. FLAPW-B and Interpolated Absorption Coefficients for $In_{0.75}Ga_{0.25}As$

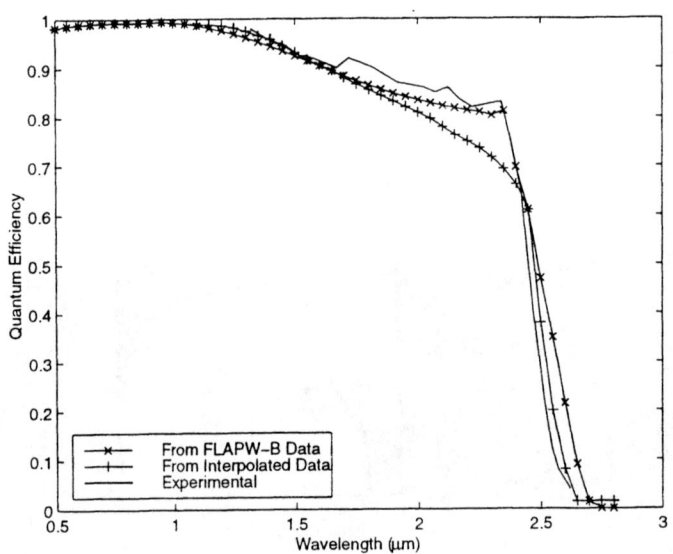

FIGURE 7. Internal QE for 0.52-eV TPV Cell

An experimental internal QE curve for the 0.6-eV ERI cell also is shown (shifted in wavelength to that of a 0.52-eV cell). As can be seen, the band edge response of an actual TPV device is more accurately reproduced with the calculated optical properties.

OPTICAL PROPERTIES OF DOPED MATERIALS

Two important effects of dopants on semiconductor optical properties are the change in optical band gap (Moss-Burstein Shift) and the smearing of the absorption edge, both of which strongly affect the near-band-edge QE of a TPV device. The Moss-Burstein Shift is most appreciable for n-doped semiconductors owing to the much lower effective mass of conduction band electrons relative to valence band holes. This effect can reasonably be modeled in the rigid band approximation, provided that one has available a reliable empirical relationship for electronic band gap narrowing. However, attempts to model band edge smearing by adding a discrete impurity band to an unperturbed host band structure probably are misguided. FLAPW calculations on both the LDA and sX-LDA level of theory show that the valence wave functions of selenium (n-type) and zinc (p-type) substitutional impurities both hybridize strongly with the wave functions of the host crystal. This is true even for a 64-atom cell LDA calculation in which the impurities in question were effectively noninteracting. Thus, it is concluded that impurity and host states remain hybridized even in the dilute limit. It follows that the impurity contribution to the energy density of states indicated for the 64-atom unit cell should span virtually the same energy range as it would in the dilute limit and should scale linearly in magnitude with concentration. On this basis, it is reasonable to approximate the energy density of states and also the dielectric function of a dilutely doped semiconductor as an appropriately weighted average of heavily doped and pure host contributions, both of which can be calculated from first principles. Both calculations must be performed using the dilute limit lattice parameters and the DOS from both calculations must be calculated and integrated with high precision to obtain accurate Fermi Levels corresponding to the dilute limit dopant concentration. For this purpose a tetrahedral k-point sampling scheme has been developed with local mesh refinement in critical areas of the Brillouin zone.

CONCLUSIONS

1. Nonlocal screened exchange corrections provide a large, systematic improvement to the theoretical description of semiconductor excitation spectra and optical properties.

2. The FLAPW-B computational code and its associated optical properties module is a practical tool for informing electronic device simulations and predicting material performance.

3. $In_xGa_{1-x}As$ compositions near x=0.75 are predicted to show strong tendencies toward high local symmetry and short-range order, which may in turn promote compositional segregation over longer length scales.

4. Absorption edge smearing is not modeled accurately by adding a discrete "impurity band" to a rigid host band. An approach is required that explicitly includes the effects of hybridization between impurity and host crystal wave functions.

REFERENCES

1. H. F. Jansen and A. J. Freeman, "Total-Energy Full-Potential Linearized Augmented Plane-wave Method for Bulk Solids: Electronic and Structural Properties of Tungsten," *Phys. Rev. B* **30**, 561 (1984).

2. W. Wolf, E. Wimmer, S. Massida, M. Posternak, and C. B. Geller, "Electronic Excitation Spectra from All-Electron Non-Local Screened Exchange Density Functional Theory," presented at the American Physical Society March Meeting, Los Angeles, CA, March 19, 1998.

3. C. B. Geller, W. Wolf, E. Wimmer, and W. Mannstadt, "Energy Band Gaps and Optical Spectra of Pure and Zn-, Se- and Rare Earth-Doped III-V Semiconductors," presented at the American Physical Society March Meeting, Los Angeles, CA, March 19, 1998.

4. A. Seidl, A. Görling, P. Vogl, and J. A. Majewski, *Phys. Rev. B* **53**, 3764 (1996).

5. M. Levinshtein, S. Rumyantsev, and M. Shur (eds.), *Handbook Series on Semiconductor Parameters, Vol. 1,* Singapore: World Scientific, 1996.

6. A. Smith and K. F. Brennan, *Solid State Electronics* **39**, #11, 1659-68 (1996).

Optimisation of InGaAsP Quantum Well Cells for Hybrid Solar-Thermophotovoltaic Applications

Carsten Rohr*, James P. Connolly, Keith W.J. Barnham,
Ian Ballard, Paul R. Griffin, Jenny Nelson,
Chris Button[†], Janice Clark[†]

*Experimental Solid State Physics, Blackett Laboratory,
Imperial College of Science, Technology and Medicine,
London SW7 2BZ, U.K.* * Email: c.rohr@ic.ac.uk
[†] *EPSRC III-V Facility, University of Sheffield, Sheffield, S1 3JD, U.K.*

Abstract. We discuss quantum well cells (QWCs) in the quaternary system $In_{1-x}Ga_xAs_yP_{1-y}$ lattice-matched to the InP substrate ($x \approx 0.47y$) for the use in solar and thermophotovoltaic (TPV) applications. The deep lattice-matched wells of up to $y = 1$ ($In_{0.53}Ga_{0.47}As$) can be incorporated without any strain. The effective band-gap for absorption in these quaternary QWCs can be tailored (up to $\sim 1.7\mu m$) to produce the ideal band-gap for a given blackbody or selective-emitter spectrum while retaining a comparatively high efficiency for the solar spectrum. This has a great potential for higher-efficiency cells, especially for hybrid solar-TPV applications. We present the results of a new model for the external quantum efficiency of InGaAsP QWCs. The model calculates the spectral response of multi-layer InGaAsP photovoltaic cells with and without quantum wells. It is in very good agreement with the experimental spectral response of InGaAsP QWCs with a variety of designs and therefore is an important tool for the optimisation of these cells. Besides improvements in quantum well geometry, processing and material quality, the high efficiencies we obtain can be even further increased by additional features such as the use of back surface reflectors. We also investigate how these cells perform under several different illuminations such as solar and blackbody spectra as well as narrow-band selective-emitter spectra from erbia and ytterbia comparing them to lattice-matched InGaAs monolithic interconnected modules. We conclude that the InGaAsP QWC system is a very promising candidate for high-efficiency photovoltaic cells for hybrid solar-TPV applications.

INTRODUCTION

The fundamental limit to the efficiency of a photovoltaic (PV) cell results from the broad energy spectrum of the radiation and the single, fixed band-gap of a conventional PV cell. The choice of material for a PV cell is a trade-off between maximising current output with a lower band-gap and voltage output with a higher band-gap semiconductor.

In the Quantum Well Cell (QWC) higher efficiencies can potentially be achieved by introducing multi-quantum wells into the intrinsic region of a p-i-n diode [1]. The enhancement in short-circuit current (I_{SC}) can be greater than the loss in open-circuit voltage (V_{OC}), and results in significantly improved cell efficiencies for band-gaps greater than optimum [2]. The V_{OC} is between that of p-i-n devices made of the well material and that made of the base cell and is higher than expected from the change in the effective band-gap alone [3]. The improved performance is due to the high current collection efficiency from the quantum wells (QWs) even in forward bias, which is attributed to efficient thermal escape of the carriers out of the QWs [4,5].

Thermophotovoltaics (TPV) is the use of PV cells to convert heat radiation, e.g. from the combustion of fossil fuels, possibly from fuel cells, or by co-generation using waste heat, into electricity. TPV is an old idea [6] currently undergoing a renewed interest due to recent advances in low band-gap cells and selective emitters [7]. The energy spectrum is often reshaped using selective emitters such as the rare earth oxides ytterbia and erbia which absorb infrared radiation and re-emit in a narrow band [8]. The re-emitted radiation may be efficiently converted to electric power using a PV cell of appropriate band-gap. For TPV low band-gaps are required since the source temperature is much lower than that of the sun but low band-gaps give low output voltages.

Advantages of InGaAsP QWCs for TPV

The quaternary InGaAsP lattice-matched to the InP substrate in QWCs is a very promising candidate to overcome commonly encountered problems of single, fixed band-gap materials such as the difficulty in spectral matching of source and cell and strain relaxation caused by lattice mismatch. With deep lattice-matched wells, the effective band-gap can be tuned, up to $\sim 1.7 \mu m$ ($In_{0.53}Ga_{0.47}As$), without introducing strain, by varying the well depth and width, to match a given spectrum.

The large enhancement in output voltage of a QWC compared to a comparable control cell made from the well material is a major advantage for TPV applications [9-11]. It also gives rise to another advantage: for the same power conversion efficiency as a conventional cell made from the well material a QWC operates at a higher voltage and hence lower current; this means series resistance effects are smaller and active areas larger. This is particularly important as the current levels in TPV applications and under concentration are high.

Finally, QWCs have superior temperature coefficients of efficiency to comparable conventional cells made from the well or the barrier material [10]. This lower susceptibility to heat makes QWCs very attractive, especially for the use in hot environments of TPV applications. It will also be of advantage under concentration; the power densities in TPV applications are expected to be ≥ 100 suns.

TABLE 1. A typical InGaAsP Quantum Well Cell (Sample A).

No. of layers	Thickness [Å]	Material	Function	Doping
1	1000	$In_{0.53}Ga_{0.47}As$	Cap	p
1	1500	InP		p
1	2500	InP	Spacer	i
60	70	InGaAsP (Q1.1)	Barrier	i
60	100	InGaAsP (Q1.6)	Well	i
1	70	InGaAsP (Q1.1)	Barrier	i
1	5000	InP		n
		InP	Substrate	n

EXPERIMENTAL

Sample Description

The $In_{1-x}Ga_xAs_yP_{1-y}$ quaternary cells are p-i-n structures where all layers are lattice-matched to the InP substrate which is equivalent to $x \approx 0.47y$. The samples were grown by metal organic vapor phase epitaxy (MOVPE). The actual fractional material composition, which is not precisely known from the growth, is determined by fitting the spectral response (SR) using the model described below.

The layer structure of a typical InGaAsP QWC is shown in Table 1 (Sample A). The p- and n-type layers are InP ($y = 0$), whereas the $\sim 1\mu m$ wide intrinsic region is made of the quaternary with a composition of $y \approx 0.36$ resulting in a band-gap of $1.15\mu m$. Incorporated into this i-region are 60 QWs of $10nm$ width made of the quaternary with a composition of $y \approx 0.86$ equivalent to a band-gap of about $1.6\mu m$, but due to the confinement the effective band-gap is close to $1.5\mu m$. A $0.2\mu m$ wide spacer region is introduced between the i- and p-region to reduce dopant diffusion into the QWs [12].

All cells were processed into $1mm$-diameter test devices with a ring contact leaving an optical window of $545\mu m$ diameter. The QWC devices were processed with and without a one-layer silicon nitride anti-reflection (AR) coating, usually optimised for about $900nm$. When an AR coating is applied, the InGaAs cap is removed.

Sample B has a thicker p-region, slightly deeper wells and an AR coating optimized for $1500nm$. Sample C has narrower QWs made of InGaAs, the spacer region is made of the quaternary of the same composition as the barrier; it has no cap, leaving the InP as front surface of this non-AR sample. Otherwise the cells are the same as Sample A (Table 1).

Spectral Response (SR)

The SR of all the cells was measured with a tungsten lamp, computer-controlled monochromator and lock-in amplifier. The absolute external quantum efficiency

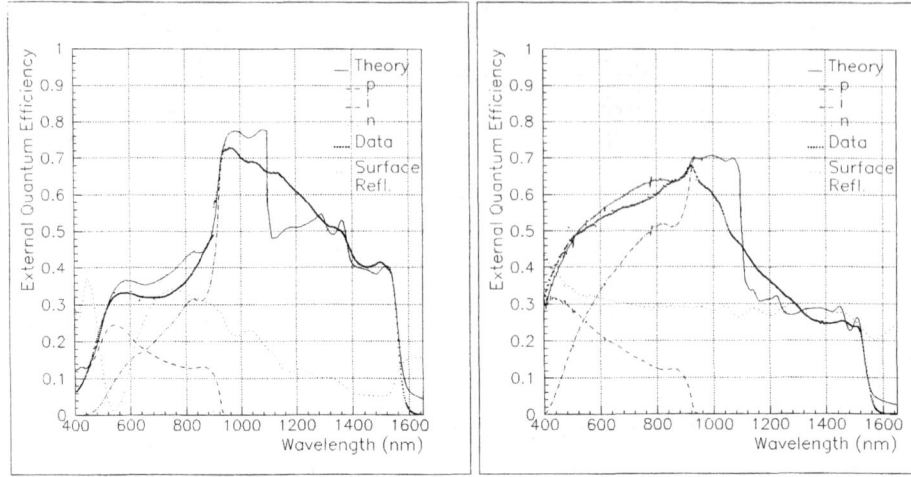

FIGURE 1. Spectral response of two different InGaAsP quantum well cells - Theory, data and reflectivity. Left: Sample B (AR coated), Right: Sample C (non-AR).

(QE) was obtained by ratioing the photocurrent of the cells with calibrated InGaAs and Silicon detectors. Typical results of very different cell designs are presented in Figure 1 (Sample B and C) and Figure 2 (Sample A) where the experimental SR is fitted with our new model described below. The QE of the QWC above $1100nm$ is generated from photon absorption in the QWs. The response falls off at longer wavelengths due to fewer energy levels in the QWs and a broad exciton peak is seen at around 1500–$1550nm$. The band-edge of the QWC is about $1550nm$. This is due to the InGaAsP well material composition, which has a bulk band-edge corresponding to longer wavelengths, and the quantisation of energy levels because of the confinement in the wells. The fall-off in QE of the QWC below $950nm$ is due to bulk and surface recombination in the InP p-layer, and in some cases due to higher reflectivity of the AR coating at smaller wavelengths.

MODELLING

Results of a new model for the external quantum efficiency of InGaAsP QWCs are presented. We find that the model is in very good agreement with the experimental results and therefore is an important tool for the optimisation of these cells.

The Model

The model calculates the spectral response of multi-layer InGaAsP devices with separate contributions from the p-, i-, and n-region. It enables us to fit the experimental results of any given cell design and is based on a model which has been

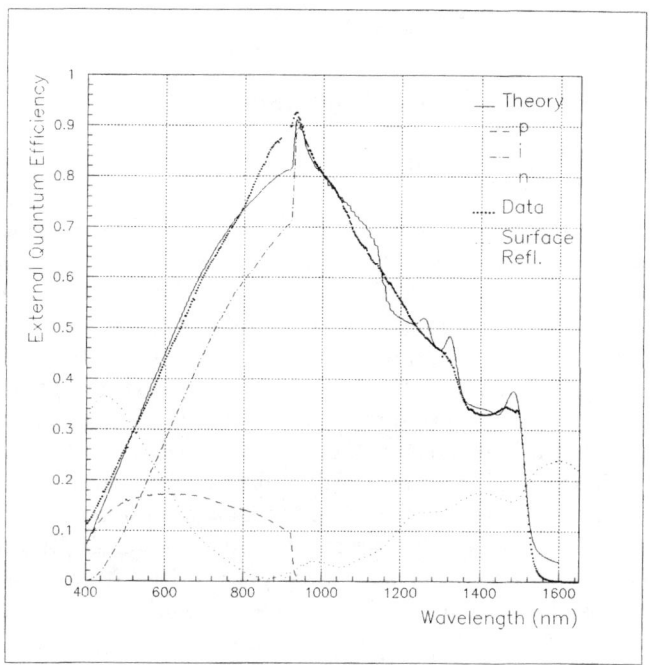

FIGURE 2. Spectral response of InGaAsP QWC Sample A (AR coated) - Theory, data and reflectivity.

developed for QWCs in the material system GaAs/AlGaAs [13], and has previously been extended to InGaAs/GaAs [14] and InGaAs/InP [12]. Now it has been further developed to deal with the quaternary system $In_{1-x}Ga_xAs_yP_{1-y}$ of any composition lattice-matched to InP.

The SR is calculated by solving the carrier transport equations at room temperature within the minority carrier and depletion approximations. The assumptions of low injection, uniform doping, and abrupt changes in composition at heterojunctions are justified for these heavily doped materials and enable an analytic solution of the transport equations to be obtained. We use measured, published [15], or modeled values for the various parameters required. The absorption is calculated by the same method as Paxman et al. [13] using published data [16], which is quite limited for InGaAsP. A self-consistent approach to determine the absorption has been applied for QWC Sample A (Table 1, Fig. 2) by using the SR of hetero-structure control cells without QWs. Fairly reliable data is available for the other optical and transport properties of InGaAsP lattice-matched to InP, thus limiting the number of fitting parameters. Experimentally measured reflectivity of the surface can be used in the model.

The model calculates the short-circuit current density for any spectrum, and assuming superposition, can also determine the efficiency if provided with the dark current data of the cell.

Fitting Data

We were able to accurately fit data of a number of cell designs with various layer thicknesses and compositions as shown in Fig. 1 and 2, which gives confidence in the model. The fit for the band-edge, the first exciton and the first continuum level is very good taking the appropriate well material composition and width. Even the second exciton can be fitted accurately. The broadness of the experimental peaks is due to non-abrupt interfaces caused by diffusion and width fluctuations. In the model one can see the sharp band-edge of the barrier material at $\sim 1100 nm$. Both the broadening of the top of the well, again due to non-abrupt interfaces, and super-lattice effects result in more energy states in the top of the well, and therefore a stronger response than predicted is observed in the region 1100–$1250 nm$. The SR is overestimated in the barrier region of 950–$1100 nm$, where the absorption is small and poorly known. Below $950 nm$ there is a big contribution from the p-layer. This region is sensitive to the p-layer thickness, the minority carrier diffusion length in the p-layer and the surface recombination velocity. The n-region contribution is nearly zero because the thick p- and i-layers absorb nearly all the incident light.

Optimisation

The model predicts that the SR and hence the current performance of an InGaAsP QWC can be improved substantially by having a thin p-region, an AR coating optimised for the appropriate wavelength, and more, deeper and wider wells. Efficiencies of up to about 40% for an Erbia-like narrow-band, 14% for a 2000K black-body and 22% for an AM1.5 spectrum were predicted [11].

Further improvements to these cells can be made by adding a back mirror enhancing the well contribution in particular, which is discussed later, and also reflecting longer wavelength radiation not absorbed by the cell back to the source.

Advanced multi-layer AR coatings appropriate to the illuminating spectrum and the SR of the cell, and using a suitable front contact grid design to balance series resistance and shading losses at higher power densities could further enhance the cell performance.

Various illuminating spectra

Black-body temperature dependence

The short-circuit current density of InGaAsP QWC Sample A as a function of black-body temperature of the illuminating spectrum is shown in Figure 3. The same behaviour is seen for all samples including homogeneous structures such as

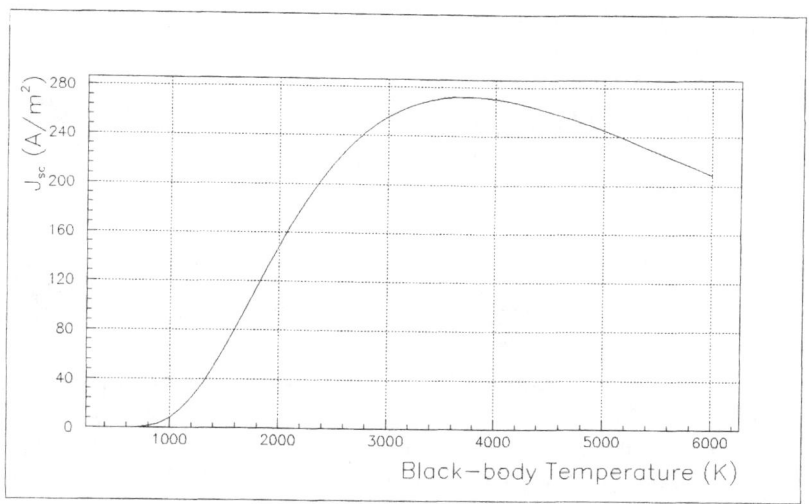

FIGURE 3. Short-circuit current density of InGaAsP QWC as a function of blackbody temperature of illuminating spectrum.

InGaAs. The linear dependence reported by Beckert *et al.* [17] is an approximation for the temperature range of about 1200–2500K.

Comparison with InGaAs monolithic interconnected modules (MIM)

We compare our InGaAsP QWC Sample A with the lattice-matched InGaAs MIM [18] ($E_g = 0.74eV$, non-AR, with back surface reflector), a state-of-the-art technology and one of the best bulk InGaAs/InP TPV cells, under typical TPV or concentration conditions of 100 suns intensity and assuming a grid shading of 5%. For comparison purposes the reflectivity of the AR coating of our QWC was replaced by an InGaAs reflectivity. A back surface reflector is an integral part of MIM technology and also particularly useful for QWCs as it enhances the well contribution significantly. The effect of such a mirror is simulated by doubling the light pass through the wells. Techniques for modelling a back mirror have been successfully applied previously [19].

The short-circuit current density was calculated by

$$J_{SC} = intensity \times \int spectrum(\lambda)\, SR(\lambda)\, \frac{q}{hc}\, \lambda\, d\lambda \qquad (1)$$

where *spectrum* is the normalised illuminating spectrum and λ the wavelength. The integral is taken from $\lambda = 500nm$ as the SR data for the InGaAs MIM was only available for $\lambda \geq 500nm$. Therefore the experimental J_{SC} should be greater— hence the quoted efficiencies are underestimates. The efficiency η is calculated from J_{SC} and the experimental dark current density with a grid shading of 5%.

TABLE 2. Comparison of predicted *efficiencies* of non-AR InGaAsP QWC and InGaAs MIM with back surface reflector under various spectra at 100 suns and 5% shading.

Spectrum	InGaAsP QWC	InGaAs MIM
1500K black-body	3.1%	3.5%
2000K black-body	7.5%	7.5%
3200K black-body	14%	12%
Ytterbia-like [10]	26%	17%
Erbia-like [10]	29%	27%

The results for various illuminating spectra normalised to 100 suns power density are summarised in Table 2.

In many cases, the higher SR of the InGaAs MIM is more than off-set by the lower dark current, i.e. better voltage performance, of the mirrowed InGaAsP QWC. As one can see in Figure 4, the InGaAsP QWC has a higher band-gap and therefore a better dark current. For black-body spectra in the range of 1500–2000K, typical TPV temperatures, the efficiencies of the InGaAsP QWC and the InGaAs MIM is expected to be comparable. Higher black-body temperatures, for examples 3200K, are favourable for the QWC. In spectra of lower black-body temperatures the InGaAs MIM is better because of the higher band-edge of 1650–1700nm for bulk InGaAs compared to 1500–1550nm for the QWC. With deeper and/or wider wells this problem could be overcome.

With narrow-band selective emitters such as Ytterbia and Erbia, which are sim-

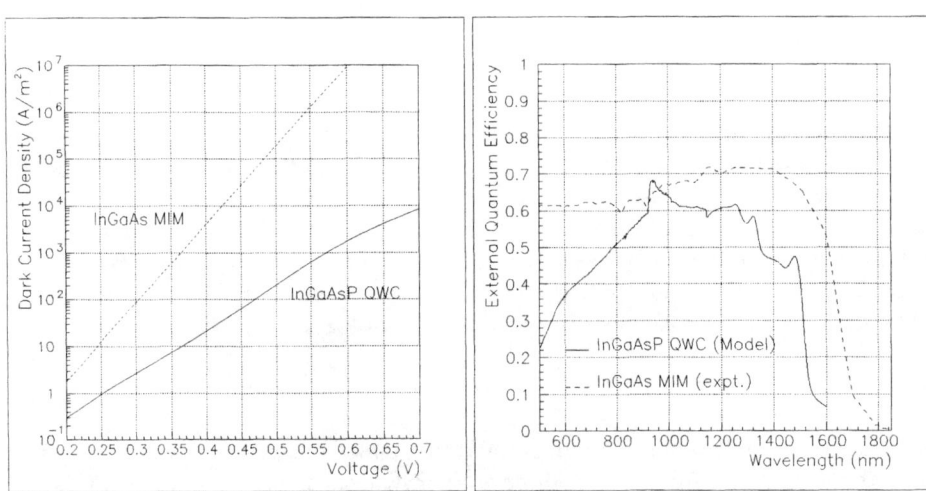

FIGURE 4. Comparison of InGaAsP QWC and InGaAs MIM. Left: Dark current, Right: Spectral response (both with back mirror, non-AR).

ulated by using narrow-band filters of $950nm$ and $1500nm$ respectively [10], the QWC has significant advantages over the InGaAs MIM. In the case of the Erbia-like spectrum the difference is less pronounced because the spectrum does not completely overlap with the SR of this QWC; again, slightly deeper or wider wells would increase the QWC advantage considerably.

Solar Spectrum

The InGaAsP QWC Sample A retains a comparatively high efficiency for the solar spectrum, calculated over the whole wavelength range. In the non-AR case, with a mirror on the back, we calculate an efficiency of 7.7% for an air mass 1.5 global solar spectrum (AM1.5) and under 100 suns concentration $\eta = 11.8\%$. With an AR coating optimised for about $900nm$, $\eta = 11.0\%$ for AM1.5 and $\eta = 16.6\%$ under 100 suns concentration.

CONCLUSIONS

We have shown that, with our newly extended model, we can accurately fit the SR of a variety of QWC in the InGaAsP quaternary system with various layer types, layer thicknesses and compositions. We are able to calculate the short-circuit current density for any given spectrum, and given the dark current calculate the expected efficiency.

The better SR of an InGaAs MIM [18] compared to our InGaAsP QWC can be more than off-set because of a significantly lower dark current. We expect the QWC to have a similar or superior performance under most illuminating spectra, such as black-body temperatures above about 1800K and narrow-band emitter like Erbia and Ytterbia, at 100 suns concentration.

Under an AM1.5 solar spectrum the AR coated QWC (Sample A) with a mirror on the back is expected to reach efficiencies of about 11% and almost 17% under 100 suns concentration.

In summary, the InGaAsP QWC system is a very promising candidate for high-efficiency PV cells, possibly in a MIM configuration, for hybrid solar-TPV applications.

ACKNOWLEDGEMENTS

We are grateful to EPSRC, The Institute of Physics, The Greenpeace Environmental Trust, The British Council and Tata Ltd. (India) for financial support, and Navid Fatemi for helpful information about InGaAs MIMs.

REFERENCES

1. Barnham, K. W. J., and Duggan, G., *J.Appl.Phys* **67**, (1990) 3490–3493.
2. Barnham, K., Ballard, I., Barnes, J., Connolly, J., Griffin, P., Kluftinger, B., Nelson, J., Tsui, E., and Zachariou, A., *Applied Surface Science* **113/114**, (1997) 722–733.
3. Barnham, K. W. J., Connolly, J., Griffin, P., Haarpaintner, G., Nelson, J., Tsui, E., Zachariou, A., Osborne, J., Button, C., Hill, G., Hopkinson, M., Pate, M., Roberts, J. S., and Foxon, T., *J.Appl.Phys* **80** (2), (1996) 1201–1206.
4. Nelson, J., Paxman, M., Barnham, K. W. J., Roberts, J. S., and Button, C., *IEEE Journal of Quantum Electronics* **29**, (1993) 1460–1467.
5. Zachariou, A., Barnes, J., Barnham, K. W. J., Nelson, J., Tsui, E. S. M., Epler, J., and Pate, M., *J.Appl.Phys* **83** (2), (1998) 877–881.
6. Aigrain, P., The thermo-photovoltaic converter, 1956. In a series of lectures, École Normale Supérieure, Paris and Dept. of Elec. Eng., MIT.
7. Coutts, T. J., and Fitzgerald, M. C., *Scientific American* **9**, (1998) 90–95.
8. Coutts, T. J., Wanlass, M. W., Ward, J. S., and Johnson, S., in *Proc. 25th IEEE PV specialists conf.*, 1996 pp. 25–30.
9. Griffin, P., Ballard, I., Barnham, K., Nelson, J., Zachariou, A., Epler, J., Hill, G., Button, C., and Pate, M., *Solar Energy Materials and Solar Cells* **50**, (1998) 213–219.
10. Griffin, P., Ballard, I., Barnham, K., Nelson, J., Zachariou, A., Epler, J., Hill, G., Button, C., and Pate, M., in *Proc. Thermophotovoltaic Generation of Electricity: Third NREL Conference*, 1997 pp. 411–422.
11. Rohr, C., Connolly, J. P., Barnham, K. W. J., Griffin, P. R., Nelson, J., Ballard, I., Zachariou, A., Button, C., and Clark, J., in *Proc. 2nd World PV Energy Conversion Conf.*, 1998 To be published.
12. Zachariou, A., Barnham, K. W. J., Griffin, P., Nelson, J., Button, C. C., Hopkinson, M., Pate, M., and Epler, J., in *Proc. 25th IEEE PV specialists conf.*, 1996 pp. 113–116.
13. Paxman, M., Nelson, J., Braun, B., Connolly, J., Barnham, K. W. J., Foxon, C. T., and Roberts, J. S., *J.Appl.Phys* **74**, (1993) 614–621.
14. Barnes, J., Ali, T., Barnham, K. W. J., Nelson, J., and Tsui, E. S. M., in *Proc. 12th European PV solar energy conf.*, 1994 pp. 1374–1377.
15. Adachi, S., *Physical Properties of III-V Semiconductor Compounds: InP, InAs, GaAs, GaP, InGaAs, and InGaAsP*, New York Chichester: John Wiley & Sons, 1992. A Wiley-Interscience publication.
16. Adachi, S., *Properties of indium phosphide*, no. 6 in EMIS datareviews series, London: INSPEC, 1991 p. 408.
17. Beckert, R., Broman, L., Jarefors, K., Marks, J., and Bücher, K., in *Proc. 2nd World PV Energy Conversion Conf.*, 1998 To be published.
18. Fatemi, N. S., Wilt, D. M., Jenkins, P. P., Weizer, V. G., Hoffman, Jr., R. W., Murray, C. S., Scheinman, D., Brinker, D., and Riley, D., in *Proc. 26th IEEE PV specialists conf.*, 1997 pp. 799–804.
19. Connolly, J., *Modelling and Optimising GaAs/AlGaAs MQW solar cells*, Ph.D. thesis, Imperial College, University of London, 1997.

A TPV System Model and an Analysis of TPV System Efficiency

Jeffery L. Gray and Ali M. El-Husseini

School of Electrical and Computer Engineering
Purdue University, West Lafayette, IN 47907

Abstract. The components of a TPV system model include a model for the optics of the system, a detailed numerical model of the TPV cell, and a thermodynamic model from which the overall system efficiency can be computed. These three components are coupled and a complete model of the TPV system requires a self-consistent treatment of all three components simultaneously. The optical model includes all the optical elements between, and including, the TPV emitter and the TPV cell. These elements include, for example, heat shields, spectral selectors, and antireflective coatings on the TPV cell. Ultimately, this model provides the information needed to compute the optical generation rate within the TPV cell (used by the TPV cell model). The thermodynamic component enforces conservation of energy and determines the operating temperature of the TPV cell. In addition, some simple analytic results are presented and used to illustrate some TPV system design tradeoffs.

INTRODUCTION

A complete TPV system model should include a model for the TPV emitter, a model for the optical components, a thermodynamic element, and a detailed model of the TPV cell. These model components are coupled and a complete TPV model requires simultaneous treatment of all the model components.

To determine the TPV system efficiency, one must define what that efficiency is. Here, the details of the TPV emitter are not modeled and so the efficiency of the conversion of the fuel into the light radiated from the emitter is not considered. The TPV system model discussed here assumes the spectral content of the light from the TPV emitter is known and that some fraction of the photons not absorbed by the TPV cell are returned to the TPV emitter. Thus, for the purposes of this paper, the TPV system efficiency is defined as

$$\eta_{sys} = \frac{P_{out}}{P_E - P_{ret}} \qquad (1)$$

where P_{out} is the net power produced by the TPV cell, P_E is the radiation power emitted by the TPV emitter, and P_{ret} is the radiation power returned to the TPV emitter.

A detailed numerical device model for the TPV cell is presented elsewhere in these proceedings (1). This paper will focus primarily on modeling the optics of the TPV system with particular attention on calculating the optical generation rate inside the TPV cell. The thermodynamic model component will also be discussed relative to its role in computing the TPV cell operating temperature.

TPV OPTICS

The optics of a TPV system is complicated by the fact that light is emitted from the TPV emitter at a variety of angles. This is in contrast to a typical PV system in which the light is incident at a single angle. This influences the design of any optical selectors must be accounted for in calculating the optical generation rate within the TPV cell.

The Optical Generation Rate

We shall first address the calculation of the optical generation rate inside the TPV cell, which is, along with the determination of the net radiation returned to the TPV emitter, the main function of the optical model. It is useful to consider three separate idealized TPV geometric arrangements: spherical, in which the light source is a point surrounded by a sphere of TPV cells; cylindrical, in which the light source is a line surrounded by a cylindrical arrangement of TPV cells; and planar, in which both the TPV emitter and the TPV cells are infinite parallel planes. It is plain that the circumstances for a typical TPV arrangement of a cylindrical emitter surrounded by TPV cells lies somewhere between the idealized cylindrical and planar geometric arrangements.

While an analytic formula for the generation rate, $G(x)$, cannot, in general be derived, such a formula is possible if suitable simplifications are made. If reflections are ignored, an analytic formula can be derived which provides some useful insights.

Figure 1 shows the planar geometric arrangement of a blackbody emitter parallel to and absorbing medium. For a "black" surface, there exists no preferential direction of emission. If the space between the blackbody and the absorber is non-absorbing, then light impinging on the absorber does so with the same intensity as the light emitted from the blackbody and also exhibits no preferential direction of incidence. Thus, it will suffice to consider the point **P** located at the surface of the absorber as a point source equivalent to the point source **P'** on the blackbody emitter, assuming reflections are ignored. If light transmitted from a surface element **P** in Figure 1, the power transmitted through any surface intersecting the same solid angle, such as the surface elements dA and dA'. This is the basis of Lambert's cosine law

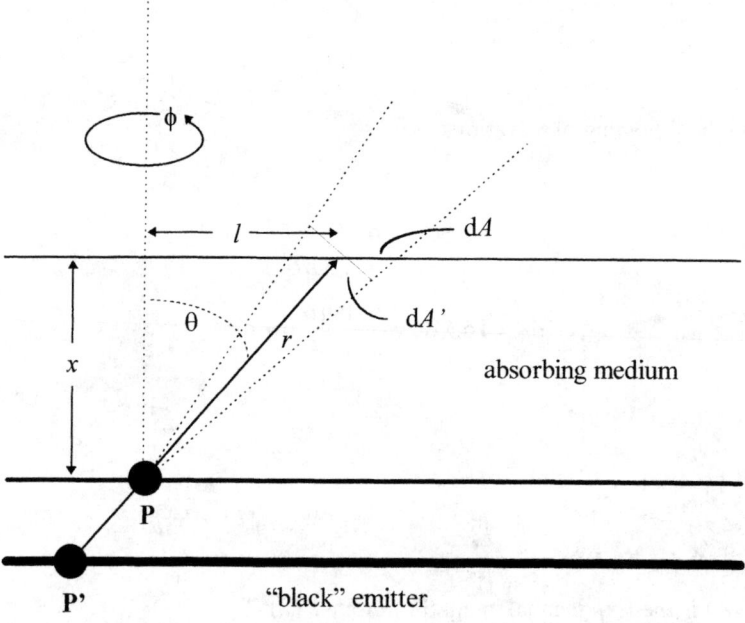

FIGURE 1. Planar TPV arrangement illustrating geometric factors.

$$B(\theta) = B_0 \cos\theta \qquad (2)$$

where the brightness, B_0, is the power impinging on the surface dA' per square centimeter per unit solid angle (assuming no absorption) and θ is angle between dA and dA'. B_0 is, itself, independent of θ. Lambert's cosine law explains why spherical, cylindrical, and other irregularly shaped blackbodies appear to the eye as flat objects.

To compute $G(x)$, we must find the photon flux arriving at the parallel plane, A, at a distance x from the surface of the absorbing medium. Since both the emitter and absorber are assumed to be uniform, the photon flux at any point on A will be independent of l and ϕ. Thus, it will be sufficient to integrate the contribution from the point **P** onto the plane A.

Let us define N_0 as the number of photons per second radiating from point **P** on the surface of the absorbing medium. By Lambert's cosine law and Beer's law of absorption, the flux density impinging on the surface element dA is

$$B(r,\theta) = \frac{N_0 \cos\theta\, e^{-\alpha r}}{2\pi r^2} \tag{3}$$

Making the following observations

$$r = \frac{x}{\cos\theta}$$
$$l = r\sin\theta = x\tan\theta \tag{4}$$
$$dA = l\, d\phi\, dl = \frac{x^2 \tan\theta}{\cos^2\theta} d\theta\, d\phi$$

and integrating, we get

$$N(x) = N_0 \int_0^{\pi/2} e^{-\alpha x/\cos\theta} \sin\theta\, d\theta = N_0\, E_2(\alpha x) \tag{5}$$

where $E_2(\alpha x)$ is the Exponential integral of order 2 (2)

$$E_n(z) = \int_1^{\infty} t^{-n} e^{-zt} dt \tag{6}$$

Assuming that the photon flux density emitted from the blackbody is Φ_0 (i.e. the points **P'** are evenly distributed) and using

$$G(x) = -\frac{d\Phi}{dx} \tag{7}$$

we get

$$\Phi(x) = \Phi_0\, E_2(\alpha x) \tag{8}$$

and

$$G(x) = \alpha\, \Phi_0\, E_1(\alpha x) \tag{9}$$

Similar relationships can be derived for both the spherical and cylindrical geometric arrangements with the additional assumption that the thickness of the absorber (TPV cell) is much less than the distance between the TPV emitter and the TPV cell.

For the cylindrical geometric arrangement, we get

$$\Phi(x) = \Phi_0 S_2(\alpha x) \tag{10}$$

and

$$G(x) = \alpha \Phi_0 S_1(\alpha x) \tag{11}$$

where S_1 and S_0 are radiation functions (3)

$$S_n(z) = \frac{2}{\pi} \int_1^\infty \frac{t^{-n} e^{-zt} dt}{\sqrt{t^2 - 1}} \tag{12}$$

For the spherical geometric arrangement, we get

$$\Phi(x) = \Phi_0 e^{-\alpha x} \tag{13}$$

and

$$G(x) = \alpha \Phi_0 e^{-\alpha x} \tag{14}$$

which is equivalent to a conventional PV system with a normal angle of incidence. In Figure 2, $\Phi(x)$ is plotted for each of the geometric arrangements. Note that light is absorbed much more quickly in both the planar and cylindrical arrangements as compared to the spherical arrangement.

As was noted earlier, analytic expressions are not possible for the planar and cylindrical arrangements if reflections are to be accounted for. Since light is emitted (and impinges) with no directional preference, it can easily be shown that the angular distribution of the photon flux density for the planar geometric arrangement is

$$\Phi(\theta) = \Phi_0 \sin \theta \tag{15}$$

and it is independent of θ for the cylindrical geometric arrangement. Accounting for reflections and the spectral distribution of the light, the photon flux density for the planar geometric arrangement can be computed numerically and can be written as

$$\Phi(x, \lambda) = \Phi_0(\lambda) \int_0^{\pi/2} (1 - R(\theta, \lambda)) e^{-\alpha x / \cos \theta} \sin \theta \, d\theta \tag{16}$$

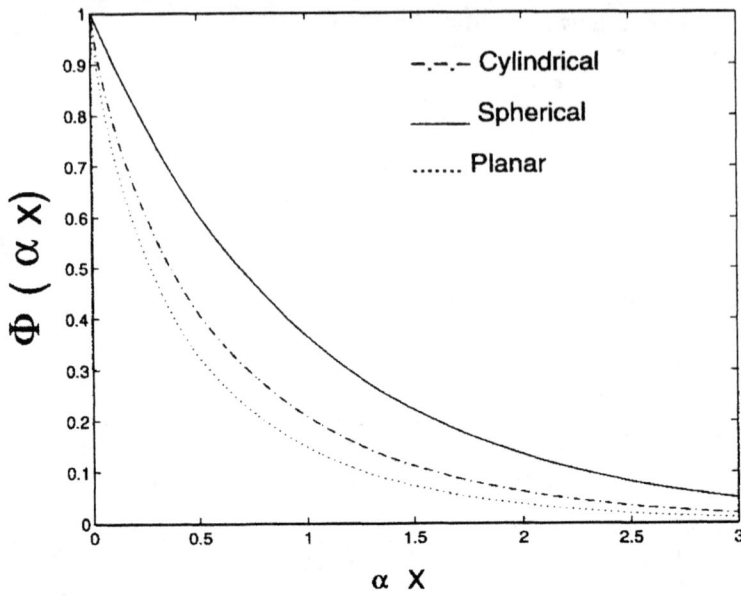

FIGURE 2. Photon flux inside TPV cell for planar, cylindrical, and spherical emitters.

Radiation Transfer

Figure 3 shows a schematic of some hypothetical optical components in a TPV system. Note that this is a coupled system and calculation of the net energy transferred requires knowledge of the optical properties of all the components and the solution of a system of equations.

Optical Selection

It is possible to return unused and/or inefficiently used photons to the TPV emitter. Some methods for accomplishing this include back surface reflectors on the TPV cells, plasma filters, and multi-layer optical filters. The multiple angles of incidence involved complicates the design of multi-layer filters since the transmission properties will depend on the light's angle of incidence. This is illustrated in Figure 4.

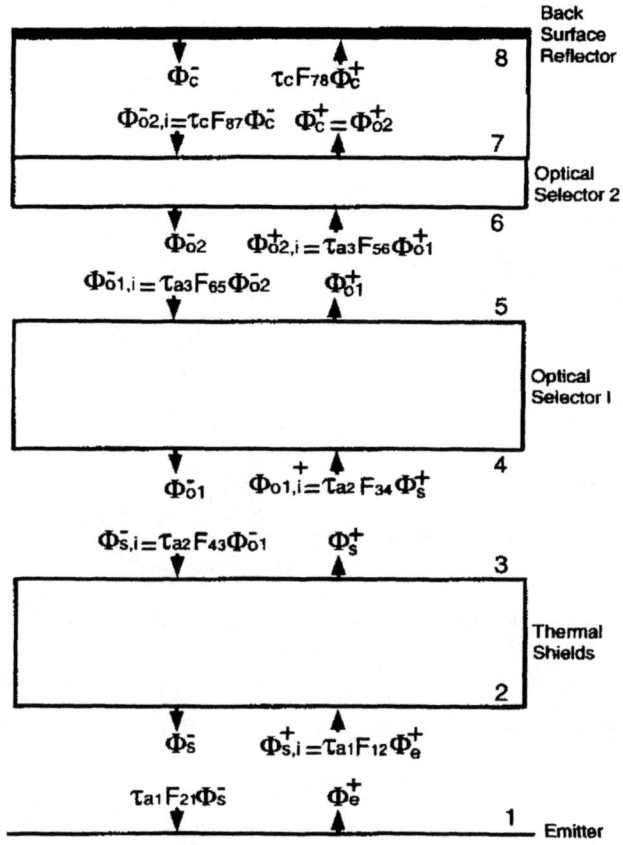

FIGURE 3. Radiant transfer schematic.

THERMODYNAMIC MODEL

The main objectives of the thermodynamic model is to calculate the TPV cell operating temperature and to account for thermal losses. This is accomplished through the application of energy conservation and modeling the flow of energy through the system via radiation, conduction, and convection. Passive cooling (via a heat sink) and reasonable assumptions for a GaSb TPV system predict a cell temperature of about 340 Kelvin.

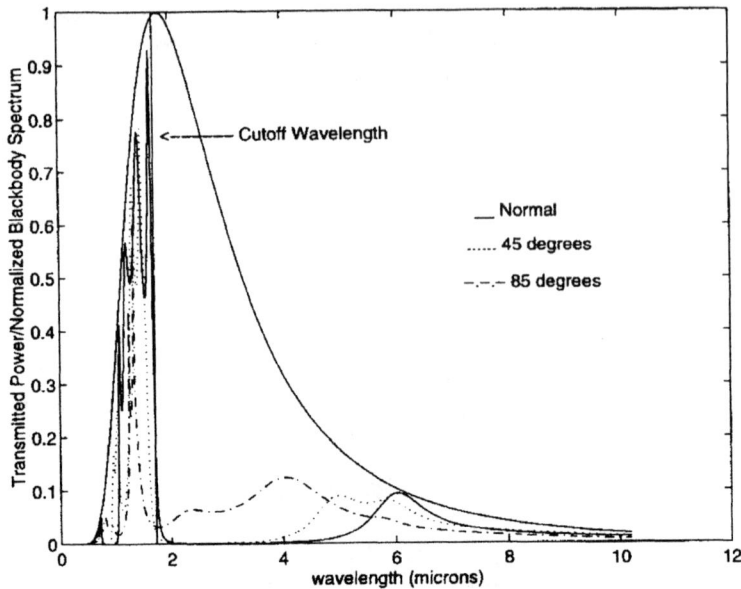

FIGURE 4. Transmission through multi-layer filter.

The waste heat that must me dissipated by the TPV cell can be written as

$$Q_{waste} = P_{abs} - P_{out} \tag{17}$$

where P_{abs} is the power absorbed by the TPV cell and P_{out} is the electrical power delivered by the TPV cell. Since P_{ou} depends of the cell operating point, it is clear the cell operating temperature will be lowest at the maximum power point of the TPV cell.

TPV SYTEM ANALYSIS

It is possible to relate the TPV cell band gap and the temperature of a blackbody TPV emitter so the power available for TPV conversion is maximized (4). This done by maximizing

$$P_{max} = E_g \int_{E_g}^{\infty} p_E(E,T_E) \frac{dE}{E} \tag{18}$$

This results in a simple relationship between the blackbody temperature, T_E, and the optimum TPV cell band gap

$$E_{g,opt} = 1.87 \times 10^{-4} T_E \tag{19}$$

One aspect of TPV system design that is different from PV system design is that efficiency and output power are independent of each other. In fact, there is a trade-off between the system efficiency, η_{sys}, as defined in Equation (1) and the TPV cell output power, P_{out}. This is illustrated in Figures 5 and 6. Each figure assumes that an ideal band pass optical selector is used that passes all radiation with energies between E_g, the TPV cell's optical band gap, and E_u, the upper cutoff energy of the filter. All other radiation is returned to the TPV emitter. In Figure 5, it can be seen that the system efficiency (assuming an idealized TPV cell with 100% collection efficiency, unity fill factor, and $V_{OC} = E_g/q$) approaches 100% when $E_u = E_g$. However, the output power falls to zero in that case.

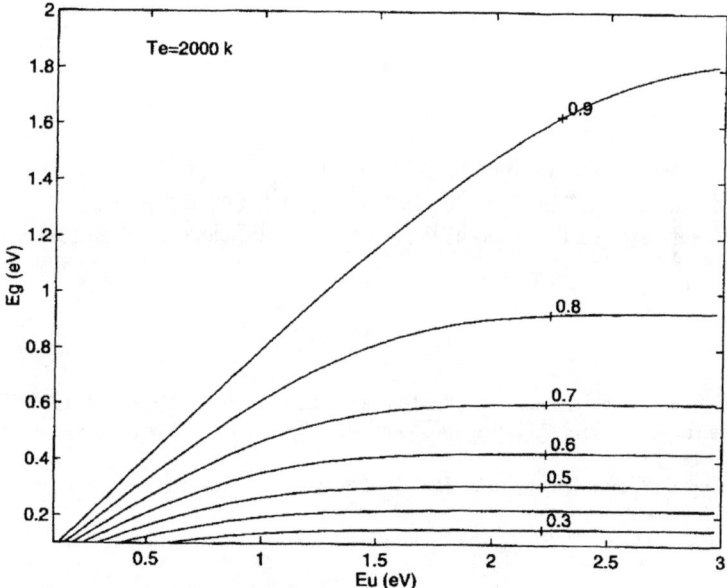

FIGURE 5. System efficiency contours.

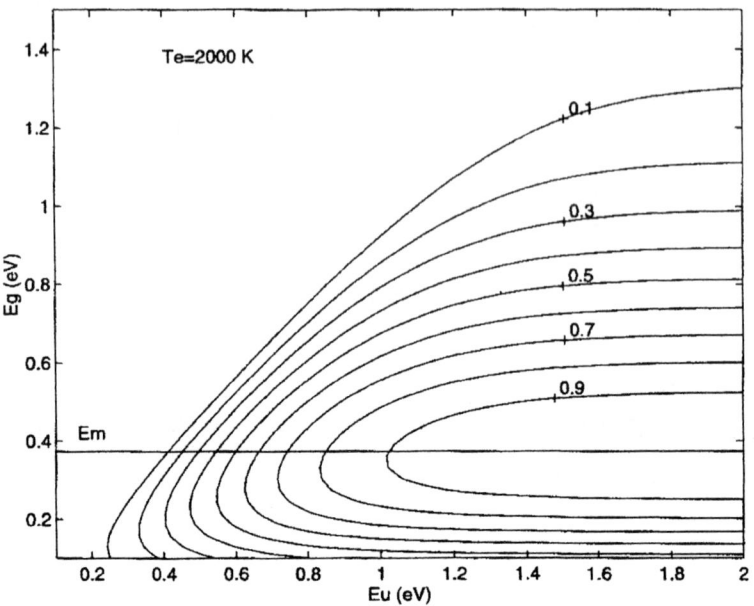

FIGURE 6. Output power contours.

SUMMARY

A TPV system model has been developed that includes an optical model, a thermodynamic model, and a detailed numerical device model. Simple analytic modeling has illustrated the tradeoff between system efficiency and output power.

REFERENCES

1. Gray, J. L. and El-Husseini, A. M., "A Numerical Semiconductor Device Model for TPV Cells," presented at the Fourth NREL Thermophotovoltaic Generation of Electricity Conference, Denver, CO, October 1998.
2. Abromowitz, M. and Stegun, I. A., *Handbook of Mathematical Functions*, Dover Publications, Inc., p. 227, 1972.
3. Siegel, R. and Howell, J. R., *Thermal Radiation and Heat Transfer*, McGraw Hill, pp. 1046-1050, 1992.
4. El-Husseini, A. M. and Gray, J. L., "Numerical Modeling of Thermophotovoltaic Cells and Systems," *Conference Proceedings of the 26th IEEE Photovoltaic Specialists Conference*, pp. 959-962, Anaheim, CA, 1997.

Efficiency and power density potential of thermophotovoltaic systems using low bandgap photovoltaic cells

Andreas Heinzel, Joachim Luther, Gunther Stollwerck, Matthias Zenker

Fraunhofer Institute for Solar Energy Systems ISE
Oltmannsstr. 5, 79100 Freiburg, Germany

Abstract. In this paper, an updated evaluation of the efficiency and power density potential of TPV technology is presented. We focus on combustion-driven systems using low bandgap photovoltaic (PV) cells. In a geometry-independent approach, we analyze (i) the thermodynamical limit, (ii) an idealized and (iii) a realistic model, extrapolating the current state of technology. Special attention is paid to the performance of the optical system. For a system containing a propane burner, GaSb PV cells and a radiator at 1500 K, we estimate that efficiencies of the order of 10 % and power densities of 1 W/cm^2 are achievable with a sufficient development effort. In addition to this theoretical analysis, the 5 W TPV test facility at Fraunhofer ISE is described.

INTRODUCTION

The renewed interest in thermophotovoltaic (TPV) energy conversion justifies a new evaluation of the TPV efficiency and power density potential. In this paper, we will take a three-step approach. To evaluate the potential of TPV, the upper limits of efficiency and power density are discussed first, taking into account only losses due to fundamental thermodynamical principles. In the next step, inherent properties of real materials are considered, but an ideal technology is assumed, providing an efficiency that can be seen as the asymptote of an optimization effort. Finally, we examine realistic devices, extrapolating the current state of technology. We consider that the resulting performance values are attainable within the next few years if a sufficient development effort is undertaken. The discussed system consists of a gas burner, a broadband or selective radiation source, an optional optical filter and low-bandgap photovoltaic (PV) cells. Specific geometries of TPV systems will not be examined here. We assume that radiator, filter and PV cells are flat and of equal area and that no radiation is lost elsewhere in the system.

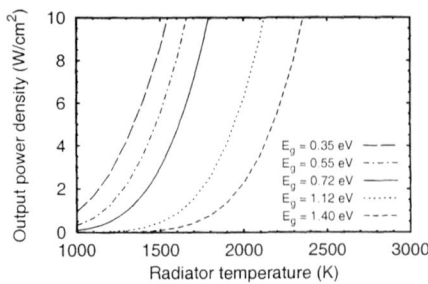

Figure 1. Maximum power density of a PV cell (thermodynamical limit) as a function of radiator temperature. The bandgaps have been chosen to correspond to the PV cell materials InAs (0.35 eV), InGaAs (0.55 eV), GaSb (0.72 eV), Si (1.12 eV), and GaAs (1.4 eV).

TPV CELLS

From thermodynamical calculations, the following expression for the upper limit of the TPV cell efficiency can be derived (1–3):

$$\eta_{PV}(V, E_g, T_{rad}, T_c) = \frac{e \cdot V \left(\int_{E_g}^{\infty} \left[\frac{E^2}{e^{\frac{E}{kT_{rad}}} - 1} - \frac{E^2}{e^{\frac{E-e \cdot V}{kT_c}} - 1} \right] dE \right)}{\int_{E_g}^{\infty} \left[\frac{E^3}{e^{\frac{E}{kT_{rad}}} - 1} - \frac{E^3}{e^{\frac{E-e \cdot V}{kT_c}} - 1} \right] dE} \quad (1)$$

where E is the photon energy, k Boltzmann's constant, T_{rad} the temperature at the radiating surface, T_c the PV cell temperature, E_g the cell bandgap, and V the cell voltage.

The efficiency increases from 50 % to 70 % as the cell bandgap increases from 0.35 eV to 1.4 eV, but shows no strong dependence on the radiator temperature in the range between 1000 K and 3000 K (3). The maximum PV cell power output density as a function of radiator temperature and cell bandgap is shown in Figure 1. The output power density reaches much higher values for smaller bandgap energies, which is a motivation for using PV cells with a lower bandgap than silicon.

GaSb cells

GaSb cells have shown to be relatively simple to fabricate with high quality (4–6), and will be taken for the remainder of this paper. A radiator temperature of 1500 K, which may be reached with real radiators, will be assumed.

The idealized cell contains a homogeneously p-doped emitter, a back surface field and a perfectly passivating and totally transparent window layer on the front side. A perfectly pure material is assumed. We take into account the GaSb-specific

Table 1. PV cell efficiency, output power density and the cell parameters fill factor (FF), short circuit current (j_{sc}), and open circuit voltage (V_{oc}) for the thermodynamical limit, an idealized and a realistic GaSb cell under the spectrum of a 1500 K blackbody radiator with an ideal optical edge filter at 1.73 µm (300 K) resp. 1.76 µm (330 K).

	η_{PV} (%)	P_{MPP} (W/cm²)	FF (%)	j_{sc}(A/cm²)	V_{oc}(V)
thermodyn. limit	60.5	3.03	83.3	5.86	0.622
idealized GaSb cell	48	2.3	81.4	5.3	0.54
realistic cell (300 K)	43	1.8	80.8	4.3	0.51
realistic cell (330 K)	38	1.6	77.8	4.4	0.46

values for radiative recombination Auger recombination and bandgap narrowing due to doping, which are discussed in detail in (7). We use a two-dimensional PV cell simulator, which is described in (8), to calculate the cell performance for given cell structure and input radiation. An optimization of the structure is performed for a 1500 K blackbody spectrum and 300 K cell temperature (3).

For a more realistic approach, we assume additional recombination by impurities in the semiconductor material and a finite surface recombination velocity (7). Additionally, a realistic front side reflection is considered, as will be discussed below. A diffused doping profile for the emitter in a GaSb wafer without back surface field is assumed, as in currently fabricated GaSb cells (9). As it is not realistic to operate a TPV cell at 300 K, the cell is simulated at a typical operating temperature of 330 K also. The bandgap variation with cell temperature is taken into account. For 300 K, $E_g = 0.72$ eV, corresponding to a wavelength $\lambda_g = 1.73$ µm.

For comparison, we have compiled the cell performance for the different degrees of idealization discussed above, assuming that all radiation with a wavelength $\lambda > \lambda_g$ is returned to the radiator by an ideal edge filter. The photons emitted by radiative recombination are considered to be lost for the idealized and realistic cells. The values for cell efficiency, output power density, and the cell parameters are listed in Table 1. The increased photogenerated current for the realistic cell at 330 K, due to the temperature-induced shift of the optical absorption edge, is overcompensated by the decrease of V_{oc}. The calculated power densities are about 100 times higher than the corresponding values for silicon PV cells under unconcentrated sunlight.

OPTICAL SHAPING

The quality of the optical system is evaluated defining a figure of merit, consisting of efficiency and output power density of a reference PV cell when illuminated by the output spectrum of the optical system (see Fig. 2). We define the reference power density as follows:

$$P_{ref} = V_{MPP}(j_{ph} - j_{rec}(V_{MPP})) \tag{2}$$

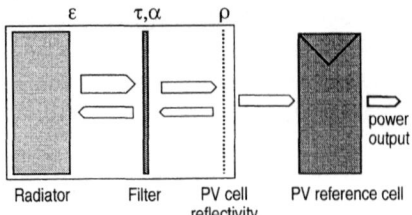

Figure 2. Schematic representation of the optical system evaluation. Multiple reflections are calculated by the standard ray-tracing method (10,11). The summations over the spectrum are carried out numerically between 0.2 μm and 10 μm. The optical characteristics of the components are taken at normal incidence.

where j_{ph} is the photogenerated current density, assuming a quantum efficiency of 100 %. For j_{rec}, the minimal value is given by thermodynamical calculations (1–3). The reference efficiency is defined as

$$\eta_{ref} = \frac{P_{ref}}{P_{rad}} \qquad (3)$$

where P_{rad} is the net radiation power density extracted from the radiator, i.e. the sum of the radiation absorbed in the optical filter and the PV cell.

Two different types of radiators are discussed for TPV, namely broadband and selective radiators. The following broadband radiators will be considered: A black body as reference, a greybody radiator with an emissivity of 0.95 as idealized approximation, and a SiC radiator. We will also examine three selective radiators: an idealized Erbium oxide (Er_2O_3) radiator, a cobalt-doped magnesia ceramic (16),

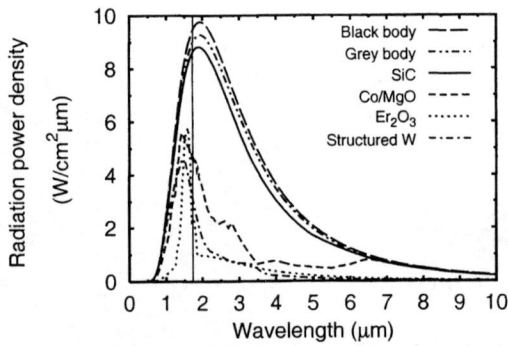

Figure 3. Radiation power density of several radiators: A SiC radiator (12), an idealized Er_2O_3 radiator (13), where $\varepsilon = 0.1$ has been assumed outside the radiation band (14), Co-doped magnesia, where a large emissivity has been taken for $\lambda > 5\,\mu m$ (15) and structured tungsten (calculated). A blackbody and a greybody spectrum ($\varepsilon = 0.95$) have been superposed for reference. The radiator temperature is 1500 K.

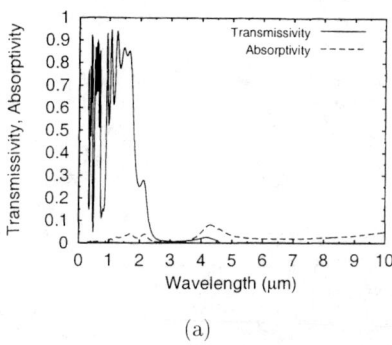
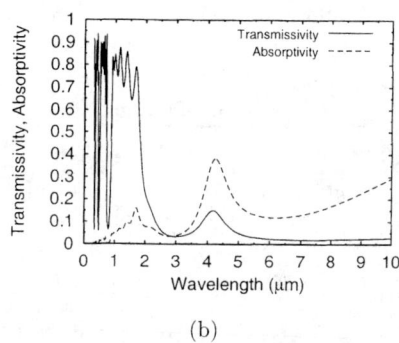

Figure 4. Calculated transmission and absorption characteristics of (a) an idealized and (b) a realistic plasma-interference filter. For the idealized filter, we assume an electron mobility of 1000 cm^2/Vs and a carrier concentration of $2.4 \cdot 10^{22}$cm^{-3} (3,19). For the realistic filter, we take a mobility of 150 cm^2/Vs (20) and a carrier concentration of 1.6×10^{22}cm^{-3}. Four interference layers, consisting of TiO$_2$ and MgF$_2$ alternately, are intended to suppress the absorption at the plasma edge.

and an emissivity characteristics resulting from our theoretical emissivity calculations for a submicrostructured tungsten surface (18). The power density spectra of the selective radiators are shown in Figure 3.

The reference efficiencies and power densities for the broadband and selective radiators, together with a blackbody spectrum with a cut-off at λ_g, are summarized in table 2. For the selective radiators, the structured tungsten exhibits the best efficiency and the second best power density. However, tungsten is not stable in air for the temperatures of interest, and would therefore have to be operated in a non-oxidising or evacuated environment. The Co-doped magnesia gives the highest power density and the lowest efficiency, while the Erbium oxide radiator is more selective for a lower power output. The broadband radiators show a poor efficiency, but a high power density.

An optical filter is another means for optical shaping. We compare an edge filter, totally transmitting for $\lambda < \lambda_g$ and totally reflecting for $\lambda > \lambda_g$, to an idealized and a realistic plasma-interference filter. Their calculated absorption and transmission characteristics are shown in Figure 4. The reference efficiencies and power densities for these filters, in combination with a blackbody radiator, are shown in table 2. Due to the lower mobility, the realistic filter shows a much higher absorption and a lower transmission than the idealized filter. The decrease in output power density is less pronounced than the decrease in efficiency. The absorptance peak slightly above the transition region is caused by the antireflective property of the interference filter in this range. To achieve a better filter quality, plasma filters with a higher electron mobility or different optical filter techniques will have to be investigated.

Table 2. Reference efficiency and power density for different broadband and selective radiators, and for a blackbody radiator in combination with different filters. The calculation is done for a radiator temperature of 1500 K and a PV cell bandgap of 0.72 eV.

		η_{ref} (%)	P_{ref} (W/cm^2)
broadband radiators	black body	11	3.0
	grey body ($\varepsilon = 0.95$)	11	2.9
	SiC (12)	11	2.8
selective radiators	black body, cut-off at 1.73 μm	58	3.0
	Er$_2$O$_3$ (idealized)	26	1.1
	Co-doped MgO (16)	17	1.8
	structured W (calculated)	30	1.6
black body with filter	none	11	3.0
	edge (1.73 μm)	58	3.0
	idealized plasma + interference	36	2.4
	realistic plasma + interference	23	2.2

The radiation impinging on the PV cell is partly reflected by the front or back side, which has therefore to be considered as a part of the optical system. For the idealized case, we assume that in the spectral range where $\lambda < \lambda_g$, zero reflectivity on the active area can be reached with a good approximation by a multilayer antireflection coating. For $\lambda > \lambda_g$, back side reflection is possible due to the low cell thickness (3). For the realistic cell, we take the measured reflectivity of a cell fabricated at Fraunhofer ISE with an antireflective coating of anodic oxide (5).

Table 3 shows the reference efficiency and power density for the radiators discussed above, in combination with the idealized and realistic filters and cell re-

Table 3. Evaluation of the optical system: Reference efficiency η_{ref} (%) and output power density P_{ref} (W/cm^2) for combinations of the radiators, filters and cell reflectivities discussed in the text. As broadband radiator, we have taken a grey body ($\varepsilon = 0.95$) for the idealized and SiC for the realistic case.

	no filter idealized cell refl.	idealized filter ideal. cell refl.	no filter realistic cell refl.	realistic filter real. cell refl.
grey (idealized)/ SiC (realistic)	47 %	42 %	14 %	20 %
	2.7 W/cm^2	2.2 W/cm^2	2.3 W/cm^2	1.7 W/cm^2
Er$_2$O$_3$	41 %	39 %	25 %	23 %
	1.0 W/cm^2	0.93 W/cm^2	0.95 W/cm^2	0.73 W/cm^2
Co/MgO	43 %	40 %	19 %	22 %
	1.7 W/cm^2	1.5 W/cm^2	1.6 W/cm^2	1.2 W/cm^2
structured W	43 %	41 %	29 %	28 %
	1.6 W/cm^2	1.3 W/cm^2	1.4 W/cm^2	1.1 W/cm^2

flectivities. In the idealized case, the high cell reflectivity for wavelengths above and the high absorptivity for wavelengths below the bandgap nearly reach perfect edge filter characteristics. The broadband radiators without filter show the highest values for efficiency and power density. For the realistic case, selective radiators show better values if no filter is used. However, the filter provides a large increase in efficiency for the broadband radiators. So we can conclude that, depending on the importance of a high efficiency or a large power density for the considered application, a selective radiator (e.g. the structured tungsten) without a filter, or a broad band radiator (SiC) with a filter could be chosen.

HEAT SOURCE

To estimate the burner-radiator efficiency, the combustion in the burner and the heat transfer to the radiator have been modeled. It turns out that the latter process is highly geometry-dependent, so that a generalization similar to the infinite flat plate model used for the radiation flow analysis in the TPV system is not possible. The TPV burner-radiator efficiency η_{BR} is defined as the ratio of P_{rad} and the incoming chemical energy flow. The incoming combustion air is assumed to be preheated to combustion temperature T_{comb} by the exhaust gas. The combustion temperature is determined from the heat transition properties of the radiator. The combustion heat is then computed, and the efficiency calculated (3). The calculations have been carried out for propane in the idealized and the realistic case. Due to increasing exhaust losses and to endothermal side-reactions, the efficiency η_{BR} decreases with increasing combustion temperature. Therefore, a good heat transfer and a low value of P_{rad} are favourable for an effective radiator heating.

SYSTEM

Table 4 summarizes the properties of the three system layouts and lists the calculated efficiencies and power densities. We would like to point out that the results obtained are subject to uncertainties originating from the fact that the calculations are partly based on estimated or extrapolated data. Furthermore, effects due to angular-dependend optical characteristics or system geometry have not been taken into account. Nevertheless, we estimate the relative uncertainty in the results to be below 20 %. The challenge in reaching TPV system efficiencies of the order of 10 %, as predicted in this paper, is to minimize the loss mechanisms that we did not consider, namely geometry-dependent radiation and heat conduction losses, and additional losses in the burner, especially in the preheating facility which we assumed to be of high quality.

Table 4. Compilation of the parameters for the three considered system layouts: (1) upper limit of system performance: hydrogen-oxygen burner with preheating of fuel and oxygen, lossless heat transfer, edge filter at 1.73 μm, thermodynamical limit cell; (2) real materials with idealized technology: propane burner with idealized heat transfer, greybody radiator, idealized GaSb cell with idealized back side reflectivity; (3) realistic system: propane burner with realistic heat transfer, SiC radiator with filter or selective structured tungsten radiator, realistic cell at 330 K.

	1. upper limit	2. idealized system	3. realistic system	
burner	$H_2 + O_2$	propane+air	propane+air	
radiator	blackbody	greybody (ε=0.95)	SiC	struct. W
filters	edge (1.73 μm)	none	plasma + interf.	none
PV cells	thermodyn. limit	GaSb (idealized)	GaSb (realistic)	
T_{cell} (K)	300	300	330	
T_{comb} (K)	1500	1500	2415	2080
η_{BR}	97 %	92 %	70 %	83%
P_{rad} (W/cm^2)	5.01	6.0	8.3	4.9
η_{filter}	100 %	–	59 %	–
η_{PV}	60.5 %	37 %	24 %	21 %
P_{el} (W/cm^2)	3.03	2.2	1.2	1.0
η_{TPV}	58.7 %	34 %	10 %	17 %

TPV TEST FACILITY

A TPV test system was realized in our laboratory (Fig. 5). The heat is generated by a premixed 2 kW propane-air burner. No exhaust gas heat recuperation is used. The radiator consists of a high-temperature resistant metallic alloy, the emissivity of which decreases from about 0.65 at a wavelength of $\lambda = 1.5\,\mu$m down to 0.33 at $\lambda = 8\,\mu$m. The PV module contains 32 GaSb cells with an active area of 1.66 cm^2 each. A quartz window is separating radiator and cells. The highly reflective walls of the system cavity and the PV module are water cooled. Burner, cells and the system have been developed and fabricated at our institute (21,5,6). As the components are easily exchangeable, testing of different kinds of filters, radiators and cells is possible.

The system has shown stable operation during an accumulated operation time of several days. Currently, an electrical power output of 5 W is generated at a radiator temperature of 1230 K. A measured I-V characteristics of the GaSb module in operation is shown in Figure 5. This test system is not optimized for high efficiency or high power output. Energy flow measurements indicate that 1.5 kW are lost in the burner, and that geometry-induced radiation losses are also large (21).

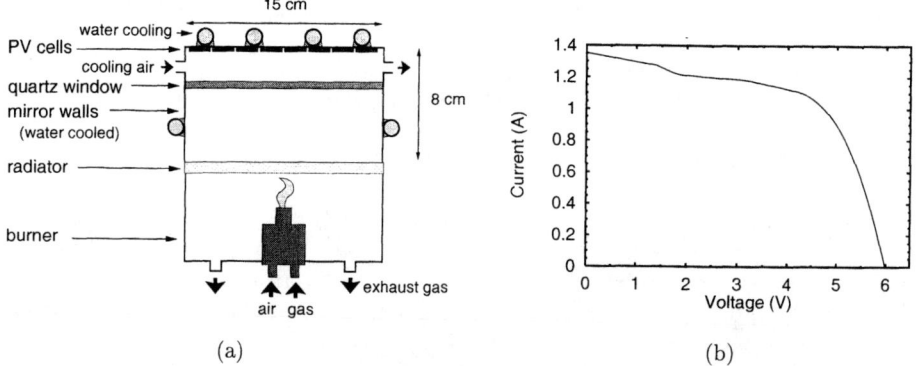

Figure 5. (a) Schematic drawing of the TPV test system at Fraunhofer ISE. (b) Measured I-V characteristics of the GaSb module in the TPV system at a radiator temperature of 1230 K.

SUMMARY

Efficiency and power density potentials of TPV conversion using a propane burner, a 1500 K radiator and low bandgap GaSb cells have been investigated with idealization degrees from the upper thermodynamical limit to realistically feasible devices. While the upper efficiency limit for this configuration is 59 %, our calculations show that efficiencies in the 10 % range and a power density of 1 W/cm^2 can realistically be attained. If sufficient development efforts are undertaken, this may be realized within a few years.

A 5 W TPV test facility has been set up and tested in our laboratory.

ACKNOWLEDGEMENTS

We would like to thank Andreas Bett, Oleg Sulima and Wolfgang Graf for fruitful discussions.

REFERENCES

1. Würfel, P., *Journal of Physics C: Solid State Physics* **15**, 3967 (1982).
2. De Vos, A., *Endoreversible thermodynamics of solar energy conversion*, Oxford University Press, 1992.
3. Zenker, M., Heinzel, A., Stollwerck, G., and Luther, J., Submitted to Journal of Applied Physics (1998).
4. Fraas, L., Ballantyne, R., Samaras, J., and Seal, M., "Electric power production using new GaSb photovoltaic cells with extended infrared response", In Coutts and Benner (22), pp. 44–53.

5. Bett, A., Keser, S., Stollwerck, G., and Sulima, O., "Large-area GaSb photovoltaic cells", In Coutts et al. (23), pp. 41–53.
6. Bett, A., Keser, S., Stollwerck, G., and Sulima, O., "Recent progress in developing of GaSb photovoltaic cells", in *Proceedings of the 14th European Photovoltaic Solar Energy Conference, Barcelona*, 1997.
7. Stollwerck, G., Sulima, O., and Bett, A., Submitted to Journal of Applied Physics (1998).
8. Sterk, S. and Glunz, S., *Simulation of semiconductor devices and processes*, volume 5, Springer Verlag Wien, 1993.
9. Bett, A., Keser, S., Stollwerck, G., Sulima, O., and Wettling, W., "GaSb - based (thermo)photovoltaic cells with Zn diffused emitters", in *Proceedings of the 25th IEEE Photovoltaic Specialists Conference*, pp. 133–136, 1996.
10. Siegel, R. and Howell, J., *Thermal radiation heat transfer*, Hemisphere Publishing Corporation, 1992.
11. White, D. and Hottel, H., "Important factors in determining the efficiency of TPV systems", In Coutts and Benner (22), pp. 425–454.
12. Touloukian, Y. and DeWitt, D., *Thermophysical properties of matter*, volume 8: Thermal radiative properties of nonmetallic solids, Plenum Press, New York, 1972.
13. Lowe, R., Chubb, D., and Good, B., "Radiative performance of rare earth garnet thin film selective emitters", In Coutts and Benner (22), pp. 291–297.
14. Chubb, D. and Lowe, R., *Journal of Applied Physics* **74**, 5687 (1993).
15. Sova, R., Linevsky, M., Thomas, M., and Mark, F., *Johns Hopkins APL Technical Digest* **13**, 368 (1992).
16. Ferguson, L. and Fraas, L., "Matched infrared emitters for use with GaSb TPV cells", In Coutts et al. (23), pp. 169–180.
17. Heinzel, A. et al., "Surface relief grating structures in tungsten for use as a selective emitter in thermophotovoltaic systems", in *Proceedings of the EOS Topical Meeting in Diffractive Optics*, 1997.
18. Heinzel, A., Boerner, V., Gombert, A., Wittwer, V., and Luther, J., "Microstructured tungsten surfaces as selective emitters", This conference.
19. Charache, G. et al., "Electrical and optical properties of degenerately-doped n-type $In_xGa_{1-x}As$", In Coutts et al. (23), pp. 215–226.
20. Graf, W., Fraunhofer Institute for Solar Energy Systems, 1998, personal communication.
21. Gabler, H., Hein, M., and Zenker, M., "A propane-fueled thermophotovoltaic energy converter using low-bandgap photovoltaic cells", in *Proceedings of the 2nd World Conference on Photovoltaic Solar Energy Conversion, Wien*, 1998.
22. Coutts, T. and Benner, J., editors, *Proceedings of the first NREL conference on thermophotovoltaic generation of electricity, Copper Mountain*, New York, 1994, The American Institute of Physics.
23. Coutts, T., Allman, C., and Benner, J., editors, *Proceedings of the third NREL conference on thermophotovoltaic generation of electricity, Colorado Springs*, The American Institute of Physics, 1997.

SESSION 2:
MONOLITHIC INTERCONNECTED MODULES (MIMS)

Square Cones for TPV; Experiments and Computer Simulations

Lars Broman*, Kenneth Jarefors[†], Eva Lindberg*, and Jörgen Marks*

*Solar Energy Research Center, Högskolan Dalarna, SE 781 88 Borlänge, Sweden, lbr@du.se
[†]Hörnell Elektrooptik AB, SE 780 41 Gagnef, Sweden, kenneth.jarefors@hornell-speedglas.se

Abstract. The efficiency of a TPV generator is very dependent on selective optical properties of the chain emitter-filter-TPV cell-reflective backing. If a selective reflective edge filter using multiple reflections in dielectric layers is employed, the slope of the edge depends on the incidence angle interval. A special geometry, consisting of a double square cone between the (square) emitter and the (square) cell array, which narrows this angle interval significantly, has been constructed and investigated experimentally as well as with ray tracing analysis.

INTRODUCTION

Optical efficiency is an important part of total TPV efficiency. Many different attempts to reach higher optical efficiency have been made, such as using selective emitters, selectively transmitting filters, TPV cells with a high back surface reflectance, or any combination of those. However the optical efficiency is still a weak link in all TPV systems.

We have constructed an Excel® document, following ideas from Per Broman (1) in which three variables can be varied individually: the emitter temperature T_e, the TPV device bandgap E_g, and the edge filter reflectance ρ. They all appear graphically; see Figure 1. By manipulating the scroll bars, the temperature can be varied from 300 K to 7000 K, the bandgap from 0,10 to 3,00 eV, and the filter efficiency from 0 to 100 %. For any combination of variable values, the fraction η_0 of the incident radiative energy that is available for the photovoltaic process is calculated and given. Figure 1 presents an example. A diagram that can be manipulated is given at www.du.se/ekos/serc/research/tpv2/ (2).

The present diagram is made for a filter that has a sharp edge between transmission and reflection, and with constant reflection for all wavelengths above the edge. Soon we will have included a more realistic filter in the diagram, but for the time being, a sloping edge will be approximated with a decreased average reflectance.

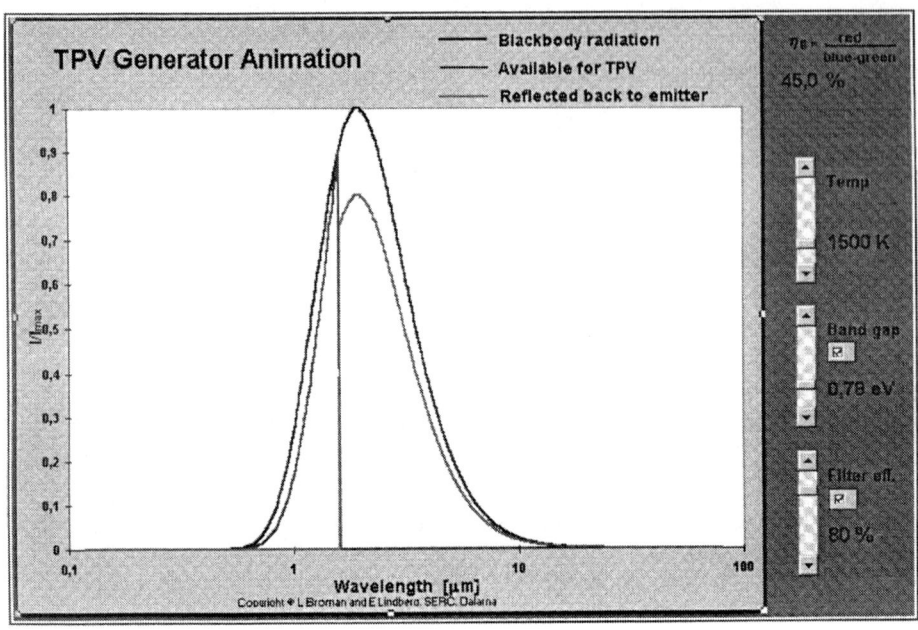

FIGURE 1. Diagram in Excel, presenting the situation with a blackbody emitter at 1500 K, a TPV cell with bandgap 0,78 eV, and a filter with 100 % transmittance below 0,78 eV and 80 % reflectance above 0,78 eV. In this example, 45 % of the radiation energy is available for the photovoltaic process.

OPTICAL GEOMETRY

This work concentrates on the possibility of increasing the optical efficiency by using a special geometry, and using this geometry in combination with interference filters. Interference filters can be designed to get a very good spectral selectivity, but they are more or less sensitive to incidence angles. Therefore we designed a geometry that limits the angle distribution of the radiation incident on the filter, at the same time as it keeps the radiation concentrated on a relatively small cell area. Earlier studies of internally reflecting truncated square cones in solar photovoltaic applications (3) have shown interesting properties that could make them very well suited also for TPV applications.

The importance of limiting the range of incidence angles is clearly demonstrated by Figure 2. This figure shows the transmittance curves for an edge filter, designed using the program TFCalc (4) for an incidence angle of 45°. As is obvious from this figure, the edge is shifted substantially when the angle of incidence is changed, and the slope gets more shelving when the range of incidence angles increases.

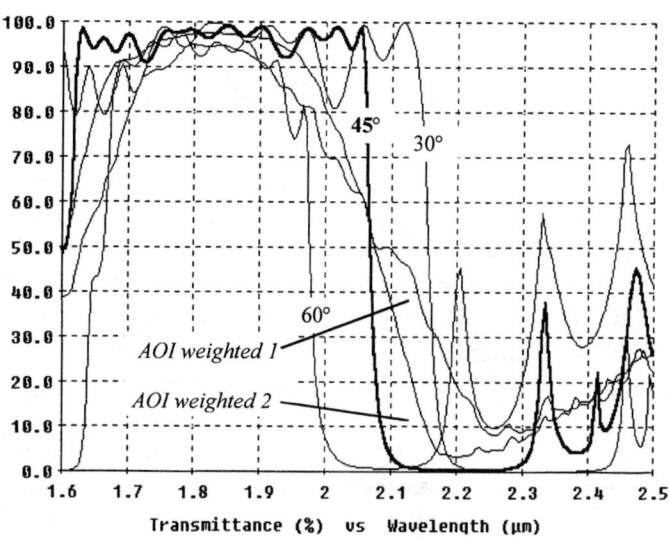

FIGURE 2. Transmittance curves for an edge filter designed for an incidence angle of 45°. The numbers indicate the angle of incidence for which the different curves are valid: 30°, **45°**, and 60°, respectively. The curve *AOI weighted 1* is the result if angles of incidence are in the range 45° ± 45° (i. e. 0° - 90°). *AOI weighted 2* is for the range 45° ± 20° (i. e. 25° - 65°).

The geometry considered in this study assumes that the emitter is a square slab (which e. g. can be made of a ceramic material with channels for hot exhaust gases from a combustion chamber). Instead of placing interference filters and TPV cell arrays close to the sides of the emitter, these sides are covered with square cones, consisting of four flat mirrors. The other end of the cone is covered by the filter, which thus is larger than the emitter. Another cone, similar to the first one, directs radiation towards the TPV cell array, which has the same size as the emitter. Figure 3 shows one side of the setup.

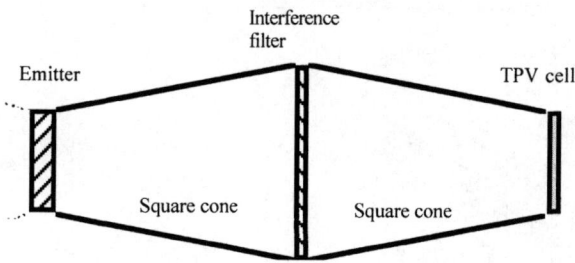

FIGURE 3. Basic double cone design.

EXPERIMENTAL SETUP AND MODEL MEASUREMENTS

Model measurements were performed in an optical bench with the following components: A beam of light was taken from a spectrophotometer with inbuilt monochromator and chopper. The beam passed a square adjustable diaphragm, and an $F = 85$ mm lens. The light illuminated a 5 cm × 5 cm square, covered with sintered $BaSO_4$. This square could serve as a model for an Lambertian emitter, since the intensity of the reflected light closely followed a cosine curve; see Figure 4 below. The square (as well as the diaphragm) could be rotated in a plane perpendicular to the beam. The spectrophotometer's detector was placed on an arm, rotating around a horizontal axis through the center of the square and perpendicular to the beam, and equipped with an angle meter. The amplified signal and the angle θ were fed into a computer program (built on LabWindows®).

For a first investigation, a cone with 30° acceptance angle and linear concentration 2 according to PV application criteria (2) was chosen; with a bottom side of 5 cm, this gave a top side of 10 cm and a height of 13 cm. The variation of intensity was measured at angles parallel to the bottom sides as well as with the surface and cones rotated 22,5° and 45°. All three measurements gave very similar results, so only the results from the first measurement are given in Figure 4. Rather than presenting the intensity, this figure shows the intensity integrated over all directions ϕ with the same θ, which integral is proportional to the total power of all rays exciting at θ.

Figure 4 contains three curves: No. 1 gives the measured power angular distribution for the "emitting" surface. A filter placed close to such an emitter will thus receive radiation over a broad interval centered at 45°. No. 2 gives the measured power angular distribution from the top of a first cone. This is where we suggest that the edge filter should be placed, and here it will receive radiation over a much narrower angle interval centered at 20°. No. 3, finally, gives the angular distribution at the bottom of a second cone. This curve looks somewhat strange, but at the time we do not have a good explanation to this.

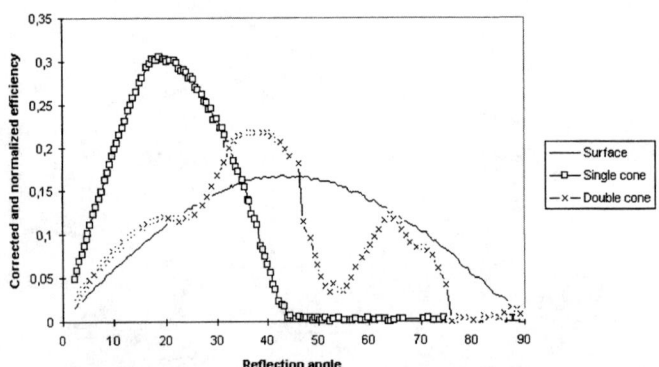

FIGURE 4. Measured power distribution from ① an "emitting" Lambertian surface, ② from the same surface covered with one cone, and ③ from the same surface covered with the double cone in Figure 3.

The reflectance ρ of the coated aluminum tin that was used in the model experiment was also measured at incidence angles θ from 30° to 80° and at wavelengths λ from 400 to 1000 nm (in the model experiment, λ = 650 nm was used). For λ ≥ 500 nm, r ≈ 0,8 for all θ:s.

COMPUTER SIMULATIONS

A ray tracing program has also been written and applied to this geometry. To date, we have been able to simulate what happens when a flat emitter radiates diffuse radiation and this surface is covered with a square cone with arbitrary emitter size, length, and filter side. As seen from Figure 5, the measured and simulated angular distributions agree very well.

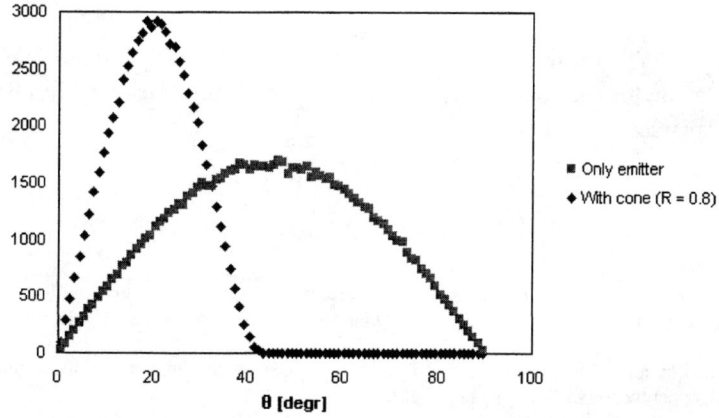

FIGURE 5. Simulated angular distribution (number of rays) of power from Lambertian emitter without cone and with cone having a reflection coefficient R = 0.8.

DISCUSSION

Both measurements and computer simulations show that the range of incidence angles onto the filter is strongly limited by the cone geometry. As an example, a cone geometry with bottom side 5 cm, top side 10 cm and length 13 cm reduces the angular spread from a Lambertian surface from ±45° to ±20°. As shown above, this fact steepens the transition slope of the edge filter considerably.

Another factor of importance is the magnitude of losses due to non-unit reflectance ρ of the cones. Summation of the radiated power in the ray tracing analysis showed that all emitted power also reaches the filter if $\rho = 1,0$, but only 74 % when $\rho = 0,8$.

In the continued work at SERC, we plan to develop the dynamic document to include filter and cell characteristics; to do model measurements on different cone designs; and to develop and use the ray tracing program to cover also the double cone. Then we will be prepared to do real measurements with a thermal radiation emitter, a double cone, an edge filter (constructed and made in collaboration with Hörnell Elektrooptik), and a water-cooled array of TPV cells.

Preliminary accounts of the double cone have recently been presented elsewhere (5, 6).

ACKNOWLEDGEMENTS

The present study was in part financed by Sparbanksstiftelsen Dalarna. We also thank Per Broman for teaching us how to make a dynamic document and Jon Back for making the ray tracing program.

REFERENCES

1. Broman, P., *Dynamic Documents®*, Broman Planetarium, pbr@planetarium.euromail.se (1998).
2. Broman, L. and Lindberg, E., *TPV Generator Animation*, www.du.se/ekos/serc/research/tpv2/ (1998).
3. Broman, L., Rönnelid, M., Binder, B., and Lindberg, E., Use of Nonimaging Concentrators for Moderate Concentration of Sunlight onto PV Cells, *Proc 9th European PV Solar Energy Conf., Freiburg, Germany*, pp 234-239 (1989).
4. *TFCalc®*. A thin film design program marketed by ARSoftware, 8201 Corporate Drive, Landover, Maryland 20785, USA (1995).
5. Jarefors, K., Broman, L., and Marks, J., Optical Interference Filters in Thermophotovoltaic Applications, *Optical Interference Coatings, Topical Meeting in Tucson, Az, USA* (June 1998).
6. Broman, L., Jarefors, K., Marks, J., and Lindberg, E., Enhancing the Optical Efficiency of a Thermophotovoltaic Generator Using Square Cones in Combination with Interference Filters, *Proc. 2nd World Conference and Exhibition on Photovoltaic Energy, Vienna, Austria* (6-10 July 1998, in print).

High-Performance, Lattice-Mismatched InGaAs/InP Monolithic Interconnected Modules (MIMs)

Navid S. Fatemi[1], David M. Wilt[2], Richard W. Hoffman, Jr.[1],
Mark A. Stan[1], Victor G. Weizer[1], Phillip P. Jenkins[1], Osman S. Khan[1],
Christopher S. Murray[3], David Scheiman[2] and David Brinker[2]

[1]*Essential Research, Inc., Cleveland, Ohio*
[2]*NASA Lewis Research Center, Cleveland, Ohio*
[3]*Bettis Atomic Power laboratory, West Mifflin, Pennsylvania*

Abstract. High performance, lattice-mismatched p/n InGaAs/InP monolithic interconnected module (MIM) structures were developed for thermophotovoltaic (TPV) applications. A MIM device consists of several individual InGaAs photovoltaic (PV) cells series-connected on a single semi-insulating (S.I.) InP substrate. Both interdigitated and conventional (*i.e.*, non-interdigitated) MIMs were fabricated. The energy bandgap (E_g) for these devices was 0.60 eV. A compositionally step-graded InPAs buffer was used to accommodate a lattice mismatch of 1.1% between the active InGaAs cell structure and the InP substrate. 1x1-cm, 15-cell, 0.60-eV MIMs demonstrated an open-circuit voltage (Voc) of 5.2 V (347 mV per cell) and a fill factor of 68.6% at a short-circuit current density (Jsc) of 2.0 A/cm^2, under flashlamp testing. The reverse saturation current density (J_o) was 1.6×10^{-6} A/cm^2. J_o values as low as 4.1×10^{-7} A/cm^2 were also observed with a conventional planar cell geometry.

INTRODUCTION

Monolithic interconnected module (MIM) devices for thermophotovoltaic (TPV) applications have been under development for the past several years **[1-6]**. In a MIM, small area InGaAs photovoltaic (PV) cells are connected in series monolithically, on a semi-insulating (S.I.) InP substrate. This results in the formation of a module with a very desirable power profile, *i.e.*, a high-voltage/low-current configuration. There are other advantages to a MIM device as compared to conventional planar devices. A prominent advantage of MIM is its ability to efficiently recuperate (recycle) the incident non-convertible infrared (IR) radiation.

CP460, *Thermophotovoltaic Generation of Electricity: Fourth NREL Conference*
edited by T. J. Coutts, J. P. Benner, and C. S. Allman
© 1999 The American Institute of Physics 1-56396-828-2/99/$15.00

Optical recuperation is mandatory to achieve high efficiencies in a TPV system. Conventionally, separate front surface filtering elements, such as dielectric stack bandpass, or plasma, or tandem filters are used in front of a PV cell to accomplish optical recuperation. These elements, however, attenuate the useful in-band radiation reaching the PV cell. As a result, the cell's output power density is diminished. They not only add complexity to the design of a TPV system, but they also tend to absorb both in-band and out-of-band radiation.

An alternative approach to optical recuperation is the use of a back surface reflector (BSR). A highly IR-reflective material such as gold can be used on the back side of the semi-insulating InP wafer to accomplish this task. As shown schematically in Figure 1, the fabrication of MIM requires the use of a semi-insulating InP substrate. Unlike doped InP substrates, S.I. InP is transparent to infrared (IR) radiation. Radiation with wavelengths greater than the device bandedge wavelength (*i.e.*, $\lambda g \sim 2$ µm) can be reflected back to the TPV radiator via the BSR. This reflected energy can be recuperated by the radiator, thus reducing the energy input to the heat source.

In fact, spectral utilization (SU) factors greater than 70% have been measured for a MIM with a BSR, and a bandgap of 0.60 eV [7]. The SU factor is defined as the above-bandgap absorbed energy in the active device layers divided by the total absorbed energy in the structure. This SU factor is greater than what has been observed with the most advanced filtering options currently available (*i.e.*, SU of <70%) [8]. The use of a BSR in a MIM configuration, therefore, eliminates the need to use separate front surface filters.

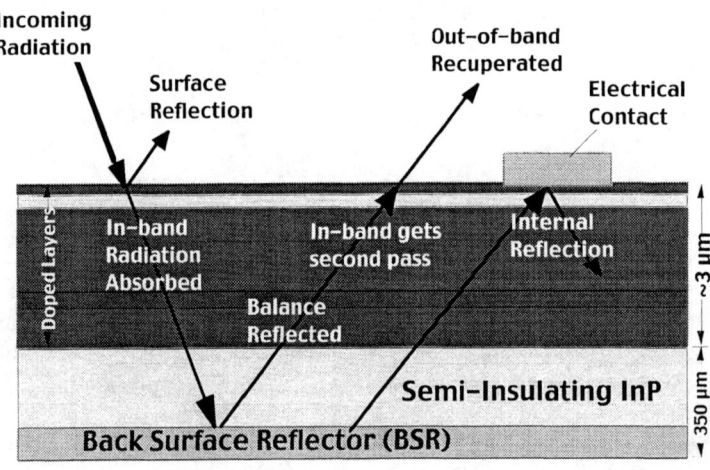

Figure 1.—Optical recuperation (recycling) in a MIM structure.

Another advantage to a MIM design is that both the negative and positive electrical connections are fabricated on the top side of the module, thereby simplifying the array design, interconnection, and thermal management. The completed device may be soldered directly onto the array substrate/heat sink without having to provide electrical isolation. Individual MIMs can then be connected in series or parallel configurations by welding or bonding of metallic interconnect ribbons to adjacent busbars. A photograph of an array, comprised of twelve 1x1-cm conventional MIMs, interconnected by this process is shown in Figure 2.

Figure 2.—Photograph of an array of twelve 1x1-cm MIMs on a heatsink.

In a TPV system, the bandgap of the PV device is ideally selected to match the peak intensity of the radiator spectral power density at the operating temperature. At low to moderate radiator temperatures (*i.e.*, 1200-1500 K), the PV cells used normally have bandgap values of 0.74 eV or smaller. The most commonly studied devices are the lattice-matched InGaAs-on-InP and GaSb cells, with bandgaps of 0.74 and 0.73 eV, respectively. The spectral response of the PV cells with lower bandgaps, however, are better matched to the graybody spectral power density profile for low to moderate radiator temperatures. Examples of cells with lower bandgaps are lattice-mismatched InGaAs-on-InP and lattice-matched InGaAsSb-on-GaSb.

In Figure 3, for example, the spectral response for two high-quality InGaAs/InP MIM devices with bandgaps of 0.74 and 0.60 eV are overlaid on top of the spectral power density curve for a blackbody at a temperature of 1500 K. As shown in the figure, the response of the lattice-mismatched device with $E_g = 0.60$ eV is far better matched to the peak intensity of the blackbody power curve. As a result, in the example shown, the output current of the 0.60-eV device is more than 80% greater than the output current for the 0.74-eV device.

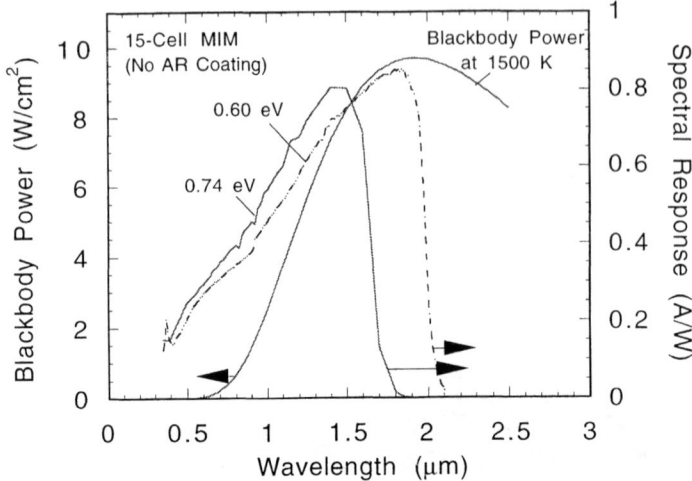

Figure 3.—Measured spectral response data for 15-cell InGaAs/InP MIMs, with Eg = 0.60 eV and 0.74 eV (with no anti-reflection coating), and the blackbody spectral power density curve at 1500 K.

We have fabricated and tested lattice-mismatched p/n InGaAs/InP MIMs, grown by organo-metallic vapor phase epitaxy (OMVPE). Both interdigitated [4] and conventional (non-interdigitated) [1-3, 5] MIMs were developed. These structures had a bandgap of 0.60 eV. High performance MIMs were developed using a compositionally step-graded InPAs buffer to accommodate a lattice mismatch of 1.1% between the active InGaAs cell structure and the InP substrate. In what follows, we will present data regarding the electrical performance of these devices.

MIM DESIGN AND STRUCTURE

Two MIM geometries were fabricated and tested. The first was the conventional design that has been under development jointly by Essential Research and NASA Lewis, and also independently by Spire Corporation, for the past several years [1–3, 5–6]. The second was the interdigitated design that was under development at NASA Lewis and refined independently at the National Renewable Energy Laboratory (NREL) [4].

The 1x1-cm and 2x2-cm conventional MIM designs had 15 and 30 cells interconnected in series, respectively. Each individual cell was approximately 550 μm wide and either 1 or 2 cm long. The 1x1-inch (2.54x2.54-cm) interdigitated design had 23 cells interconnected in series. Each individual cell was approximately 975 μm wide and 2.54 cm long. A photograph of several conventional MIMs as processed on 2-inch InP wafers are shown in Figure 4.

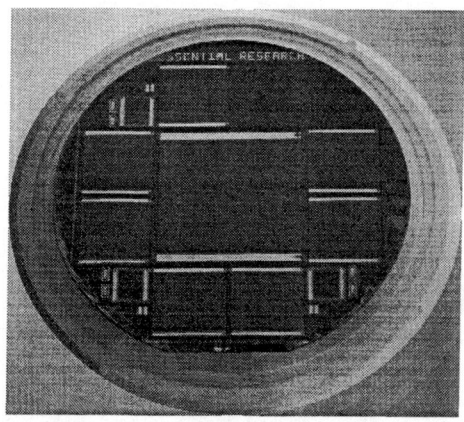

Figure 4.—Photograph of a 2-inch diameter Semi-insulating InP wafer with one 30-cell (2x2-cm) and seven 15-cell (1x1-cm) conventional MIMs.

The conventional and interdigitated MIMs had very similar structures. A schematic cross-sectional view of a MIM structure is shown in Figure 5. The advantage of the interdigitated MIM is that the design allows for the use of a thinner, lower doped lateral conduction layer (LCL). Typically, the LCL layer for the interdigitated structure was 0.25 μm thick with a carrier concentration in the low 10^{18} cm^{-3} range. In contrast, the LCL layer for the conventional structure was 1.0 μm thick with a carrier concentration in the low 10^{19} cm^{-3} range. Thinner and lower doped semiconductor layers show lower free-carrier absorption (FCA). Lower FCA results in more efficient optical recuperation via the BSR [2, 5].

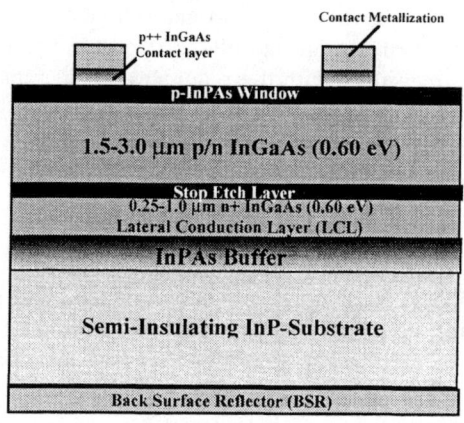

Figure 5.—Schematic cross-sectional view of InGaAs/InP MIM.

The advantage of the conventional design is that it is less sensitive to the minority carrier diffusion length in the base region. The base layer of the device can be thinner than optimum in this design (*i.e.*, 1.5 μm versus 3.0 μm). As a result, the in-band photons impinging on the device are not completely absorbed in the first pass through the device. Therefore, the unabsorbed portion of the usable photons can reflect off the BSR and have a second pass through the active regions of the device. This process forces the carrier generation to occur closer to the p/n junction. The high doping level in the InGaAs LCL layer effectively increases the optical bandgap of this layer (i.e., the Burnstien-Moss shift) to allow the shorter wavelength convertible radiation to pass through to the BSR with negligible absorption.

Recent Monte-Carlo photonic modeling suggests that the processing techniques used can have a significant effect on the amount of photons that can be recuperated at the radiator. Four areas of particular interest are the specular/diffused reflectance of the BSR, light trapping due to surface features (i.e., isolation trenches), the reflectance of the electrical metallization at the metal-semiconductor interface, and IR absorption in the anti-reflective coating.

RESULTS

External quantum efficiency (QE) and high-intensity illumination current versus voltage (I-V) measurements were performed to characterize the MIMs. We will present data for p/n InGaAs/InP 0.60-eV lattice-mismatched conventional and interdigitated MIMs, as well as some data for conventional planar (i.e., non-MIM) one-junction devices

External Quantum Efficiency (QE) Data

Initially, planar one-junction conventional 0.60-eV p/n InGaAs/InP cells were fabricated and tested. Excellent current collection over a wide range of wavelengths (i.e., 0.45-2.0 µm) was measured with these devices. This data is shown in Figure 6. Note that the cell did not have an anti-reflection (AR) coating. The estimated internal quantum efficiency was near unity near the bandedge. This result demonstrates the effectiveness of the buffer layer, grown between the InP substrate and the active regions of the 0.60-eV InGaAs device, to limit the propagation of threading dislocations from the buffer-InP interface to the top active layers of the device.

The QE data for both conventional and interdigitated MIM structures were similar to the data measured for the planar cells. The main difference between the planar cell structure and the MIM structure was that the latter had an extra LCL layer. The QE plots for conventional and interdigitated MIM structures are illustrated in Figure 7. Note that the MIMs did not have any AR coating.

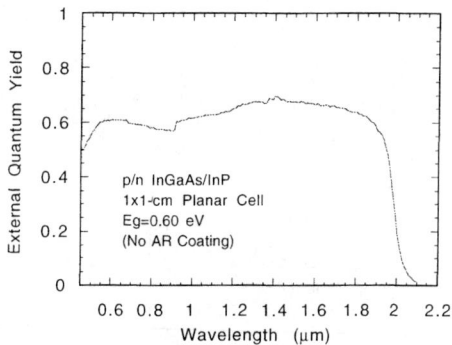

Figure 6.—External quantum efficiency data for a conventional planar cell with Eg = 0.60 eV (No AR coating).

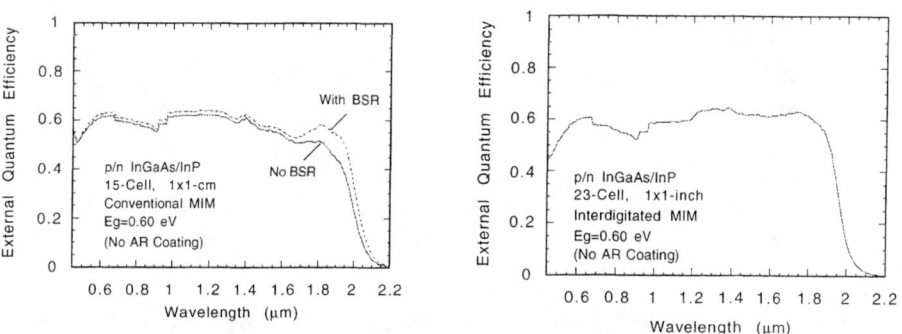

Figure 7.—External quantum efficiency data for a conventional MIM structure (left), and an interdigitated MIM structure (right), with Eg = 0.60 eV (No AR coating).

The QE data for the conventional MIM also shows the effect of the BSR in improving the current collection, especially near the bandedge. As mentioned earlier, this is due to increased absorption in the active layers, during the second pass, after reflecting off the BSR.

Current-Voltage (I-V) Data

The MIMs were tested under high-intensity illumination, using a large-area pulsed solar simulator (LAPSS), to assess their performance under simulated operating conditions. The results for the variation in the open-circuit voltage (Voc) and fill factor (FF) with the short-circuit current density (Jsc), for 0.60-eV conventional and interdigitated MIMs are given in Figures 8 and 9, respectively.

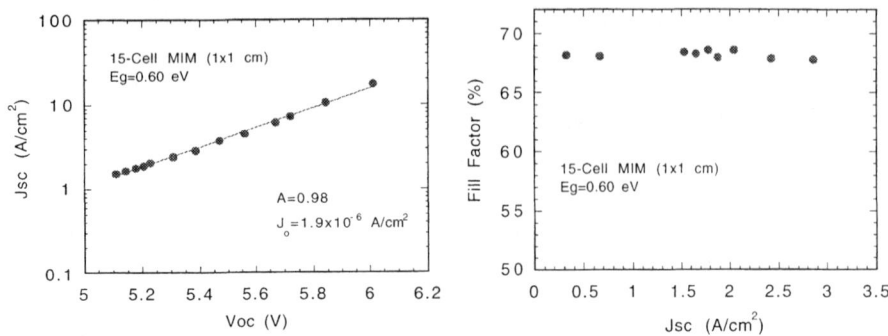

Figure 8.—Variation of Voc (left) and FF (right) with Jsc for a conventional 15-cell InGaAs/InP MIM, with Eg = 0.60 eV.

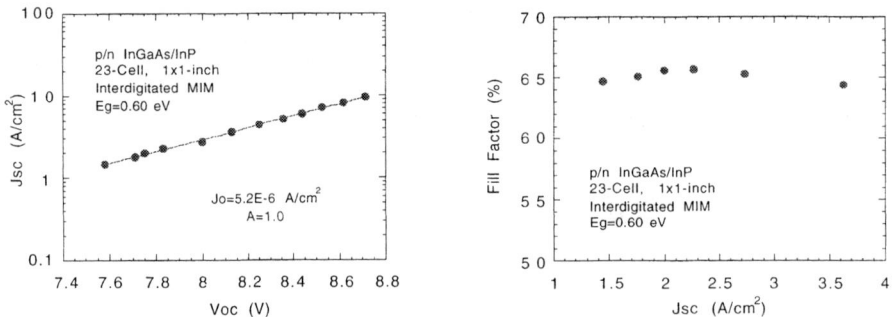

Figure 9.—Variation of Voc (left) and FF (right) with Jsc for a 23-cell interdigitated InGaAs/InP MIM, with Eg = 0.60 eV.

As shown in the figures, the diode ideality factor (A) for both structures was unity. The reverse saturation current density (Jo) was calculated to be in the range of 1.9–5.2x10^{-6} A/cm^2. This range of Jo values is more than an order of magnitude lower than the best results reported in the literature for an InGaAs/InP cell with Eg = 0.60 eV [9] We have also observed Jo values, as low as 4.1x10^{-7} A/cm^2 for 1x1-cm planar (non-MIM) p/n InGaAs/InP 0.60-eV cells.

For both device structures, they remained at or above 65% for the short-circuit current densities up to about 3 A/cm^2. The results for the I-V data presented suggest that the buffering technique used to grow the 0.60-eV lattice-mismatched InGaAs on InP was effective in preventing the majority of the dislocations created at the cell-substrate interface from threading up into the active regions of the cell structure.

From the QE and I-V data presented above, we calculated the expected electrical performance of the MIMs in a TPV system. The graybody radiator was assumed to have an emissivity of 0.9 and a view factor of 0.9. The graybody radiator temperature was varied from 1100 to 1500 K, and the cell temperature was assumed to be constant at 25°C. For comparison, the performance of a conventional MIM with a bandgap of 0.74 eV was also calculated.

The short-circuit current density, Jsc, was calculated by integrating the measured spectral response times the spectral graybody power density over the wavelength range of 0.35 to 2.1 µm. The illuminated I-V data measured by the LAPSS was then used to determine Voc and FF at the Jsc point calculated above. The output power was the product of Voc, FF, and Isc. The results for a 1x1-inch 23-cell interdigitated (Eg = 0.60 eV) and two 1x1-cm 15-cell conventional MIMs (Eg = 0.60 and 0.74 eV) are shown in Table I.

TABLE 1.—Electrical performance of p/n InGaAs/InP MIMs under simulated TPV conditions.

MIM Device	Eg (eV)	Radiator Temperature (K)	Jsc (A/cm^2)	Voc (V) / Voc per Cell (V)	Fill Factor (%)	Power (W) / Power Density (W/cm^2)
Interdigitated (2.54x2.54 cm)	0.60	1100	0.66	7.30 / 0.317	63.5	0.62 / 0.13
		1300	2.26	7.83 / 0.340	65.7	2.36 / 0.51
		1500	5.83	8.43 / 0.367	61.0	6.10 / 1.31
Conventional (1 x 1 cm)	0.60	1100	0.69	4.69 / 0.313	68.1	0.12 / 0.15
		1300	2.36	5.26 / 0.351	68.3	0.45 / 0.57
		1500	6.00	5.64 / 0.376	64.9	1.15 / 1.46
Conventional (1 x 1 cm)	0.74	1100	0.27	5.70 / 0.380	77.1	0.063 / 0.079
		1300	1.13	6.30 / 0.420	76.1	0.28 / 0.36
		1500	3.30	6.66 / 0.444	75.0	0.86 / 1.10
		1700	7.69	6.95 / 0.463	71.3	2.00 / 2.54

The data in the above table shows that relatively large electrical output power, in the range of 0.45-1.15 watts, can be obtained with 0.60-eV, 1x1-cm MIMs at moderate radiator temperatures (1300-1500 K). The better matching of the 0.60-eV versus 0.74–eV device response to the blackbody peak radiation curve (see Figure 3), and the high quality of the lattice-mismatched devices, result in significantly higher output power for the lower bandgap devices. In the temperature range of 1300-1500 K, for example, the 0.74-eV MIMs produced only about 62-75% of the power that the

0.60-eV MIMs produced. It should be noted that the structure of the MIMs with Eg = 0.60 eV was optimized for operation with a radiator temperature of about 1300 K, whereas the structure of the MIMs with Eg = 0.74 eV was optimized for operation with a radiator temperature of 1500 K.

We believe that the data presented in Table 1 are practical values that we can expect to observe under actual TPV conditions. Specifically, in a separate experiment we have demonstrated that our calculated data for the 0.74-eV MIMs are in close agreement with the measured experimental data taken under actual TPV conditions. An array of 12, 0.74-eV, 1x1-cm MIMs (see Figure 2) was coupled to a combustion heated SiC radiator operated at 1300 K, resulting in measured Voc and Isc values 6% higher and 2% lower than the calculated values, respectively.

SUMMARY

The results presented in this work are summarized below:

1. We have fabricated and tested lattice-mismatched p/n InGaAs/InP monolithic interconnected modules (MIMs), with a bandgap of 0.60 eV. These devices had a gold back surface reflector (BSR) that is used to recuperate non-absorbed photons at the radiator, increasing TPV system efficiency.

2. Both interdigitated and conventional MIM designs were fabricated. The 1x1-cm and 2x2-cm conventional MIMs had 15 and 30 cells interconnected in series, respectively. The 1x1-inch (2.54 x 2.54-cm) interdigitated MIMs had 23 cells interconnected in series.

3. The external quantum efficiency data for the lattice-mismatched structures showed very good collection efficiency. The internal QE was estimated to be near unity close to the bandedge.

4. MIM devices were tested under high-intensity illumination. The diode ideality factor (A) and the reverse saturation current density (J_o) for the MIMs with Eg = 0.60 eV were near or at unity, and $1.9\text{-}5.2\times10^{-6}$ A/cm^2, respectively. Lower J_o values (i.e., 4.1×10^{-7} A/cm^2) were also observed for the planar (non-MIM) structures. The fill factor values remained at or above 65% for short-circuit current densities up to 3 A/cm^2.

5. From the measured QE and I-V data, the expected output power for the 0.60-eV devices, when coupled to a graybody radiator (emissivity of 0.9 and view-factor of 0.9) was calculated. Relatively large power densities, in the range of 0.45–1.15 W, were calculated for a radiator temperature range of 1300-1500 K.

REFERENCES

[1] N. S. Fatemi, D. M. Wilt, P. P. Jenkins, V. G. Weizer, R. W. Hoffman, C. S. Murray, D. Scheiman, D. Brinker, and D. Riley, "InGaAs Monolithic Interconnected Modules (MIMs)," Twenty Sixth Photovoltaic Specialists Conference (PVSC), p. 799, 1997.

[2] D. M. Wilt, N. S. Fatemi, P. P. Jenkins, V. G. Weizer, R. W. Hoffman, R. K. Jain, C. S. Murray, and D. Riley, "Electrical and Optical Performance Characteristics of 0.74 eV p/n InGaAs Monolithic Interconnected Modules, " Third NREL TPV Conference, 1997.

[3] N.S. Fatemi, D.M. Wilt, P.P. Jenkins, R.W. Hoffman, V.G. Weizer, C.S. Murray, and D. Riley, "Materials and Process Development for the Monolithic Interconnected Module (MIM) InGaAs/InP TPV Devices", Third NREL TPV Conference, 1997.

[4] J. S. Ward, A. Duda, M. W. Wanlass, J. J. Carapella, X. Wu, R. J. Matson, T. J. Coutts, and T. Moriarty, "A Novel Design for Monolithically Interconnected Modules (MIMs) for Thermophotovoltaic Power Conversion," Third NREL TPV Conference, p. 227, 1997.

[5] D. M. Wilt, N. S. Fatemi, P. P. Jenkins, R. W. Hoffman, G. A. Landis, and R. K. Jain, "Monolithically Interconnected InGaAs TPV Module Development," Twenty-fifth Photovoltaic Specialists Conference (PVSC), 1996.

[6] S. Wojtczuk, "Comparison of 0.55 eV InGaAs Single-Junction vs. Multi-Junction TPV Technology," Third NREL TPV Conference, 1997.

[7] M. B. Clevenger, C. S. Murray, and D. R. Riley, "Spectral Utilization in Thermophotovoltaic Devices," Materials Research Society (MRS) Fall Meeting, Boston, MA, December 1997.

[8] Data presented by Knolls Atomic Power Labs (KAPL) in a review meeting held in Schenectady, NY, January 1998.

[9] G. W. Charache, J. L. Egley, L. R. Danielson, D. M. Depoy, P. F. Baldasaro, B. C. Campbell, S. Hui, L. M. Fraas, and S. J. Wojtczuk, "Current Status of Low-Temperature Radiator TPV Devices", Twenty-fifth IEEE PVSC, p. 137, 1996.

High-Performance, 0.6-eV, $Ga_{0.32}In_{0.68}As/InAs_{0.32}P_{0.68}$ Thermophotovoltaic Converters and Monolithically Interconnected Modules

M. W. Wanlass, J. J. Carapella, A. Duda, K. Emery, L. Gedvilas, T. Moriarty, S. Ward, J. D. Webb, and X. Wu

National Renewable Energy Laboratory (NREL), 1617 Cole Blvd., Golden, CO, 80401, U. S. A.

C. S. Murray

Bettis Atomic Power Laboratory, P. O. Box 79/ZAP08D, West Mifflin, PA, 15122, U. S. A.

Abstract. Recent progress in the development of high-performance, 0.6-eV $Ga_{0.32}In_{0.68}As/InAs_{0.32}P_{0.68}$ thermophotovoltaic (TPV) converters and monolithically interconnected modules (MIMs) is described. The converter structure design is based on using a lattice-matched $InAs_{0.32}P_{0.68}/Ga_{0.32}In_{0.68}As/InAs_{0.32}P_{0.68}$ double-heterostructure (DH) device, which is grown lattice-mismatched on an InP substrate, with an intervening compositionally step-graded region of $InAs_yP_{1-y}$. The $Ga_{0.32}In_{0.68}As$ alloy has a room-temperature band gap of ~0.6 eV and contains a p/n junction. The $InAs_{0.32}P_{0.68}$ layers have a room-temperature band gap of ~0.96 eV and serve as passivation/confinement layers for the $Ga_{0.32}In_{0.68}As$ p/n junction. $InAs_yP_{1-y}$ step grades have yielded DH converters with superior electronic quality and performance characteristics. Details of the microstructure of the converters are presented. Converters prepared for this work were grown by atmospheric-pressure metalorganic vapor-phase epitaxy (APMOVPE) and were processed using a combination of photolithography, wet-chemical etching, and conventional metal and insulator deposition techniques. Excellent performance characteristics have been demonstrated for the 0.6-eV TPV converters. Additionally, the implementation of MIM technology in these converters has been highly successful.

INTRODUCTION

TPV converters that use low-band-gap, epitaxial $Ga_xIn_{1-x}As$ alloys grown on InP substrates have been under investigation by a number of groups in recent years [1]. Additionally, substantial progress has been made in developing MIMs from similar TPV converter structures [1]. In this paper, we describe recent progress at NREL in the development of both discrete converters and MIMs with a band gap of 0.6 eV.

CP460, *Thermophotovoltaic Generation of Electricity: Fourth NREL Conference*
edited by T. J. Coutts, J. P. Benner, and C. S. Allman
© 1999 The American Institute of Physics 1-56396-828-2/99/$15.00

Low-band-gap (0.5–0.6 eV) $InAs_yP_{1-y}/Ga_xIn_{1-x}As/InAs_yP_{1-y}$ DH TPV converters have been under development at NREL since 1993. The DH converter concept is based on using high-band-gap layers of $InAs_yP_{1-y}$ to effect passivation and minority-carrier confinement when applied to the front and back surfaces of a low-band-gap $Ga_xIn_{1-x}As$ p/n-junction absorber. A full description of this concept, and the experimental demonstration of its utility, are given elsewhere [2]. The DH structure provides higher photovoltaic (PV) performance because it reduces minority-carrier recombination and, thus, results in both a higher quantum efficiency and lower reverse-saturation current density for the converter. Similar DH designs have been used to produce high-efficiency solar PV converters in III-V materials that have high surface-recombination velocities (e.g., solar cells based on the lattice-matched combinations $Ga_{0.52}In_{0.48}P/GaAs$, $Al_xGa_{1-x}As/GaAs$, and $InP/Ga_{0.47}In_{0.53}As$). At 300 K, $Ga_{0.47}In_{0.53}As$ is lattice matched to InP and has a band gap of 0.74 eV. Therefore, TPV converters using $Ga_xIn_{1-x}As$ epilayers grown on InP require mismatched epitaxy in order to produce the lower band gaps (0.5–0.6 eV) necessary for optimal conversion efficiency. The problems that arise from the lattice mismatch between the substrate and the converter epilayers are mitigated by including intervening compositionally graded layers in the converter structure. In the past, continuous, linearly graded regions of $Ga_xIn_{1-x}As$ have been used to produce converters with excellent performance characteristics [2, 3]. In the recent work described here, we have found that properly designed $InAs_yP_{1-y}$ step-graded regions give even better results.

To a good approximation, the band gap (E_g) of $Ga_xIn_{1-x}As$ as a function of x at 300 K is given by the following expression [4],

$$E_g(x) = (0.555x^2 + 0.505x + 0.356)eV, \quad (1)$$

which yields x = 0.35 for E_g = 0.6 eV. Our experience with $Ga_xIn_{1-x}As$ epilayers shows that this value of x is somewhat high; experimentally, we have found that x ≈ 0.32 is the correct value to attain E_g = 0.6 eV (2067 nm). The lattice mismatch (M) between a $Ga_xIn_{1-x}As$ epilayer and an InP substrate is given precisely by [2]

$$M(Ga_xIn_{1-x}As / InP) = \left(\frac{-0.378 + 0.810x}{11.927 - 0.405x}\right). \quad (2)$$

Thus, M = -1.01% for x = 0.32, where the minus sign indicates that the $Ga_xIn_{1-x}As$ epilayer experiences biaxial compression, a condition that maintains the integrity of the $Ga_xIn_{1-x}As$ (i.e., no microcracking within the epilayer). The lattice-matching condition for epilayers of $InAs_yP_{1-y}$ and $Ga_xIn_{1-x}As$ is expressed exactly by [2]

$$y = 1 - 2.143x. \tag{3}$$

This equation yields $y \approx 0.32$ for $x \approx 0.32$. We have found close lattice-matching between the $Ga_xIn_{1-x}As$ and $InAs_yP_{1-y}$ epilayers to be of paramount importance, particularly when $InAs_yP_{1-y}$-graded layers are used. Otherwise, microcracks in the epistructure have been observed, which occur, presumably, as a result of the strain between the two ternaries. The band gap of $InAs_yP_{1-y}$ at 300 K is best expressed as [5]

$$E_g(y) = (0.320y^2 - 1.315y + 1.351)eV, \tag{4}$$

which gives $E_g = 0.96$ eV for $y = 0.32$.

The above parameters define the 0.6-eV $InAs_{0.32}P_{0.68}/Ga_{0.32}In_{0.68}As$ DH TPV converter design concept. Further details of the actual converter structures are outlined in the sections that follow. It should be noted that the basic $InAs_yP_{1-y}/Ga_xIn_{1-x}As$ DH converter design can be used to achieve a wide range of band gaps; recently, we demonstrated high-performance 0.5-eV converters using this concept. In the remainder of this paper, we describe the results obtained from fabricating 0.6-eV discrete converters and MIMs.

EPITAXIAL GROWTH and DEVICE FABRICATION PROCEDURES

The APMOVPE process used in this work employs an NREL-built gas-handling system and a specially designed, vertically oriented reactor vessel [6] that yields highly uniform epilayers. InP substrates (prepared with an "epi-ready" surface), which were oriented (100) 2° toward (110), were loaded directly into the reactor for growth as received from the vendor. Discrete converters were grown on polished/etched S-doped InP substrates, and MIM structures were grown on double-side-polished (DSP), Fe-doped InP substrates. Ethyldimethylindium, triethylgallium, arsine, and phosphine were used as the primary reactants, and diethylzinc and hydrogen sulfide were used to dope epilayers p- and n-type, respectively. Epitaxial growth was performed in a purified hydrogen ambient at a temperature of 620°C. The deposition rates were 0.1 µmmin^{-1} for $Ga_{0.32}In_{0.68}As$ and 0.07 µmmin^{-1} for $InAs_{0.32}P_{0.68}$. A detailed description of the APMOVPE system and growth process is given elsewhere [2].

Conventional procedures were used to process devices. Selective wet-chemical etching was performed using 3 H_3PO_4 : 4 H_2O_2 : 1 H_2O to dissolve $Ga_{0.32}In_{0.68}As$ and concentrated HCl to dissolve $InAs_yP_{1-y}$. Electroplated Au was used to form the metallizations on the discrete converters. Our general approach to fabricating MIMs has been described previously [7]. The insulator used in MIM fabrication is SiO_2 (~200 nm) deposited by a chemical-vapor-deposition process. The metallization used for the MIMs is an electron-beam-evaporated multi-metal stack consisting of Cr (30 nm), Pd (20 nm), Ag (4 µm), and Pd (20 nm). The optimized back-surface reflector (BSR), which is applied to the back surface of MIM converters, consists of MgF_2 (130 nm), Ti (2.5 nm), and Au (150 nm). The characteristics of front and back electrical contacts used in the MIMs are discussed in a companion paper given at this conference [8].

DISCRETE CONVERTERS

Discrete converters were fabricated using the epistructure shown in Figure 1. From the substrate up, the basic components of the structure include an $InAs_yP_{1-y}$ step-graded layer, the 0.6-eV DH converter layers, and a contact layer to facilitate contacting the front surface. Details of the composition, doping level, and thickness for each of the component layers in the structure are given in the diagram. Design rules derived from a previous study of lattice-mismatched $GaAs_{1-z}P_z$/GaAs structures [9] were used to guide the design of the graded layer used in the present TPV converters. The $InAs_yP_{1-y}$ graded layer consists of 10 equal compositional steps, which results in a lattice mismatch of -0.1% at each step interface. The grade is terminated with a layer of constant composition ($InAs_{0.32}P_{0.68}$) that is 1 µm in thickness. This thicker layer allows the final network of misfit dislocations arising from the last compositional step to evolve fully, leaving only a relatively low density of threading dislocations to propagate into the active converter layers above. As shown later, this grading technique results in converter layers with excellent electronic characteristics. For the reasons that follow, the total thickness of the $Ga_{0.32}In_{0.68}As$ p/n junction layers of the DH converter (2.3 µm) is about half the thickness required for complete absorption of above-band-gap-energy photons. Firstly, the converter structure is designed for eventual use as a MIM, where a BSR is available to reflect unabsorbed near-band-gap-energy photons back through the converter layers for a second pass. Secondly, thinner p/n-junction layers result in lower reverse-saturation current densities, which, in turn, result in higher converter voltages (~18 mV higher for half the thickness, at 300 K). Finally, thinner layers require less growth time, which is a potential manufacturing advantage. A range of thicknesses (0.1–0.3 µm) have been used for the p$^+$-Ga$_{0.32}$In$_{0.68}$As emitter layer in the DH converter.

The microstructure of the 0.6-eV converters was characterized using atomic-force microscopy (AFM), cross-sectional transmission electron microscopy (XTEM), and electron-beam-induced current (EBIC) techniques. To the naked eye, the surface of the structure appears to have a fine, orthogonal, biaxial crosshatched morpology, which is typical of single-crystal, lattice-mismatched III-V epilayers grown on substrates with a near-(100) orientation [9]. Detailed examinations of the surface topography by AFM show that a three-dimensional ripple morphology is superimposed on the crosshatch. The root-mean-square roughness of the surface is ~23 nm. XTEM examinations show that the structures exhibit extensive misfit dislocation networks in the vicinity of the compositional steps in the graded region. At the top of the grade, threading dislocations were observed to propagate into the DH converter layers above. Plan-view EBIC was performed on discrete devices to assess the degree of minority-carrier recombination at dislocations within the converter layers. The EBIC studies showed that carrier recombination was not occurring to any measurable degree; the threading dislocations could not be resolved as dark lines or spots. The above result has been corroborated by minority-carrier lifetime measurements, as we have observed minority-carrier lifetimes of several µs in undoped 0.6-eV DH structures [10]. The above results indicate that the grading technique used in the 0.6-eV DH structure produces converter layers with superior properties; the usual deleterious effects of threading dislocations are almost nonexistent.

Spectral quantum efficiency (QE) and reflectance (R) data for a discrete 0.6-eV DH converter without an antireflection coating (ARC) are given in Figure 2. The figure shows R, active-area absolute external quantum efficiency (AAAEQE), and internal quantum efficiency (IQE) data for a representative high-performance 0.6-eV converter. The IQE data show that the carrier collection efficiency over the bulk of the response range is excellent (~90%–100%), a result of excellent bulk-material quality and low interfacial recombination at the outer edges of the p/n-junction absorber. Again, these results corroborate the abovementioned observations made with EBIC and carrier lifetime studies. However, near the band edge of the converter, the IQE rolls off, which is a result of the thin absorber layers discussed previously. As shown later, the near-band-edge QE improves markedly when the converter structure is used in a MIM configuration with a BSR.

In Figure 3, current-voltage data for a representative discrete 0.6-eV converter under high-intensity illumination are given. The typical photovoltaic performance parameters [open-circuit voltage (V_{oc}), short-circuit current density (J_{sc}), and fill factor (FF)] are listed. For a band gap of 0.6 eV, the values of FF (73.4%) and the voltage factor (defined as V_{oc}/E_g, 0.59 V/eV) are excellent. Typical ranges for the diode ideality factors and reverse-saturation current densities for these converters [derived from $V_{oc}(J_{sc})$ analyses] are 0.95-1.05, and $5\text{-}10 \times 10^{-7}$ Acm^{-2}, respectively, at a temperature of 25°C. These data illustrate that high-performance TPV converters are realized using the aforementioned design concepts.

MONOLITHICALLY INTERCONNECTED MODULES

As shown in Figure 4, the converter structure grown for MIM fabrication is virtually identical to the structure used for discrete converters, with four exceptions: 1) The InP substrate is DSP and semi-insulating (SI) (i.e., Fe-doped and transparent to sub-band-gap-energy photons); 2) The $InAs_yP_{1-y}$ compositionally graded layer is undoped (making it transparent to sub-band-gap-energy photons); 3) A 0.3-μm-thick n+/p $InAs_{0.32}P_{0.68}$ cell isolation diode (CID) is included as part of the 1-μm-thick $InAs_{0.32}P_{0.68}$ layer above the graded layer; 4) The thickness of the p+-$Ga_{0.32}In_{0.68}As$ emitter layer is fixed at 0.1 μm to reduce free-carrier absorption of sub-band-gap-energy photons (empirically, we have found that p-type layers have a higher sub-band-gap-energy, free-carrier absorption coefficient). At this thickness, the p+-$Ga_{0.32}In_{0.68}As$ emitter layer has a sheet resistance of ~1600 ohms per square. We have investigated CIDs for MIMs for the last 2 years. In the present structure, the CID serves two important functions, as follows: 1) Electrical isolation of the component cells in the MIM is achieved by etching a trench through the CID n+/p junction with the reverse-bias characteristic of the CID preventing inter-cell shunting; 2) The depth of the isolation trench is minimized because the CID is disposed directly beneath the DH converter. This feature improves the characteristics of the cell interconnections on the MIM. In fact, the n+-$InAs_{0.32}P_{0.68}$ layer of the CID is designed to serve several important functions, as follows: 1) It is the back-surface confinement layer for the DH converter; 2) It works as a stop-etch layer for device processing, which ensures correct placement of the back contact within the MIM structure; 3) It is also the back-contact, lateral-conduction layer for the MIM — the sheet resistance of the layer is ~50 ohms per square, which results in negligible Joule losses; 4) It is the emitter layer for the CID. In an effort to reduce free-carrier absorption of sub-band-gap photons in the MIM structure, the p-$InAs_{0.32}P_{0.68}$ layer of the CID is grown just thick enough to support the space-charge region. Overall, the CID technique has been extremely helpful in implementing MIM technology in these lattice-mismatched converter structures.

Spectral AAAEQE data (no ARC) are shown in Figure 5 for a MIM structure both with and without a BSR. The improvement in the near-band-gap response is quite significant. These data demonstrate that the BSR serves the intended purpose of improving the long-wavelength QE of optically thin DH converters.

Current-voltage data for a representative 0.6-eV, four-cell MIM operated under high-intensity illumination are given in Figure 6. The value of V_{oc} for the MIM (1.546 V) gives an average V_{oc}/cell of 0.387 V, which yields a voltage factor of 0.64 V/eV. These data also show that the CID provides excellent electrical isolation between the component cells in the MIM. The FF value (66.1%) can probably be improved somewhat through an optimization of the epistructure/metallization design. If we assume that the IQE data shown in Figure 2 are a good approximation to the AEQE of this MIM with an optimized ARC applied, then the data shown in Figure 6 correspond to operation of the MIM under an ideal blackbody radiator at a temperature of 1027°C. Under these conditions, the maximum power density generated by the MIM is 1.07 Wcm^{-2}, which is an excellent level of performance.

High-reflectance BSRs are necessary for efficient recuperation of sub-band-gap-energy photons in order to achieve high TPV system-conversion efficiencies. Full details of BSR optimization and fabrication are given in a companion paper at this conference [8]. Additionally, sub-band-gap energy free-carrier absorption within the MIM structure must be as low as possible to maintain efficient recuperation. Long-wavelength, spectral total reflectance data for an as-grown, 0.6-eV MIM structure (no metallization) with an optimized BSR are shown in Figure 7. The data show that high reflectances (80%–90%) are achieved for wavelengths ranging from 2 to 12 µm, which indicates that the BSR has a high reflectance and that the free-carrier absorption of sub-band-gap energy photons is low.

CONCLUSION

A 0.6-eV, $InAs_{0.32}P_{0.68}/Ga_{0.32}In_{0.68}As$ DH TPV converter structure has been described that employs an $InAs_yP_{1-y}$ step-graded layer to produce converter layers with superior electronic properties. The DH design yields TPV converters with high quantum efficiencies and low reverse-saturation current densities. High-performance, 0.6-eV discrete converters and MIMs have been demonstrated with this structure.

ACKNOWLEDGMENTS

We would like to thank Dr. Richard Ahrenkiel, Kim Jones, Alice Mason, Rick Matson, and Dr. Helio Moutinho, all of NREL, for their technical support.

REFERENCES

1. See Proc. of the first three NREL Conferences on Thermophotovoltaic Generation of Electricity, Pub. by American Institute of Physics (AIP), Woodbury, NY, (AIP Conf. Proc. 321 (1994), 358 (1995), and 401 (1997)).
2. M. W. Wanlass, "Development and Characterization of Low-Band-Gap $Ga_xIn_{1-x}As$ Thermophotovoltaic Energy Converters," PhD thesis (Physics), University of Wales, College of Cardiff (1997).
3. M. W. Wanlass, J. S. Ward, K. A. Emery, M. M. Al-Jassim, K. M. Jones, and T. J. Coutts, "$Ga_xIn_{1-x}As$ Thermophotovoltaic Converters," *Solar Energy Materials and Solar Cells*, 41/42, pp. 405-417, 1996.
4. P. Bhattacharya, ed., *Properties of Lattice-Matched and Strained Indium Gallium Arsenide*, INSPEC EMIS Datareviews Series, Short Run Press Ltd., Exeter, UK, 1993, p. 7.
5. O. Madelung, M. Schulz, and H. Weiss, eds., *Landolt-Bornstein Numerical Data and Functional Relationships in Science and Technology*, Group III, Vol. 17, Subvol. a, Springer-Verlag, Berlin, 1982, p. 340.
6. M. W. Wanlass, "Reactor Design for Uniform Chemical-Vapor-Deposition-Grown Films Without Substrate Rotation," U.S. Patent No. 4,649,859, 3/17/87.
7. J. S. Ward, A. Duda, M. W. Wanlass, J. J. Carapella, X. Wu, R. J. Matson, T. J. Coutts, T. Moriarty, C. S. Murray, and D. R. Riley, "A Novel Design for Monolithically Interconnected Modules (MIMs) for Thermophotovoltaic (TPV) Power Conversion," *Proc. 3rd NREL Conf. on Thermophotovoltaic Generation of Electricity, Colorado Springs, CO, USA (1997)*, AIP Conf. Proc. 401, pp. 227-236.
8. X. Wu, A. Duda, J. J. Carapella, S. Ward, and M. W. Wanlass, "A Study of Contacts and Back-Surface Reflectors for 0.6-eV $Ga_{0.32}In_{0.68}As/InAs_{0.32}P_{0.68}$ Thermophotovoltaic Monolithically Interconnected Modules," this conf.
9. M. W. Wanlass, "Effects of Lattice Mismatch on the Photovoltaic Performance of $GaAs_{0.74}P_{0.26}$ Shallow Homojunctions," MS thesis (Physics) (T-3905), Colorado School of Mines, Golden, CO, USA (1990).
10. R. K. Ahrenkiel, R. Ellingson, S. Johnston, M. W. Wanlass, and J. J. Carapella, "Recombination Lifetime of $In_xGa_{1-x}As$ Alloys as a Function of Doping Density," this conf.

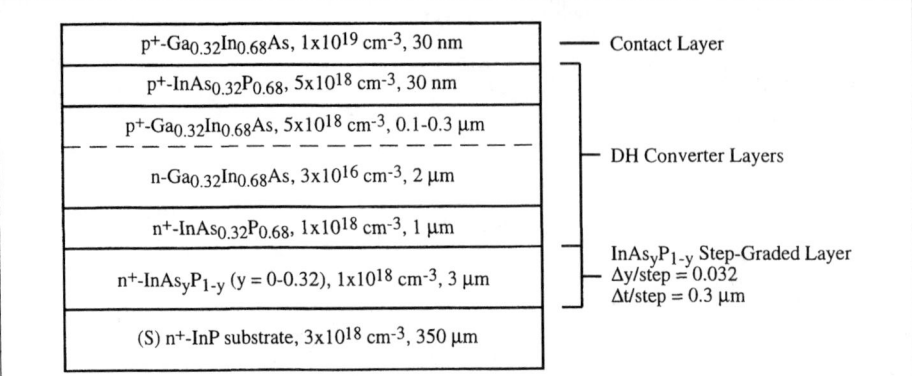

FIGURE 1. Schematic cross-sectional diagram of the epistructure used for the fabrication of discrete 0.6-eV TPV converters.

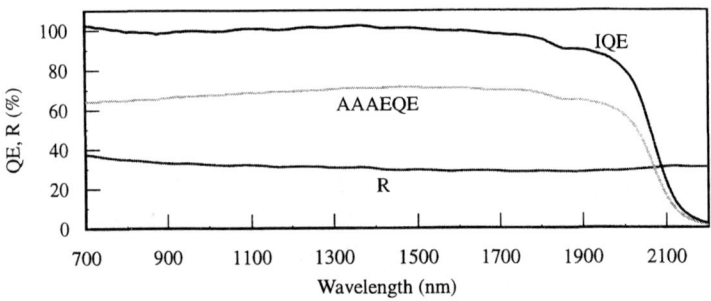

FIGURE 2. Spectral quantum efficiency and reflectance data for a discrete 0.6-eV TPV converter.

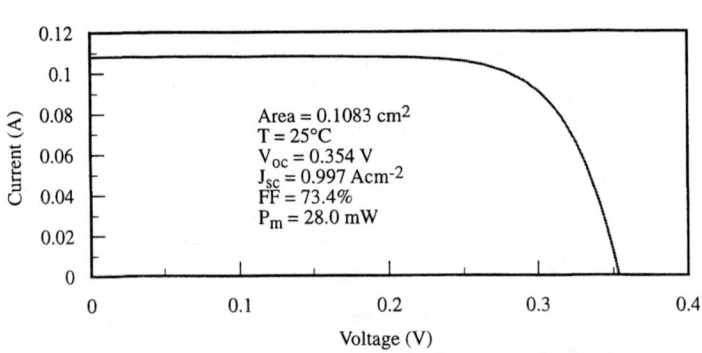

Area = 0.1083 cm^2
T = 25°C
V_{oc} = 0.354 V
J_{sc} = 0.997 Acm^{-2}
FF = 73.4%
P_m = 28.0 mW

FIGURE 3. Current-voltage data for a discrete 0.6-eV TPV converter under high-intensity illumination.

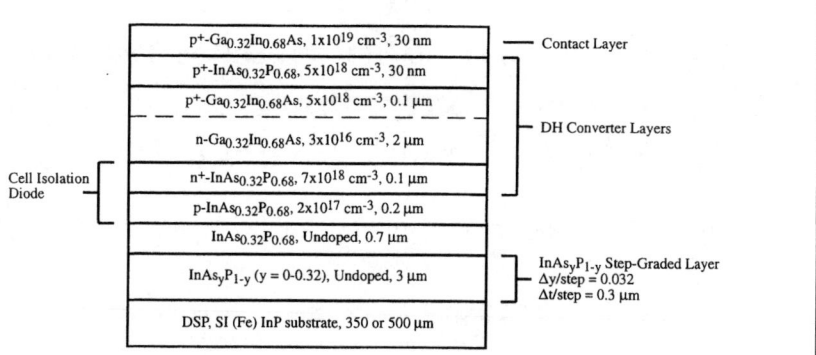

FIGURE 4. Schematic cross-sectional diagram of the epistructure used for 0.6-eV MIM fabrication.

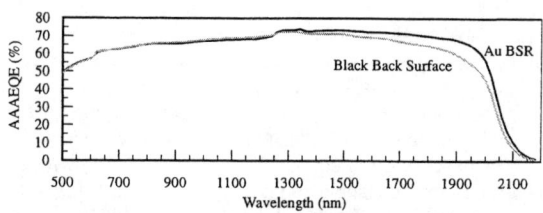

FIGURE 5. Spectral AAAEQE data for a 0.6-eV MIM both with and without a BSR.

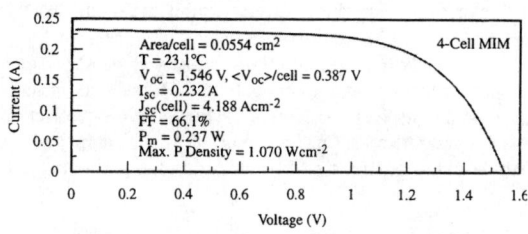

FIGURE 6. Current-voltage data for a 0.6-eV MIM under high-intensity illumination.

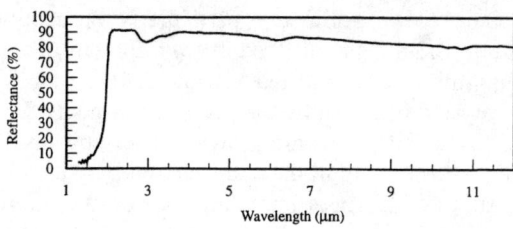

FIGURE 7. Long-wavelength spectral reflectance data for a 0.6-eV MIM converter structure with a BSR.

Growth and Properties of InGaAs/FeAl/InAlAs/InP Heterostructures for Buried Reflector/Interconnect Applications in InGaAs Thermophotovoltaic Devices

S.A. Ringel,* R.N. Sacks,* L. Qin,* M.B. Clevenger,[†] and C.S. Murray[†]

Department of Electrical Engineering, The Ohio State University, Columbus, OH 43210
[†]*Bettis Atomic Power Laboratory, West Mifflin, PA 15122*

Abstract. InGaAs thermophotovoltaic cells are of extreme promise for high efficiency, low bandgap TPV conversion. In the monolithic interconnected module scheme, an InGaAs lateral conduction layer (LCL) is used for the series connection between TPV cells results in undesirable free carrier absorption, causing a tradeoff between series resistance and optical absorption losses in the infrared. A potential alternative is to replace the LCL with an epitaxial metal layer that would provide a low-resistance interconnect while not suffering from free carrier absorption. The internal metal layer would also serve as an efficient, panchromatic back surface reflector, providing the additional advantage of increased effective optical thickness of the InGaAs cell. In this paper, we present the first results on the growth and development of buried epitaxial metal layers for TPV applications. High quality, single crystal, epitaxial Fe_xAl_{1-x} layers were grown on InAlAs/InP substrates, having compositions in the range $x = 0.40$-0.80. Epitaxial metal layers up to 1000Å in thickness were achieved, with excellent uniformity over large areas and atomically smooth surfaces. X-ray diffraction studies indicate that all FeAl layers are strained with respect to the substrate, for the entire composition range studied and, surprisingly, for all thicknesses. The FeAl layers exhibit excellent resistance characteristics, with resistivities from 60μohm-cm to 100μohm-cm being measured for layers having thickness greater than 200 Å. This closely matches bulk FeAl resistivity values, indicating that interface scattering has a negligible effect on lateral conductivity. Reflectance measurements show that the FeAl thickness must be at least 1000Å to achieve >90% reflection in the infrared. Overgrowth of InGaAs and AlInAs on FeAl surfaces was achieved, with layer quality depending strongly on As overpressure, possibly due to In segregation to the FeAl surface.

INTRODUCTION

Thermophovoltaic (TPV) cells are particularly susceptible to free carrier absorption (FCA) due to the intense infrared energy content of typical TPV thermal sources. In the monolithic interconnected module (MIM) configuration, FCA has been dealt with to a very large degree by the use of semi-insulating InP substrates on which low bandgap InGaAs TPV cells are grown.[1] The semi-insulating nature of the substrates mitigates FCA losses within the substrate; and, by coating the back surface of the substrate with Au, provides an efficient reflector of the infrared photons back to the heat source for recuperation. The MIM structure with a Au back surface reflector has already demonstrated outstanding TPV performance, which is now well documented.[2] One of the important issues remaining to be addressed is optimization of the lateral conduction layer (LCL), which is needed to provide the series interconnection between adjacent TPV cells. The typical LCL consists of heavily n-doped InGaAs, and its thickness must be such that resistance losses are minimized. Hence, the use of relatively thick (tenths of microns or more) n^+ InGaAs as the LCL can lead to significant FCA. The ideal LCL would be one in which ultra-low resistance is possible and infrared radiation is either completely transmitted through to the backside of the wafer or is reflected back to the heat source by the LCL directly, with negligible absorption. With the latter in mind, we have been investigating buried epitaxial metal layers as possible LCL's for future MIM TPV cells.

In this paper, we describe an initial investigation of Fe_xAl_{1-x} layers grown epitaxially onto InP substrates and their suitability for eventual integration within InGaAs TPV cell structures. The use of such a metal layer may potentially demonstrate several advantages: (1) ultra-low resistance for the purpose of interconnections, (2) very high reflectivity throughout the infrared to reduce or eliminate absorption, and (3) very high reflectivity for above-bandgap photons which provides an effectively increased optical pathlength for the TPV cell. Hence, an epitaxial metal interconnect/reflector layer incorporated into a TPV cell above the substrate would theoretically behave as a near-ideal interconnect while having the added advantage of broad band reflectivity. Of course, there are some significant materials issues that must be overcome for this to become reality, as InGaAs/FeAl/InP-type structures represent a challenging integration of dissimilar materials. However, the notion of developing epitaxial semiconductor/metal heterostructures is not new, and has been explored recently for applications ranging from metal base transistors to magnetic devices.[3] In general, the great potential of epitaxial semiconductor-metal heterostructures as the basis for new classes of devices has motivated significant efforts to identify the most promising metal compounds for pairing either with GaAs or InP due to the dominance of these semiconductors in many device technologies.[3] For InP (and also GaAs), the transition metal-group III alloys have been identified as particularly promising for the fabrication of buried metal layer structures. This is due to the thermodynamic stability of these compounds with InP and GaAs, the fact that these compounds have dominant cubic phases, and, since the lattice parameter of many transition metal-group III compounds are approximately half that of InP (and GaAs), the potential to achieve lattice matching at the metal/semiconductor interfaces ($2a_{metal} \sim a_{semiconductor}$). Specifically, the compound Fe_xAl_{1-x} has been investigated for pairing with InP since the Fe_xAl_{1-x} lattice constant is slightly less than half that of InP (for x = 0.5), resulting in a total misfit between $2a_{FeAl}$ and a_{InP} of 0.9%.[3-6] However, in contrast to early work which focussed on the problem of achieving smooth, continuous, *thin* (< 100Å) layers for applications such as metal base transistors, TPV cells will require much thicker (up to 1000Å or more) layers to achieve the desired conductivity and reflectance properties.

EXPERIMENTAL DETAILS

Fe_xAl_{1-x}/III-V heterostructures were grown using a multi-chamber MBE cluster, comprised of separate III-V MBE and electron-beam evaporation chambers connected by ultra high vacuum transfer tubes. Hence, combinations of Fe_xAl_{1-x} and III-V layers could be grown while maintaining a pristine surface for high quality heteroepitaxy. III-V MBE growth was performed in standard fashion using conventional solid sources. Fe_xAl_{1-x} deposition was achieved by co-evaporation using dual e-beam evaporators. Iron and aluminum fluxes were independently controlled by an electron impact emission spectroscopy (EIES) controller/monitor so that the FeAl composition could be precisely tuned in each run. This technique is based on the measurement of light intensity emitted from the evaporant atoms when bombarded with electrons. Each element emits photons at characteristic wavelengths, and with the use of a beam splitter and filters, the contribution from each evaporant source can be simultaneously determined, with the emission intensity being proportional to the evaporant beam flux. We investigated Fe_xAl_{1-x} compositions of $x = 0.40 - 0.80$, so that a range of lattice constants could be explored for suitability with InGaAs/InP cell structures.[3,4] Deposition rates were adjusted to be within 0.75 to 1.0 Å/sec. Fe_xAl_{1-x} layers were grown to thicknesses between 200Å and 1200Å, sufficient for evaluation of both optical reflectance and lateral conductivity. Based on earlier work, a range of substrate temperatures from 100° to 300°C for Fe_xAl_{1-x} epitaxy was investigated, with best results occurring for a 200°C substrate temperature, consistent with earlier reports.[6] A 3000Å III-V buffer was grown on the (001) InP substrates in the III-V growth chamber prior to UHV transfer into the metallization chamber for Fe_xAl_{1-x} epitaxy to provide a high quality surface for Fe_xAl_{1-x} nucleation. For this purpose, we used either $In_{0.53}Ga_{0.47}As$ or $Al_{0.52}In_{0.48}As$ buffers lattice-matched to the underlying InP substrate. Both were successful in providing a good surface for FeAl epitaxy, but AlInAs was chosen as standard since it was anticipated that the presence of Al (instead of Ga) on the surface prior to Fe_xAl_{1-x} deposition may encourage better Fe_xAl_{1-x} nucleation. Hence, the nominal structure investigated was FeAl/AlInAs/InP. Reflection high energy electron diffraction (RHEED) was used to monitor the quality of the growing films at various stages of growth, and Auger electron spectroscopy (AES) was performed *in-situ* by transferring the wafer into an attached UHV analysis chamber equipped with AES and sputtering capabilities. A variety of *ex-situ* measurements were used to assess correlations between growth parameters and final film properties. For the remainder of the paper, we will refer to Fe_xAl_{1-x} as FeAl.

RESULTS AND DISCUSSION

Epitaxial growth of FeAl on InP substrates

Figure 1 shows a representative series of RHEED patterns obtained on the surface of a typical AlInAs buffer immediately prior to FeAl deposition, and after

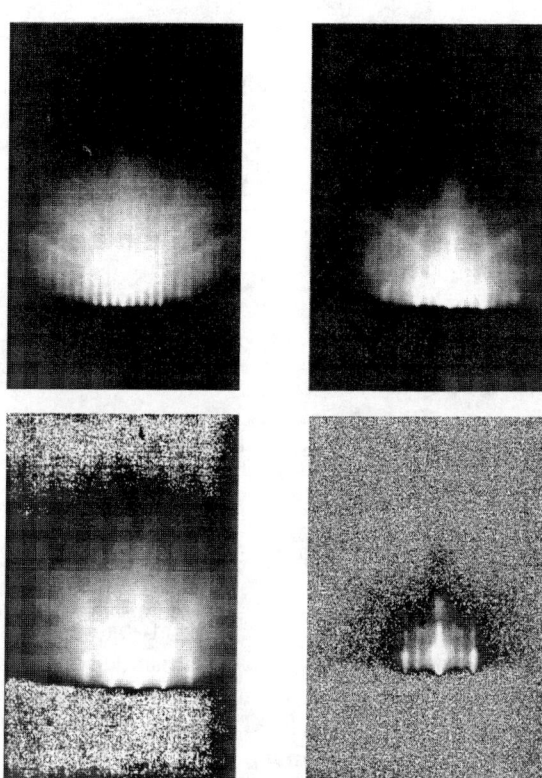

FIGURE 1. RHEED patterns of (a) AlInAs/InP, and of FeAl surfaces for FeAl (45% Fe) thicknesses of (b) 200Å, (c) 500Å and (d) 1000Å. Similar patterns were observed for %Fe = 40% - 80%

FeAl deposition as a function of layer thickness for a single FeAl composition (x = 0.45). The substrate temperature in each case was 200°C. As seen in the figure, the AlInAs surface displays the expected two-fold reconstruction pattern. However, significant differences are seen after various stages of FeAl deposition. In general, FeAl exhibits dominant diffraction streaks with a much wider spacing than the AlInAs diffraction streaks, consistent with an FeAl lattice constant that is approximately half that of AlInAs. Occasionally we observe strong four-fold reconstruction on the FeAl surface, and this is seen clearly in the 200Å and 1000Å thick FeAl RHEED patterns in Figure 1. There has been very limited information published on the surface structure of FeAl layers to date, and this information is vital for successful III-V epitaxy on the FeAl surfaces. Nevertheless, the fact that all RHEED patterns display streaky features with only minimal degradation as the FeAl layer thickness is increased is indicative of high quality single crystal epitaxy. The RHEED results were further substantiated by AFM measurements from which we derived an RMS roughness of 3Å to 4Å for various FeAl layers (identical to the starting InP substrates), featureless Nomarski micrographs, and featureless SEM images, with no evidence of pinholes.

One issue that did arise in evaluating growth properties of the FeAl layers came from Auger analysis performed *in-situ* in our attached UHV analysis chamber. Figure 2a shows an Auger spectrum of the surface of a 500Å FeAl layer grown on AlInAs/InP, indicating a large indium peak. To determine the source of the In within the FeAl layer, and to see whether the indium was distributed throughout the layer or was "surface-riding" as a result of initial segregation on the FeAl growth front, Auger depth profiling was performed. The In signal was observed to disappear after argon ion sputtering the FeAl surface for 30 seconds, enough to remove the first few monolayers of FeAl. This indicates that indium likely segregates to the FeAl surface

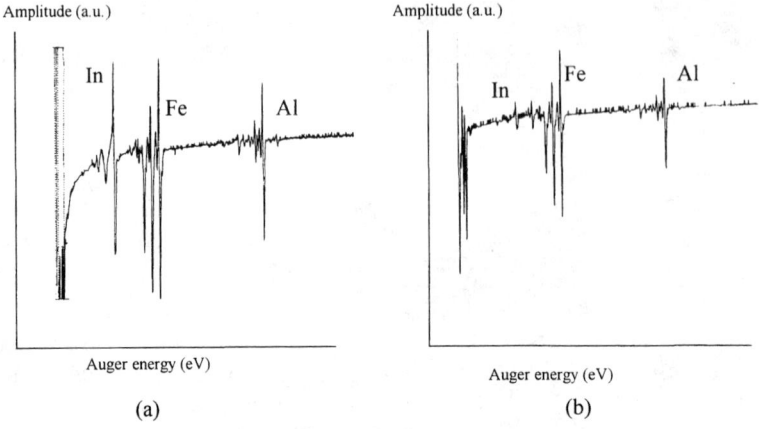

FIGURE 2. In-situ AES spectra for the surface of FeAl layers grown on (a) AlInAs/InP and (b) AlAs/AlInAs/InP, indicating the suppression of indium for the latter.

initially and that subsequent growth of FeAl blocks any further solid-state diffusion after the initial surface segregation. To confirm this more quantitatively, we conducted a series of AES measurements made on test samples consisting of calibrated doses of indium deposited onto polycrystalline FeAl layers grown on silicon substrates. These measurements indicate that the indium on the surface of the FeAl/AlInP/InP samples corresponded to a coverage of approximately 1-2 monolayers. The presence of indium on the FeAl surface may complicate III-V overgrowth and lattice-matching at the interface by forming an InAs layer upon exposure to arsenic. Hence, we investigated methods to eliminate the presence of indium. The growth of a 10 monolayer (ML) AlAs barrier layer at 500 °C on the AlInAs/InP structure prior to FeAl deposition greatly suppresses, but does not eliminate the In segregation. This is seen in figure 2b. The indium signal can be completely suppressed by growth of a 20 ML AlAs layer, by reducing the growth temperature of the 10 ML AlAs layer to 400 °C, or by simply depositing the FeAl below 100 °C without any AlAs layer. However, in each of the last three cases, the FeAl layers demonstrated poor RHEED patterns and DCXRD spectra. Hence, a 10 ML thick, strained AlAs layer was incorporated into the FeAl structures (FeAl/AlAs/AlInAs/InP) for subsequent studies.

The long-term technological goal of the project is to complete InGaAs TPV cell growths onto the FeAl surfaces. Hence, the thermal stability of the FeAl layers from the viewpoint of both indium suppression and also of surface crystallinity is important, since III-V epitaxy will occur a few hundred degrees higher than the FeAl deposition. RHEED patterns were monitored as a function of in-situ annealing for $x = 0.4 - 0.8$. For compositions of $\sim 0.4 - 0.6$, the RHEED pattern became spotty above 350 °C. However, high Fe content layers with $x = 0.7 - 0.8$ did not degrade until temperatures in excess of 550 °C were reached. Moreover, AES measurements made on 80% iron samples grown using a 10 ML AlAs barrier, as described above, did not show any increase in indium surface coverage. We conclude from this that high Fe content layers with AlAs barriers should be suitable for InGaAs overgrowth, which typically occurs at temperatures above ~ 500 °C in MBE.

Bulk crystalline properties of epitaxial FeAl/AlInAs/InP

High-resolution x-ray diffraction measurements were used to assess the crystalline quality of the bulk FeAl layers and to examine how the lattice-mismatch between FeAl and AlInAs/InP varies with nominal changes in composition. Surprisingly, all FeAl layers which showed good RHEED patterns also gave high resolution x-ray diffraction patterns (rocking curves) that indicated that the layers were fully strained. An example is given in Figure 3, which shows the (004) rocking curve of a sample with the structure: 1,000Å $Fe_{0.75}Al_{0.25}$ / 10mL AlAs / 2,900Å (In,Al)As / InP. Also shown in this figure is a simulated rocking curve. The (In,Al)As buffer is too closely lattice-matched to the InP substrate to be seen in this rocking curve, and the 10mL AlAs layer is too thin to be seen under these conditions. Thus, the only peaks visible are those from the FeAl layer and the (In,Al)As+InP combination. Note the presence of Pendellosung oscillations around the FeAl peak. These oscillations are only observed for layers with high crystalline quality. It should be pointed out that this peak is actually the (002) reflection of the FeAl layer, not the (004), since the a_0 of the metal is ~0.5 that of the semiconductor. Rather than attempting to simulate a FeAl layer, the simulation was actually done using a 1,000Å thick layer of (In,Al)As with the In content chosen to give an a_0 matching that of the observed peak, assuming it is fully strained. This was done simply to approximate the

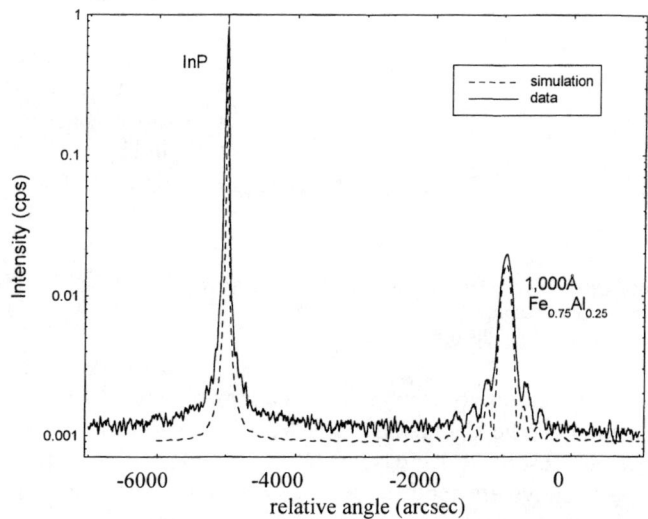

FIGURE 3. Experimental and simulated (004) x-ray rocking curves of 1000Å FeAl (75% Fe) layer on AlInAs/InP.

peak width, shape, and height to be expected from such a layer. The only error that should result from this approximation is in the predicted peak height due to the different scattering factors for FeAl and (In,Al)As. As seen in Figure 3, the simulated peak width and shape, including the presence and spacing of the Pendellosung oscillations, is in excellent agreement with the measurement, implying the layer is fully strained. However, it was expected that the large mismatch (~1.6%) coupled with the thickness (1,000Å) of this layer should instead have resulted in almost complete relaxation, with dislocations generated by the relaxation causing extreme broadening of the peak, wiping out the Pendellosung oscillations. Clearly, this was not observed, and our conclusion that the FeAl layers are fully strained was confirmed by reciprocal space mapping using triple axis XRD.

The variation of $2a_0$ with alloy content for the Fe_xAl_{1-x} layers is shown in Figure 4. The lower curve is the lattice parameter perpendicular to the surface (in the growth direction) extracted directly from the separation of the InP and metal x-ray peaks, with no correction for tetragonal distortion. The upper curve shows an approximate correction for tetragonal distortion assuming the Poisson ration (v) is 0.333. This upper curve would represent the $2a_0$ of a relaxed metal layer with no strain (i.e. - the bulk value). The lattice parameter for x = 0.50 is very close to the bulk FeAl value.[4]

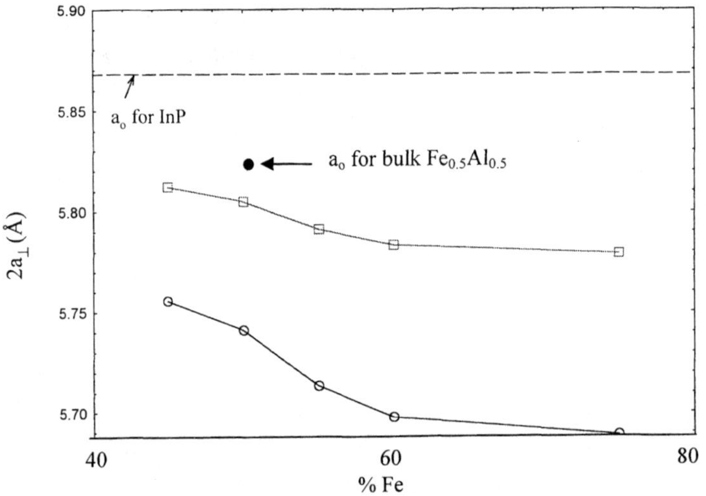

FIGURE 4. Variation of perpendicular FeAl lattice parameter (in the growth direction) with composition. Open circles assume the layer is fully relaxed and open squares are corrected values for tetragonal distortion.

Increasing the iron content beyond 50% initially is seen to decrease the lattice parameter, consistent with the reported behavior for bulk FeAl crystals.[4] However, as the iron content approaches 70%, we continue to observe a decrease in lattice parameter whereas the lattice parameter for bulk FeAl alloys are reported to pass through a minimum and increase.[4] These differences could be due to a number of factors acting alone or in combination: the much lower temperature of our depositions compared to bulk synthesis, the difference between layer-by-layer deposition and bulk sintering, and the possible mediating effects of the surface-segregating indium layer.

Electrical and optical properties of epitaxial FeAl layers

For TPV applications, the FeAl layer must simultaneously demonstrate excellent conductivity and reflectivity, while maintaining high crystalline quality to support device-quality InGaAs TPV cell overgrowth. To date, there have been no reports on the effect of either FeAl composition or layer thickness on either of these critical properties for device applications and it may be expected that both will have large impacts on these properties. Resistivities were obtained for a number of FeAl/AlAs/InAlAs/InP samples by four point probe measurements, where the Fe/Al ratio ranged from 0.4 to 0.8 and the FeAl layer thickness was varied from 200Å to

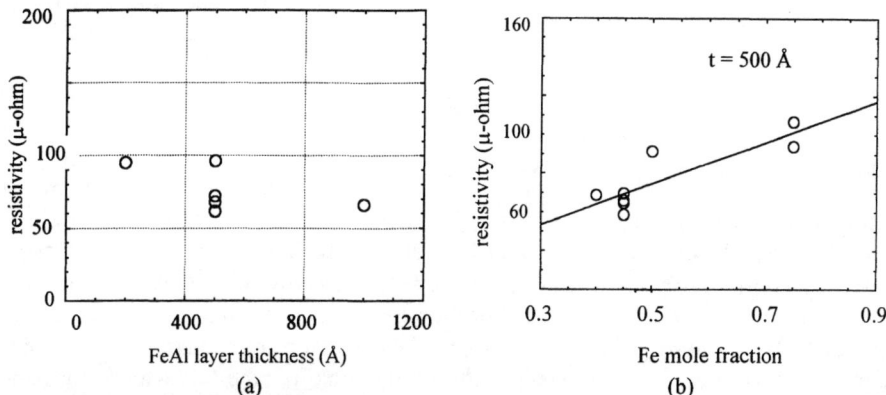

FIGURE 5. FeAl resisitivity as a function of (a) layer thickness for %Fe = 45 – 55% and (b) composition for 500Å thick layers.

1000Å. Figure 5a shows the resistivity results as a function of FeAl layer thickness. As seen, the resistivity of all FeAl layers were in the range of ~50µohm-cm to 100µohm-cm with no clear dependence on layer thickness, indicating that the lateral conductivity of the FeAl epilayers is not limited by interface scattering. This is further substantiated by the close agreement between these measurements and the resistivity values reported for bulk FeAl crystals (80µohm-cm to 90µohm-cm at room temperature).[7] Note that the resistivity of the FeAl is orders of magnitude lower than that of heavily doped InGaAs, demonstrating that FeAl layers have the potential of providing improved electrical interconnects with minimized parasitic losses.

It is of interest to determine how the FeAl resistivity may depend on composition, since adjusting the Fe:Al ratio allows some tuning of the lattice constant layer resistivity on composition. Figure 5b shows the FeAl resistivity to increase slightly with iron content, increasing from ~60μohm-cm to ~100μohm-cm as the iron mole fraction is increased from 0.45 to 0.75. This trend is consistent with an expected increase in resistance for iron-rich layers. This weak dependence is clearly advantageous from the viewpoint of optimizing the lattice match with the surrounding semiconductor layers since the lattice constant, which does depend on composition, can be essentially adjusted independently of any effect on resistivity. Thus, from the electrical viewpoint, FeAl layers exhibit considerable promise.

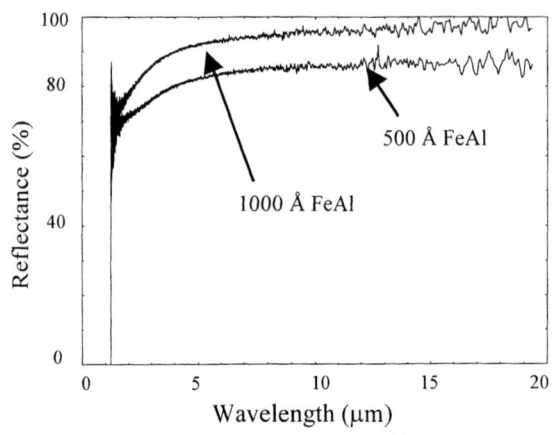

FIGURE 6. Front-side reflectance for FeAl layers.

However, for TPV applications, the total reflectance of the FeAl layers is also critical. Figure 6 shows a comparison of reflectance data measured from the surface of bare $Fe_{0.50}Al_{0.50}$ layers having thicknesses of 500Å and 1000Å, grown on AlAs/ InAlAs/InP. As expected, an increase in reflectance is observed as the FeAl layer thickness is increased. Reflectance values were found to maximize in excess of 90% for wavelengths between 4μm and 20μm. However, all films exhibited a loss of reflectance at shorter wavelengths with the reflectance decreasing continuously with decreasing wavelength below 4μm. We attribute this reduction at shorter wavelengths to absorption by the iron transition metal component in the FeAl.[8] This absorption can be mitigated by use of a front surface filter. These experimental results are consistent with predictions of the optical performance modeled from band structure calculations for FeAl.[8,9]

InGaAs overgrowth on FeAl/AlAs/AlInAs/InP: preliminary findings

Once a degree of confidence was established for the FeAl epitaxy, a series of samples were transferred back into the III-V growth chamber after FeAl epitaxy for InGaAs overgrowth. Since the FeAl lattice is strained with respect to the InP substrate, the in-plane lattice constant of the FeAl layer will be matched to 0.74 eV bandgap InGaAs overgrowth. Initial overgrowth exhibited spotty RHEED patterns, indicative of polycrystalline growth. The ability to nucleate reasonable quality InGaAs on the FeAl surface was found to depend strongly on the presence of arsenic. By reducing the arsenic exposure on the FeAl surface, streaky RHEED patterns could be obtained during nucleation. The reason for this is not yet clear and more work must be done, especially from the viewpoint of FeAl surface stability in the presence of a

III-V flux at growth temperatures. Some preliminary work did indicate that the FeAl surface stability improved significantly for higher iron compositions, discussed earlier, and we plan to conduct future overgrowth studies on these higher Fe-content layers.

CONCLUSIONS

FeAl epitaxial layers were explored to investigate their potential as buried interconnect/reflector layers in InGaAs TPV cells. Successful FeAl epitaxy was achieved up to 1000Å in thickness on AlInAs/InP and AlAs/AlInAs/InP substrates. RHEED and DCXRD demonstrate high crystalline quality for compositions of $x = 0.40 - 0.80$. All FeAl layers were fully strained with respect to the InP substrate. AFM and SEM measurements indicate that the FeAl surfaces have monolayer smoothness, enhancing the chances for good quality InGaAs heteroepitaxy on these layers. Auger indicates that indium from the AlInAs buffer layer rides the surface of the FeAl growth front, and by inserting an AlAs layer prior to FeAl epitaxy, the presence of indium is dramatically reduced. The FeAl layers exhibit electrical properties consistent with bulk crystal values, having resistivities in the 50µohm-cm to 100µohm-cm range. Optical reflectance in excess of 90% was achieved from 4µm to 20µm for 1000Å thick FeAl layers, which combined with the ultra-low resistivity, imply that a 1000Å thick FeAl layer will be optimum for eventual TPV applications. The presence of iron in the FeAl layer was explained to be responsible for a roll off of reflectance below 4µm. To mitigate this effect, a front surface filter could be employed. InGaAs epitaxy on FeAl/AlAs/AlInAs/InP structures was also demonstrated in preliminary studies, and InGaAs quality was found to depend on arsenic exposure at the FeAl surface.

ACKNOWLEDGEMENTS

This work was supported by Bettis Atomic Power Lab under contract no. 73-911801.

REFERENCES

1. D.M. Wilt, N.S. Fatemi, P.P. Jenkins, R.W. Hoffman Jr., G.A. Landis and R.K. Jain, *Proc. 25th IEEE Photovoltaic Spec. Conf.*, 43, 1996.
2. N. S. Fatemi, D.M. Wilt, P.P. Jenkins, V.G. Weizer, R.W. Hoffman Jr., C.S. Murray, D. Scheiman, D.Brinker and D. Riley, *Proc. 26th IEEE Photovolt. Spec.Conf.*, 799, 1997.
3. T. Sands, C.J. Palmstrom, J.P. Harbison, V.G. Keramidas, N. Tabatabaie, T.L. Cheeks, R. Ramesh and Y. Silberberg, *Materials Science Reports* 5, 99, 1990.
4. A. Taylor and R.M. Jones, *J. Phys. Chem. Sol.* 6, 16, 1958.
5. C.P. Wang, F. Jona, N.R. Leason, D.R. Strongin and P.M. Marcus, *Surf. Sci.* 298, 114, 1993.
6. A.M. Wowchak, J.N. Kuznia and P.I. Cohen, *J. Vac. Sci. Tech.* B7, 733, 1989.
7. W.B. Muir, J.I. Budnick and K. Raj, *Phys. Rev. B.* 25, 726 (1982).
8. C. Koenig and M. A. Khan, *Phys. Rev. B* 27, 6129, 1983.
9. M.B. Clevenger, C.S. Murray, S.A. Ringel, R.N. Sacks, L. Qin, G.W. Charache and D.M. Depoy, *Proc. 4th Conf. on Thermophovoltaics*, in press, 1998.

n/p/n Tunnel Junction InGaAs Monolithic Interconnected Module (MIM)

David M. Wilt
NASA Lewis Research Center, Cleveland, OH 44135

Christopher S. Murray
Bettis Atomic Power Laboratory, West Mifflin, PA 15122

Navid S. Fatemi and Victor Weizer
Essential Research, Inc., Cleveland, OH 44122

The Monolithic Interconnected Module (MIM), originally introduced at the First NREL thermophotovoltaic (TPV) conference, consists of low-bandgap indium gallium arsenide (InGaAs) photovoltaic devices, series interconnected on a common semi-insulating indium phosphide (InP) substrate. An infrared reflector is deposited on the back surface of the substrate to reflect photons, which were not absorbed in the first pass through the structure. The single largest optical loss in the current device occurs in the heavily doped p-type emitter. A new MIM design (pat. pend.) has been developed which flips the polarity of the conventional MIM cell (i.e., n/p rather than p/n), eliminating the need for the high conductivity p-type emitter. The p-type base of the cell is connected to the n-type lateral conduction layer through a thin InGaAs tunnel junction. 0.58 eV and 0.74 eV InGaAs devices have demonstrated reflectances above 90% for wavelengths beyond the bandgap (> 95% for unprocessed structures). Electrical measurements indicate minimal voltage drops across the tunnel junction (< 3 mV/junction under 1200K-blackbody illumination) and fill factors that are above 70% at current densities (J_{sc}) above 8 A/cm^2 for the 0.74eV devices.

INTRODUCTION

In thermophotovoltaic (TPV) energy conversion, a radiator is heated to incandescence and a photovoltaic device is placed in view of the radiator to convert the radiant energy into electrical energy. Research in TPV has been renewed recently due to the development of new radiator, filter and photovoltaic cell technologies [1]. Most current efforts in TPV research have concentrated on using front surface spectral control elements such as selective emitters (radiators) [2] or greybody radiators combined with plasma, dielectric or dipole filters [3,4] in order to improve system efficiency. The front-surface spectral control approach generally produces systems with lower power densities (W/cm^2).

A different approach involves the use of rear-surface spectral controls. Using this technique, the entire radiant output from the radiator is incident upon the photovoltaic (PV) device, thereby providing higher output power densities. Photons that the PV device is unable to convert pass through the cell structure and reflect off of a rear reflector back to the radiator for recuperation. Researchers have developed TPV cells that utilize low-doped substrates and reflective rear contacts to provide photon recuperation [5,6].

CP460, *Thermophotovoltaic Generation of Electricity: Fourth NREL Conference*
edited by T. J. Coutts, J. P. Benner, and C. S. Allman
© 1999 The American Institute of Physics 1-56396-828-2/99/$15.00

A Monolithic Interconnected Module, or MIM, TPV cell design, originated at NASA Lewis Research Center [3], takes a different approach to rear surface spectral control. The device consists of series-connected indium gallium arsenide (InGaAs) devices on a common, semi-insulating indium phosphide (InP) substrate. An infrared reflector is deposited on the rear surface of the InP substrate to reflect photons back toward the front surface of the cell. This provides a second pass opportunity for photons capable of being converted by the cell. In addition, long wavelength photons are returned to the radiator for "recuperation", improving the system efficiency.

The MIM design offers several advantages. Firstly, small series-connected cells provide high voltages and low currents, reducing I^2R losses. In addition, the small size of the cells permits an array to be comprised of series/parallel strings rather than a single series-connected string of larger cells. This should improve the reliability of the TPV module since the failure of a single cell would not debilitate the entire array. Also, the cell size and distribution may be easily adjusted to minimize the losses associated with radiator non-uniformity (i.e., variation in view factor, temperature, etc.).

The MIM design maximizes output power density since losses associated with front-surface spectral controls are eliminated. The lack of front surface spectral control represents a significant simplification of TPV system design and thermal management, as there are no filters to cool. In addition, the rear surface of the device is not electrically active, therefore the cell may be directly bonded to the substrate/heat sink without concern for electrical isolation. This greatly simplifies the array design and improves the thermal control of the cells. Lastly, photons that are weakly absorbed have the possibility of multiple passes through the cell structure. This feature is particularly important for lattice-mismatched devices, where poor minority carrier diffusion length can be partially offset by making the cell thin, forcing the carrier generation to occur closer to the p/n junction.

Although the MIM design has many beneficial attributes, there are limitations. The optical efficiency of the conventional MIM design [7] and interdigitated MIM design, developed by NREL [8], suffer from free carrier absorption losses in the heavily doped p-type InGaAs emitter. The trade off between optical and electrical losses forces a balance of emitter thickness and doping level. To date, most devices have utilized heavily doped ($>1 \times 10^{19}$ cm^{-3}) emitters of approximately 0.1μm in thickness. This layer accounts for the majority of the optical losses in the MIM for unprocessed structures. Recent Monte Carlo photonic modeling, being presented at this conference [9], suggests that the processing techniques used can also have a significant affect on the amount of photons that can be recuperated at the radiator. Four areas of particular concern are the specular/diffuse reflectance of the BSR, light trapping due to surface features (i.e., isolation trenches), the reflectance of the electrical metallization at the metal semiconductor interface, and IR absorption in the anti-reflective coatings.

CELL DESCRIPTION

In order to optimize the optical efficiency of the MIM device, it is desirable to minimize the thickness of heavily doped p-type material. P-type material has exhibited ~20x higher free carrier absorption compared to similarly doped n-type InGaAs. Converting the photovoltaic device polarity from p on n (p/n) to n/p allows the use of n-type material for the emitter. In addition to reducing the optical losses, majority carriers in the n-type emitter have approximately 25x higher mobility than p-type material, leading to lower resistive and grid shadowing losses.

It is also desirable to use n-type material for the lateral conductor layer (LCL) in order to minimize electrical and optical losses. This layer conducts the collected carriers to the interconnect (in the case of the conventional MIM design) or to the closest back contact (in the case of the interdigitated MIM). The use of p-type material for this layer imposes significant optical and electrical losses. Therefore, we employ a thin InGaAs tunnel junction (TJ) to connect the p-type base to the n-type LCL. The complete device structure for 0.74eV InGaAs is shown in Figure 1. For use in lattice-mismatched InGaAs, a suitable buffer layer would be inserted between the substrate and the active device. The use of the n/p/n TJ MIM structure is applicable to both the conventional and interdigitated MIM configurations.

0.1μm n++ InGaAs	Contact Layer
0.1μm n+ InP	Front Window
0.3μm n+ InGaAs	Emitter
1.5μm p InGaAs	Base
0.1μm p+ InP	Back Window
500Å p++ InGaAs	Tunnel Junction
500Å n++ InGaAs	
0.1μm n+ InP	Etch Stop
1.0μm n++ InGaAs	Lateral Conduction Layer
InP:Fe	Substrate

FIGURE 1. Test structure for 0.74 eV InGaAs n/p/n TJ MIM (conventional).

TUNNEL JUNCTION DEVELOPMENT

Successful development of the tunnel junction (TJ) MIM requires the development of a thin, high conductance tunnel junction. Test TJ devices were fabricated using a low-pressure Organo-Metallic Vapor Phase Epitaxy (OMVPE) reactor. Tunnel junction structures were produced using both 0.74 eV and 0.60 eV InGaAs. 150 μm diameter test diodes were fabricated for characterization with TJ layer thickness' from 200Å to 1 μm. Figure 2 shows the I-V characteristic of a 0.74 eV InGaAs tunnel junction with 0.5 μm layers. The TJ demonstrated excellent resistivity ($R_{max} = 3.7 \times 10^{-4}$ ohm-cm^2) and current carrying capability ($J_p = 1900$ A/cm^2). Tunnel junctions fabricated from 0.6 eV InGaAs also demonstrated excellent I-V characteristics ($J_p = 700$ A/cm^2, $R_{max} = 5.6 \times 10^{-4}$ ohm-cm^2). Perfect 0.6 eV devices (QE = 1) fabricated using these tunnel junctions could expect to lose 2.4 mV/junction under a 1200K blackbody illumination (view factor = 1).

Next, tunnel junction test structures were grown with the appropriate over-layer thickness of InGaAs. This was done so that the effects of an extended temperature soak on dopant distribution within the tunnel junction were appropriately considered. Three test n/p 0.74 eV cell structures were fabricated and tested. Two of them n/p/n structures that had tunnel junctions with layer thickness of 1000Å and 500Å. The third was a control structure, which had the same n/p cell layers, but deposited on a p-type substrate. The I-V tests of these structures were identical, indicating that the tunnel junction was not negatively impacting the device operation.

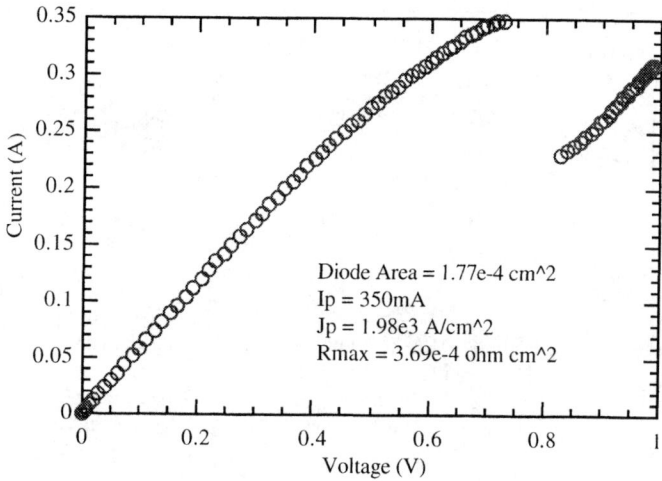

FIGURE 2. I-V characteristic of tunnel junction fabricated from 0.74 eV InGaAs.

0.74 eV n/p/n MIM Characterization

Verification of the n/p/n cell concept was accomplished by fabricating the test device structure shown in figure 1. Figure 3 shows the variation of fill factor with current density for this structure compared to several conventional p/n structures. The n/p/n device was processed using our standard contact design for p/n MIM devices. All of the devices had the same LCL thickness and doping level. The differences in FF can be attributed to variations in the emitter type, doping level and thickness.

Measurements were taken in a flash simulator with the cell at room temperature. As shown in the data, there is very little reduction in FF as the current density is increased for the n/p/n device, whereas the p/n structures all show significant degradation with increasing intensity. For operation of this device under a 1200K radiator, it is expected that the current density would not exceed 0.9 A/cm^2. Thus, there is ample opportunity to reduce layer thickness', doping levels and grid coverage without sacrificing electrical performance. This optimization was not performed because the bandgap of this device is too high for the illumination source of interest.

Figure 4 shows the reflectance measurements for an n/p/n device structure (NAS468 - not a processed device) similar to that shown in Figure 1. The principle differences are a reduction in the thickness of the tunnel junction layers from 500 Å to 100 Å, and a reduction in the LCL thickness from 1.0 μm to 0.15 μm. There is also a reduction in the doping level of the LCL from 3×10^{19} cm^{-3} to the low 10^{18} cm^{-3} range. This structure is suitable for fabrication into an interdigitated MIM device.

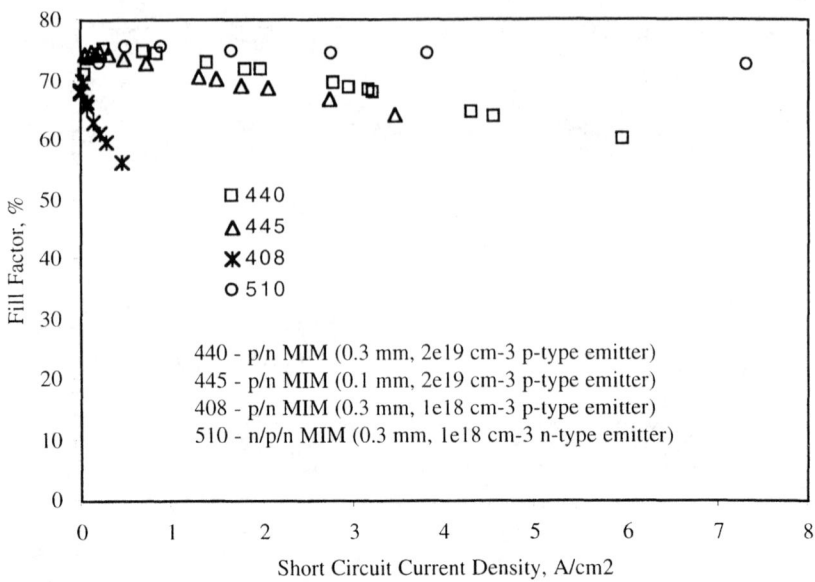

FIGURE 3. Variation in fill factor with current density for the n/p/n 0.74 eV MIM with the architecture shown in Figure 1.

For comparison, a conventional p/n MIM structure reflectance is also plotted (NAS445). There are two features of significance in this figure, the elimination of the plasma absorption peaks at approximately 6 μm and 9 μm and the increase in reflectance in the short wavelength region. The elimination of the plasma peaks can be attributed to the interdigitated cell design with it's lower doped LCL. This feature is not specific to the n/p/n design. The increase in the short wavelength reflectance from ~0.88 to ~0.95 is a direct result of the elimination of the heavily doped p-type emitter through the n/p/n design.

Table 1 lists the short circuit current density, FF, spectral utilization (Fu) and the product of FF and Fu for the five (5) devices previously described under a variety of blackbody temperatures. The spectral utilization is a ratio of the useful energy absorbed in the emitter and base divided by the total energy absorbed. Also shown is the change in the FF*Fu product compared to the n/p/n interdigitated device design. The FF*Fu product captures both the electrical efficiency and optical efficiency of the MIM device in a single, evenly weighted factor. The p/n conventional MIM devices all show significantly lower combined efficiency, particularly at the lower emitter temperatures. The difference between the conventional and interdigitated n/p/n MIM is smaller, particularly at the higher temperature emitters.

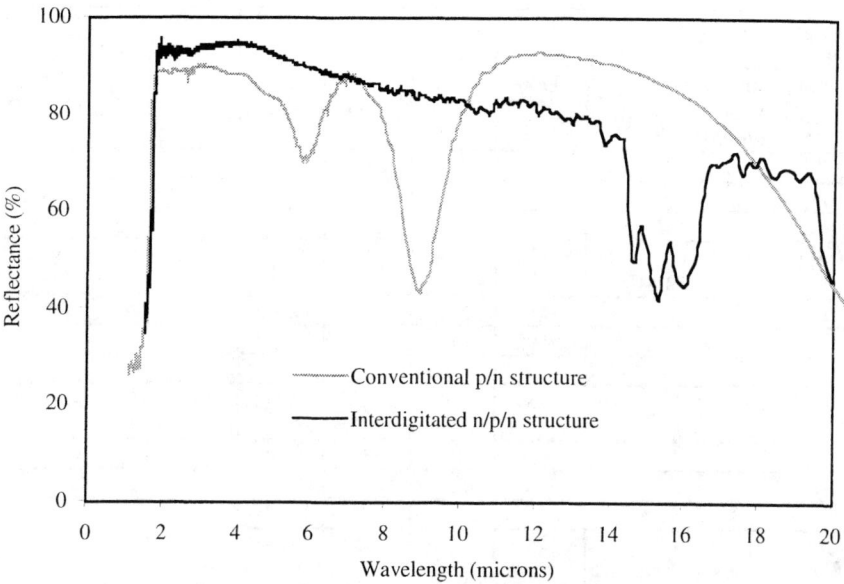

FIGURE 4. Reflectance comparison of conventional p/n and interdigitated n/p/n structures.

0.6 eV n/p/n CELL DEVELOPMENT

As was stated previously, the goal of this development effort is to develop a low bandgap MIM device suitable for operation with a low temperature radiator. After demonstration of the 0.74 eV n/p/n device, efforts shifted to development of 0.6 eV and lower bandgap n/p/n structures. These device structures, being lattice mismatched to the InP substrate, use a proprietary buffer layer which has demonstrated reverse saturation current densities of $<4 \times 10^{-7}$ A/cm^2 in p/n 0.6 eV MIMs.

Several test structures have been fabricated, both conventional and interdigitated n/p/n MIM designs. These devices have utilized 200Å tunnel junction layers. The data from these structures indicates that resistance in the thin tunnel junction does not limit the FF. A typical interdigitated n/p/n design is shown in Figure 5.

Figure 6 compares the reflectance test data from a 0.58 eV n/p/n structure, as grown material (as shown in Figure 5), to a processed cell from the same material. The cell includes a dual layer anti-reflective (AR) coating (ZnS/MgF) and Cr/Au contact metallization. The structure has a gold BSR layer only. The data demonstrates a reduction in the reflectance of the cell beyond the InGaAs bandedge. This difference is due to several factors. The Cr adhesion layer for the front surface metallization reduces the reflectance in the 2-4 μm region. Overlying that is an absorption peak near 3 μm associated with water absorbed in the AR coating. The difference in reflectance shrinks to insignificant levels by approximately 7 μm.

MIM	Trad °F	Jsc A/cm²	FF	Fu	Fu*FF	% Change Relative to 468
440, 0.3 mm, 2x10¹⁹ p-type emitter p/n MIM	1750	0.68	0.746	0.197	0.147	-97%
	2250	2.85	0.69	0.389	0.268	-65%
	2500	4.65	0.63	0.472	0.297	-64%
445, 0.1 mm, 2x10¹⁹ p-type emitter - p/n MIM	1750	0.68	0.731	0.232	0.17	-71%
	2250	2.85	0.67	0.443	0.297	-49%
	2500	4.65	0.6	0.526	0.316	-54%
408, 0.3 mm, 1x10¹⁸ p-type emitter - p/n MIM	1750	0.68	0.5	0.272	0.136	-13%
	2250	2.85	-	0.499	-	
	2500	4.65	-	0.583	-	
510, 0.3 mm, 1x10¹⁸ n-type emitter - n/p/n MIM	1750	0.68	0.754	0.296	0.223	-30%
	2250	2.85	0.744	0.508	0.378	-17%
	2500	4.65	0.740	0.584	0.432	-13%
468, 0.1 mm, 1x10¹⁸ n-type emitter - n/p/n MIM	1750	0.68	0.754	0.384	0.29	-
	2250	2.85	0.744	0.596	0.443	-
	2500	4.65	0.740	0.660	0.488	-

Table 1. Summary of Fill Factor and Spectral Utilization Data for 0.74 eV p/n and n/p/n MIMs.

0.1μm n++ InGaAs	Contact Layer
500Å n+ InPAs	Front Window
0.3μm n+ InGaAs	Emitter
3.0μm p InGaAs	Base
500Å p+ InPAs	Back Window
200Å p++ InGaAs	⎫ Tunnel Junction
200Å n++ InGaAs	⎭
500Å n+ InPAs	Etch Stop
0.3μm n+ InGaAs	Laterial Conduction Layer
InPAs	Buffer Layer
InP:Fe	Substrate

Figure 5. Low bandgap (< 0.74 eV) n/p/n interdigitated MIM test structure.

Figure 7 shows the external quantum efficiency for the 0.58 eV n/p/n interdigitated MIM device described above. The device has a dual layer AR coating and Au back surface reflector. This data is from the first attempt at producing this device. Continued development is expected to improve this encouraging start.

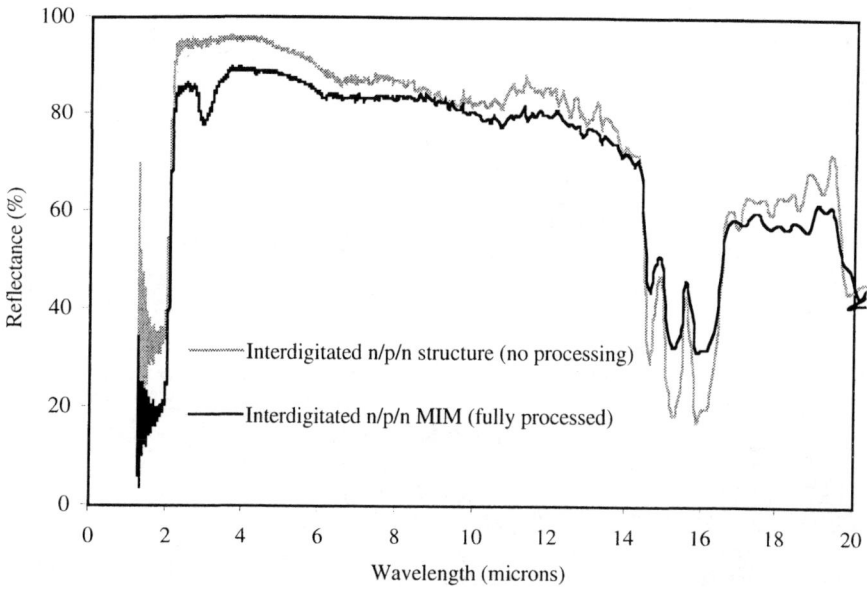

FIGURE 6. Comparison of 0.58 eV Cell and Structure Reflectance.

CONCLUSION

A new tunnel junction MIM design has been described (pat. pend.) which offers the potential for higher power density and efficiency compared to p/n conventional or interdigitated MIM devices. Both 0.74eV conventional and 0.58eV interdigitated n/p/n MIM devices have been demonstrated with encouraging results. The 0.58eV and 0.74 eV InGaAs devices have demonstrated reflectance's above 90% for wavelengths beyond the bandgap (> 95% for unprocessed structures).

Electrical measurements indicate minimal voltage drops across the tunnel junction (< 3 mV/junction under 1200K-blackbody illumination) and fill factors that are above 70% out to current densities (J_{sc}) above 8 A/cm^2 for 0.74eV devices. The emphasis now shifts to optimizing the devices through reduction in tunnel junction layer thickness' (currently at 200 angstroms), active cell region optimization, grid design and new processing strategies. Care must be exercised in designing a fabrication process since these processes can seriously degrade the optical efficiency of the completed MIM device.

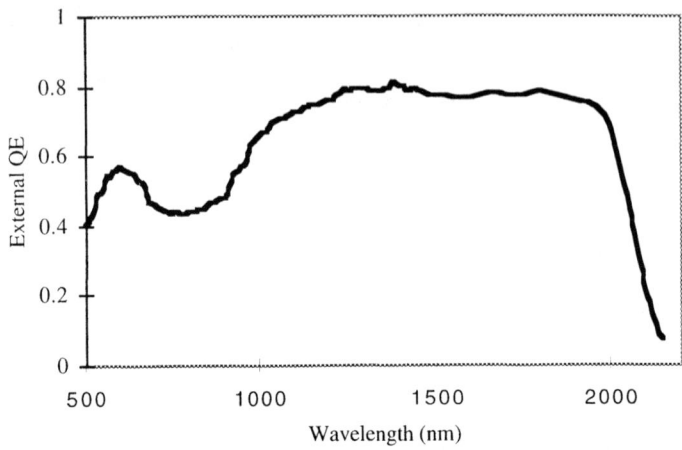

FIGURE 7. External QE of 0.58 eV n/p/n Interdigitated MIM.

REFERENCES

1) Proc. First NREL Conf. On TPV Generation of Electricity, AIP 321, 1994.
2) Chubb, D.L., "Reappraisal of Solid Selective Emitters," Proc. 21st IEEE Photovoltaic Specialists Conference, pp. 1326-1342, 1990.
3) Wilt, D.M., et al., "InGaAs PV Device Development for TPV Power Systems," 1st NREL Conf. On TPV Gen. Of Elect, pp. 210, AIP 321, 1994.
4) Horne, W.E., et al., "IR Filters for TPV Convertor Modules", Proc. 2nd NREL Conf. On TPV Gen. Of Elect, pp. 35, AIP 358, 1995.
5) Charache, G.W., et al., "Thermophotovoltaic Devices Utilizing Back Surface Reflector for Spectral Control," Proc. 2nd NREL Conf. On TPV Gen. Of Elect, pp. 191, AIP 358, 1995.
6) Iles, P.A. and Chu, C.L., "TPV Cells with High BSR," Proc. 2nd NREL Conf. On TPV Gen. Of Elect, pp. 361, AIP 358, 1995.
7) Wilt, D.M., et al., "Electrical and Optical Performance Characteristics of 0.74eV p/n InGaAs Monolithic Interconnected Modules," Proc. 3rd NREL Conf. On TPV Gen. Of Elect, pp. 237, AIP 401, 1997.
8) Ward, J.S., et al., "A Novel Design for Monolithically Interconnected Modules (MIMs) for Thermophotovoltaic Power Conversion," Proc. 3rd NREL Conf. On TPV Gen. Of Elect, pp. 227, AIP 401, 1997.
9) Ballinger, C.T., et al., "Monte Carlo Analysis of a MIMs Device with a Back Surface Reflector," Proc. 4th NREL Conf. On TPV Gen. Of Elect, AIP, 1998.

Monte Carlo Analysis of a Monolithic Interconnected Module with a Back Surface Reflector

*C. T. Ballinger, *G.W. Charache, and **C.S. Murray

*Lockheed Martin Corporation, Schenectady, NY
**Bettis Atomic Power Laboratory, West Mifflin, PA

Abstract: Recently, the photon Monte Carlo code, RACER-X, was modified to include wavelength dependent absorption coefficients and indices of refraction. This work was done in an effort to increase the code's capabilities to be more applicable to a wider range of problems. These new features make RACER-X useful for analyzing devices like Monolithic Interconnected Modules (MIMs) which have etched surface features and incorporates a back surface reflector (BSR) for spectral control. A series of calculations were performed on various MIM structures to determine the impact that surface features and component reflectivities have on spectral utilization. The traditional concern of "cavity photonics" is replaced with "intra-cell photonics" in the MIM design. Like the cavity photonic problems previously discussed [1], small changes in optical properties and/or geometry can lead to large changes in spectral utilization. The calculations show that seemingly innocuous surface features (*e.g.*, trenches and grid lines) can significantly reduce the spectral utilization due to the non-normal incident photon flux. Photons that enter the device through a trench edge are refracted onto a trajectory where they will not escape. This leads to a reduction in the number of reflected below bandgap photons that return to the radiator and reduce the spectral utilization. In addition, trenches expose a lateral conduction layer in this particular series of calculations which increase the absorption of above bandgap photons in inactive material.

INTRODUCTION

A photon Monte Carlo code, RACER-X, has been developed and proven to be a useful tool in analyzing and guiding in-cavity thermophotovoltaic (TPV) experiments. RACER-X predictions are routinely within 10% of the measured values for heat absorbed and short circuit current, which is an improvement over the last publication [1]. The code has been modified to model both wavelength dependent refraction and absorption so that it is applicable to a wider range of problems. These new features make RACER-X useful for evaluating designs like Monolithic Interconnected Modules (MIMs) with a back surface reflector (BSR) [2-4].

TPV devices that use a BSR for spectral control must simultaneously balance below bandgap absorption losses with series resistance losses by manipulating the free carriers in each device layer [5]. Hence, the optimal design must consider the combination of photon and electron transport through the multiple layers in the device. The 0.6 eV InGaAs interdigitated MIMs design [6], which is the focus of this study, had been optimized assuming near-normal incidence for photons upon a featureless surface [7]. However, this report demonstrates that optimizing a MIM with a BSR must consider photon angle of incidence effects, detailed surface features, and refraction.

CP460, *Thermophotovoltaic Generation of Electricity: Fourth NREL Conference*
edited by T. J. Coutts, J. P. Benner, and C. S. Allman
© 1999 The American Institute of Physics 1-56396-828-2/99/$15.00

MONTE CARLO MODEL DESCRIPTION

Geometric Description

The analysis in this report is based upon the concept of an 0.6 eV InGaAs interdigitated MIM however, the conclusions are generally applicable to any TPV device that uses a BSR for spectral control. Several TPV diode geometries were constructed and analyzed in an effort to quantify the impact particular geometric features have on spectral utilization and below band gap reflectivity. MIMs typically require etched trenches on the front surface along with metal contacts. Figure 1 shows the macro-structure of the MIM including the BSR, lateral conduction layer (LCL), and substrate. Notice that the model includes a radiator that is parallel to the top of the TPV surface. Perfectly reflecting boundary conditions were imposed on the four sides of this model to simulate an infinite plate with the MIM structure. Figure 2 shows the details on the surface of the MIM where the photonically important features are the trenches and the grid lines. The back contact trenches in a MIM using an interdigitated grid design are dovetail etched such that they undercut the diode as shown in Figure 2. The isolation trenches are etched as "V-notch" trenches to allow smooth metallization between diodes.

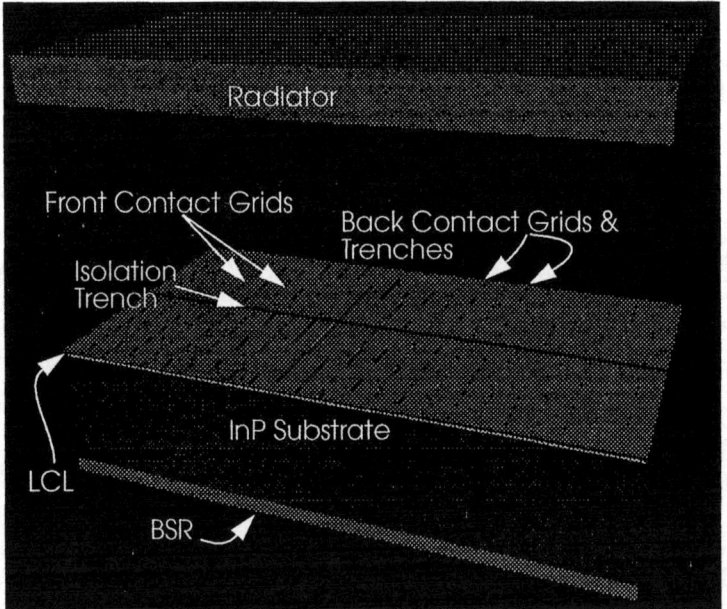

FIGURE 1: Interdigitated MIM macro-structure.

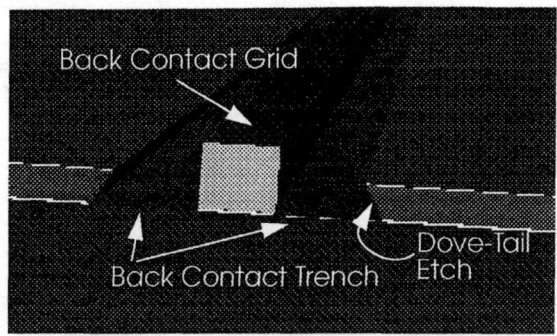

FIGURE 2: Dove-tail back contact trench model.

Several RACER-X models were constructed to isolate the effects of trenches, grids, BSR reflectivity, specularity, diode thickness, and dopant level. This was accomplished by constructing models with and without grids, trenches, etc. and comparing the results. Table 1 provides details of dimensions for individual components used in the computational model.

Optical Properties

Important material properties include wavelength dependent reflectivities and both the real and imaginary parts of the refractive index. Table 2 lists the optical properties of the particular features that were used in the Monte Carlo model. The above bandgap absorption cross-sections were simply modeled with published values of absorption coefficients for binary III-V semiconductors and linear combinations of binary semiconductors for ternary III-V materials [8]. The InGaAs n-type below bandgap absorption coefficients were derived from a Drude model [9], while the InGaAs p-type absorption coefficients were taken from Wilt *et al.* [2]. The degree of specularity is nearly as important as the reflectivity, as is discussed later, so, the percent specularity is also included.

MONTE CARLO RESULTS

A series of RACER-X calculations was performed based on a 1750°F radiator and the aforementioned material and geometric description. The radiator was assumed to be a blackbody with a constant 90% emissivity and a diffuse angular emission distribution, (Lambertian). Table 2 shows some selected results from the RACER-X calculations as a function of the geometric and material properties.

The results of most interest are below bandgap reflectively and spectral utilization since they provide useful insight into the performance of the TPV system. Of course, the ideal situation is to have 100% of the below bandgap photons returned to the radiator. Notice that small changes in the below bandgap reflectivity result in large changes to the spectral utilization.

Component	Geometry / Properties
Grid Lines (Gold)	5 μm wide, 4 μm high spaced 100 μm back-to-front contact distance 95% reflective on exposed sides, 0-90% reflective ohmic contact (100% specular)
Back Contact Trench	16 μm wide, 2.1 μm deep dove-tail etch 30° from normal (some models use perpendicular)
Isolation Trench	30 μm wide, 6.25 μm deep v-notch etch 30° from normal (some models use perpendicular)
InP substrate	500 μm thick, $\alpha_{<Eg} = 0$ cm^{-1}, n = 3.6
N-type Graded Layers N < 1e15 cm^{-3}	t = 2.0 μm, $\alpha_{<Eg} = 0$ cm^{-1}, n = 3.6
Lateral Conduction Layer N-type 0.6 eV InGaAs N = 2e18 cm^{-3}	t = 0.15 μm, $\alpha_{<Eg} = 580$ cm^{-1}, n = 3.6
Base N-type 0.6 eV InGaAs N = 1e17 cm^{-3}	t = 2.5 μm, $\alpha_{<Eg} = 0$ cm^{-1}, n = 3.6
Emitter P-type InGaAs P = 2e19 cm^{-3}	t = 0.1 μm, $\alpha_{<Eg} = 2560$ cm^{-1}, n = 3.6
Anti-reflection coating	3 layer MgF$_2$/ZnS/MgF$_2$ - wavelength and angular dependent reflectivity generated with commercially available TFCalc^{TM}.
Radiator (Poco-Graphite)	10% reflectivity (100% diffuse)

TABLE 1 - Model Geometry and Optical Properties (t = thickness, $\alpha_{<Eg}$ = spectrum weighted below bandgap absorption coefficient, n = real part of index of refraction).

These results demonstrate a number of interesting features for an incident photon flux with a Lambertian distribution, including:

- High impact of either front (grids and trenches) or back surface non-specularly.
- Minimal impact of trench geometry (i.e. perpendicular vs. dove tail / V-notch) (cases 9, 12).
- Minimal impact of ohmic contact reflectivity on spectral utilization (cases 9-11).

TABLE 2 - Racer-X Results with 0.6 eV Interdigitated MIM and a 1750 F Radiator.

CASE #	Geometric Schematic	BSR Characteristic	Grid Property	Photon Incident Angular Distribution	Spectral Utilization/ Below Bandgap Refletiviity
1	No surface features	100% reflectivity 100% specular	N/A	Perpendicular to surface	77.9% / 95.4 %
2	No surface features	100% reflectivity 100% specular	N/A	Lambertian	77.2% / 95.3%
3	No surface features	98% reflectivity 98% specular	N/A	Lambertian	71.8% / 93.3%
4	Grids only	100% reflectivity 100% specular	95% reflectivity 80% ohmic contact reflectivity	Lambertian	72.1% / 94%
5	Grids only	100% reflectivity 100% specular	95% reflectivity 90% ohmic contact reflectivity	Lambertian	73.4% / 94.5%
6	Perpendicular Trenches Only	100% reflectivity 100% specular	N/A	Lambertian	68.3% / 93.7%
7	Perpendicular Trenches Only	100% reflectivity 100% specular	N/A	Perpendicular to surface	75.2% / 95.6%

CASE #	Geometric Schematic	BSR Characteristic	Grid Property	Photon Incident Angular Distribution	Spectral Utilization/ Below Bandgap Refletiviity
8	Perpendicular Trenches Only	98% reflectivity 98% specular	N/A	Lambertian	55.4% / 88.4%
9	Perpendicular Trenches & Grids	100% reflectivity 100% specular	95% reflectivity 80% ohmic contact reflectivity	Lambertian	60.3% / 90.9%
10	Perpendicular Trenches & Grids	100% reflectivity 100% specular	95% reflectivity 90% ohmic contact reflectivity	Lambertian	60.5% / 91.0%
11	Perpendicular Trenches & Grids	100% reflectivity 100% specular	95% reflectivity 0% ohmic contact reflectivity	Lambertian	59.6% / 90.8%
12	Dove-tail/V-notch Trenches & Grids	100% reflectivity 100% specular	95% reflectivity 80% ohmic contact reflectivity	Lambertian	59.8% / 90.8%
13	Dove-tail/V-notch Trenches & Grids	98% reflectivity 98% specular	95% reflectivity 80% ohmic contact reflectivity	Lambertian	55.0% / 88.5%

TABLE 2 - Racer-X Results with 0.6 eV Interdigitated MIM and a 1750 F Radiator.

ANALYTIC RESULTS

The RACER-X results show that the spectral utilization is surprisingly sensitive to trenches, grids, and BSR characteristics including specularity. As a consistency check of the results, some analytic calculations were performed.

Spectral Utilization Upper Bounds

The key metric for photonic economy is the spectral utilization which is given by,

$$Spectral\ Utilization = \frac{\sum_{i=1}^{N_{diodes}} \int_E EP_i(E)dE}{\sum_{i=1}^{N_{diodes}} \int_0^\infty EP_i(E)dE + \sum_{j=1}^{N_{inactive}} \int_0^\infty EP_j(E)dE} \quad (1)$$

where $P_i(E)dE$ probability of a photon with energy in dE about E being absorbed in region i. N_{diodes} number of diodes, and $N_{inactive}$ number of inactive areas (e.g. busbars, LCL, BSR, etc.).

Thus, the spectral utilization is the above bandgap energy absorbed in the diodes divided by all the energy absorbed in the entire device. Another number that is often cited as a measure of the performance is the below bandgap reflectivity. The spectral utilization is related to the below bandgap reflectivity (R_{bg}), since we can assume that any photons that was not reflected must have been absorbed. Approximately 16% of the total energy of a 1750°F black body radiator is above the 0.6 eV bandgap energy. Assuming that all of the above bandgap energy is absorbed in a device, the spectral utilization at 1750°F is given by,

$$Spectral\ Utilization = \frac{0.16 \frac{A_{diode}}{A_{total}}}{0.84(1 - R_{bg}) + 0.16} \quad (2)$$

Figure 3 shows a plot of spectral utilization versus below bandgap reflectivity for the idealized case, where there is no inactive area and all the above bandgap energy is absorbed. These calculated spectral utilizations represent an upper bound since they are idealized. Notice that the slope of the curve is steep for high reflectivities. For example, a 95% below bandgap reflectivity results in a spectral utilization of 79%, while a reduction in reflectivity to 92% results in spectral utilization of 70%. As one might expect, it does not take much to change the reflectivity by a few percent.

The upper bound spectral utilizations are consistent with the RACER-X calculations in all cases. Although this adds credence to the RACER-X results, a more detailed analytic calculation is given below.

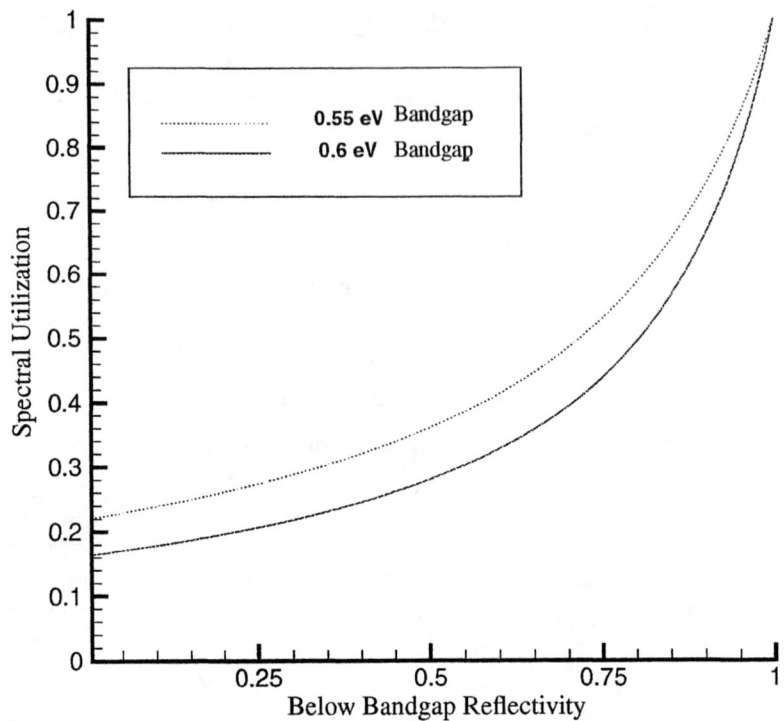

FIGURE 3: Upper bound spectral utilization versus below band-gap reflectivity.

Analytic Estimation of Spectral Utilization

Another reality check for RACER-X is an analytic estimation of the spectral utilization for the case where there are perpendicular-walled trenches without grids and a 1750°F spectrum of photons incident normal over the entire surface. This problem is simple enough to approximate analytically based on the geometry and the cross sections. Spectrum weighted absorption coefficients were calculated for above and below bandgap energies using the material properties listed in Table 1. These were used to calculate the "survival" probability which is simply the expected attenuation based on the absorption coefficient and material thickness [Table 1]. These survival probabilities are per pass. Given a single layer device mounted to a nonabsorbing insulator and back surface reflector with a single pass survival probability, p, yields a net return of, $(1-p) + (1-p)p$, after two passes.

For this analytic calculation, there are no grid lines and the trench walls are perpendicular. The trenches consume approximately 10% of the surface area so the spectral utilization can be approximated as,

$$\text{Spectral Utilization} = \frac{P_{live} P_{ABG} \overline{E}_{ABG} P_1}{P_{live} P_{ABG} \overline{E}_{ABG} P_1 + P_{dead} P_{ABG} \overline{E}_{ABG} P_2 + P_{live} P_{BBG} \overline{E}_{BBG} P_3 + P_{dead} P_{BBG} \overline{E}_{BBG} P_4} \quad (3)$$

where, P_{live} = the probability of a photon hitting "live" material, the diode, P_{dead} = the probability of a photon hitting "dead" material, the trench, P_{ABG} = the probability of a photon being born above the bandgap energy, P_{BBG} = the probability of a photon being born below the bandgap energy, $\overline{E}_{ABG}$ = the average energy of an above bandgap photon (at birth), $\overline{E}_{BBG}$ = the average energy of a below bandgap photon (at birth), P_1 = the net probability that an above bandgap photon that hits the diode is absorbed in live material, P_2 = the net probability that an above bandgap photon that hits the trench is absorbed in dead material, P_3 = the net probability that a below bandgap photon that hits the diode is absorbed in live material, and P_4 = the net probability that an above bandgap photon that hits the trench is absorbed in dead material.

The net absorption probabilities require knowledge of the survival probability, and two passes are always assumed. Any photon that hits the diode has the opportunity to go through the p-layer, the n-layer, the lateral conduction layer, and then reverse to go back through the layers. This leads to some complicated net absorption probabilities as given by,

$$P_1 = \left(1 - P_P^{ABG}\right) + P_P^{ABG}\left(1 - P_N^{ABG}\right) + P_P^{ABG} P_N^{ABG}\left(1 - P_{lcl}^{ABG}\right)$$
$$+ P_P^{ABG} P_N^{ABG} P_{lcl}^{ABG}\left(1 - P_{lcl}^{ABG}\right) + P_P^{ABG} P_N^{ABG} P_{lcl}^{ABG} P_{lcl}^{ABG}\left(1 - P_N^{ABG}\right)$$
$$+ P_P^{ABG} P_N^{ABG} P_{lcl}^{ABG} P_{lcl}^{ABG} P_N^{ABG}\left(1 - P_P^{ABG}\right) \quad (4)$$

Other probabilities are derived in similar fashion, resulting in a set of probabilities listed in Table 3. Given these probabilities, the spectral utilization is calculated to be 74.8% with trenches. Eliminating the "trench" terms in the spectral utilization equation (terms 2 and 4 in the denominator of Eqn. 3) the spectral utilization is 77.1%. These numbers are in good agreement with the RACER-X calculations of 75.2% and 77.9%, respectively. The slight inconsistencies could be due to the anti-reflective coating that is modeled in the Monte Carlo code and not in the analytic calculation. Furthermore, the below bandgap reflectivity can be easily calculated as an area weighted fraction,

$$R_{BBG} = 0.9(1 - P_3) + 0.1(1 - P_4) = 94.9\% \quad (5)$$

which again is in good agreement with the RACER-X value of 95.3%.

Variable	Value	Variable	Value
P_1	98.6%	P_P^{ABG}	91.0%
P_2	24.83%	P_P^{BBG}	97.5%
P_3	5.50%	P_N^{ABG}	14.9%
P_4	1.792%	P_N^{BBG}	100%
P_{ABG}	6%	P_{live}	90%
P_{BBG}	94%	P_{dead}	10%
$\overline{E}_{ABG}/\overline{E}_{BBG}$	3.22		

TABLE 3 - Prababilities utilized in analytic calculation

The RACER-X calculations compare well to the analytic calculations for simple geometries. Analytic calculations for non-normal incident photons and/or more complex geometries are nearly impossible given the added complexity of refraction and light trapping.

ANALYSIS OF RESULTS

Refraction and Light Trapping

Refraction in a device with a back surface reflector becomes an important issue because some photons can get sent off at an angle where they get internally reflected and cannot escape. This is analogous to light trapping in numerous optoelectronic devices: solar cells [10], light emitting diodes [11], and quantum well infrared photodetectors (QWIPs) [12]. The refracted angle, θ_2 (with respect to the surface normal), can be calculated based on the incident angle, θ_1, and the index of refraction in both materials, N_1 and N_2, respectively using Snell's law,

$$N_1 \sin\theta_1 = N_2 \sin\theta_2. \qquad (6)$$

The refracted angle can never be greater than $90°$ from the normal, which poses a problem when traveling from high index to low index. In this case, there are a range of incident angles that result in total internal reflection since refraction beyond $90°$ is impossible. The typical index of refraction for these TPV materials is 3.6, while that of air is 1.0. Photons that travel from the MIM device towards an air (vacuum) interface must hit the surface at an angle less than 16° from the surface normal to be refracted, otherwise they are reflected.

Non-specular reflection from the BSR reduces the overall reflectivity of a device because once a photon has an angle greater than 16° from the normal, it is internally reflected. Based on a Lambertian distribution for diffuse scattering, the probability of scattering within 16° of the surface normal is given by,

$$P_{0-16} = \int_0^{16} \sin\theta \cos\theta \, d\theta = 0.038. \tag{7}$$

Thus, only 3.8% of the diffusely scattered photons are within 16° of the surface normal.

Another way a photon can become internally reflected is by entering the MIM via a trench edge. Figure 4 shows a schematic of the incident and refracted ray when it hits a trench. Trenches act as a black absorber for photons since effectively all that enter, get absorbed.

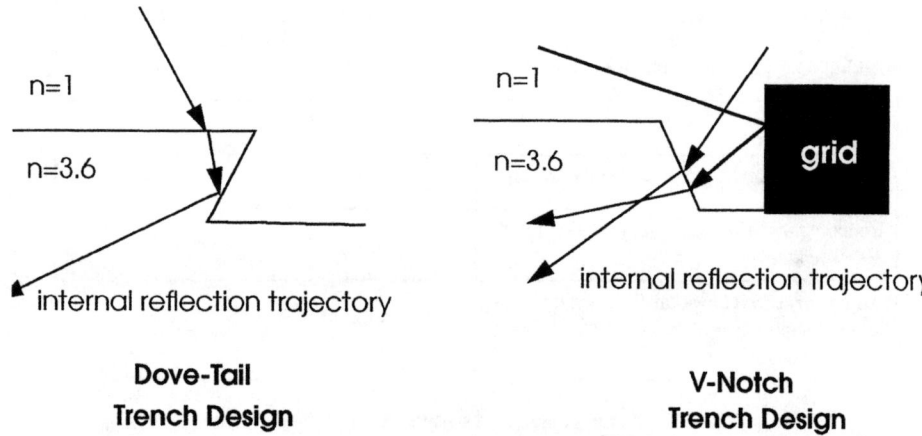

FIGURE 4 - Refraction through a trench

The calculations show that the trenches and grids have a large impact on the spectral utilization. Trenches reduce the spectral utilization in two ways. First, photons entering a trench edge have little hope of escaping due to refraction and subsequent internal reflection. Secondly, above bandgap photons that enter via the trench bottom are much more likely to be absorbed in the LCL than photons entering the diode. Along with the trench refraction issues, the BSR specularity and reflectivity are important. A non-specular scattering event from the BSR almost always guarantees a trapped photon which, again, greatly affects the spectral utilization.

Quantifying the Effect of Surface Features

RACER-X was used to determine the sensitivities to the various surface features and the results are summarized in Table 4. Trenches have the greatest negative impact on the spectral utilization. The effect of these trenches is magnified when coupled with grid lines. In the interdigitated MIM design that was analyzed, the grids in the back contact trench are taller than the trench itself, which tends to scatter more photons into the walls of the trench [Fig. 5]. Clearly, reducing the trench width will have a positive impact on the spectral utilization without adversely affecting the electrical properties of the device.

The calculations suggest that even with no trenches, the spectral utilization will be approximately 65% for a MIM with a 98% reflective, 98% specular BSR, and grid lines. Thus ensuring that a 100% specular back surface is critical.

MIM/BSR Feature (the starting point for all differences is a featureless MIMs with a perfect BSR)	% Reduction on Spectral Utilization	% Reduction on Below Band-Gap Reflectivity
Adding perpendicular trenches	-11.5%	-1.7%
Adding Dove-tail and V-notch trenches	-12.4%	-2.0%
Adding grids	-4.9%	-1.0%
Adding grids and Dove-tail/V-notch trenches	-22.5%	-4.8%
Reducing the BSR specularity to 98% (still 100% reflective)	-7.0%	-2.2%
Adding grids and Dove-tail/V-notch trenches and reducing the BSR reflectivity and specularity to 98%	-28.7%	-7.2%

TABLE 4: Racer-X predicted sensitivities for Interdigitated MIM device with properties listed in Section II.

CONCLUSIONS

A detailed photonic analysis is necessary when optimizing a TPV device that utilizes a back surface reflector for spectral control, especially when dealing with a non-normally incident photon flux which is expected in prototypical environments. Surface features like trenches that occupy a small fraction of the surface area can have an enormous impact on the device efficiency.

In addition, spectral characterization of fully fabricated MIMs is required which include multiple angle of incidence reflection measurements and total power absorption measurements. As this analysis demonstrated, measurement of near-normal incidence on unprocessed MIM structures, over-estimates spectral utilization in a cavity environment. These measurements are difficult to perform accurately due to the non-specular nature of a processed MIM and the high accuracy required for values of reflection greater than 90%. Thus, in-cavity heat absorption measurements may prove to be a more useful measure of spectral utilization [1].

This analysis has also highlighted some improvements that may be made to existing MIM designs that will help mitigate the reduction in spectral utilization of a fully processed MIM, these include:

- Minimize trench area. This will require tighter lithographic tolerance to prevent diode shorts.
- Plating of metallization to mushroom over trenches.
- Increase base thickness to increase above bandgap absorption in the active diode and not in the lateral conduction layer. This will occur at the expense of a decreased open circuit voltage.
- Increase the bandgap of the lateral conduction layer to prevent absorption of above bandgap photons.
- Lower the bandgap of the active layers or increase radiator temperature. This will decrease the sensitivity of the spectral utilization by decreasing the fraction of below bandgap photons.
- Ensure a highly specular back surface reflector. A 2% reduction in specularity reduced spectral utilization by 7%.
- Utilize a front surface interference filter to reflect near bandedge (2-5 micron) below-bandgap photons.

REFERENCES

[1] C.K., Gethers, Ballinger, C.T., Postlethwait, M.A., DePoy, D.M., and Baldasaro, P.B., "TPV Efficiency Predictions and Measurements for a Closed Cavity Geometry", The Third NREL Thermophotovoltaic Conference, AIP Conference Proceeding, **401**, 471 (1997).

[2] D.M. Wilt, N.S. Fatemi, P.P. Jenkins, R.W. Hoffman, G.A. Landis, and R.K. Jain, "Monolithically Interconnected InGaAs TPV Module Development," 25th PVSC, 43 (1996).

[3] D.M. Wilt, N.S. Fatemi, P.P. Jenkins, V.G. Weizer, R.W. Hoffman, R.K. Jain, C.S. Murray, and D.R. Riley, "Electrical and Optical Performance Characteristics of 0.74 eV p/n InGaAs Monolithic Interconnected Modules," The Third NREL Thermophotovoltaic Conference, AIP Conference Proceeding, **401**, 237 (1997).

[4] S. Wojtczuk, "Comparison of 0.55 eV InGaAs Single Junctions vs. Multijunction TPV Technology," The Third NREL Thermophotovoltaic Conference, AIP Conference Proceeding, **401**, 205 (1997).

[5] G.W. Charache, D.M. DePoy, P.F. Baldasaro and B.C. Campbell, "Thermophotovoltaic Devices Utilizing a Back Surface Reflector", The Second NREL Thermophotovoltaic Conference, AIP Confer-

ence Proceeding, **358**, 339 (1996).

[6] J.S. Ward, A. Duda, M.W. Wanlass, J.J. Carapella, Z. Wu, R.J. Matson, T.J. Coutts and T. Moriarty, "A Novel Design for Monolithically Interconnected Modules (MIMs) for Thermophotovoltaic Applications," The Third NREL Thermophotovoltaic Conference, AIP Conference Proceeding, **401**, 227 (1997).

[7] M.B. Clevenger, C.S. Murray, and D.R. Riley, "Spectral Utilization in Thermophotovoltaic Devices," Presented at the Fall Meeting of the MRS Conference (1997).

[8] J. Borrego, M. Zierak and G. Charache, "Parameter Extraction for TPV Cell Development," The First NREL Thermophotovoltaic Conference, AIP Conference Proceeding, **321**, 371 (1994).

[9] G.W. Charache, D.M. Depoy, J.L. Egley, R.J. Dziendziel, M.J. Freeman, P.F. Baldasaro, B.C. Campbell, P.R. Sharps, M.L. Timmons, R.E. Fahey, K. Zhang, and J.M. Borrego, "Electrical and Optical Properties of Degenerately-Doped N-type InGaAs, The Third NREL Thermophotovoltaic Conference, AIP Conference Proceeding, **401**, 215 (1997).

[10] C.B. Honsberg and A.M. Barnett, "Light Trapping in Thin Film GaAs Solar Cells, 21st PVSC, 262 (1991).

[11] D.A. Vanderwater, I.H. Tan, G.E. Hofler, D.C. Defevere and F.A. Kish, "High Brightness AlGaInP Light Emitting Diodes," Proc. IEEE **85**, 1752 (1997).

[12] C.J. Chen, K.K. Choi, W.H. Chang and D.C. Tsui, "Two-color Corrugated Quantum Well Infrared Photodetector for Remote Temperature Sensing," Appl. Phys. Lett., **72**, 7 (1998).

SESSION 3: RADIATORS AND OPTICAL CONTROL

HIGH TEMPERATURE OPTICAL PROPERTIES OF THERMOPHOTOVOLTAIC EMITTER COMPONENTS

Daniel E. Pierce[1] and Guido Guazzoni[2]

[1]William Paterson University, Wayne New Jersey
[2]U.S. Army CECOM, Ft. Monmouth, New Jersey

Abstract. The hemispherical transmissivity, reflectivity, and emissivity were determined at high temperatures for various types of quartz glass and borosilicate glass. From the information, the behavior of these materials in filtering 1300°C blackbody radiation was calculated. The results show, that depending on the glass, the losses of radiation usable for photovoltaic conversion (<1.8μm) can be minimized, while filtering of unusable radiation (>1.8μm) maximized. In addition, high temperature optical parameters were determined for sapphire, translucent polycrystalline sapphire (TPS), and alumina. This information was used to consider these materials as substrates for selective emitters. Both sapphire and TPS have low emissivity, at high temperatures (>1000°C), extending to >4μm. Combining this property with the application of selective emitters, in principle, results in a more efficient emitter, limiting the generation of unusable radiation. Optical properties of erbium oxide bulk and film materials were measured and combined with literature values of absorptivity to determine what thickness of rare earth oxide would make an effective emitter.

INTRODUCTION:

The purpose of this investigation was twofold. First, the high temperature optical properties of uncoated bulk filter elements were determined. In a thermophotovoltaic (TPV) system, these elements will be placed between an emitter and photovoltaic (PV) cells. Commonly, quartz glass is used for this purpose. There exist however, many kinds of quartz and optical glasses, each with its own optical characteristics. The materials investigated included four types of quartz glass and one of optical glass. Spectroscopic measurements were made at high temperature and used in calculating the glasses effectiveness as filter elements in transmitting or re-emitting radiation useful in a PV power conversion. The second purpose, was to measure the high temperature optical properties of several forms of alumina (Al_2O_3), which could serve as emitter components such as the substrate or host for rare earth emitters.[1,2] The information was used to conceptually design an emitter using translucent polycrystalline sapphire as a substrate.

When a piece of quartz glass or other material transparent to the near infrared is placed between an emitter and PV cell, the material also acts as a heat shield. The material prevents direct contact of hot combustion gases and reduces heat transferred via thermal convection to the PV cells. In addition, the material filters radiation primarily through absorption of the infrared radiation outside its transmission window. The filter becomes hot and will re-emit radiation characteristic of the material's temperature and emissive nature. Temperatures on the order of 700°C to 800°C can be expected for filters in close proximity to an emitter operating at high temperature (e.g. 1300°C). It is widely known that quartz has high transmission and very little absorption in the visible and near-infrared regions. The specific details however, of the transmissivity, emissivity, reflectivity, and in particular, how high temperature effects these properties, are not generally considered.

In the schematic diagram in Figure(1), the emitter is shown at T_1 and the filter at T_2. For the sake of discussion the T_1 will be 1300°C, while the filter temperature, T_2, will be variable. The emitter has a total radiance of $R_1(T_1)$, with the spectral components $R_1(\lambda,T_1)$, which are defined as $R_1(\lambda,T_1) = R_{bb}(\lambda,T_1) * \varepsilon_1(\lambda,T_1)$, where R_{bb} indicates blackbody radiance and ε the emissivity. Whatever the nature of $R_1(\lambda,T_1)$, (i.e. SiC, rare earth oxide, etc.), the radiation will be filtered according to the materials spectral tansmissivity; $R_1(\lambda,T_1) * \tau_2(\lambda,T_2)$, and reflectivity; $R_1(\lambda,T_1) * \rho_2(\lambda,T_2)$. When the filter comes to equilibrium at some temperature, it will emit radiation hemispherically, in both directions a quantity $R_2(\lambda,T_2)$, which is defined as $R_2(\lambda,T_2) = R_{bb}(\lambda,T_2) * \varepsilon_2(\lambda,T_2)$. The total radiance that is projected in the direction of the photovoltaic cells will then be,

$$R_{PV} = \sum_\lambda R_{bb}(\lambda,T_1) * \varepsilon_1(\lambda,T_1) * \tau_2(\lambda,T_2) + R_{bb}(\lambda,T_2) * \varepsilon_2(\lambda,T_2) \quad \text{Eq.(1)}$$

EXPERIMENTAL:

The apparatus and technique for measuring high temperature optical properties has been discussed elsewhere.[3,4,5] The transmission and reflection spectra are determined using the radiation of a blackbody at 900°C. The blackbody and sample are placed at the two foci of a large parabolic mirror so that the blackbody radiation is hemispherically focussed onto the sample surface. The sample itself reflects and intrinsically emits radiation from the front surface, which is directed into a Fourier transform infrared (FTIR) spectrometer, and spectra were taken from 0.8μm to 20μm. The two components (reflection and emission) are separated and measured by chopping the blackbody source. Likewise, the radiation from the back of the sample is composed of both transmitted and emitted, which again is separated by chopping the blackbody source. The sample is heated by directing the hot gases of a propane torch to the backside of the sample. By adjusting the flame intensity and nearness to the sample, the temperature could be controlled over a wide range (limited here to about 1000°C). The sample spectral emissivity is determined by Eq.(2), while the sample temperature determined by Eq.(3). By dividing the measured spectral radiance by the measured spectral emissivity (in Eq.(2)), the temperature is calculated by adjusting the variable T_2, in the right hand side of Eq.(3), until the best fit is obtained.

$$\varepsilon_2(\lambda,T_2) = 1 - (\tau_2(\lambda,T_2) + \rho_2(\lambda,T_2)) \qquad \text{Eq.(2)}$$

$$\frac{R_2(\lambda,T_2)}{\varepsilon_2(\lambda,T_2)} = R_{bb}(\lambda,T_2) \qquad \text{Eq.(3)}$$

Samples of four types of quartz glass and one of borosilicate glass were obtained commerically and all were 1.00mm thick double side polished opitical windows. Sapphire samples were 2.00mm thick double side polished optical windows. Translucent polycrystalline sapphire[6] (TPS) was obtained as double side polished windows and unpolished smooth surface windows, both 1.00mm thick. Tubular TPS material, used commercially in sodium vapor lamp manufacture was also obtained which had very smooth (polished-like) inner and outer surfaces. The alumina material was rough polished, 1.00mm thick opaque plate material. Erbium oxide pellet was 2.00mm thick and formed by pressing and sintering pure finely powdered Er_2O_3. Erbium oxide films were deposited by plasma spray onto borosilicate glass substrates, with varying plasma power levels and oxygen gas flows.

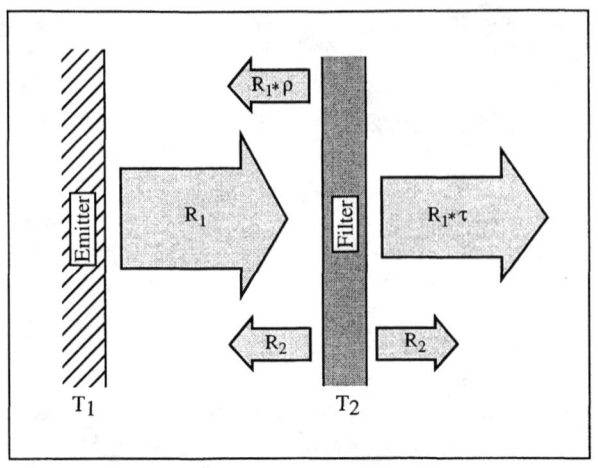

FIGURE 1. Interactions between emitter and filter

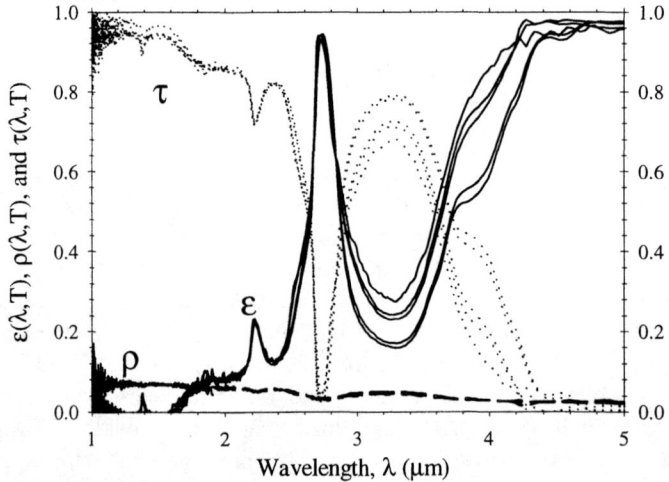

FIGURE 2. Transmissivity, emissivity and reflectivity of quartz glass sample-SA, at the temperatures 180, 270, 550, 640, 820°C

RESULTS AND DISCUSSION:

In Figure(2), the $\tau(\lambda,T)$, $\rho(\lambda,T)$, and $\varepsilon(\lambda,T)$ values of the first type of quartz glass is shown for five different temperatures, 180, 270, 550, 640, and 820°C. This sample type is identified as SA. The reflectivity is low and not significantly affected by temperatures in the 0.5-5μm range. The transmission is high up to about 1.7μm, decreasing gradually after this point and then more rapidly to a transmission minimum, occurring at slightly above 2.7μm. This minimum is due to the presence of water in the material in the form of hydroxyls.[7,8] Within the bulk of the quartz glass, water reacts with Si-O-Si linkages to form Si-OH + HO-Si, as well as other related forms, all having characteristic absorption[7,8]. The region leading up to the 2.7μm band is not strongly affected by temperature, however, after this point, transmission is strongly affected. The transmission window decreases with temperature while emission increases.

Of the four quartz glasses studied, three showed very different optical properties. The emissivity at 800°C of these three along with the borosilicate glass are shown in Figure(3). The borosilicate glass, identified as BSG, shows the characteristic hydroxyl band previously discussed for the SA sample, which is shown again in Figure(3). The sample SA is the best representative of commonly available quartz glass. The other quartz glasses, SI, and SB emitted no measurable radiation until 3.0μm and 3.5μm respectively, at 800°C. These materials are presumably manufactured in the absence of water such as in the formation of $SiO_2(s)$ by reaction of $SiF_4(g)$ and $O_2(g)$. Not shown, is sample SC, which appears much like SI, but with a small narrow peak at 2.7μm. The presence of the -OH bands in some samples is of concern in the near infrared region, especially when considering the possibility of high performance PV cells working out beyond 2μm.

The measured data is useful in understanding the net radiation from an emitter under various filtering conditions. The effects of using either SA or SI type quartz to filter a blackbody radiating at 1300°C is considered in Figure(4). In the figure, the blackbody radiation is given in the uppermost curve. The next lower curve is the same radiation but after filtering through a SI filter at 800°C. The remaining two lower curves are the result of filtering the blackbody radiation through a single SA filter at 800°C (next to lowest curve), and filtering through the combination of two SA filters, one at 800°C and the other at 450°C (lowest curve). A comparison of SI with SA in the near infrared up to about 1.6μm shows them to be equivalent After 1.6μm, the SI is superior due to higher transmission. The total

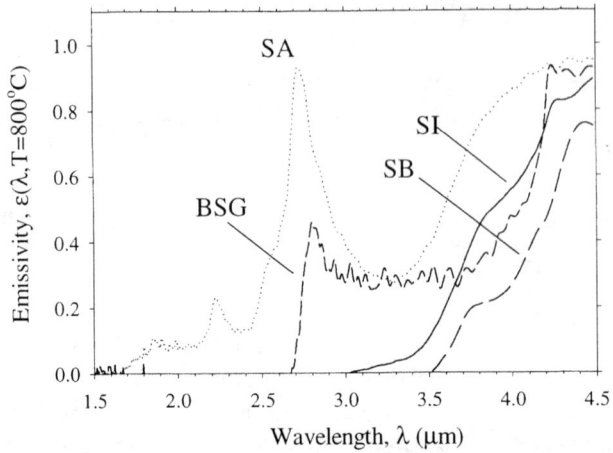

FIGURE 3. Emissivity of three types of quartz filters (SA, SB, and SI) and borosilicate glass (BSG), all at 800°C.

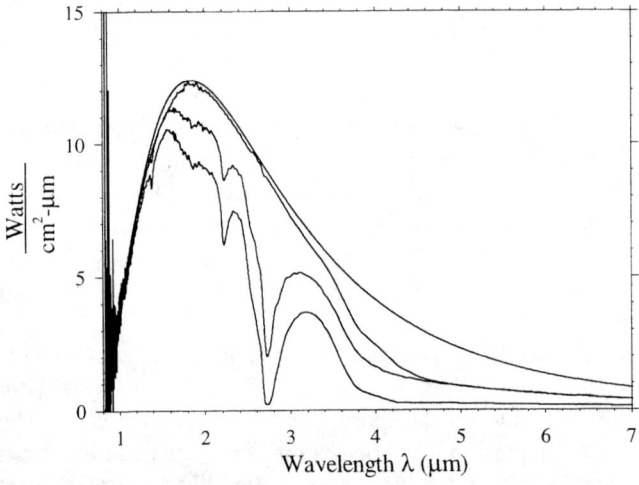

FIGURE 4. Spectra in descending order represent, 1300°C blackbody radiation, after one SI filter (800°C), one SA filter (800°C), and two SA filters (800°C and 450°C)

radiation from the blackbody in the 0.8-20μm range of the experiment, was 34.7 Watts/cm^2. This could be divided into 8.1 Watts/cm^2 and 26.6 Watts/cm^2, for the regions 0.8-1.8μm and 1.8-20μm respectively. The net radiation after filtering through SI is 7.8 Watts/cm^2 and 20.8 Watts/cm^2 in the same regions, and for SA, 7.5 Watts/cm^2 and 16.2 Watts/cm^2. In transmission of radiation that is usable for PV conversion, SI is slightly better than SA. In the region of unusable energy however, the higher transmission of SI means a greater radiation load to the PV cells and from this point of view SA is better. If we then consider the effect of having two filters present, one at 800°C and the other at 450°C, the unusable energy to the cells is much reduced using two filters, while the usable energy is still high for both. The results are summarized in Table(1). In particular, the table shows that combination of two SA filters maintained 88% of the usable energy while the unusable was reduced to only 38% of the blackbody source.

TABLE 1

Blackbody/Filter#1 / Filter#2 1300°C / 800°C / 450°C	0.8-1.8μm		1.8-20μm	
	W/cm^2	R_{net}/R_{bb}	W/cm^2	R_{net}/R_{bb}
R_{bb}	8.1	1.00	26.6	1.00
R_{bb} /SI	7.8	.96	20.8	0.60
R_{bb} /SA	7.5	.93	16.2	0.47
R_{bb} /SI/SI	7.6	.94	17.6	0.51
R_{bb} /SA/SA	7.1	.88	10.1	0.38

Another important application of high temperature optical properties is in the emitter itself. Here the properties of translucent polycrystalline sapphire (TPS) are considered. This material is also known as translucent polycrystalline alumina as well as by various trade names[6]. It is closely related to both sapphire; a single crystal material, and high purity alumina; a fine-grained ceramic material. All three of these materials have the composition Al_2O_3. Each has very good high temperature stability with melting points around 2000°C, and high resistance to chemical attack. Sapphire is commonly used as an optical window due to its high transmission range extending from the ultraviolet to the mid- infrared. Alumina on the other hand has poor transmission, due largely to light scattering from the microscopic Al_2O_3 particles, which results in a highly diffuse reflective material.

As a single crystal, sapphire is difficult to process into useful shapes and sizes. However, in recent years sapphire tubes of 1.5 in. diameter and greater, have been made by the "edge-defined" growth method[9]. Tubes and large plates of sapphire are commercially available but expensive. Alumina on the other hand can be manufactured into any number of shapes and sizes inexpensively. With respect to manufacturing, TPS bears resemblance to alumina and can be processed into a variety of shapes with some difficulty. In particular, TPS requires a final high temperature (>1800°C) annealing step in a reducing environment which fuses the material, making it translucent.

An application that serves to illustrate the unique thermal, optical, and chemical properties of TPS, is the high pressure sodium vapor lamps used in street and highway lighting.[6] TPS tubes serve as a sealed container for high temperature sodium vapor, transmitting the radiation emitted from within. The invention of TPS was a key step in the development of these lamps. In principle, TPS can be made relatively inexpensively. Geometries with dimensions different from that needed for sodium vapor lamps however, are not easily obtained.

A comparison of the transmissivity spectra of sapphire, TPS, and alumina is made in Figure(5). Sapphire and TPS are transparent out to about 7μm. The TPS transmission is rather diffuse and is accompanied by as much as a 15% reflectivity. The hemispherical measurement technique used is well suited for accurately determining diffuse properties. It should also be noted, that though Figure(5) suggests TPS is more transmissive than sapphire between 5.2μm and 7.0μm, this is due mainly to the thicker sapphire sample thickness (2mm versus 1mm). Alumina exhibits diffuse transmissivity out to about 5μm accompanied by substantial diffuse reflectivity.

Figure(6) shows the emissivity behavior of sapphire and TPS at about 1000°C. Both curves reveal a peak at about 4.2μm due to the emission of very hot CO_2 produced in the intense propane flame directed onto the sample. At high temperature, sapphire and TPV retain their characteristic transmissivity to at least 4μm. Sapphire has no measured emissivity out to 4μm, while TPS has an emissivity of 0.15, characteristic of a greybody. This greybody behavior is presumed to be the result of the formation of optical cavities within the grain structure of the TPS material. As a result of their low emittance, neither sapphire nor TPS have any value as emitters themselves, however they can serve as substrates for emitting

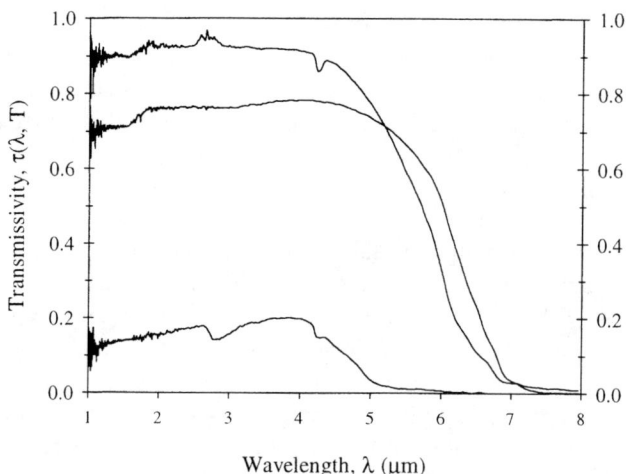

FIGURE 5. Transmissivity of three types of Al_2O_3 material at 1000°C. In descending order (at 1μm), sapphire (2.0mm thick), TPS (1.0mm), and alumina (1.0mm).

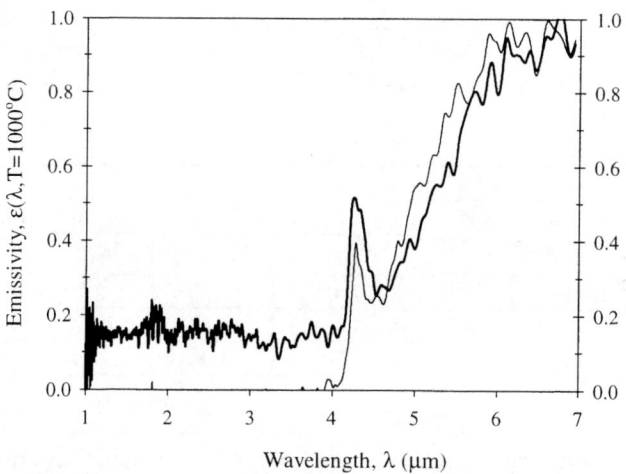

FIGURE 6. Comparison of high temperature emissivity at 1000°C. Sapphire is shown as the lighter line and TPS as the darker. (Note: CO_2 emission is evident at 4.2μm)

coatings such as rare earth oxide films. Their use as substrates would have the advantage of minimizing emission of unusable radiation while providing a mechanically strong, thermally and chemically stable substrate for a selective emitter. Previous experiments with selective emitter coatings deposited on silicon carbide[10], found that after repeated thermal cycles, the coating failed. The chemical and physical properties of Al_2O_3 bear a much greater resemblance to the rare earth oxides (of the formula M_2O_3), than to SiC or SiO_2. This suggests that films of greater stability toward thermal cycling could be made using sapphire or TPS as substrates.

In Table(2), the radiance of a 1300°C blackbody is compared with that of both sapphire and TPS at the same temperature. The blackbody produces 34.7 W/cm^2 in the 0.8μm to 20μm range, with components; 7.1W/cm^2 in the 0.8-1.8μm range and 27.5W/cm^2 above this. Using 1.8μm as an upper wavelength limit for PV power conversion, sapphire at 1300°C, emits 3.8W/cm^2 unusable and no radiation usable for photovoltaic conversion. TPS on the other hand emits 7.7 W/cm^2 unusable and 1.2 W/cm^2 usable. The main consideration here is minimizing the unusable radiation, since the usable radiation can be controlled by application of selective emitters. For example using TPS rather than a blackbody emitter reduces the unused energy to a fraction (0.28) of the backbody, a savings of 72% (almost 20 W/cm^2). Sapphire reduces this energy to a small fraction (0.14), a savings of 86%. These energy savings should translate into increased fuel efficiency. Decreased losses should also result in increased temperature.

TABLE 2				
Emitter 1300°C	0.8-1.8μm		1.8-20μm	
	W/cm^2	R_{net}/R_{bb}	W/cm^2	R_{net}/R_{bb}
Blackbody	7.1	1.00	27.5	1.00
TPS	1.2	0.17	7.7	0.28
Sapphire	0	0.00	3.8	0.14

Early experiments with self-supporting monolithic erbium oxide structures, found cracking within a limited number of thermal cycles at temperature typical of a practical thermophotovoltaic system.[10] An emitter based on a coated TPS substrate is schematically represented in Figure(7). The selective emitter film is shown on the flame side at the temperature T_2, while the substrate temperature vary across its thickness reaching T_3 at the exiting side. Here, TPS or another suitable substrate would support the rare earth oxide. The film itself could be a mixture of several rare earth oxides co-deposited or deposited in layers. Routes of application include plasma spray or sol-gel, or rare earth ions might be incorporated directly within the substrate. The film emissivity, which depends on thickness reaches full saturation only at impracticably large thickness. Certain peaks within a selective emission band are stronger emitters and these approach saturation at more practical thickness. To demonstrate this phenomenon, the selective emission of a 2mm thick erbium oxide pellet is compared with that of 120μm and 85μm thick plasma deposited films are shown in Figure(8). Clearly the peak at 1.53μm is approaching a saturation value much faster than the peak for example, at 1.64μm or 1.40μm. From the data, 100μm may be set (somewhat arbitrarily), as a useful erbium oxide film thickness.

An attempt to estimate useful rare earth film thickness was made by considering the molar absorptivity of the various rare earth ions in solution.[11] In solution, the rare earth ion is surrounded by a coordination sphere of oxygen in water molecules. This is not unlike the oxygen coordination experienced by ions in an oxide. The molar absorptivity of Er^{3+} measured in solution had a peak value $e(\lambda \approx 1.5\mu m) = 2.1$ L/mole-cm. At roughly the same wavelength, Pr^{3+} has $e(\lambda \approx 1.5\mu m) = 4.7$ L/mole-cm. From these absorptivity values (e), the oxide density (d), formula weigh t (FW), film thickness (t), and stoichiometric coefficient of the rare earth (for M_2O_3 this value is 2), the absorbed light can be determined at a given wavelength (assuming reflection is small). In Eq.(4), transmissivity is first calculated, where R_o is the incident radiation. From the transmission, the absorptivity (a) and emissivity (ε) can be estimated (assuming low reflectivity).

$$\tau(\lambda,T) = \frac{R(\lambda,T)}{R_o} = 10^{-e*t*c} = 10^{-e*t*1000*(2d/FW)} \qquad \text{Eq.(4)}$$

$$\varepsilon(\lambda,T) \approx a(\lambda,T) \approx 1 - \tau(\lambda,T) \qquad \text{Eq.(5)}$$

From the data, it appears that Pr^{3+} is a much better candidate than Er^{3+}, based on its higher absorptivity. The Pr^{3+} ion also has stronger and wider absorption band than Er^{3+}, which in principle translates to greater usable emission in the solid state. The results are summarized in Table(3). Aside from the possible advantages of Pr_2O_3, the stability of Pr_2O_3 is of

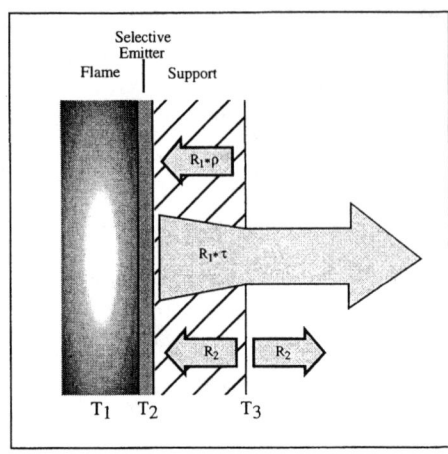

FIGURE 7. Drawing of thermophotovoltaic emitter using an infrared transparent material (e.g. TPS) as a substrate for selective emitter film

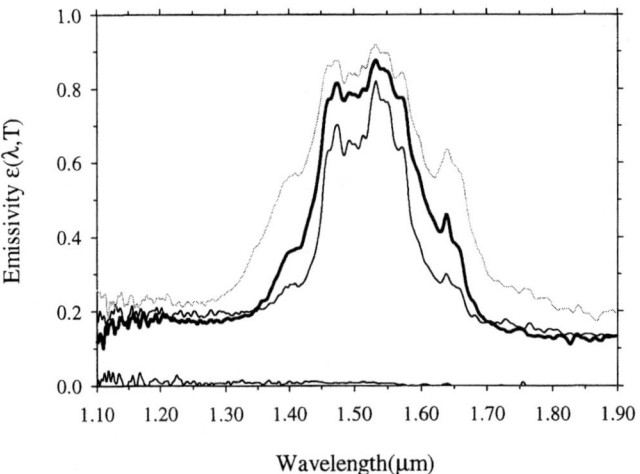

FIGURE 8. Emissivity of erbium oxide pellet (2.0mm thick, upper line), compared with plasma spray deposited erbium oxide films of approximately 120μm and 85μm thickness (next two lower line). The substrate itself is shown at the very bottom near zero.

concern and little appears to be known about the material. Another possibility is to incorporate Pr^{3+} ion into an infrared transparent material such as TPS or translucent polycrystalline yttria $(Y_2O_3)^{12}$ in order to stabilize the ion. If the equivalent of a 50μm film of Pr_2O_3 were contained within a 2.0mm thick substrate it would amount to 2% by volume of the material. It seems reasonable to expect that significant Pr^{3+} ion could be substituted for Al^{3+} ion in Al_2O_3 or for Y^{3+} ion in Y_2O_3. These possibilities can only be considered further through a more detailed study.

TABLE 3				
Rare Earth	$e(\lambda_{peak})$ L/mole-cm	Absorptivity at given film thickness		
		t=20μm	t=50μm	t=100μm
Er^{3+}	2.1	0.35	0.66	0.89
Pr^{3+}	4.7	0.60	0.90	0.99

CONCLUSIONS:

The high temperature optical properties of various glass materials was measured and the information used in calculating their efficiency as filters which transmit usable and block unusable radiation from an emitter. The use of two quartz shields was found to reduce the unusable radiation load to the PV cells from a 1300°C blackbody to a fraction, (0.38) of that emitted by the blackbody itself, while maintaining a high fraction (0.88) of the radiation usable for photovoltaic conversion. The property was very dependent on the type of material and the presence of hydroxyl in the glass.

Translucent polycrystalline sapphire (TPS) was shown to have a low emission of radiation and a high infrared transparency out to >4μm, even at high temperature. With a blackbody and a TPA substrate both at 1300°C, the TPA emits only 14% as much radiation as the blackbody over the 1.8-20μm wavelength region. TPA has strength and chemical stability at high temperature and can be fabricated into emitter structures relatively easily (e.g. tube structures). TPS and other infrared transparent materials such as translucent polycrystalline yttria can be used as substrates or hosts for rare earth selective emitters. The net effect would be to minimize the unusable radiation load at the photovoltaic cell while providing a stable support for a selective emitter.

From measurements of erbium oxide bulk and thin film materials and from the literature information on praseodymium Pr^{3+} ion, the useful thickness of Er_2O_3 and Pr_2O_3 films was estimated to be about 100μm and 50μm respectively.

ACKNOWLEDGEMENTS:

The authors would like to express our gratitude to Professor Ravindra of New Jersey Institute of Technology and his team at the NJIT spectral emissometry facility, especially Dr. O.H.Gokce and M.Babladi for their assistance. The assistance of GTS for summer support and travel support for D.E.P. is also gratefully acknowledged.

REFERENCES:

(1) Guazzoni,G.E., Applied Spectroscopy **26**, 60-65, 1972
(2) Chubb,D.L. and Lowe,D.L., Journal of Applied Physics **74**, 5687-5692 , 1993
(3) Markam,J.R.,Kinsella,K.,Carangelo,R.M.,Brouillette,C.R.,Carangelo,M.D.,Best,P., and Solomon,P.R., Review of Scientific Instruments **64**, 2515-2522, (1993).
(4) Ravindra,N.M.,Abedrabbo,S.,Chen,W.,Tong,F.M.,Nanda,A.K.,and Speranza,T. IEEE Transactions on Semiconductor Manufacturing **11**, 30-39, 1998
(5) Abedrabbo, S., *Emissivity Measurements and Modelling of Silicon Related Materials and Related Structures,* Ph.D. Thesis, New Jersey Institute of Technology 1998.
(6) Burke,J.E., Material Research Society Bulletin **21**, 61-68, 1996.
(7) Brice,J.C. and Cole,A.M., Proc. 32nd Annual Symp. Freq.Control, 1-10, 1978.
(8) Sawyer,B., IEEE Trans. On Ultrasonics, Ferroelectrics, and Frequency Control 34, 558-565, 1987
(9) Schwab,M.L., Saphikon Inc., private communication.
(10) Guazzoni, G., USArmy CECOM, private communication
(11) Carnall,W.T. and P.R.Fields, Advances in Chem. Ser. **71**, 1967.
(12) Klocek, P., *Handbook of Infrared Optical Materials*, 355-356, Marcel Dekker, New York 1991.

Microstructured Tungsten Surfaces as Selective Emitters

A. Heinzel, V. Boerner, A. Gombert, V. Wittwer, J. Luther

Fraunhofer Institute for Solar Energy Systems ISE
Oltmannsstr. 5, 79100 Freiburg, Germany

Abstract. This paper presents the optical properties of periodic microstructured tungsten surfaces used as selective emitter materials. Theoretical calculations are done using rigorous diffraction theory, and the influence of the grating period and the grating depth on the absorptance of tungsten gratings is investigated. Microstructured tungsten surfaces are fabricated using a holographic process to produce a photoresist mask. Subsequently a reactive ion etching process or a chemical etching process transfers the structure into tungsten. The absorptance of the tungsten gratings is measured and compared with the theoretical predictions of the rigorous diffraction theory.

INTRODUCTION

To improve the efficiency of thermophotovoltaic systems, the emitted radiation spectrum has to be matched to the sensitivity of the photovoltaic cell. In our system we use a low bandgap GaSb photocell (bandgap energy at 0.73 eV, corresponding a wavelength of 1.7 µm). The long wavelength part of the radiation (above 1.7 µm) of a greybody emitter can not be converted by the PV-cell (Fig.1). Radiation management can be realised by using a bandpass filter or a selective emitter as radiation source. The most common selective emitter materials are rare earth metal oxides. The disadvantage of this type of selective emitter is the high evaporation rate and therefore a short lifetime.

In this paper we describe a new type of selective emitter that consists of a microstructured tungsten surface. Tungsten is very stable at high temperatures and the evaporation rate is only a few nanometers a year at an operation temperature of 1700K in vacuum. At high temperatures tungsten oxidizes in air. So it is necessary to operate in vacuum or in a protective gas environment.

The wavelength selective behaviour of the emittance of tungsten makes it favourable for use in combination with a low bandgap cell, especially with a GaSb photovoltaic cell. This selectivity is caused by a low emittance in the infrared region above a wavelength of 2 µm, which increases towards shorter wavelengths.

Optical resonances are well known from diffractive gratings. We use this effect to increase the emittance in a limited wavelength range. In this way the emittance of the

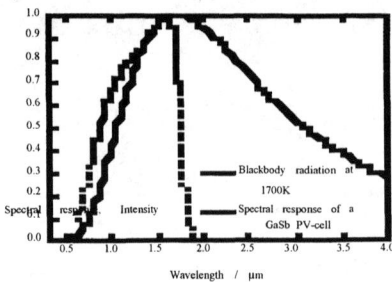
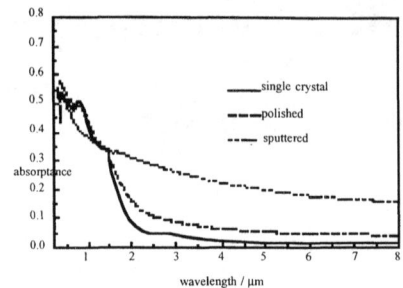

FIGURE 1. Normalized spectral response of a GaSb PV-cell in comparison to the blackbody radiation at 1700K

FIGURE 2. Absorptance of different tungsten samples

periodic microstructured tungsten can be adapted to the spectral response of the photovoltaic cell.

THEORETICAL CONSIDERATIONS

Fig. 2 shows the absorptance of different plane tungsten surfaces. A low absorptance means a low emittance according to Kirchhoff's law. The values of single crystal tungsten are taken from (1), whereas the absorptance of the polished and the sputtered tungsten was measured. In the far infrared wavelength range, high reflectivity and therefore low absorptance is caused by exciting the conducting electrons. At a wavelength of about 2.0 μm, interband transitions lead to absorption bands.

Because of the polycrystallinity, the conductivity of sputtered tungsten is weak and therefore the metal shows higher absorptance in the infrared region, compared to the other tungsten samples.

Considering that the emitter should have a low emittance at longer wavelengths than the bandgap of the photovoltaic cell, sputtered tungsten is not suitable as selective emitter material.

Microstructuring of the metal surface can enhance the emittance in a limited wavelength range. In order to calculate the optical properties of these metal surface gratings, especially in the resonance domain, a rigorous diffraction theory is needed. There are different methods; among these the rigorous coupled wave analysis (RCWA) is able to solve grating problems for gratings with arbitrary profiles and optical constants without changing the algorithm. In the RCWA the electromagnetic field is expanded into a fourier series in the modulated region. This leads to a matrix eigenvalue problem. The dimension of the matrix depends on the number of fourier coefficients. In our simulation we used an algorithm, that reduces the dimension of the matrix eigenvalue problem significantly (2, 3). A grating with a continous profile is divided into a stack of

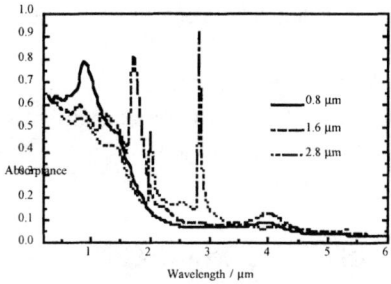 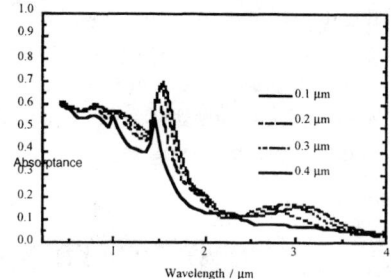

FIGURE 3. Absorptance of tungsten gratings with different periods, the grating depth is 0.3 µm

FIGURE 4. Absorptance of tungsten gratings with different depths, the grating period is 1.4 µm

binary gratings. A recursive matrix algorithm is used to match the electromagnetic fields at the interfaces between each slice (4). The disadvantage of the RCWA is a long computing time, especially for deep three-dimensional gratings of materials with high refractive indices.

In this section we will discuss the influence of the grating parameters, which are the period and depth, on the optical properties of tungsten surface relief gratings.

As basic input, we chose the refractive indices of single crystal tungsten. With the RCWA the spectral absorptance of a grating with a given profile is calculated. As mentioned, the emittance at a given wavelength is equal to the absorptance at the same wavelength.

In the calculations (Fig. 3, Fig. 4) the shape of the structure was described by periodically arranged tungsten cylinders on a plane tungsten surface, which is a reasonable approximation of real structures like that shown in fig. 6.

Fig. 3 shows the dependence of the absorptance of a microstructured tungsten surface on the grating period. The structure depth and the shape was fixed. If the first diffracted order can propagate in the metal, i. e. if the wavelength is in the range of the grating period, there is a strong enhancement of the absorptance and therefore of the emittance. The height and width of this resonance peak depends strongly on the wavelength-dependent conductivity which increases towards longer wavelengths. The higher the conductivity of the material is, the higher and narrower gets the resonance peak. For producing selective emitters, the best results are expected if the resonance is in the range of the onset of the interband transitions.

Variation of the structure depth shows, that a depth between 0.3 µm and 0.4 µm is sufficient (Fig. 4). In this calculation the period and shape of the grating was fixed. Increasing the depth does not lead to a remarkably higher absorptance peak. The second absorptance peak at a wavelength of about 3.0 µm is caused by interference effects between the upper side of the grating and the tungsten substrate. With a deeper grating structure it shifts towards longer wavelengths. When the grating profile becomes smoother, this peak is expected to decrease.

As the conclusion of this section, the best grating parameters for a periodic microstructured tungsten surface as selective emitter material in combination with a GaSb-cell are a period of 1.4 µm and a structure depth from 0.3 µm to 0.4 µm.

FABRICATION OF TUNGSTEN GRATINGS

We fabricated surface structures in tungsten by a two step process. In order to obtain an etching mask, the metal surface was first coated with a standard positive photoresist. The resist was exposed by a holographic process. Two coherent beams from an argon ion laser source interfered. A single wavelength of 363.8 nm was used for the exposure. In order to get grating structures, which are periodic in two directions, the sample was turned between two exposures. After development we obtained a photoresist mask. The structure was transfered into the metal using a reactive ion etching process with SF_6 as reactive gas, or a wet chemical etching process. The advantage of a holographic process is the possibility to structure even rough surfaces. With this method large areas can be structured homogenously.

EXPERIMENTAL RESULTS

In order to compare the theoretical predictions with measured data we first used sputtered tungsten films on glass substrates. These films have a smooth surface and therefore show weak scattering. The refractive indices of sputtered tungsten films were measured by ellipsometry (Fig. 5). The refractive indices differ strongly from the single

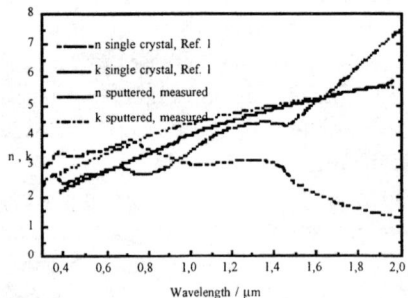

FIGURE 5. Real part (n) and imaginary part (k) of the optical constants of single crystal tungsten (from Ref. 1) and sputtered tungsten (measured)

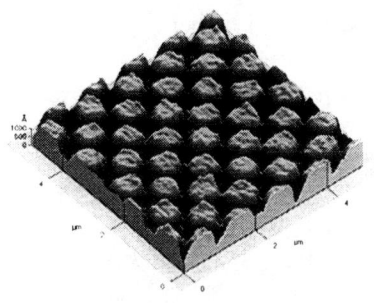

FIGURE 6. SPM image of a microstructured sputtered tungsten film

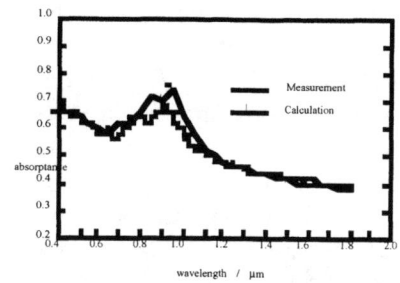

FIGURE 7. Measured and calculated absorptance of the tungsten sample of Fig. 4

crystal values, given in (1). The profile of the microstructured tungsten sample was taken from scanning probe microscopy (SPM) measurements (Fig. 6). This structure was produced in a wet chemical etching process. It has a grating period of 0.8 µm and a structure depth of 0.15 µm. The measured grating profile was used as input for the RCWA calculations. The theoretical results fit well with the experimental data (Fig. 7). Since the optical properties of sputtered tungsten are not suitable for thermophotovoltaic applications, we next structured polished tungsten sheets. Because of their still rough surface, it was not possible to determine the optical constants by ellipsometry. From Fig. 2 we can assume, that the optical constants are closer to the single crystal than to the measured sputtered tungsten values.

We used a 0.7 µm thick photoresist mask with a period of 1.4 µm and achieved a structure depth of 0.3 µm by reactive ion etching. Fig. 8 shows an SPM image of the structured tungsten surface. Fig. 9 shows the measured absorptance of this sample before and after etching. The maximum absorptance is reached at the expected wavelength of about 1.6 µm. The increase of the absorptance at this wavelength is about 0.4. The surface roughness of this polished tungsten sample explains the high absorptance of 0.17 at longer wavelengths. Other samples show a smoother surface and

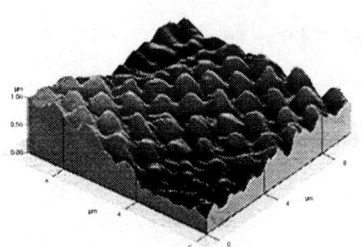

FIGURE 8. SPM image of a microstructured polished tungsten sample

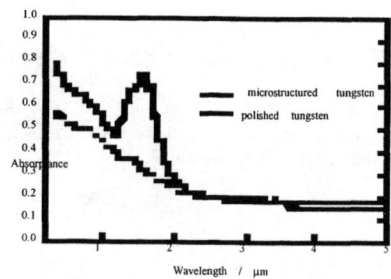

FIGURE 9. Measured absorptance of a polished tungsten sample before and after structuring

therefore a lower absorptance (Fig.2). We can expect that a higher selectivity of the microstructured samples can be achieved by using plane tungsten surfaces.

As a first stability test we heated this tungsten sample up to 1700K for several hours in vacuum. After that treatment we could not observe any change in the surface structure.

First measurements of the angular dependent absorptance of microstructured tungsten showed a decreasing absorptance with higher angles in the spectral range above 1.2 µm. Thus it is expected, that the emittance is highest in the direction normal to the surface.

CONCLUSION

Microstructured tungsten surfaces show a good selective behaviour which makes them suitable for emitters in thermophotovoltaic applications. The theoretical calculations based on a rigorous diffraction theory promise a high emittance in a specified wavelength range without an enhancement of the low infrared emittance of plane metal substrates. Initial experiments with periodically structured sputtered tungsten films show a good agreement with theoretical predictions. In experiments with polished tungsten as substrate material we achieved a peak emittance of 0.75 at a wavelength of about 1.6µm with an infrared emittance lower than 0.2. With smoother polished tungsten samples we expect an infrared emittance lower than 0.1, according to the theoretical calculations.

A first durability test indicates that these surface microstructures are stable at temperatures of 1700K.

Therefore we believe, that microstructured tungsten can be used as an emitter material for high efficiency thermophotovoltaic devices. We think, that the technical problems to build a vacuum system are not too high if one takes into account the advantages of such systems, i.e. the lack of heat transfer.

REFERENCES

1. Palik, E. D., *Handbook of Optical Constants of Solids,* Orlando: Academic Press, 1985, pp. 357-367
2. Peng, S., Morris, G.M., *J. Opt. Soc. Am.* A **12** (5), 1088-1096 (1995)
3. Li, L., *J. Opt. Soc. Am.* A **14** (10), 2758-2767 (1997)
4. Chateau, H., Hugonin, J.-P., *J. Opt. Soc. Am.* A **11** (4), 1321-1331 (1994)

Thermal Spray Approach for TPV Emitters

Christopher J. Crowley, Nabil A. Elkouh, and Patrick J. Magari

Creare Incorporated, P.O. Box 71, Hanover, NH 03755

Abstract. We have fabricated selective emitters and systematically investigated their performance in comparison with a gray body (SiC-based composite) emitter. The two key elements of our approach are (1) using thermal-spraying to manufacture thin-film selective emitters, and (2) exploring a variety of oxide materials, including rare-earths (erbia, thulia, and holmia) and cobalt-doped spinel. We fabricated the emitters by plasma-spraying coatings at thicknesses ranging from 15 to 250 microns. The spectral emittance was measured as a function of wavelength for each of the various types of emitters. Our plasma-sprayed, composite emitters are able to operate at temperatures as high as 1760 K. We find that the selective emission characteristics of the oxides are retained through the plasma-spray process. The emittance of the supporting components in the composite emitter structure is the key to the performance of these selective emitters. We investigated the use of reflective metal layers and various substrates (alumina, yttria, and silicon carbide) to optimize the performance. The various emitters were tested in a small, approximately one-dimensional TPV system which simulates prototypical operating conditions. The system is electrically heated, incorporates 5 cm × 5 cm emitters, and uses a 0.70 eV GaSb PV cell. That 1 cm^2 PV cell incorporates an integral filter. This system allows the cell output power, total power emitted, and conversion efficiency for various emitters to be compared under similar conditions. For this particular system, the power conversion efficiency with selective emitters on yttria substrates is comparable to the gray body emitter. Output power density and cooling load are smaller for the selective emitters, but might be improved. These results are specific to this TPV test configuration. Other geometries, PV cells, or filters may result in different conversion efficiency, power output, or cell cooling loads.

INTRODUCTION

In this paper we report on:

- The plasma-spray fabrication approach,
- Spectral emittance characteristics, and
- Thermophotovoltaic (TPV) performance

of a new configuration for selective emitters for TPV. The fabrication approach that we have used for our selective emitters – plasma-spraying of thin films – has not been reported previously. The plasma-spray fabrication approach is successful. We find that the plasma-sprayed materials:

CP460, *Thermophotovoltaic Generation of Electricity: Fourth NREL Conference*
edited by T. J. Coutts, J. P. Benner, and C. S. Allman
© 1999 The American Institute of Physics 1-56396-828-2/99/$15.00

- Retain their spectral emission characteristics through the deposition process,
- Survive prototypical operating temperatures (up to 1760 K), and
- Exhibit negligible degradation with long-term operation at high temperature.

The performance of TPV systems is highly dependent upon the combinations of PV cells, emitters, and filters selected for the system, as well as the specific system geometry. We have compared the performance of the selective emitters in a one-dimensional geometry, with a 0.70 eV band-gap GaSb PV cell and a specific dielectric filter. Our results are specific to that configuration. Behavior with other cells in other configurations may be predicted using the emissive spectra that we present here.

Under these TPV test conditions, we find that the best selective emitter of the rare-earth oxides is a combination of erbia (Er_2O_3) and thulia (Tm_2O_3). Compared with a single oxide, the mixture provides a stronger emission and hence useful power within the operating band of the PV cell. At previous NREL conferences, JX Crystals (Issaquah, Washington) has reported (1) a cobalt-doped spinel "matched emitter" which has selective emission characteristics. With their assistance in providing materials, we prepared an emitter which at present is comparable to the rare-earth oxide emitters, but might be superior with further improvement.

BASIC EMITTER CONFIGURATION

Figure 1 illustrates the emissive characteristics of various rare-earth oxides. These spectra are for the oxides in their original form, which is powders of various particle sizes. This figure shows that the oxides each have a characteristic wavelength at which the peak emittance occurs. That characteristic wavelength depends upon the particular metal (M) in the oxide, which has the general chemical composition of M_2O_3. For example, the peak for erbia is at a wavelength of about 1.5 microns.

Figure 2 illustrates the emissive characteristics of spinel (Al_2MgO_4) doped with cobalt. Ferguson and Fraas (1) have investigated cobalt-doped alumina (Al_2O_3), magnesia (MgO), and spinel. The doped oxide has selective emission characteristics similar to the rare-earth oxides, with a peak at a wavelength of about 1.7 microns. This material generally exhibits a broader peak than the rare-earth oxides.

Ideally, we would like to have a structure which provides exactly, and only, the emissive characteristics of the raw materials as shown above. This would provide all of the emitted power in the usable band of the PV cell, and very little outside of that band, for very high conversion efficiency. Guazzoni (2) has shown that fabricating monolithic structures of these materials is difficult, because they cannot survive thermal stresses at the high operating temperatures. In addition, they are not optically thick, so out-of-band radiation is transmitted from the heat source through the emitter structure. That significantly reduces the conversion efficiency. In practice, a substrate

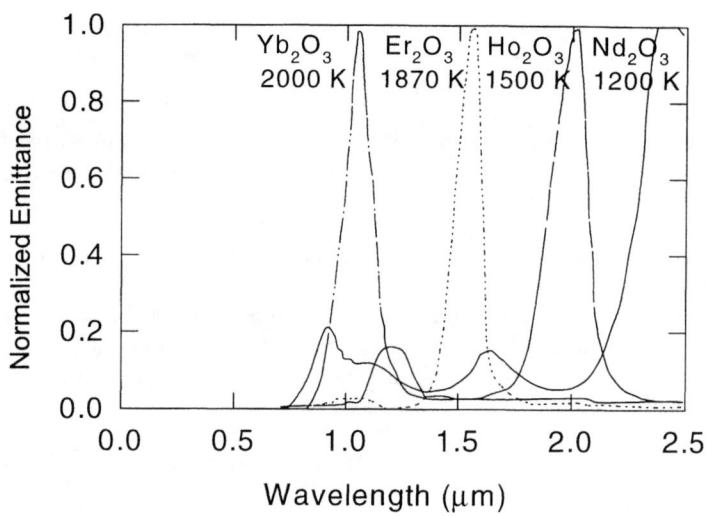

FIGURE 1. Emissive Characteristics of Rare-Earth Oxides

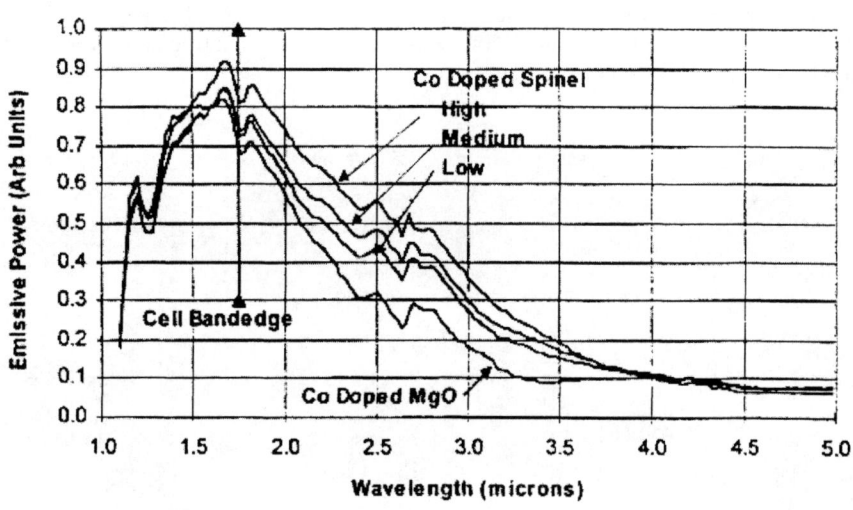

FIGURE 2. Emissive Characteristics of Cobalt-Doped Spinel (1)

is required to support the selective emitter material. As summarized by Chubb *et al.* (3), approaches such as using woven fibers and doping crystals with the rare-earth oxides have produced selective emitters.

We plasma-sprayed thin coatings of oxides with selective emission characteristics on substrates. Three configurations of the plasma-sprayed selective emitter (Figure 3) have been tried on this project:

- A coating applied directly to a silicon carbide (SiC) based composite substrate,
- A coating applied directly to an oxide (alumina or yttria) substrate, with a reflective metal coating on the <u>back</u> face of the substrate, and
- A coating applied directly onto a reflective metal layer which has in turn been deposited on the <u>front</u> face of an oxide substrate.

The second configuration is the most successful, for reasons that we discuss in this paper.

Selective emission characteristics will be superimposed on the basic behavior of the substrate. A coating applied directly to a silicon carbide substrate (first configuration) must be optically thick. This is difficult to achieve, because it requires coatings possibly millimeters thick. Figure 4 shows the emissivity of an SiC-based composite, Dow Corning Sylramic S200 CMC, without an oxide coating. (For convenience, we will hereafter refer to this sample as the "SiC-based" emitter.) With a thin (<50 microns), plasma-sprayed coating of the oxides applied to this substrate, the spectral emittance is the same as the substrate. Thicker coatings alter the spectral emittance, as discussed later.

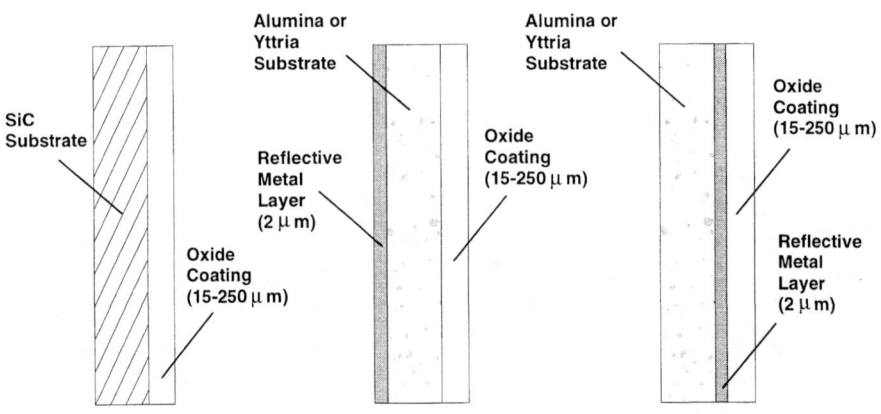

FIGURE 3. Emitter Configurations

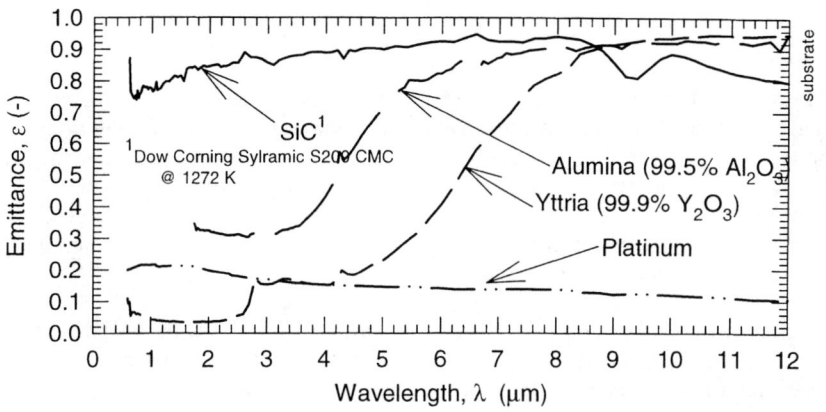

FIGURE 4. Emissivity of Various Substrates

A thin (2 micron) layer of highly reflective, refractory metal sputter-deposited on the surface of the substrate produces a very small emissivity at all wavelengths, as seen in Figure 4 for platinum. It blocks any emissive behavior of the substrate. This approach cannot be used with silicon carbide substrates because the metals (platinum or rhodium) react with the silicon carbide at elevated temperature to form silicides. The approach can be used with oxide (alumina or yttria) substrates. However, we found that plasma-spraying onto the metal layer adversely affects the emissivity of the composite structure. Furthermore, the oxides themselves exhibit "lattice vibration" emissions, leading to high emissivity above a "cut-off" wavelength similar to the oxide substrate materials. (See below.)

We also investigated a configuration with a thin coating of metal on the back face a substrate which has high transmissivity and low absorptivity in the operating wavelength range of the PV cell, so that the selective emitter dominates in this range. The emissive characteristics of alumina and yttria in this configuration are also illustrated in Figure 4. Note that these materials have "lattice vibration" emissions, resulting in a "cut-off" in the emittance at about 3.5 and 6 microns, respectively. These long-wavelength emissions contribute to the out-of-band (non-useful) power transmission, reducing the overall conversion efficiency and increasing the cooling load. Generally, a filter can recycle a significant portion of these emissions.

The results presented in the remainder of the paper are for substrates in the configuration with the metal coating on the back face and the plasma-sprayed coating on the front face of the substrate.

PLASMA-SPRAY PROCESS

In the plasma-spray process (Figure 5) particles of the coating material are entrained into a gas flow and melted in a hot plasma generated by an electrical arc. The melted particles are directed onto the surface to be coated. At the surface, the particles re-solidify as they are cooled. The plasma-spray gun is passed over the surface a number of times, depositing material in a swath which is several microns thick and centimeters wide each pass. Layers from 15 to 250 microns thick can be built up on the substrate in just a few minutes.

Using an in-house Praxair 3702 Plasma System, we prepared the sample emitters listed in Table 1. Erbia and cobalt-doped spinel coatings were prepared directly. Raw material of the cobalt-doped spinel was provided by JX Crystals, and the powder was sorted to usable particle sizes at Creare. Thulia (Tm_2O_3) and holmia (Ho_2O_3) powders were mixed with erbia prior to spraying, in ratios of about 2:1 and 3:1, respectively. Erbia is the dominant component in the mixtures. Emitter samples were prepared at a size up to 5 cm × 5 cm.

Characterization tests included transmission electron microscope (TEM) examination which confirmed that the crystal structure was the same as in the original powders after plasma-spraying. Via measurements of the mass and thickness of the deposited coatings, we estimate that the density of the plasma-sprayed coatings is 80% to 100% of the original.

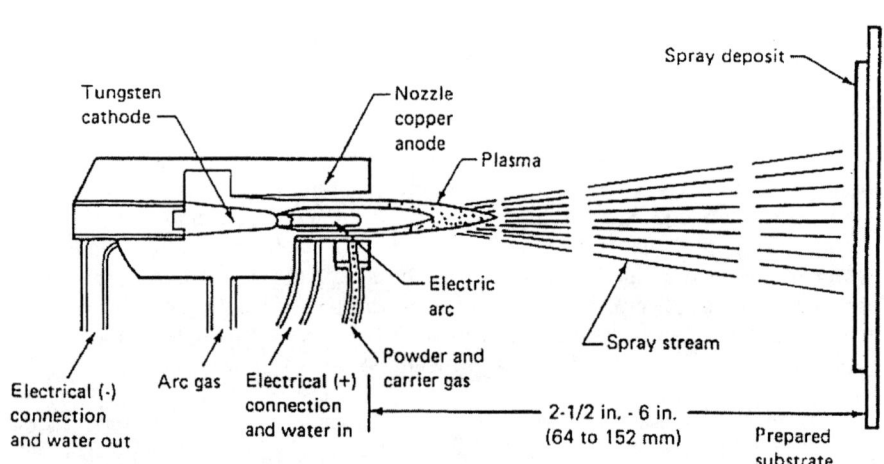

FIGURE 5. Illustration of the Plasma Spray Process

TABLE 1. Sample Emitters Prepared

Plasma Sprayed Coating (μm)	Substrate (mm)	Metal (μm)	Sample ID	Max. Temp. (Heat Rate)
Erbia (15-100)	99.5% Alumina (2.54)	Platinum (2)	R1	1760 K (5-16 K/min)
Erbia+Thulia (50-80)	99.5% Alumina (2.54)	Platinum (2)	R2	1760 K (5-16 K/min)
Erbia+Holmia (80)	99.5% Alumina (2.54)	Platinum (2)	R7	1470 K (10 K/min)
Co-Doped Spinel (70)	99.5% Alumina (2.54)	Platinum (2)	SP1	1475 K (10 K/min)
Erbia (50)	99.9% Yttria (3.18)	Platinum (2)	R3	1370 K (6 K/min)
Erbia+Thulia (100)	99.9% Yttria (3.18)	Platinum (2)	R4	1480 K (6 K/min)
(none)	SiC-Based* (0.381)	(none)	S1	1450 K (100 K/min)
Erbia (100-250)	SiC-Based (0.381)	(none)	R6	1230 K (100 K/min)
Co-Doped Spinel (125-210)	SiC-Based (0.381)	(none)	SP3	1400 K (100 K/min)

*All SiC substrates are Dow Corning silicon carbide based Sylramic S200 CMC.

SPECTRAL EMITTANCE

Emissive spectra of the samples were measured in a High-Temperature Spectral Emittance Facility at McDermott Technologies, Inc. (Alliance, Ohio). Samples up to 5 cm × 5 cm square are heated between 800 K and 1600 K in a high-temperature furnace. Typical temperatures are 1100-1200 K for the emittance measurements. Three photodetectors measure light intensities from 0.6 to 14.4 microns. A thermopile detector measures total emittance directly. The system is calibrated daily with a blackbody radiation source. Spectral emittance measurements are within $\varepsilon \pm 0.05$. In the range of interest for the selective emission characteristics, readings are typically obtained at wavelength increments of 0.020 microns, at a band-pass of 0.0024 microns from 0.6 to 1.1 microns wavelength and a band-pass of 0.0098 microns from 1.1 to 2.4 microns wavelength.

The emissive spectra of various oxides on alumina substrates are shown in Figure 6. Selective emission characteristics representative of each oxide appear as expected, showing that the selective emission characteristics are retained through the plasma-spray coating process. For the rare-earth oxides, erbia (R1) shows the

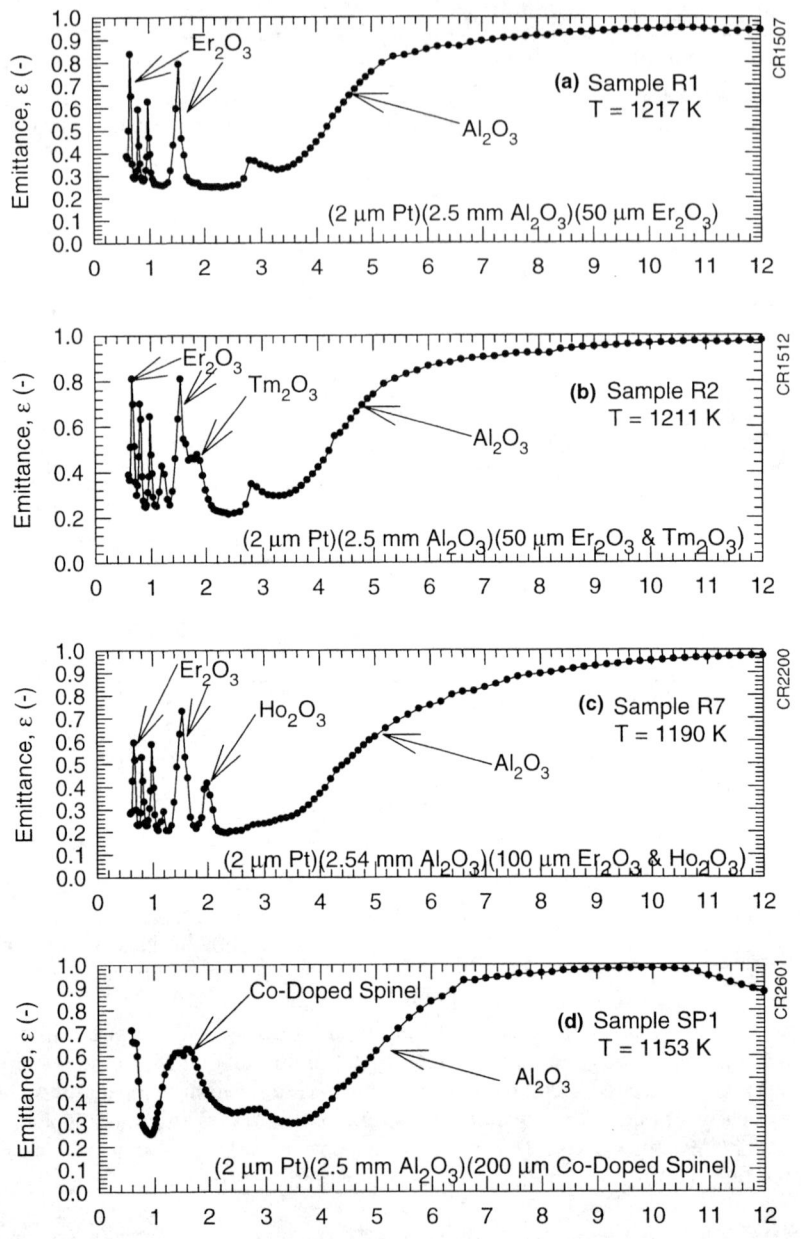

FIGURE 6. Spectral Emittance Characteristics of Plasma-Sprayed Oxides on Alumina Substrates

expected primary peak emissivity ($\varepsilon \cong 0.7$) at a wavelength of about 1.5 microns. Lesser peaks at other wavelengths (0.64, 0.80, and 1.0 microns) are also typical of erbia (3). This sample is representative of all others containing erbia. Independent emittance tests using facilities at NASA Lewis Research Center (3) show similar selective emittance at temperatures up to 1760 K. The emittance peak at 1.5 microns is slightly lower ($\varepsilon \cong 0.6$) in those measurements, possibly due to temperature gradients in the coating. Erbia coatings 50-100 microns thick show little difference in emittance spectra, while samples with thinner coatings (15 microns) clearly show a decrease in the peak emittance. So, coatings as thin as 50 microns can be used. This is important since thinner coatings tolerate thermal stresses better.

Emitters with thulia (R2) and holmia (R7) in combination with erbia are of interest because of the potential for additional emission in the useful band of the PV cell. In Figure 6 additional peaks in the emission are observed at 1.7 microns and 2.0 microns, which are consistent with thulia and holmia, respectively. The peak emissivity of $\varepsilon \cong 0.4$ to 0.5 is lower than for the erbia. The lower value for the second oxide may be due either to interference with the erbia, or the lower percentage of thulia or holmia in the mixed oxide. With the present materials it is not possible to spray coatings of thulia and holmia separately, nor is possible to increase the percentage of thulia or holmia.

The cobalt-doped spinel (SP1) also exhibits selective emission. There is a broad emittance band, with peak of $\varepsilon = 0.64$ in the emittance at about 1.7 microns. The peak emittance is somewhat lower than the peak of $\varepsilon \cong 0.8$ reported earlier by JX Crystals (1). (See Figure 2.) Since the selective emission characteristics may be very sensitive to the doping level or grain size, this may indicate that further work is required to adjust the raw material for the plasma-spray approach.

Because the cut-off wavelength for emittance due to lattice vibrations occurs at longer wavelengths for yttria than for alumina, we also investigated the use of this substrate material. Figure 7 shows that the higher cut-off is the main difference in spectral emittance between these samples (Figure 7, R3 and R4) and the corresponding samples on alumina (Figure 6, R1 and R2). The emittance in the wavelength range of 2 to 4 microns is also lower. The yttria substrates tend to be more fragile than the alumina substrates, with a greater chance of failure due to thermal or mechanical stresses.

Figure 8 shows that application of very thick (200 microns) erbia (R6) or cobalt-doped spinel oxides (SP3) onto silicon carbide also provides selective emission. Selective emittance of the oxides is evident at 1.5 and 1.7 microns, respectively. However, the emittance at wavelengths above 2 microns is still substantially higher than for the rare-earth oxide or the cobalt-doped spinel on alumina (Figure 6, samples R1 and SP1).

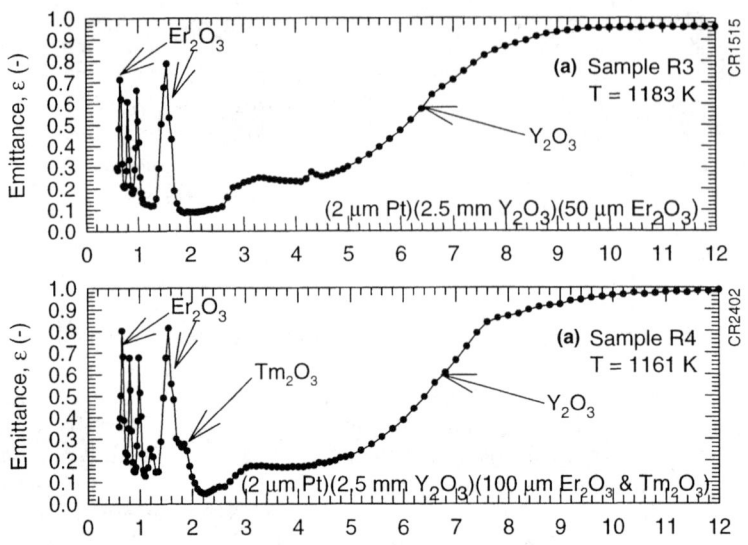

FIGURE 7. Spectral Emittance Characteristics of Plasma-Sprayed Oxides on Yttria Substrates

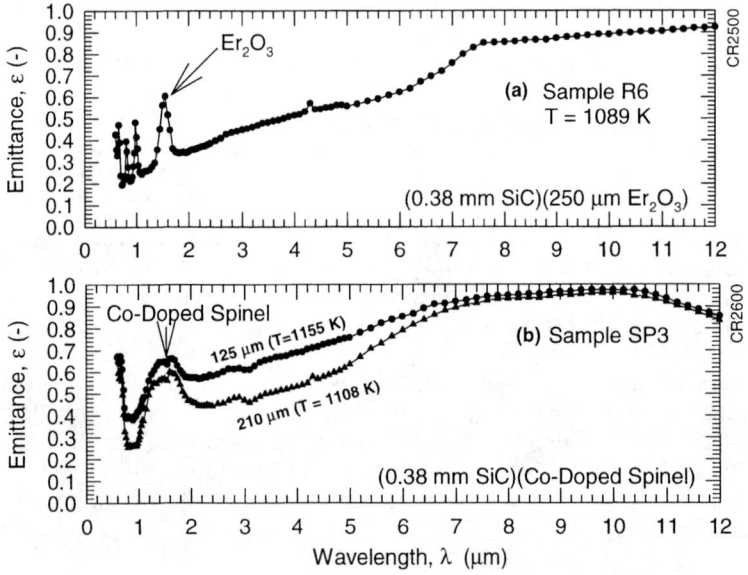

FIGURE 8. Spectral Emittance Characteristics of Plasma-Sprayed Oxides on Silicon Carbide Substrates

BENCH-TOP TPV TESTS

To test our selective emitter samples under similar, prototypical operating conditions, we used a bench-top TPV system. A small test apparatus which incorporates electrical heaters, an aperture, and a photovoltaic cell was fabricated for this purpose. See the related paper by Fraas et al. (5) in this conference proceedings for a description of this facility and test procedures. Key features of the facility include:

- Approximately a one-dimensional geometry,
- Silicon carbide heaters capable of 1600 K temperatures,
- Emitter test samples at 5 cm × 5 cm size,
- An aperture 1.25 cm in diameter to view the samples,
- A 0.70 eV GaSb cell (1 cm^2)
- A thin-film, dielectric filter[1] integral with the PV cell, and
- An optical pyrometer for determining the surface temperature of the emitter at wavelengths of 8 to 11 microns.

Except for the pyrometer, the test apparatus was supplied by JX Crystals. This apparatus provides a means to subject the selective emitter samples to prototypical temperatures, heat fluxes, and temperature gradients. It also permits quantitative comparisons of conversion efficiency and output power densities for the various selective emitters and serves as an integral check of the spectral emittance discussed above.

To characterize the conversion efficiency, the emitter is first heated to the desired temperature without the PV cell in place. After the emitter reaches the desired temperature, a block with the mounted PV cell is slid up to the aperture. The current vs. voltage characteristics of the PV cell are determined in a fast sweep. The sweep is made within a few seconds. These results are used to determine the short-circuit current (I_{sc}) and the open-circuit voltage (V_{oc}). The short-circuit current depends upon the temperature of the emitter, the type of emitter, and the temperature of the PV cell. The open-circuit voltage depends on the temperature of the PV cell, and is typically about 0.45 V. The current and voltage measurements, along with the fill factor (FF) for the PV cell, determine the output power ($P_{out} = I_{sc}V_{oc}FF$). From the mass of the block, its specific heat capacity, and the rate of temperature rise of the block, the total heat input to the block (P_{in}) is calculated. The block heats up over several minutes. To first order, the conversion efficiency is the ratio of these two numbers, P_{out}/P_{in}.[2]

---

[1] The cut-off wavelength at the edge of the band-gap for this GaSb cell is approximately 1.8 microns.

[2] A correction is made to the total heat input to the cell. The test as described above is repeated with a gold-coated, brass block which does not have an active PV cell installed. The heatup of this block occurs by conduction and convection rather than radiation, so this corrects the total input power for non-radiative sources of heat transfer. The correction typically amounts to about 25% of the total input power.

TPV Results

Figure 9 illustrates the TPV results for various oxides on alumina substrates. Refer to Figure 6 for the spectral emittance characteristics of the oxide samples in these figures. Conversion efficiency, PV cell power output, and cooling load for various emitters are shown for this specific test configuration. The limiting case of blackbody ($\varepsilon = 1$ at all wavelengths) is used as a reference in these comparisons.

For the emitters with alumina substrates, the erbia/thulia REO combination and cobalt-doped spinel have the highest conversion efficiency and power density. At a temperature of 1400 K, the conversion efficiency is approximately 9-10% for all of these emitters. This is a little more than half of the theoretical limit for a blackbody emitter in this configuration. The output power density is a little less than one-half of the value of a blackbody emitter. The output power is consistent with the emittance values shown in Figure 6 for the wavelengths (<1.8 μm) at which the PV cell produces useful power. The effect of the greater integrated area under the emittance spectrum of the erbia/thulia combination (Figure 6b) compared with the erbia-only emitter (Figure 6b) can be seen in the TPV test as a slightly (~1.1 times) greater output power. The effect of the holmia in the erbia/holmia combination does not show up in the output power because the peak in the emittance at 2 μm due to the holmia is outside the useful wavelength range for this PV cell (<1.8 μm). With the selective emitters on alumina substrates, the cooling load on the PV cell is about two-thirds of the value with a blackbody emitter.

The conversion efficiencies with the selective emitters on yttria substrates (Figure 10) are larger than the corresponding emitters on alumina substrates (Figure 9). At a temperature of 1400 K, the conversion efficiency on yttria substrates is about 12%, or about three-fourths of the conversion efficiency expected with a blackbody emitter. This compares with a 9-10% conversion efficiency on the alumina substrates (Figure 9). The larger conversion efficiency is due to a lesser cooling load with the yttria substrate – about half of the blackbody limit instead of two-thirds. This is primarily an effect of the low emittance of the yttria substrate compared with the alumina substrate above 1.8 microns, as shown in Figure 4 and Figure 6a (Sample R1) compared with Figure 7a (Sample R3). The power output is the same for alumina and yttria substrates because the emissive behavior at wavelengths less than 1.8 microns is the same. The advantage of the lower emittance of the yttria substrate is less apparent with the integral filter on the PV cell than it would be without it.

Figure 11 illustrates the TPV results with an SiC-based substrate. An uncoated SiC-based emitter has a conversion efficiency of about 12% at 1400 K, or about three-fourths of the blackbody limit. The corresponding output power density (0.75 W/cm^2) is about three-fourths of the blackbody. This power output is still about 20% smaller than we expected based upon the spectral emittance (Figure 4), suggesting that the emittance of the SiC-based emitter may be lower at higher temperatures. Decreasing emittance of SiC-based emitters with temperature would be consistent with other published data (4). The cooling load on the PV cell/filter approaches the blackbody emitter.

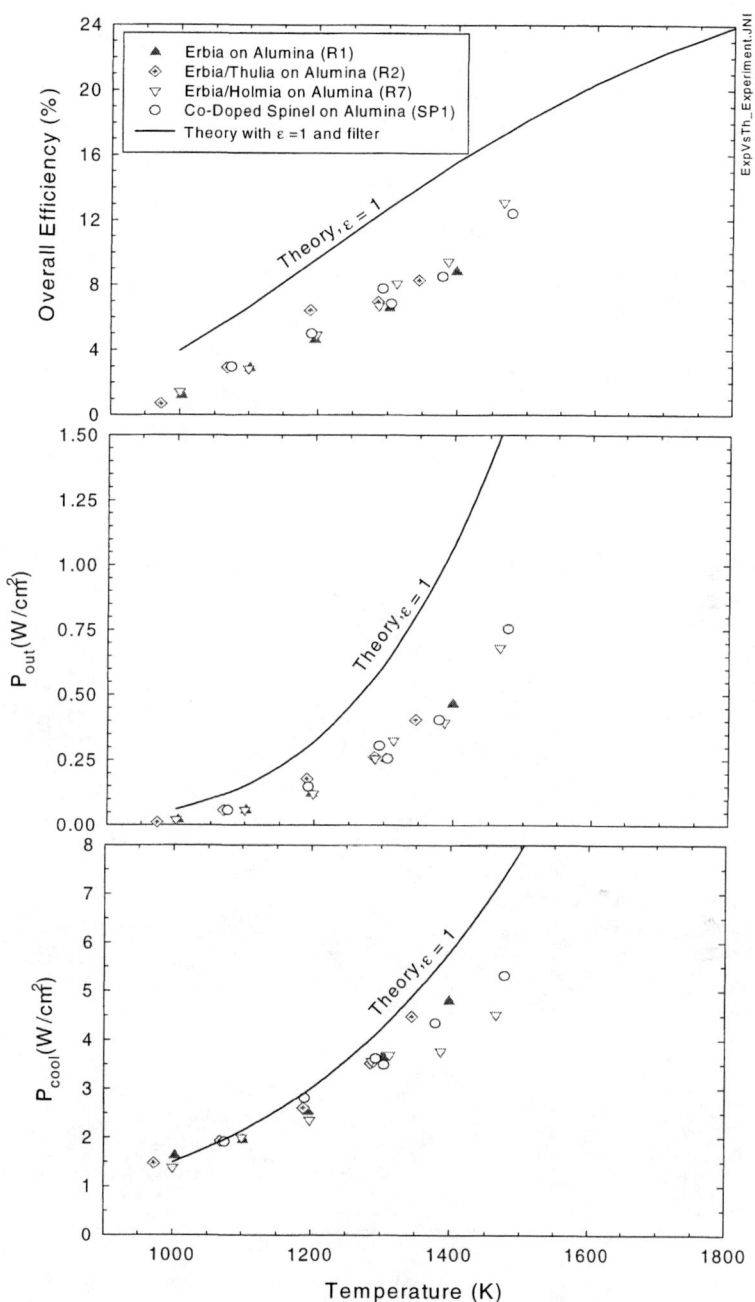

FIGURE 9. TPV Results for Emitters on Alumina Substrates

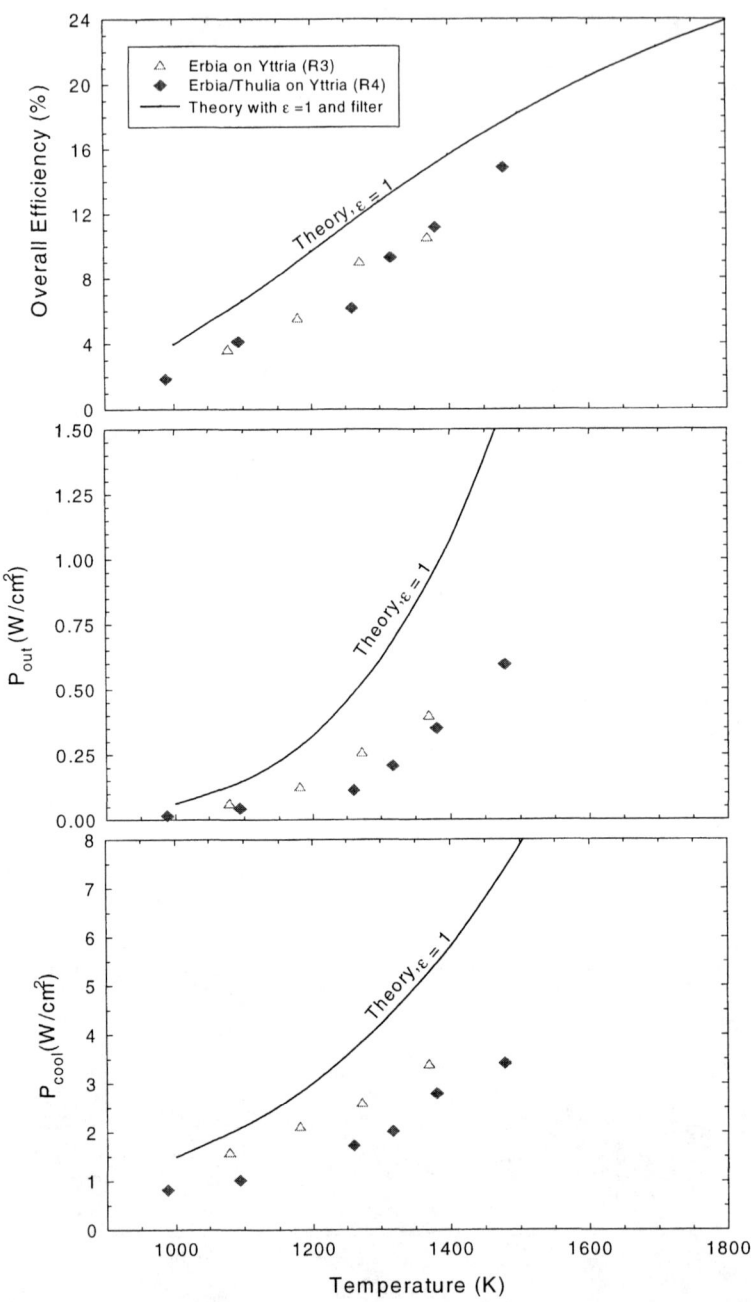

FIGURE 10. TPV Results for Emitters on Yttria Substrates

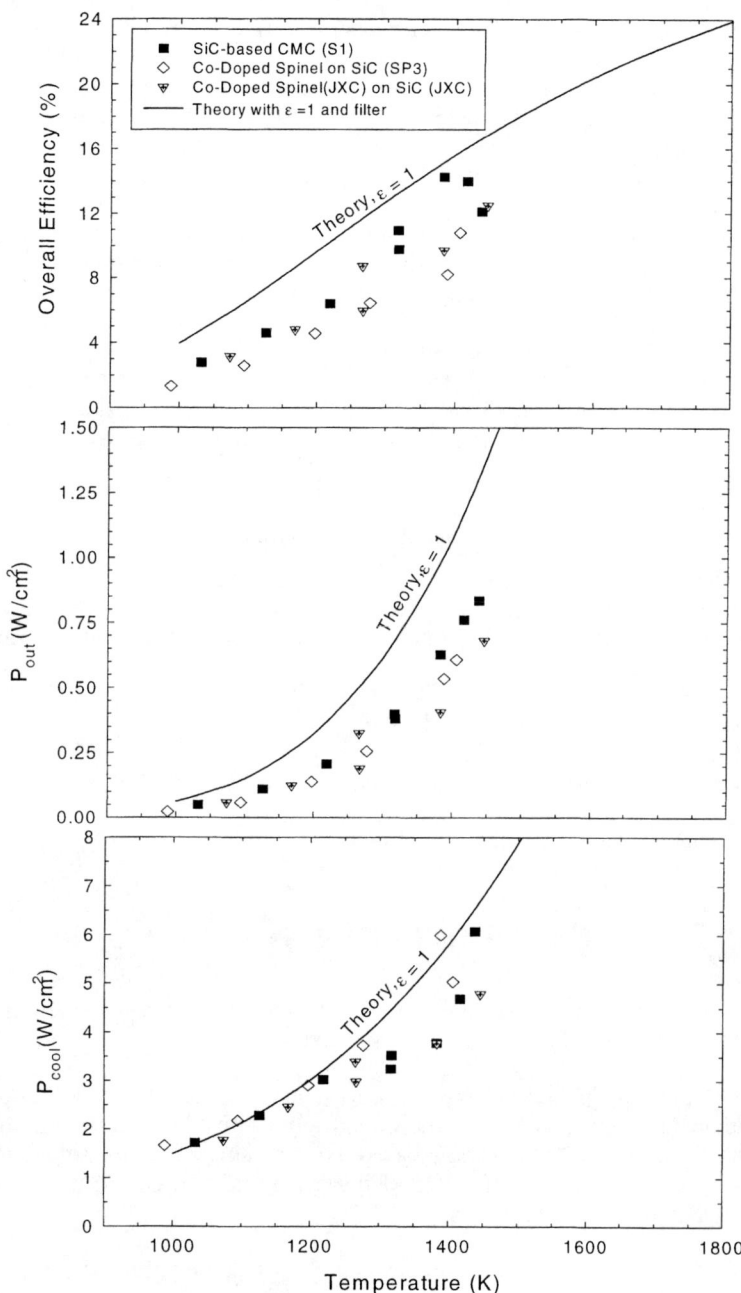

FIGURE 11. TPV Results for Emitters on Silicon Carbide Substrates

We previously discussed that thicker plasma-sprayed coatings are needed to mask the emissive behavior of the SiC-based substrate. See Figure 8. Figure 11 shows the TPV results for one of our plasma-sprayed emitters, as well as a matched emitter provided by JX Crystals. The matched emitter from JX Crystals was prepared by their proprietary coating process, rather than plasma-spraying. Both have thick ($\cong$200 μm) oxide coatings on the SiC-based substrate. The conversion efficiency of these emitters with cobalt-doped spinel on SiC-based substrates is similar to the emitters on alumina substrates. That conversion efficiency is somewhat less than the uncoated SiC-based emitter in this configuration, because the output power density is slightly less while the cooling load is about the same.

It remains to be determined whether the thicker coatings can survive operating temperatures higher than 1500 K at which these emitters have operated in the bench-top TPV tests. In general, the thermal stresses are greater with thicker coatings and there is a greater probability of mechanical failure at high temperatures.

Endurance Testing

Selective emitter samples are undergoing life tests. In these tests, the emitters are subjected to a temperature of 1500 K in a furnace for long periods of time and periodically removed, when:

- TPV performance is measured again,
- The samples are weighed to quantify any mass loss or gain, and
- The spectral emittance measurements are repeated.

Negligible changes have been detected after 1,000 hours.

CONCLUSIONS

- Plasma-spraying is a viable, rapid, inexpensive approach to fabrication of TPV emitters with selective emission characteristics.

- The plasma-sprayed selective emitters survive prototypical TPV operating temperatures (1100 K to 1760 K) and temperature gradients. The thin-film configuration of the plasma-sprayed oxide coatings and the use of substrates with fairly similar coefficients of thermal expansion enable the emitters to tolerate the thermal stresses.

- With the particular TPV test configuration used, a combination of rare-earth oxides (erbia and thulia) on an yttria substrate approaches the conversion efficiency of a silicon carbide tested. The power density and cooling load are about a factor of 0.75 times the values with a silicon carbide emitter.

- With the particular TPV test configuration used here, a cobalt-doped spinel oxide (from JX Crystals) on an alumina substrate also equals the conversion efficiency of the silicon carbide emitter. The power density and the cooling load with the cobalt-doped spinel emitter are also lower than the silicon carbide emitter. It may be possible to improve the performance of emitters with this oxide if the peak emittance can be increased from the current peak ($\varepsilon = 0.63$) to the peak $\varepsilon = 0.8$ reported with oxides prepared previously (1).

- Thick ($\cong 200$ µm) coatings of the oxides can be applied directly to SiC-based substrates to achieve similar TPV performance with this TPV test configuration up to 1500 K.

- TPV results with the selective emitters are consistent with a one-dimensional analytical model. TPV results with an uncoated SiC-based emitter are lower than predicted analytically, possibly due to decreasing emittance at high temperature.

ACKNOWLEDGMENTS

The authors gratefully acknowledge the sponsorship and technical contributions of NASA Lewis Research Center (Contract NAS3-97018). We also wish to thank: JX Crystals for providing cobalt-doped spinel oxides and hardware for the TPV test apparatus; McDermott Technologies Inc. for spectral emittance measurements, as well as SEM and EDS analysis of samples; and Essential Research for information on the performance characteristics of PV cells and discussions concerning the modeling of PV cell behavior.

REFERENCES

1. Ferguson, L. and Fraas, L.; "Matched Infrared Emitters for Use with GaSB TPV Cells"; *In the 3rd NREL Conference Thermophotovoltaic Generation of Electricity*, CP401, Benner and Coutts, eds.; pp. 169-179, 1997.

2. Guazzoni, G. E.; *Cylindrical Erbium Oxide Radiator Structures for Thermophotovoltaic Generators*, Report AD/A-001 525, Fort Monmouth, NJ: Army Electronics Command, August 1974.

3. Chubb, D. L., Pal, A. T., Patton, M. O., and Jenkins, P. P., "Rare Earth Doped High Temperature Ceramic Selective Emitters," presented at New Developments in High Temperature Ceramics, Istanbul, Turkey, August 12-15, 1998.

4. Guazzoni, G. E., and McAlonan, M., "Multifuel (Liquid Hydrocarbons) TPV Generator," *In the 3rd NREL Conference Thermophotovoltaic Generation of Electricity*, CP401, Benner and Coutts, eds., 1997, pp. 341-354.

5. Fraas, L., Groenveld, M., Magendanz, G., Hai, S., and Custard, P., "A Single TPV Cell Power Density and Efficiency Measurement Technique," this volume.

Temperature Gradient Effects in an Erbium Aluminum Garnet Selective Emitter

Brian S. Good and Donald A. Chubb
National Aeronautics and Space Administration
Lewis Research Center
21000 Brookpark Road, MS 302-1
Cleveland, OH 44135

AnnaMaria Pal
Essential Research
23811 Chagrin Blvd., Suite 280
Cleveland, OH 44122

Abstract. Spectral control through the use of selective emitters is an important means of improving the efficiency of thermophotovoltaic (TPV) systems. The thin-film selective emitters developed in our laboratory offer a number of potential advantages for use in such systems. It has long been realized, however, that there may exist relatively large temperature gradients across the thickness of these emitters in operation, and that such gradients are likely to have a detrimental impact on emitter performance. Previous efforts at modeling TPV emitter or system performance have either ignored thermal gradient effects or assumed the temperature profile to be linear (1-5). A detailed investigation of the temperature profile and subsequent effects on emitter performance has not yet been given.

In this paper, we present results of a detailed theoretical and computational study of the effects of thermal gradients, along with some other film parameters, on the performance of an erbium aluminum garnet ($Er_3Al_5O_{12}$) selective emitter. Equations for the internal energy flux within the emitter are developed, under the assumption that heat transfer occurs via conduction and radiation. One face of the emitter is assumed to remain at fixed temperature, as would be the case if a thin emitter were in contact with a massive heat bath. The other face is assumed to experience either "vacuum interface" conditions, where heat transfer from the surface occurs via radiation only, or "lossy" conditions, where additional losses via conduction and convection are included. The temperature profile across the emitter thickness is computed by requiring that energy be conserved everywhere.

We investigate the effects of the thermal gradient on the useful output power and emitter conversion efficiency. We also consider the sensitivity of the temperature gradient to the film and substrate physical properties.

INTRODUCTION

The importance of temperature gradients within a thin-film selective emitter has been recognized for some time (1), (5). It is clear that the maximum performance of such an emitter will be obtained when the emitter is at a uniform temperature, but such a condition is unlikely to exist in an emitter in operation. However, previous efforts at modeling TPV emitter or system performance have either ignored thermal gradient effects or assumed the temperature profile to be linear (1-5). A detailed investigation of the temperature profile and consequent effects on emitter performance has not yet been given.

In this paper, we present results of a detailed theoretical and computational study of the effects of thermal gradients on the performance of an erbium aluminum garnet ($Er_3Al_5O_{12}$) selective emitter. The emissive properties of this material are characterized by a major emission band at $\lambda \approx 1500$nm, smaller emission bands at $\lambda \approx 1000$nm, 800nm and 640nm, and large emission for $\lambda > 6000$nm due to lattice vibrations. Equations for the energy flux within the emitter are developed, under the assumption that internal heat transfer occurs via conduction and radiation. One face (the substrate face) of the emitter is assumed to remain at fixed temperature, as would be the case if a thin emitter were in contact with a massive heat bath. The other face (the radiating face) is assumed to be described by one of two boundary conditions. When all heat transfer at the radiating surface occurs via radiation, the "vacuum interface" condition is assumed to exist. This boundary condition is appropriate for a TPV system operating in a space environment, where the possibility of convective loss does not exist, and any conductive loss is most likely negligible. In this case, the temperature profile is computed by requiring that energy conservation be satisfied at all points within the film and on the film surface. The "lossy" interface condition, appropriate for terrestrial operation, exists when there is the possibility of conductive and/or convective energy transfer at the radiating surface. In the lossy case, the radiating surface temperature is fixed, which (along with the input power entering the film from the substrate) determines the conductive and convective losses.

We investigate the effects on emitter performance of the calculated average thermal gradient, for both vacuum and lossy interface conditions. The gradient is controlled by changing the temperature at the substrate face of the film and (for the lossy interface) at the radiating surface, and by varying the film properties. The emitter's performance is characterized by its total output power, above-bandgap ("useful") output power, wavelength-dependent emittance, total emittance, and conversion efficiency.

FILM EMITTANCE MODEL

The film geometry and fundamental quantities used in this work are displayed in Fig. 1, along with the boundary conditions relating the positive and negative-going intensities (i^+ and i^-) at the film boundaries. The input thermal flux Q_{in} passes from the substrate to the film; the detailed form of the output flux depends on the thermal boundary condition. In the vacuum interface case, heat transfer from the radiating surface to the environment occurs by radiation only, so the temperature gradient at the surface is zero. In the case of a lossy interface, there will be an additional thermal loss at the film surface, via a combination of convection and conduction, so that the gradient at the film surface will be nonzero. The film emittance model incorporates the following assumptions:

(1) The problem is one-dimensional, that is, the emitter film thickness is small compared to its lateral dimensions, and edge effects are ignored.

(2) Scattering within the film, if present, is isotropic.

(3) The boundaries are diffuse.

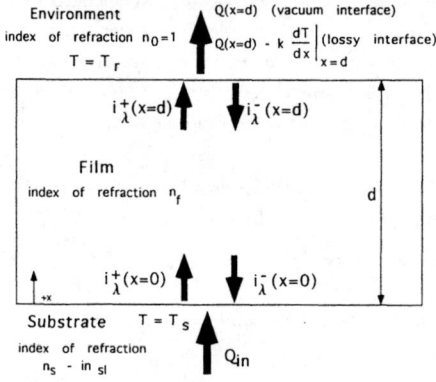

FIGURE 1. Film Emittance Model

(4) The index of refraction and the extinction coefficient of the film are independent of position within the film, but depend on wavelength.

(5) The film's environment, i.e. the thermal cavity in a TPV system, is ignored; no radiative flux impinges on the emitter from the environment.

SUMMARY OF THEORETICAL DEVELOPMENT

Equations for the net heat flux (radiative and conductive) have been derived. The derivation is lengthy, and only a brief outline will be presented here.

The general expressions for the positive and negative-going radiative intensities (as shown in Fig. 1) are obtained following Siegel and Howell (6), expressed in terms of a radiative source function. These equations can be evaluated at the film boundaries to yield $i^+(d)$ and $i^-(0)$, and the boundary conditions applied to relate i^+ and i^- at the film-vacuum and substrate-film interfaces. The expression can be simplified by applying assumption (3) above, and also assuming that the source function is independent of angle. The wavelength-dependent fluxes can be obtained from the intensities by integrating over angle. At the substrate interface, the integration is over all angles. At the film-vacuum interface, however, provisions need to be made to allow for total reflection consistent with Snell's law. The total wavelength-dependent radiative flux at the radiating surface is the signed sum of the positive- and negative-going fluxes. Energy conservation within the film requires that, in the steady state, the sum of the wavelength-integrated radiative flux and the conductive flux at each point is a constant, equal to the output power at the outer film surface plus losses, if any.

COMPUTATIONAL PROCEDURE

A computer code has been written to characterize the performance characteristics of a selective emitter, based on the previously described analytical model. The input data for the emitter model are the wavelength-dependent emitter extinction coefficient and index of refraction, the substrate index of refraction, and the film thickness and thermal conductivity. Because the optical properties of the film and substrate materials are wavelength-dependent, all computations are performed over a wavelength range of 1-10000 nm, with a wavelength interval of 1 nm. For the largest substrate temperature (2000K), a 10000 nm cutoff is reasonable, because the blackbody function is negligible at these larger wavelengths, even though the emitter material considered here has a long-wavelength high-emittance tail. For lower substrate temperatures (1200K, 1600K), however, the peak in the blackbody function occurs at longer wavelengths, with the result that the long-wavelength tail of the function is more significant. We have therefore included a long-wavelength correction term; a representative value for the long-wavelength emittance is chosen from the experimental values near 10000 nm, and this value is assumed to be appropriate for all wavelengths greater than 10000 nm. Radiation for these wavelengths is then assumed to be described by the product of this emittance and the blackbody function.

A general outline of the computation procedure employed by the code follows. Properties and temperatures are evaluated on a grid of 41 nodes that span the film thickness.

(1) An initial guess is made for the outer surface temperature. The temperature profile is initially assumed to be linear across the film thickness.

(2) The wavelength-dependent source function is computed iteratively using the current temperature profile.

(3) The wavelength-dependent thermal flux at each node is computed using the above source function.

(4) The wavelength-dependent flux is integrated over wavelength to obtain the net radiative (wavelength-independent) thermal flux at each node.

(5) The radiative net flux is used, along with the one-dimensional heat equation and the appropriate boundary condition, to compute the temperature gradient at each node. The computed gradient and the (fixed) substrate temperature are used to generate a new temperature profile.

Steps (2)-(5) are repeated until the wavelength-dependent flux converges. At this point, the total output power, above-bandgap ("useful") output power, emitter efficiency, wavelength-dependent emittance, and total emittance are computed. The useful output power is defined as the fraction of the output power whose energy is greater than the bandgap energy of the PV cell used in the system, i.e. the portion of the output power that can be absorbed by the cell. The emitter efficiency is the ratio of useful power to total power. The wavelength-dependent emittance is the ratio of the wavelength-dependent flux at the radiating surface to the blackbody function evaluated at the specified wavelength and the substrate temperature. The total emittance is the ratio of the total output power to σT_s^4; note that referencing the emittances to the substrate temperature guarantees that the emittance will not be greater than unity.

For the work described below, a set of default parameters is established, as follows. The extinction coefficient and film index of refraction are obtained from experimental measurements of an $Er_3Al_5O_{12}$ film. The substrate index of refraction is obtained from the literature, using data for platinum. The default film thickness is 0.63 mm. The default film thermal conductivity is taken to be 0.01 W/cm/K, and the default environmental emittance and reflectance are zero. The wavelength corresponding to the PV cell bandgap energy is assumed to be 1650 nm. Scattering is incorporated by assuming that the scattering coefficient is a constant fraction of the extinction coefficient; the default value for this fraction is 0.1. Substrate temperatures of 1200K, 1600K and 2000K are used.

As previously mentioned, two boundary conditions are used. When vacuum interface conditions are employed, the temperature at the radiating surface is computed self-consistently. On the other hand, when lossy conditions exist, the surface temperature must be specified in advance. In the following calculations, we specify the surface temperatures in terms of the "average gradient" $(T_s - T_r)/T_s$; values of 0.05, 0.1, 0.15 and 0.2 are used.

RESULTS AND DISCUSSION

Several sets of computer runs have been performed to investigate various aspects of selective emitter performance. The starting point for each set of runs is the default parameter set described above. For each run, data is collected on the total output power, the useful power, the selective emitter effiency, the wavelength-dependent emittance, and the total emittance.

Effects of the Temperature Gradient

The first study considers the effects of the average temperature gradient across the film thickness on emitter performance. For each of the three substrate temperatures (1200K, 1600K amd 2000K), one run was performed using the vacuum interface boundary condition, and three ($T_s = 2000K$) or four ($T_s = 1600K, 1200K$) runs were performed using the lossy boundary condition. Temperature profiles are presented in Fig. 2a-c, with useful power, total output power, efficiency and emittance shown in Figs. 3a,b, c and d.

Several trends are evident. First, the degree of curvature in the temperature profile is an increasing function of substrate temperature. This is simply a manifestation of the competition between internal radiative ($\sim T^4$) and conductive ($\sim dT/dx$) heat transfer. For lower substrate temperatures, internal conductive transfer completely dominates, while at larger substrate temperatures, this is no longer true. For $T_s = 1200K$, all temperature profiles are close to linear. For $T_s = 1600K$, the vacuum interface and 0.05 average gradient profiles exhibit noticeable deviation from linearity. Finally for $T_s = 2000K$, all the profiles exhibit curvature, though the 0.2 gradient profile is close to linear.

Based on these results, the assumption of a linear temperature profile derived from measured substrate and film surface temperatures, as was done in previous work, is reasonable as long as the substrate temperature is not too high. At higher substrate temperatures, the deviation from linearity is appreciable if the average gradient is small. However, it should be noted that, because the temperature dependence of radiative heat transfer is so strong, even a small deviation from linearity may result in significant variation in computed properties.

Note that for the 2000K substrate temperature the average gradient for the vacuum interface case is greater than 0.05; because the vacuum interface case has no convective or conductive losses at the radiating surface, it will exhibit the highest possible surface temperature. Assuming a gradient smaller than the one computed using the vacuum interface condition would yield unphysical results, so no 0.05-gradient results are presented for this substrate temperature.

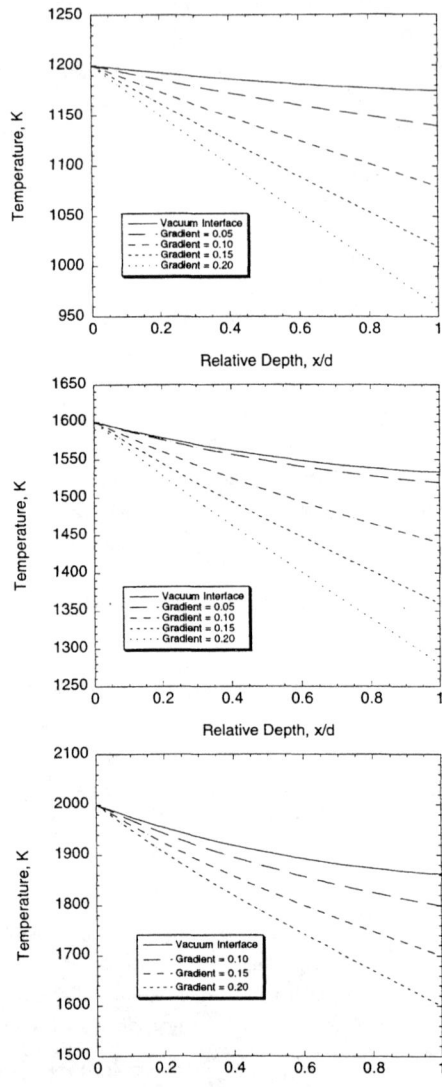

FIGURE 2a-c. Temperature profiles for substrate temperatures of (a) 1200K, (b) 1600K and (c) 2000K; Vacuum interface and average gradients of 0.05, 0.10, 0.15 and 0.20.

The emitter efficiency versus average gradient is presented in Fig. 3c. It can be seen that the efficiency decreases slightly with increasing average gradient. This behavior can be understood by considering individually the effects of the temperature gradient on the useful and total powers (Figs. 3a and 3b).

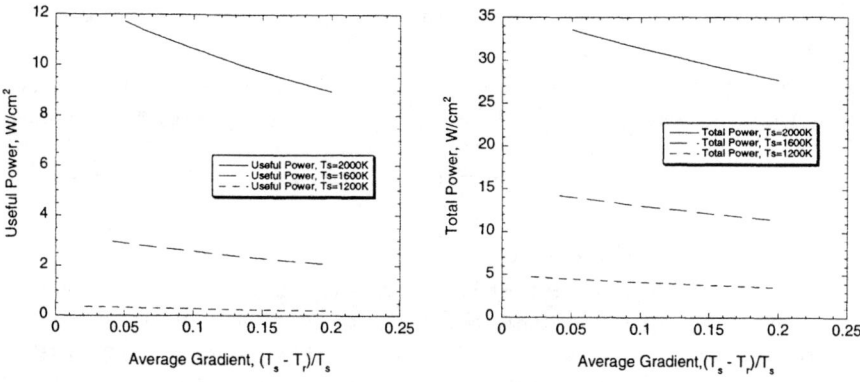

FIGURE 3a. Useful output power versus average temperature gradient.

FIGURE 3b. Total output power versus average temperature gradient.

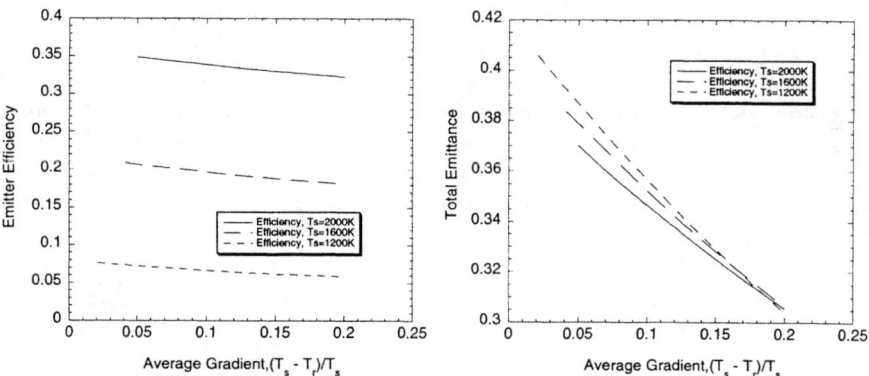

FIGURE 3c. Emitter Efficiency versus average temperature gradient.

FIGURE 3d. Total Emittance versus average temperature gradient.

The useful power consists primarily of a contribution from within the narrow emitter emission band of the film, where the extinction coefficient is large, along with a contribution from all wavelengths shorter than the PV cell cutoff wavelength, primarily from the substrate, where the extinction coefficient is smaller. The emission-band contribution will be strongly temperature-dependent, since the power radiated is proportional to T^4, and T varies significantly across the film thickness. As the average gradient increases, an increasing fraction of the emitter thickness is at a temperature substantially lower than the substrate, so that the useful power is strongly sensitive to the gradient.

The total power, on the other hand, includes the temperature-sensitive contribution from the emission band, another contribution from medium-wavelength radiation where the extinction coefficient is small, and a third contribution from long wavelengths where the extinction coefficient is large, but the contribution to the power is small (especially at low temperature). The middle-wavelength contribution is relatively insensitive to the emitter temperature, since it originates in the substrate, which is held at fixed temperature. Because the emitter extinction coefficient is small in this region of the spectrum, the film is essentially transparent and radiation from the substrate passes through the film relatively unattenuated. This contribution is therefore insensitive to the gradient, and the total power is less sensitive to the gradient than is the useful power; as the temperature gradient increases, both the useful power and the total power decrease, but the useful power decreases more strongly, so that the efficiency decreases as well.

The total emittance versus average gradient is presented in Fig. 3d. As expected, the emittance is a decreasing function of gradient; because the emittance is computed by comparing with the output of a blackbody at the substrate temperature, as the temperature across the rest of the film deviates from the substrate temperature, the total power decreases compared with that of the reference blackbody.

Effects of Film Thickness

Because the film is not a perfect emitter/absorber, the radiation emerging from the radiating surface will not be a surface phenomenon alone; the total emitted power will include components from within the interior of the film, and from the substrate. If the film were at a constant temperature across its thickness, the total emitted power would increase with film thickness and eventually saturate, at a thickness determined by the film opacity. However, for vacuum interface conditions, as the film thickness increases, the film surface temperature drops and the average gradient increases, so that the temperature-dependent portion of the radiation from the emission band will tend to decrease. The competition between these two effects means that there will be an optimum thickness which maximizes the useful power, as shown (for the vacuum interface case) in Fig. 4a. The maxima for the three temperatures are broad, but it can be seen that for $T_s = 2000K$ the maximum occurs at a film thickness of around 0.06 cm, and the optimum thickness increases with decreasing substrate temperature. The emitter efficiency (Fig. 4b) decreases with increasing film thickness, as the temperature-sensitive emission-band contribution to the power decreases with gradient, while the total power increases or stays constant.

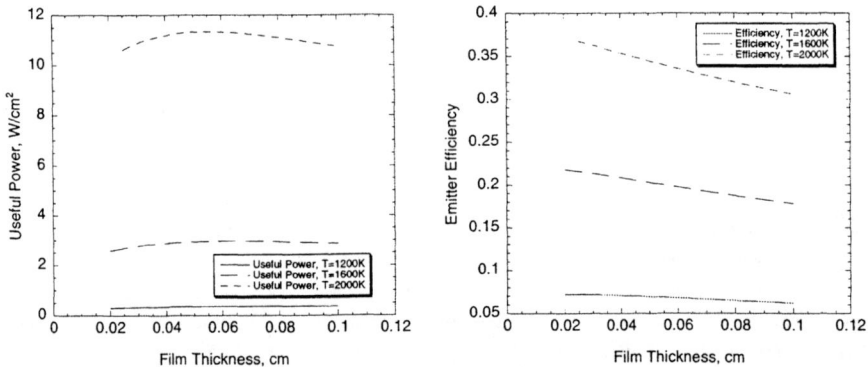

FIGURE 4a. Useful output power versus emitter film thickness.

FIGURE 4b. Emitter Efficiency versus emitter film thickness.

Effects of Thermal Conductivity

To further investigate the competition between conductive and radiative heat transfer, we next present results of a series of calculations in which the thermal conductivity k is varied from 0.005 to 1.0 W/cm/K, with all other parameters being the default ones.

The emitter efficiency and useful output power increase with increasing k for small k, and saturate at a thermal conductivity of around 0.1 W/cm/K. For large values of k, the heat transfer within the film is conductive, the temperature profile is close to linear, and the the radiating surface temperature is very close to the substrate temperature. On the other hand, when k is small, the temperature profile exhibits appreciable curvature, the radiating surface temperature is much lower than the substrate temperature, and the heat transfer is radiation-dominated, depending strongly on the opacity of the film.

It is also true that, in each case, the useful power and the efficiency are both larger for the vacuum interface than for the 10% gradient case. At large values of k, this can be understood by noting that the surface temperature is largest for the vacuum interface case. Because the two quantities are increasing functions of temperature, the vacuum interface case will exhibit larger values for these quantities than does the 10% gradient case.

Additionally, for small k the vacuum interface case exhibits greater sensitivity to k. For the 10% gradient case (i.e. with fixed radiating surface temperature), as the thermal conductivity decreases, the temperature profile shows greater curvature; the temperature everywhere but at the substrate and radiating face is below the linearized value, so that both computed quantities show reduced values. For the vacuum interface case, the preceding is true, but in addition, the surface temperature decreases with decreasing thermal conductivity, so that the decrease in the computed quantities with decreasing k is more pronounced.

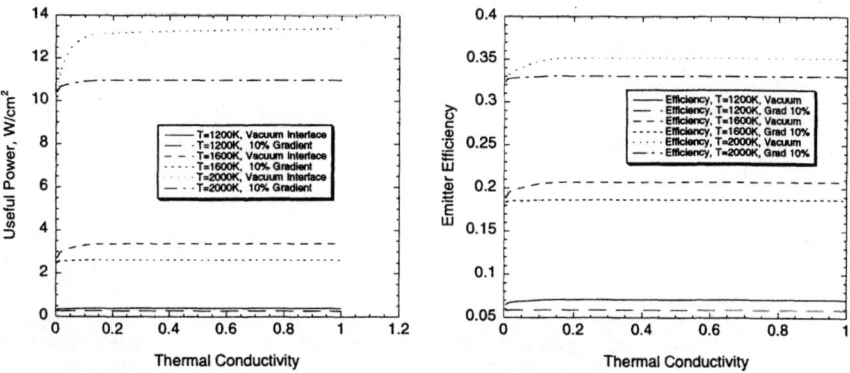

FIGURE 5a. Useful output power versus thermal conductivity. **FIGURE 5b.** Emitter Efficiency versus thermal conductivity.

Effects of Scattering

Finally, we consider the effects of changes in the scattering coefficient on emitter performance. We consider values of the scattering coefficient (expressed as a fraction of the extinction coefficient) of 0.0 to 0.9, for all three substrate temperatures, using the vacuum interface condition. The useful power decreases with increasing scattering coefficient, with a sharper decrease as the coefficient approaches unity. The efficiency at the two lowest substrate temperature exhibits a very broad maximum at a scattering coefficient of approximately 0.7 ($T_s = 1200K$) and 0.8 ($T_s = 1600K$). For ($T_s = 2000K$), the efficiency increases without a maximum for the range of scattering coefficients used here; there may be such a maximum at a still larger value of the scattering coefficient.

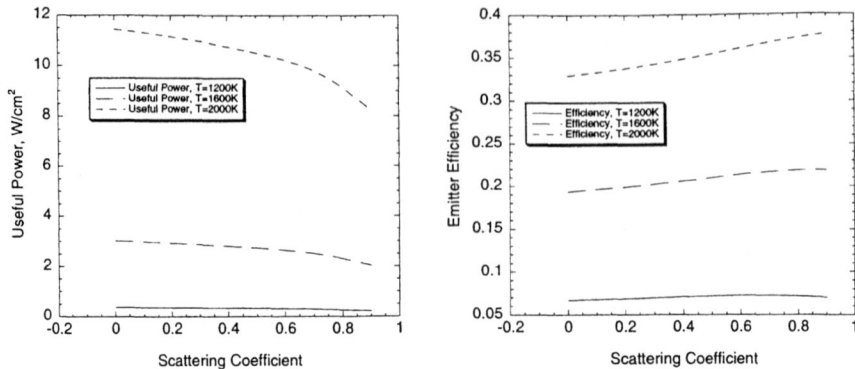

FIGURE 6a. Useful output power versus scattering coefficient.

FIGURE 6b. Emitter Efficiency versus scattering coefficient.

CONCLUSIONS

We have presented results of calculations of the performance of an erbium aluminum garnet selective emitter. For the range of physical properties investigated, we have obtained both linear temperature profiles and profiles that show appreciable deviation from linearity. The assumption of a linear profile is a reasonable first approximation; however, because the useful power output depends strongly on the temperature, if specific and detailed information about emitter performance is required, more detailed calculations are appropriate.

The useful power, total output power, emitter efficiency and total emittance are all decreasing functions of temperature gradient. The useful power exhibits a maximum at a value of film thickness that depends on substrate temperature, while the efficiency is simply a decreasing function of film thickness. The emitter efficiency and useful output power increase with increasing thermal conductivity for small values of the conductivity, and saturate at larger values. The useful power is a decreasing function of scattering coefficient, while the efficiency exhibits broad maxima for relatively large scattering coefficients.

REFERENCES

1. Good, B. S., Chubb, D. L., and Lowe, R. A., "Temperature-Dependent Efficiency Calculations for a Thin-Film Selective Emitter," in The First NREL Conference on Thermophotovoltaic Generation of Electricity, AIP Conf. Proc. 321, 1994, pp 263-275.

2. Good, B. S., Chubb, D. L., and Lowe, R. A., " Comparison of Selective Emitter and Filter Thermophotovoltaic Systems," in The Second NREL Conference on Thermophotovoltaic Generation of Electricity, AIP Conf. Proc. 358, 1995, pp 16-34

3. J. L. Gray and A. El-Husseini, "A Simple Parametric Study of TPV System Efficiency and Output Power Density Including a Comparison of Several TVP Materials," in The Second NREL Conference on Thermophotovoltaic Generation of Electricity, AIP Conf. Proc. 358, 1995, pp 3-15.

4. Good, B. S. and Chubb, D. L., " Effect of Geometry on the Efficiency of TPV Energy Conversion," in The Third NREL Conference on Thermophotovoltaic Generation of Electricity, AIP Conf. Proc. 401, 1997, pp 487-503.

5. Chubb, D. L., Good, B. S., Clark, E. B., and Zheng Chen, "Effect of Temperature Gradient on Thick Film Selective Emitter Emittance," in The Third NREL Conference on Thermophotovoltaic Generation of Electricity, AIP Conf. Proc. 401, 1997, pp 293-313.

6. Siegel, R., and Howell, J. R., *Thermal Radiation Heat Transfer*, Washington: Hemisphere Publishing, 1981, Ch. 14.

SESSION 4:
GASB-RELATED MATERIALS AND DEVICES

Bulk crystal growth of antimonide based III-V compounds for TPV applications

P.S. Dutta[*], A.G. Ostrogorsky[*], R.J. Gutmann[**]

Center for Integrated Electronics and Electronics Manufacturing
[*]Department of Mechanical Engineering, Aeronautical Engineering and Mechanics
[**]Department of Electrical, Computer Science and Engineering
Rensselaer Polytechnic Institute, Troy, New York – 12180

Abstract. In this paper, the bulk growth of crack-free GaInSb and single phase GaInAsSb alloys are presented. A new class of III-V quasi-binary semiconductor alloys $[A_{III}B_V]_{1-x}[C_{III}D_V]_x$ has been synthesized and bulk crystals grown from the melt for the first time. The present investigation is focused on the quasi-binary alloy $(GaSb)_{1-x}(InAs)_x$ (0<x<0.05) due to its importance for thermophotovoltaic applications. The structural properties of this melt-grown quasi-binary alloy are found to be significantly different from the conventional quaternary compound $Ga_{1-x}In_xAs_ySb_{1-y}$ with composition x = y. Synthesis and growth procedures are discussed. For the growth of ternary alloys, it was demonstrated that forced convection or mixing in the melt during directional solidification of $In_xGa_{1-x}Sb$ (0<x<0.1) significantly reduces cracks in the crystals.

INTRODUCTION

Semiconductor crystals with band gap in the range of 0.5 – 0.7 eV are needed for applications involving thermophotovoltaic (TPV) generation of electricity [1]. The availability of bulk substrates would significantly simplify the fabrication cycle, as the devices will be made on diffused junctions and the overall cost of the final device will be reduced. Although no elemental or binary semiconductor possesses such a band gap, this range of band gap can be realized in III-V ternary alloys like $In_xGa_{1-x}Sb$, $In_xGa_{1-x}As$ and $InAs_yP_{1-y}$. The antimonide based system (InGaSb) is preferred over the arsenic and phosphorus counterparts due to technical simplicity during growth, like extremely low partial vapor pressure of antimony [2]. Usually, melt-grown InGaSb crystals exhibit a high density of cracks due to: (1) large lattice mismatch between InSb and GaSb, and (2) segregation of InSb during growth due to the large separation between the liquidus and solidus curves in the pseudo-binary phase diagram of the GaSb-InSb system [3]. These features make the growth of single crystals of ternary alloys from melt extremely difficult. The lattice mismatch also hinders the use of binary seeds like GaSb, InAs, or InSb for single crystal growth. Therefore, ternary semiconductor crystals are currently produced in the form of thin layers ("epitaxy" or thin film growth) by non-equilibrium growth techniques (from diluted solutions and vapor phase) on binary substrates using compositionally graded buffer layers to reduce the stress in the mismatched lattice. Lattice matching to GaSb or InAs can be achieved by incorporating arsenic to form GaInAsSb quaternary alloys. Miscibility gaps in the pseudo-quaternary systems and phase separation are the main obstacles for quaternary alloy solidification from melts [4,5]. These arise mainly from differences in chemical interactions between the constituent elements in the melt. Therefore, quaternary semiconductor crystals are also produced in the form of thin layers by non-equilibrium epitaxial growth techniques either from diluted solutions or vapor phase. In this paper, we addressed the issues of crack elimination during the bulk growth of $Ga_{1-x}In_xSb$ and avoiding multiphase formation during the growth of GaInAsSb quaternary alloys. It was demonstrated that forced convection or mixing in the melt during directional solidification of bulk $In_xGa_{1-x}Sb$ ($0<x<0.1$) ternary alloys significantly reduces cracks in the crystals. Growth of single phase GaInAsSb quaternary alloys has been demonstrated from chemically associated melt [6].

EXPERIMENTAL DETAILS

Growth of GaInSb

Synthesis of $In_xGa_{1-x}Sb$ has been carried out in a multi-zone Mellen furnace [7] in 32 mm diameter silica crucibles from pre-synthesized GaSb and InSb freshly etched with CP4 etchant (CH_3COOH : HF : HNO_3 in 3:3:5 by volume). For the synthesis of GaSb and InSb, 6N pure In, Ga, and Sb were used without any chemical treatment. The unseeded directional solidification was done in flat bottom or conical tipped silica crucibles of 20 and 32 mm in diameter. Liquid encapsulation of the melt was provided by LiCl : KCl eutectic (58 mol % : 42 mol%) alkali halide salt to avoid volatilization and to reduce the probability of multiple nucleation from the crucible wall. A pre-growth baking of the charge and salt (to remove moisture) was carried out at approximately 300°C for a period of 10 - 12 hours under vacuum. After the baking, the furnace was filled with 1 atm of argon and heated to about 10°C above the melting temperature of GaSb (712°C). The synthesis was carried out by vertically raising and lowering the baffle in the melt for a period of 30-40 minutes. At the end of the synthesis cycle, the baffle was placed 1 cm away from the solid-liquid interface and the ampoule was lowered at a constant rate. Forced convection in the melt [8] was produced by (1) steady rotation of the baffle at 19 or 35 rpm, or (2) oscillatory rotation, i.e., by alternating the direction of rotation while the baffle was rotated at 19 or 35 rpm. The frequency of alternating the direction of rotation was 0.37 s^{-1}. The furnace temperature gradient near the melt-solid interface was approximately 15 °C/cm. The translation rate of the crucible was 3.3 mm/hr. After solidification, the furnace was cooled down slowly to room temperature, over a period of several hours.

Growth of GaInAsSb

The quaternary GaInAsSb alloys were synthesized from various combinations of starting elements (Ga, In, Sb) and compounded (InAs, GaSb, GaAs, InSb). Synthesis was carried out in a multi-zone Mellen furnace (vertical Bridgman set-up) in 20 and 32 mm diameter silica. Melt encapsulation was provided by boric oxide (B_2O_3) or alkali halide salts (LiCl-KCl and NaCl-KCl). Boric oxide encapsulation was found to be more satisfactory and suitable (due to extremely low vapor pressure and high viscosity) for inhibiting volatilization from the melt surface. The growth chamber was usually pressurized to slightly more than 1 atm by argon gas to prevent decomposition of the arsenic based compounds. Synthesis was done at various temperatures in the range of 712 - 945°C. After synthesis, the crucible was lowered at a constant rate of 3.3 mm/hr through a temperature of 15 – 20 °C/cm. At the end of solidification, the furnace was cooled down slowly to room temperature over a period of several hours. Crystal growth was performed without seed, either in flat bottom or conical tipped crucibles.

After the growth, the ingots (GaInSb and GaInAsSb) were sliced parallel to the growth axis to evaluate the structural and compositional properties. The composition of the grown crystals (Ga, In, Sb, and As) was evaluated by the Electron Probe Micro-Analysis (EPMA) measurements in a JEOL 733 electron microprobe set-up. The standards used were InAs, InSb, GaAs and GaSb single crystal substrates. Corrections for atomic number (Z), self-absorption (A), and fluorescence (F) effects (ZAF corrections) were performed by employing the commercial software

SCOTT-I. The error in determining the composition was in the order of 1-2% of the measured values. The microstructures of the crystals were studied through Secondary Electron Microscopy (SEM).

FIGURE 1. Longitudinally sliced $In_xGa_{1-x}Sb$ crystals grown (a) without a stationary baffle (x = 0.05, v = 3.3 mm/hr), (b) with a rotating baffle (x = 0.1, v = 3.3 mm/hr).

RESULTS AND DISCUSSION

Crack elimination in GaInSb

Fig.1a shows a typical GaInSb crystal grown from a melt containing 5 mol% InSb – 95 mol% GaSb without a baffle in the melt. Cracks typical of ternary alloys are clearly seen in this crystal. Fig. 1b shows a typical crystal grown from a melt of 10 mol% InSb – 90 mol% GaSb with a rotating baffle. Unlike the crystal in Fig. 1a, no cracks could be seen in this crystal. Moreover, the rotating baffle significantly improves the axial and radial compositional homogeneity in the grown crystal.

Growth of single phase quaternary or quasi-binary alloys

In the absence of any phase diagrams, the initial studies were focused towards studying the microstructure of quaternary $Ga_{1-x}In_xAs_ySb_{1-y}$ synthesized at various temperatures and charge preparation cycles from melts containing 20 mol% In, 80 mol% Ga, 13 mol% As, and 87 mol% Sb. This alloy composition was attempted with the aim of obtaining a band gap of ~ 0.55 eV which is necessary for specific TPV applications. In preliminary studies, spatial inhomogeneity in the crystal composition due to the multi-phase formation was observed as shown in Fig. 2(a-c). Phase separation is thermodynamically expected in quaternary systems grown from a regular solution [4]. The multi-phase formation is attributed to the presence of elemental sources.

Fig. 2a shows the microstructure of $Ga_{1-x}In_xAs_ySb_{1-y}$ synthesized at 950°C from melt containing elemental Ga, In, Sb, and compounded InAs (arsenic source). The multiple phases are formed due to the decomposition of InAs and subsequent formation of random mixed alloys with elemental Ga, In, and Sb. Fig. 2b shows typical microscopic phases observed in $Ga_{1-x}In_xAs_ySb_{1-y}$ synthesized from a melt containing elemental Ga, In, Sb, and compounded GaAs (arsenic source) at 950°C. Low synthesis temperature improves the compositional homogeneity. However, it does not fully avoid the formation of multiple phases. Fig. 2c shows inclusions in the quaternary crystals synthesized at 800°C from the same melt composition as the crystal in Fig. 2b. By correlating the microstructure of the crystals with the melt constituents, it can be concluded that the multiple phase formation arises from chemical interaction between elemental sources in the melt. Hence, they can be suppressed by synthesizing the charge from compounded sources as discussed below.

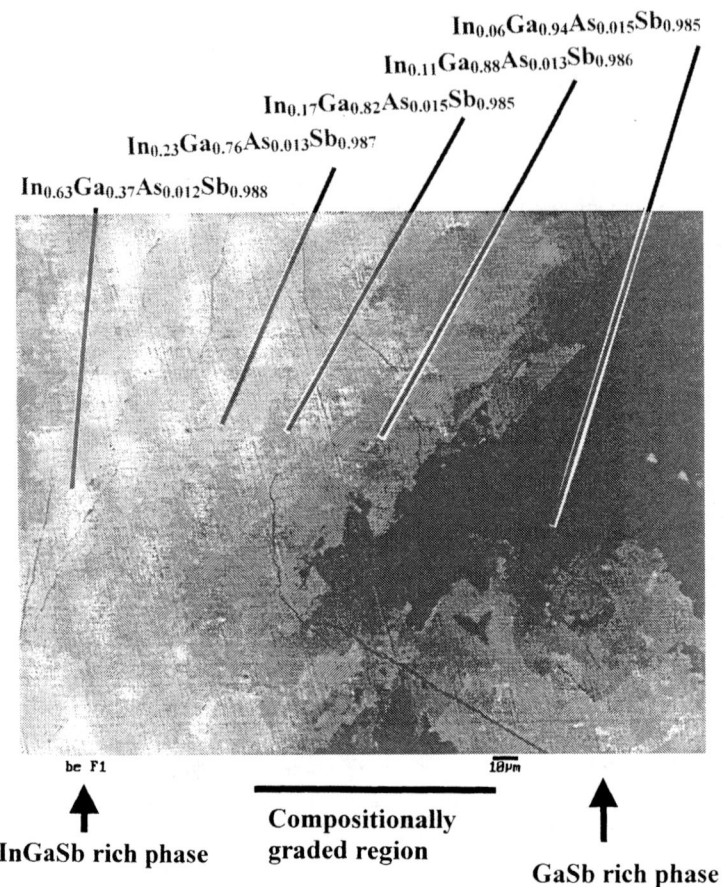

2(a)

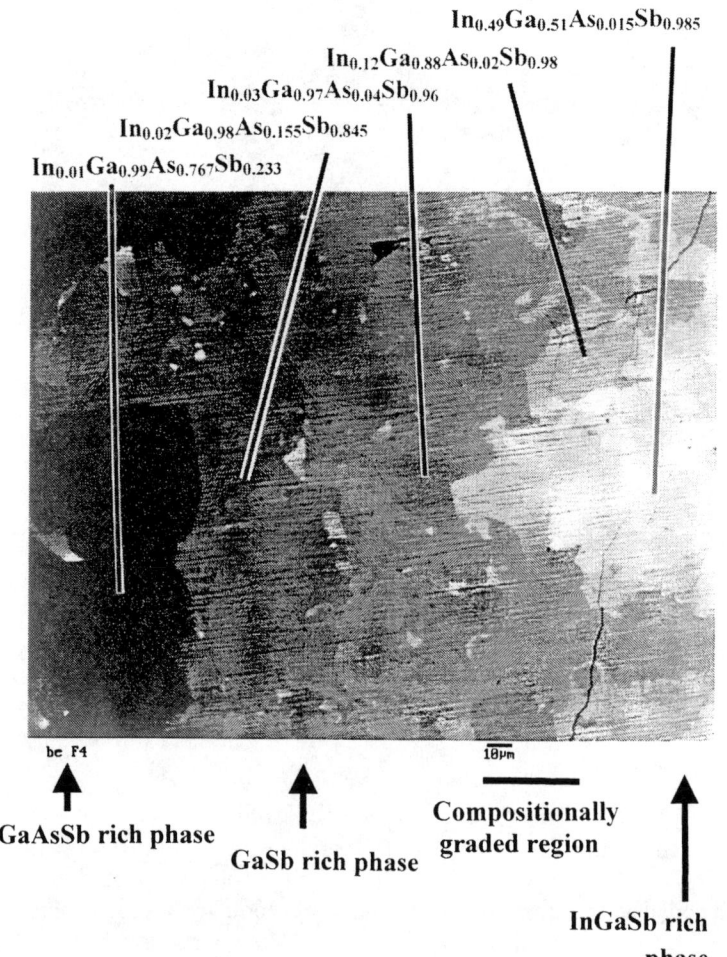

2(b)

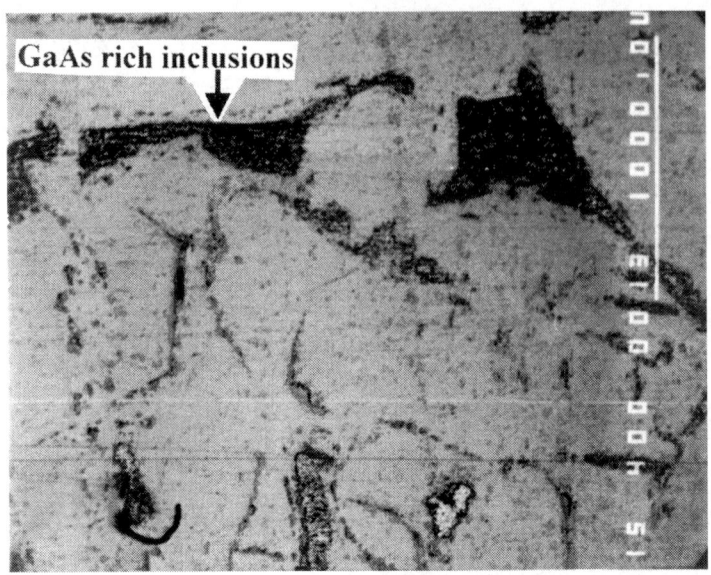

2(c)

FIGURE 2. Microstructures (SEM photomicrograph) of GaInAsSb quaternary alloys synthesized from (a) Ga, In, Sb and InAs at 950°C, (b) Ga, In, Sb and GaAs at 950°C, (c) Ga, In, Sb and GaAs at 800°C. The mol % for In, Ga, As, and Sb in the melt were 20, 80, 13, and 87, respectively.

Typical microstructure of the quasi-binary $(GaSb)_{1-x}$-$(InAs)_x$ synthesized in the 720 – 850°C temperature range from compounded GaSb and InAs is depicted in Fig. 3a [6]. From SEM studies, it is concluded that the quasi-binary crystals are completely single phase in nature, unlike the quaternary alloys. Moreover, synthesis from compounded sources significantly reduces the probability of multi-phase formation. It is also evident from the comparison of Figs. 3a and 3b, that the quasi-binary GaSb-InAs crystals do not exhibit cracks, unlike their ternary GaSb-InSb counterpart. This is due to 10 times less lattice mismatch of InAs and GaSb as compared to InSb and GaSb.

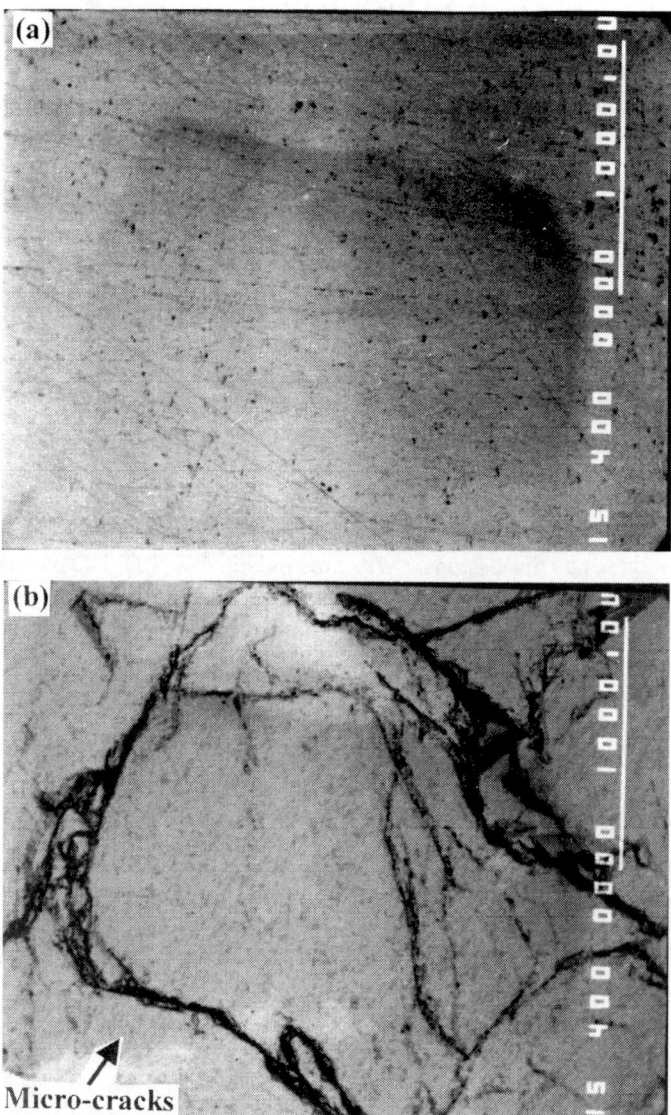

FIGURE 3. Microstructures (SEM photomicrograph) of (a) $(GaSb)_{0.97}$-$(InAs)_{0.03}$ quasi-binary alloy synthesized from GaSb and InAs, (b) $(GaSb)_{0.97}$ -$(InSb)_{0.03}$ ternary alloy synthesized from GaSb and InSb

CONCLUSION

In conclusion, the present study demonstrates that the use of forced convection during the growth of mixed alloys (InGaSb in this case) will produce spatially homogeneous composition, otherwise unattainable even by post growth thermal treatments. In turn, homogeneous composition inhibits cracking of the crystal and improves the microscopic crystalline quality. Large polycrystals of semiconductor quasi-binary alloys of $(GaSb)_{1-x}-(InAs)_x$ were grown from melt for the first time. These alloys possess better crystalline perfection (crack free and are single phase) and compositional homogeneity (close to unity segregation) than melt grown bulk ternary and quaternary alloys.

ACKNOWLEDGMENTS

The authors would like to thank Dr. Greg Charache and Mr. Greg Nichols (Lockheed Martin Inc., Schenectady, NY) for invaluable scientific information and discussions. We are indebted to Dr. David Wark (Rensselaer Polytechnic Institute, Troy, NY) for the assistance in the EPMA measurements. The growth equipment used in the present work was provided by Lucent Technologies, New Jersey.

REFERENCES

1. Coutts, T.J., Allman, C.S. and Benner, J.P., *The Third NREL Conference on Thermophotovoltaic Generation of Electricity,* AIP Conference Proceedings **401**, AIP Press, 1997.
2. Dutta, P.S., Bhat, H.L. and Kumar, V., *J. Appl. Phys.* **81**, 5821 (1997).
3. Ansara, I., Gambino, M. and Bros, J.P., *J. Crystal Growth* **32**, 101 (1976).
4. Bachmann, K.J., Thiel, F.A. and Schrieber Jr., H., *Progress in Crystal Growth and Characterization* **2**, 171 (1979).
5. Nakajima, K., Osamura, K., Yasuda, K. and Murakami, Y., *J. Crystal Growth* **41**, 87 (1977).
6. Dutta, P.S. and Ostrogorsky, A.G., *Alloys and Methods for their preparation,* (U.S. Patent pending) ; see also: Dutta, P.S. and Ostrogorsky, A.G., *Melt growth of quasi-binary $(GaSb)_{1-x}(InAs)_x$ crystals, J. Crystal Growth* (in press).
7. Dutta, P.S., Ostrogorsky, A.G. and Gutmann, R.J., "Bulk growth of GaSb and $Ga_{1-x}In_xSb$" *The Third NREL Conference on Thermophotovoltaic Generation of Electricity,* AIP Conference Proceedings **401**, Eds. Coutts, T.J., Allman C.S. and Benner, J.P., 1997, pp. 157.
8. Meyer, S. and Ostrogorsky, A.G., *J. Crystal Growth* **171**, 566 (1997).

Pseudo-closed box diffusion of Zn into InGaAsSb and AlGaSb for TPV devices

Andreas W. Bett[1], Boris Y. Ber[2], Michael G. Mauk[3],
Joseph T. South[3], and Oleg V. Sulima[4]

[1] *Fraunhofer Institute for Solar Energy Systems, Oltmannstr. 5, D-79100 Freiburg, Germany,*
  Phone: (+49) 761-4588257, Fax: (+49) 761-4588250, E-mail: bett@ise.fhg.de
[2] *Ioffe Institute, Polytechnicheskaya 26, 194021 St. Petersburg, Russia, Phone: (+7) 812-*
  2479362, Fax: (+7) 812-2471017, E-mail: ber@ntcmic.ioffe.rssi.ru
[3] *AstroPower, Inc., Solar Park. Newark, De 19716-2000, USA, Phone: (+1) 302-3660400, Fax:*
  (+1) 302-3686474
[4] *Freiburg Materials Research Center, Stefan-Meier Str. 21, D-79104, Freiburg, Germany,*
  Phone: (+49) 761-4588256, Fax: (+49) 761-4588250, E-mail: sulima@ise.fhg.de

Abstract: The first systematic study of Zn diffusion into InGaAsSb and AlGaSb is presented. The diffusion parameters, which provide good control over p-n junction depth, are determined. Anodic oxidation of $In_{0.15}Ga_{0.85}As_{0.17}Sb_{0.83}$ is investigated. InGaAsSb/GaSb photovoltaic (PV) cells were fabricated using two technologically simple methods: Liquid phase epitaxy and pseudo-closed box diffusion. The measured open-circuit voltage is the highest ever published for PV-cells with $E_g \sim 0.55$ eV.

INTRODUCTION

Epitaxial InGaAsSb and AlGaAsSb layers lattice-matched to GaSb are of large interest for TPV devices. Two technological approaches are mainly used for the growth of p-n structures in these quaternary compounds: MOVPE and LPE. We have studied another approach. LPE was used for the growth of n-type layers on GaSb, while p-layers were formed by the pseudo-closed box (PCB) diffusion method of Zn into the epitaxial n-doped layers. The PCB diffusion method applied to n-GaSb has proved to be simple, productive and reproducible [1-3]. Combination of two technologically simple methods (LPE and PCB diffusion) can provide more technological flexibility and has the potential to reduce production costs.

During the last years the PCB diffusion of Zn into n-GaSb was intensely studied. However, the Zn diffusion process in related ternary or quaternary compounds is a new field of research. Essential changes in diffusion profiles take place due to the partial substitution of Ga and Sb atoms in GaSb by In/Al and As atoms, respectively. GaSb PV cells fabricated by using the PCB diffusion method have shown a strong dependence on the diffused emitter parameters. Therefore, the exact Zn diffusion behavior should be studied in the quaternary/ternary systems before fabrication of devices.

CP460, *Thermophotovoltaic Generation of Electricity: Fourth NREL Conference*
edited by T. J. Coutts, J. P. Benner, and C. S. Allman
© 1999 The American Institute of Physics 1-56396-828-2/99/$15.00

This paper presents the first systematic study of Zn diffusion into InGaAsSb and AlGaSb. A series of Zn diffusions at different temperatures and times were performed into $In_{0.15}Ga_{0.85}As_{0.17}Sb_{0.83}$, lattice matched to GaSb, and $Al_xGa_{1-x}Sb$ (x = 0.05 - 0.78) layers. We consider AlGaSb as an intermediate step in going to AlGaAsSb lattice matched to GaSb. Zn profiles together with the composition of III- and V- elements were measured by the Secondary Ion Mass Spectroscopy (SIMS) method. The diffusion parameters, which provide good control over p-n junction depth, are determined. The very accurate etching of InGaAsSb by using of anodic oxidation is studied. These new data were used for the fabrication of InGaAsSb/GaSb PV-cells and can be used for AlGaAsSb/InGaAsSb/GaSb PV-cells.

EXPERIMENTAL

n-$In_{0.15}Ga_{0.85}As_{0.17}Sb_{0.83}$ layers on 1cm^2 (100)-oriented n-GaSb substrates were grown by LPE. The growth solution was comprised of a 4-gram melt of indium (atomic fraction: X_{In} = 0.590), antimony (X_{Sb} = 0.214) and gallium (X_{Ga} = 0.196) saturated with an InAs "float" wafer (in excess). N-type doping was effected by adding 1 mg of PbTe to the melt. The melt was equilibrated at 533°C for 1 hour and then ramped to a growth temperature of 516°C. The InGaAsSb layers were grown at 516°C for two minutes.

Pseudo-closed box diffusion (PCB) of Zn from the vapour phase into n-InGaAsSb layers was performed essentially as described in [2] for GaSb. However, in comparison with GaSb, lower diffusion temperatures were studied (360-460°C). The diffusion time was 1, 2 and 4 hours for 360-420°C diffusions and 4 hours for 440-460°C diffusions. The diffusion process was performed in a H_2-atmosphere, purified by a Pd cell, in a specially designed multi-wafer graphite boat. Separated pure Zn and Sb vapour sources were used in more than sufficient quantities to provide the saturation of vapour pressures. The design of the graphite boat ensures the uniformity of the Zn and Sb vapour pressure across the wafer surface, and thus the uniformity of the p-InGaAsSb layer depth.

$Al_xGa_{1-x}Sb$ layers (x=0.05, 0.16, 0.29, 0.35, 0.67, 0.78) on 1cm^2 (100) oriented GaSb substrates were grown by LPE. Pseudo-closed box diffusion of Zn from the vapour phase into $Al_xGa_{1-x}Sb$ layers was performed essentially as into GaSb [2]. The temperature of diffusion was 480°C as for the standard diffusion used at Fraunhofer ISE for the fabrication of GaSb TPV cells. The diffusion time was 60 and 205 minutes.

INVESTIGATION OF Zn DIFFUSION

$In_{0.15}Ga_{0.85}As_{0.17}Sb_{0.83}$

Figure 1 shows Zn concentration profiles in InGaAsSb in dependence on temperature for a 4-hour diffusion.

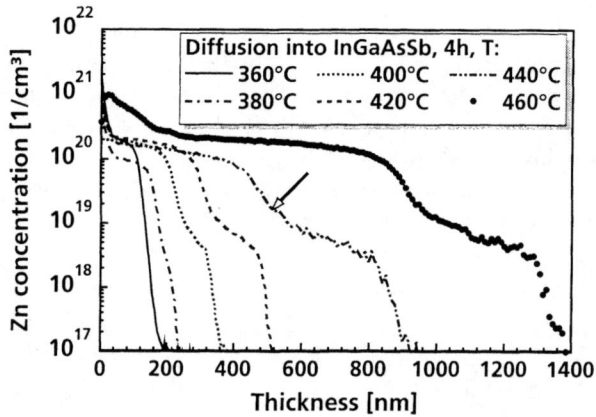

FIGURE 1. SIMS profiles of Zn concentration in $In_{0.15}Ga_{0.85}As_{0.17}Sb_{0.83}$ in dependence on temperature after a 4-hour diffusion. The arrow shows the position of the kink in the Zn concentration profile for a 440°C – diffusion.

It was found that Zn-diffusion profiles generally consist of 5 parts (except the diffusion profile at 360°C): (i) a part with a steep Zn concentration gradient near the surface, (ii) a first plateau with a relatively slow decrease of Zn concentration, (iii) a kink part (the exact position of the kink is defined as the point where the second derivative of the Zn concentration profile is at maximum [4]), (iv) a second plateau with a relatively slow change of Zn concentration and (v) a steep diffusion front. The first part was usually thinner than 50 nm, except the diffusion profile at 460°C where it was as large as 200 nm. We consider that this part of the SIMS measurements is not reliable enough. Therefore, we did not take this part into further consideration. Hereafter we define as the surface concentration the maximum concentration at the first plateau region. Similar Zn diffusion profiles were obtained in GaSb, however at higher diffusion temperature [2].

Figure 2 shows the dependence of the Zn surface concentration on the diffusion temperature for different diffusion times. It is noteworthy, that there is almost no difference in the Zn surface concentration at 360 – 460°C after 1h-, 2h- or 4h-diffusions. This shows, that the saturation of Zn concentration of InGaAsSb in this time-temperature range takes place at the level of $\approx 2 \times 10^{20} cm^{-3}$. This value is nearly the same as the one determined for GaSb after 1h – diffusion at 450-500° [2]. The fact that the surface concentration can be saturated is very important for the reproducibility and simplicity of the p-layer formation process. In general, the diffused p-layer thickness depends on the surface concentration, diffusion temperature and time. On the other hand, the surface concentration itself can vary with temperature and time. This can make the temperature and time dependencies of the p-layer thickness complicated and hardly reproducible. If one can fix the surface concentration at the saturation level, the p-layer thickness can be regulated either by diffusion temperature or diffusion time with high accuracy.

The possibility of the reproducible formation of a high Zn surface concentra-on is of practical interest for TPV. For example, the higher doping concentration leads to lower contact/sheet resistance. This can be helpful for high-current PV-cells. High surface concentration can be useful for the formation of strong built-in electrical fields caused by the gradient of doping concentration. These fields increase the effective diffusion length due to the drift of minority carriers and decrease the influence of the surface recombination.

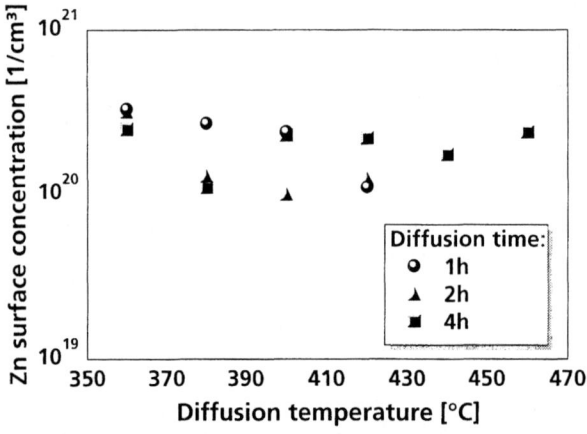

FIGURE 2. Dependence of the Zn surface concentration on the diffusion temperature for different diffusion times.

A large diffusion length is always important for the PV-cells. Therefore, a lower Zn concentration can be also advantageous. For example, a high diffusion length (up to 29 µm) was determined in low-doped InGaAsSb ($p \approx 2 \times 10^{17} cm^{-3}$) [5]. In contrast, the relatively thick first plateau region with a high Zn concentration (Fig. 1) essentially decreases the diffusion length in the PV-cell emitter. The surface concentration can be decreased if the Zn pressure in the graphite diffusion boat is reduced. The decreasing of the diffusion temperature for this purpose is not very effective. Figure 2 shows that the reduction of the diffusion temperature down to 360°C does not decrease the surface concentration. It means that the even at 360°C the Zn vapour pressure is still high enough to provide the saturation level of the Zn doping concentration. The use of diluted Zn-vapour sources (Zn-Ga melts) for the decreasing of the Zn doping concentration was studied for GaSb in [2]. It was shown that a Zn depletion in the melt and a Zn out-diffusion from the diffused layer can take place. This leads to a relatively poor reproducibility of the Zn diffusion profiles. Therefore, one has to conclude, that a reproducible formation of Zn diffused layers with low Zn concentration near the surface in GaSb or InGaAsSb is hardly possible by using the PCB method.

However, one should consider the "after-kink-part" of the Zn diffused profiles (Fig.1) which is characterized by essentially lower Zn concentration in comparison with the surface one. The Zn concentration which corresponds to the kink position for 4h-diffusions at 400-460°C is $2\text{-}3 \times 10^{19}$ cm^{-3}. Similar to the Zn surface concentration, (Fig.2) it is practically independent on the diffusion temperature.

Figure 3 shows temperature dependencies of p-n junction (for $n=3 \times 10^{17}$ cm^{-3}) and kink positions for the 4h-diffusion processes. As the diffusions at temperatures < 400°C did not lead to reproducible Zn concentration profiles (except Zn surface concentration), only the diffusion temperatures ≥ 400°C were considered.

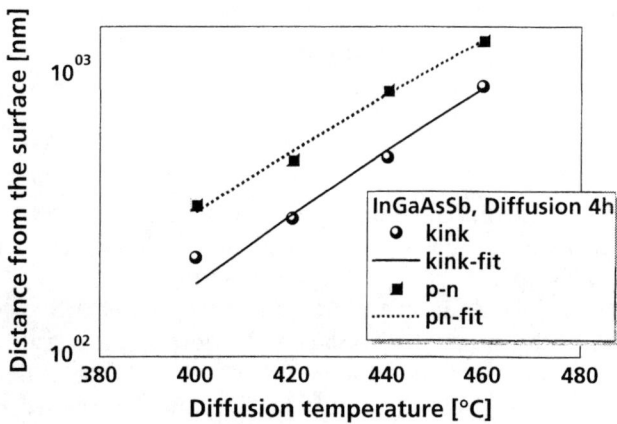

FIGURE 3. Temperature dependencies of p-n junction ($n=3 \times 10^{17}$ cm^{-3}) and kink positions for 4h-diffusions.

Fitting of the data from Fig.3 was performed by using the following equation:
$$x = A \exp(B/(T+273)),$$
where x is a corresponding distance of the p-n junction/kink from a surface (depth), T is a diffusion temperature in °C and A, B are fitting parameters.

The following fitting equations were obtained:
$$x_{p-n}(nm) = 1.1 \cdot 10^{10} \exp(-11640/(T+273)),$$
$$x_{kink}(nm) = 5.8 \cdot 10^{10} \exp(-13181/(T+273)).$$

As it was determined for GaSb [2], for certain diffusion conditions the p-n junction depth is proportional to the square root of the diffusion time. In this work, it was found that for InGaAsSb the proportionality of the p-n junction depth on the square root of the diffusion time (t) is true for diffusion temperatures T≥400°C which are longer than 2h. If this fact is considered, the fitting equations can be transformed to:

$$x_{p-n}(nm) = 6.9 \cdot 10^8 \sqrt{t} \exp(-11640/(T+273)),$$

$$x_{kink}(nm) = 3.8 \cdot 10^9 \sqrt{t} \exp(-13181/(T+273)),$$

where t is given in minutes.

With the help of these two equations and the fixed Zn surface concentration, three important points of diffusion profiles are predictable. This is extremely important for the PV-cell technology: One can regulate the thicknesses of the highly and moderately doped parts of the diffused emitter (first and second plateaus, Fig.1). Additionally, by using a precise etching technique – anodic oxidation/selective etching (see below) - one can practically stepless remove the surface layers. This provides additional possibility to decrease (also locally) the emitter thickness and the only possibility to reduce surface concentration of the emitter.

For fabrication of the first cells, investigated in this work, we used a 2-step approach consisting of Zn diffusion at 430°C (4h) and anodic oxidation of the structure. The anodic oxide in this case was also used as an antireflection coating

$Al_xGa_{1-x}Sb$

Figure 4 shows Zn concentration profiles in $Al_xGa_{1-x}Sb$ in dependence on AlSb content (x) for a 205 minutes diffusion at 480°C – the standard temperature for fabrication of GaSb TPV-cells at the Fraunhofer ISE.

One can note that the Zn profile in $Al_xGa_{1-x}Sb$ with a low AlSb content (x = 0.05) is essentially the same as in GaSb (x = 0). However, at $x \geq 0.16$ the first plateau (see discussion of profiles from Fig. 1) becomes thicker and a new part appears – a flat tail. Moreover, at $x \geq 0.35$ the kink and thus the second plateau disappear. Obviously, the p-n junction depth increases with Al content. This is due to both the increase of the first plateau thickness and the appearance of the flat diffusion tail.

Zn profiles in $Al_xGa_{1-x}Sb$ with x = 0.78 (not shown in Fig. 4) were generally much deeper, however, not reproducible neither in surface concentration, nor in thickness. It can be explained by crystal defects originating from the large lattice mismatch between $Al_{0.78}Ga_{0.12}Sb$ and GaSb.

Consideration of the above shown influence of Al on the Zn diffusion in GaSb-based materials can be useful for the design and fabrication of PV-cells containing AlGaAsSb.

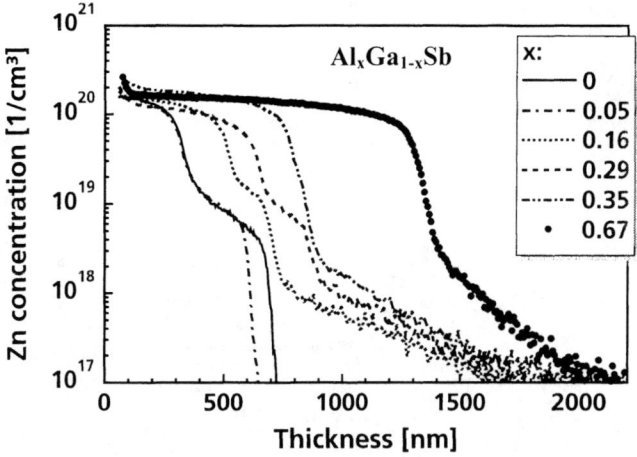

FIGURE 4. SIMS profiles of Zn concentration in $Al_xGa_{1-x}Sb$ in dependence on AlSb content (x) for a 205 minutes diffusion at 480°C.

STUDY OF ANODIC OXIDATION OF $In_{0.15}Ga_{0.85}As_{0.17}Sb_{0.83}$

Anodic oxidation in combination with selective etching of anodic oxide is an extremely accurate method for the thinning of semiconductor layers. In this work the study of anodic oxidation of 1 cm^2 n-$In_{0.15}Ga_{0.85}As_{0.17}Sb_{0.83}$ layers was performed. The anodization technique used was similar to that reported for GaAs in [6]. The thickness and the refractive index of anodic oxide layers were measured with the ellipsometer (Plasmos, SD 2000). To determine the thickness of consumed (oxidized) InGaAsSb, selective anodic oxidation was carried out. After the oxidation the oxide was selectively removed in HCl and the height of the etch step was measured by a surface profiler. The applied voltage was in the range of 20-100 V. The anodization was performed at constant current density (5 mA/cm^2) conditions. The anodization was stopped when the current density dropped down to 0.1 mA/cm^2. Figure 5 shows oxide thickness and amount of consumed InGaAsSb vs applied voltage. The consumption rate of n-$In_{0.15}Ga_{0.85}As_{0.17}Sb_{0.83}$ is 2.26 nm/V and the oxidation rate is 2.51 nm/V. The refractive index of the anodic oxide, measured at the different depth of the layer, was 1.97 ± 0.03. Our previous studies of n- and p-GaSb anodic oxidation (n,p > 3×10^{17} cm^{-3}) have shown no dependence of the oxide thickness and the amount of consumed GaSb on the type and value of the GaSb doping. Therefore, the same behaviour is assumed for InGaAsSb. This assumption is important for the fabrication process of InGaAsSb PV cells.

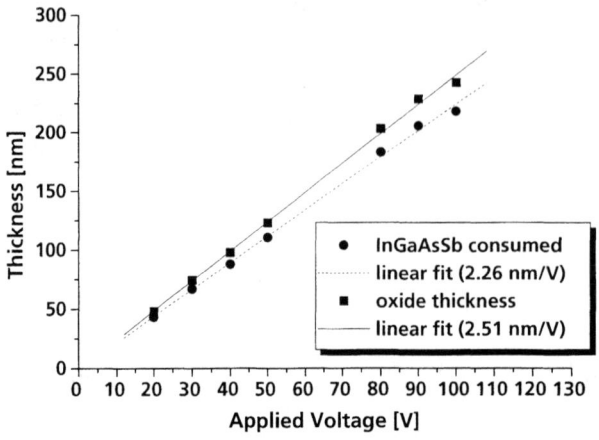

FIGURE 5. Anodic oxide thickness and amount of consumed n-$In_{0.15}Ga_{0.85}As_{0.17}Sb_{0.83}$ vs applied voltage.

FABRICATION AND INVESTIGATION OF InGaAsSb PV-CELLS

PV-Cell Fabrication

The InGaAsSb PV cell structure investigated in this work is shown schematically in Fig.6. A 5 μm thick n-$In_{0.15}Ga_{0.85}As_{0.17}Sb_{0.83}$ base layer was grown by LPE. A p-layer (emitter) was formed by Zn diffusion at 430°C during 4h. A post-diffusion thinning of the p-layer by anodic oxidation was applied. The resulting thickness of the emitter is approximately 0.5 μm. The same anodic oxide layer with a thickness of 136 nm was used as an antireflection coating. Front contacts were made of Ti/Ni/Au and back contacts of a Au-Ge/Au. The circular PV-cells (∅ 2 mm) have a contact grid consisting of a busbar and radial contact fingers and covering ≈ 25% of the cell area (area covered by the fingers only is ≈ 4%).

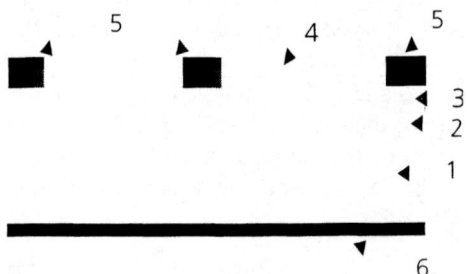

FIGURE 6. Scheme of a InGaAsSb PV cell: 1 - n-GaSb base (substrate), 2 – n-InGaAsSb LPE-grown base, 3 - diffused p-InGaAsSb emitter, 4 - antireflection coating (anodic oxide), 5 - front contacts, 6 - back contact

Measurement of I-V Curves

I-V curves of the PV-cells were measured at different illumination levels and at different cell temperatures. Figure 7 shows the dependence of the open-circuit voltage (V_{oc}) at T = 25, 50, 60°C on current density obtained in this work together with other published data [7,8].

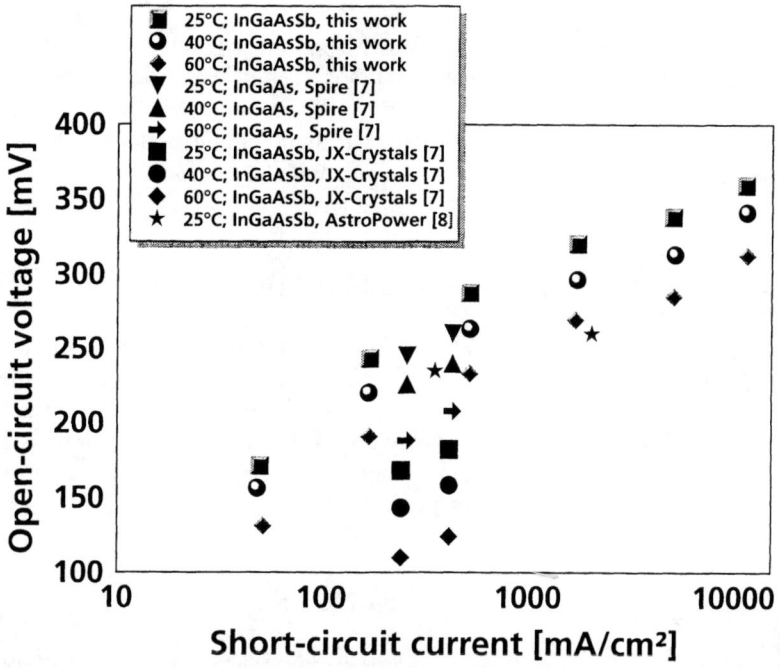

FIGURE 7. Dependence of the open-circuit voltage (V_{oc}) on current density of InGaAsSb and InGaAs PV-cells with Eg = 0.54-0.56 eV at T = 25, 50, 60°C. The shown data were obtained in this work as well as taken from Ref. [7,8].

One can see that at any temperatures the V_{oc} values of our cells exceed the values reported in other papers for cells with a similar band gap (0.54-0.56 eV). This shows that a high-quality p-n junction is formed by the diffusion process.

The temperature coefficient of V_{oc} in the current density range of 0.17 – 11.15 A/cm^2 was 1.34 – 1.55 mV/K. The highest FF achieved in our cells at 25°C was 0.64 (J=1.7 A/cm^2). The temperature coefficient of the FF at this current density was 0.11abs.%/K.

CONCLUSIONS

The pseudo-closed box (PCB) diffusion method applied to LPE-grown n-$In_{0.15}Ga_{0.85}As_{0.17}Sb_{0.83}$ has proved to be simple and reproducible. The surface concentration of Zn was found to be independent of diffusion temperature (360 - 460°C) and time (1 - 4 h). The observed value of $\approx 2 \times 10^{20}$ cm^{-3} is obviously the saturation level of Zn in $In_{0.15}Ga_{0.85}As_{0.17}Sb_{0.83}$. Analytical expressions, summarizing measured temperature/time dependences of the p-n junction and the diffusion profile kink positions, were determined. The following parameters of anodic oxidation of n-$In_{0.15}Ga_{0.85}As_{0.17}Sb_{0.83}$ were obtained: The consumption rate - 2.26 nm/V, oxidation rate - 2.51 nm/V and the refractive index of the anodic oxide - 1.97 ± 0.03. The first PV cells made by the LPE-PCB combination and anodic oxidation have an open-circuit voltage, which is the highest among published values for PV-cells with $E_g \sim 0.55$ eV.

ACKNOWLEDGEMENTS

The authors wish to thank the colleagues at Fraunhofer ISE E. Schäffer, C. Schetter and S. van Riesen for technical assistance, as well as Prof. J. Luther and Prof. W. Wettling for permanent interest in this investigation. This work was partly financed by the Federal Ministry of Education, Science, Research and Technology (BMBF), Germany (contract 0328554D). The authors are responsible for the content of this publication.

REFERENCES

1. Fraas, L.M., Avery, J.E., Martin, J., Sundaram,V.S., Girard, G., Dinh, V.T., Davenport, T.M., Yerkes, J.W., and O'Neil, M.J., *IEEE Trans: Electron. Dev.* **37**, 43-449 (1990)
2. Bett, A.W., Keser, S., and Sulima, O.V., *J. Cryst. Growth* **181**, 9-16 (1997)
3. Bett, A.W., Keser, S., Stollwerck, G., and Sulima, O.V., in: *Thermophotvoltaic Generation of Electricity: Third NREL Conference*, New York: AIP Conference Proceedings 401, 1997, pp. 41-53
4. Reynolds, S., Vook, D.W., and Gibbons, J.F., *J.Appl. Phys.*, **63**, 1052-1059 (1988)
5. Martinelli, R.U., Garbuzov, D.Z., Lee, H., Morris, N., Odubanjo, T., Taylor, G.C., and Connoly, J.C., in: *Thermophotvoltaic Generation of Electricity: Third NREL Conference*, New York: AIP Conference Proceedings 401, 1997, pp. 389-395
6. Hasegawa, H., and Hartnagel, H.L., *J. Electrochem. Soc.*, **123**, 713 - 723 (1976)
7. Carache, G.W., Egley, J.L., Danielson, L.R., DePoy, D.M., Baldasaro, P.F., and Campbell, B.C., in: Conference Record of the Twenty Fifth IEEE Photovoltaic Specialists Conference, IEEE, 1996, pp. 137-140
8. Shellenbarger, Z.A., Mauk, M.G., DiNetta, L.C., and Charache, G.W., in: Conference Record of the Twenty Fifth IEEE Photovoltaic Specialists Conference, IEEE, 1996, pp. 81-84

Interfacial Recombination in In(Al)GaAsSb/GaSb Thermophotovoltaic Cells

V. B. Khalfin, D. Z. Garbuzov, H. Lee, G. C. Taylor, N. Morris,
R. U. Martinelli, and J. C. Connolly

Sarnoff Corporation, CN 5300, Princeton, NJ 08543-5300

Abstract. We have studied efficient p-on-n homo- (InGaAsSb/GaSb) and hetero-junction (InGaAsSb/AlGaAsSb/GaSb) thermophotovoltaic (TPV) cells with respect to the recombination velocity at the cap-layer/emitter interface, S. In both cell types the open-circuit voltage, V_{oc}, and the short-circuit current, J_{sc}, have about the same sensitivity to S. The dark current, J_0, is the most sensitive of all. An examination of the essential factors in the one-dimensional minority-carrier diffusion model shows that under short-circuit conditions, photogenerated electrons diffuse rapidly away from the interface to the junction, whereas under open-circuit conditions, they remain in the emitter for a much longer bulk-recombination time, and therefore, they are more likely to recombine at the interface. A factor of 2.2 increase in S from 2.5 to 5.05 x 10^4 cm/s produces a 25-mV decrease in V_{oc}, a 12-percent decrease in J_{sc} or quantum efficiency, and a factor of two increase in J_0. This work points out the critical importance of interfacial recombination even in efficient TPV cells.

INTRODUCTION

We have characterized p-on-n InGaAsSb/GaSb thermophotovoltaic (TPV) cells grown on GaSb by molecular-beam epitaxy (1, 2). These cells exhibit external spectral quantum efficiencies of 45 percent at a wavelength of 1 µm and 55 percent at 2.25 µm, implying internal efficiencies of 82 and 95 percent, respectively. Using the measured absolute spontaneous emission of forward biased cells, the ratio of the Auger to the radiative recombination coefficients is 1.55 x 10^{-18} cm^3 for cells having p-emitters doped to 1 and 3 x 10^{18} cm^{-3} (3). Optimal p-emitter doping is about 1 x 10^{18} cm^{-3}, leading to open-circuit voltages of 300 mV at short-circuit current densities of 1 A/cm^2. The ideality factors are near unity, i. e., 1.04. Series resistances for 1 x 1 cm^2 are 9 mΩ cm^2. Fill factors of 59 percent have been measured at current densities of 2.5 A/cm^2. Such operational characteristics show that these devices are well suited for thermophotovoltaic applications.

This paper discusses an important aspect of the TPV cell's operating characteristics: the different influence of cap-layer/emitter interfacial recombination velocity, S, upon the short-circuit current, J_{sc}; the open-circuit voltage, V_{oc}; and upon the dark current density, J_0. We show that for efficient cells a change in S from 2.5×10^4 to 5.05×10^4 cm/s produces a 25-mV drop in V_{oc}, a 12-percent decrease in J_{sc}, and a factor of two increase in J_0.

TPV CELL ARCHITECTURE

Figure 1(A) shows schematically the vertical profile of the 0.55-eV TPV cell, and Fig. 1(B) shows the front-surface metallization configuration. All the cells were grown by molecular-beam epitaxy.

A 1-µm-thick n-GaSb (Te-doped to 1×10^{18} cm^{-3}) buffer layer is grown on the substrate, followed by the active layers of the cell. The n-$Al_{0.25}Ga_{0.75}As_{0.02}Sb_{0.98}$ or n-$In_{0.16}Ga_{0.84}As_{0.14}Sb_{0.86}$ base thickness is 1 µm (Te-doped to 5×10^{17} cm^{-3}). The 3-µm-thick p-$In_{0.16}Ga_{0.84}As_{0.14}Sb_{0.86}$ emitter is Be-doped to 1×10^{18} cm^{-3}. Homo- and hetero-junction structures were used in the interfacial recombination-velocity studies. Owing to the relatively wider bandgap of 0.95 eV of the n-$Al_{0.25}Ga_{0.75}As_{0.02}Sb_{0.98}$ base and depletion region, recombination occurs predominantly in the p-$In_{0.16}Ga_{0.84}As_{0.04}Sb_{0.96}$ emitter, thus facilitating the interpretation of experimental results. Both cell types exhibit comparable quantum efficiencies at high excitation levels.

Finally, a thin cap layer of p-GaSb is grown on the emitter. This layer provides a conduction-band barrier at the cell input surface that reduces the effective recombination velocity at the cap-layer/emitter interface, while absorbing little radiation. Initial devices incorporated a 0.05-µm-thick cap, Be-doped to 1×10^{19} cm^{-3}. During processing, we discovered that the photoresist developer etches the GaSb cap layer at a rate of 20 nm/min. The estimated thickness of the etched cap-layer is 20 to 30 nm, leaving open the possibility of electron tunneling to the free GaSb surface. To avoid creating such a thin cap layer, an additional 0.1 µm of GaSb was regrown on the cap layer, creating a cap layer 0.15 µm thick. During processing, 20 to 30 nm was etched, so that the final thickness was 0.12 to 0.13 µm. In some cases the first 0.05 µm of GaSb was doped to 2×10^{18} cm^{-3}, and the remaining 0.05 µm was doped to 1×10^{19} cm^{-3}, producing a built-in electric field that sweeps photogenerated electrons into the emitter. The characteristics of cells with thin (20 - 30 nm) cap layers and ones with thick (0.12 - 0.13 µm) cap layers, some of which contained a built-in field, were studied to determine the influence of interfacial recombination on J_{sc}, V_{oc}, and J_0.

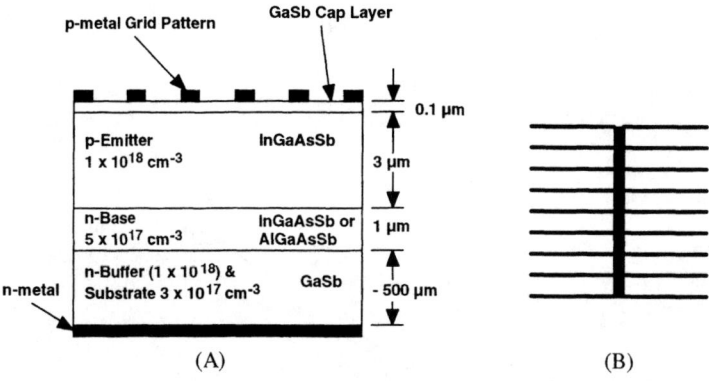

FIGURE 1. (A) The vePrtical profile of the TPV cells. (B) TPV Cell front-surface metallization pattern.

The front surface metallization pattern is an array of 10-μm-wide fingers on 100-μm centers. The fingers emanate from a central 200-μm-wide busbar. The metallization was e-beam deposited Ti(10nm)/Au(100nm)/Pt(100nm)/Au(50nm), followed by 5-μm of plated Au. This metallization pattern was applied to 4 x 4 mm² cells and to 1 x 1 cm² cells. The substrate metallization was Ni(5nm)/Au(45nm)/Ge(20nm)/Ni(15nm)/Au(200nm). The cells were In-soldered to copper submounts for testing at 15 °C.

THE INFLUENCE OF CAP-LAYER/EMITTER RECOMBINATION ON J_{SC}, V_{OC} AND J_0

Figure 2 shows the semilogarithmic J_{sc} - V_{oc} characteristics of two 4 x 4 mm² heterojunction cells, one with the thin GaSb cap layer and one with the thick cap layer without a built-in field.

Both cells were made from the same wafer. Short-circuit current densities from 0.5 to 3000 mA/cm² were generated in each cell, and the corresponding values of V_{oc} were measured. We assume the data to be described by the usual exponential dependence of J_{sc} on V_{oc}: $J_{sc} = J_0 \exp(V_{oc}/nkT)$, where J_0 is the dark-current-density parameter, n is the ideality factor, and kT is the thermal energy measured in eV. Each characteristic has two branches: one with an ideality close to 1.2 at excitation levels less than about 100 mA/cm², and the other with an ideality factor close to 1.0 at excitation levels higher than 100 mA/cm². The low-level-excitation branches presumably arise from recombination in the depletion region; those at high-excitation levels arise from recombination in the emitter bulk and cap-layer/emitter interface. Model fits to the four highest data points of each characteristic are also shown, as dashed and dot-dashed lines, along with the values of J_0 and n used; these fitted

Figure 3 shows J_{sc} for both cells as a function of the illumination level, P, in arbitrary units.

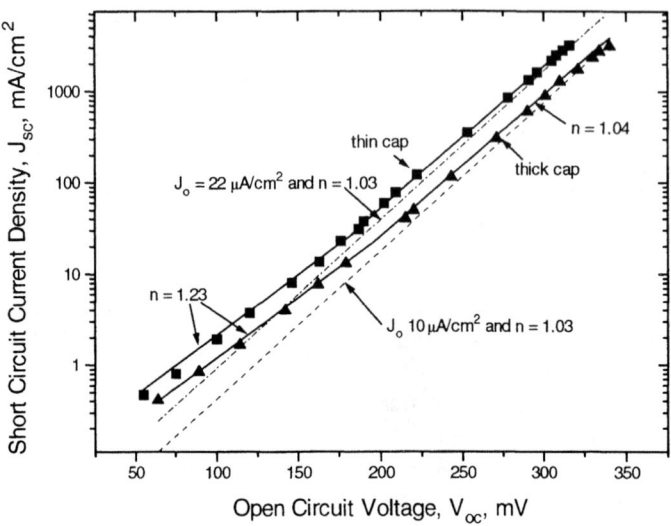

FIGURE 2. Short-circuit current density and open-circuit voltage of two heterojunction TPV cells: one with the original GaSb etched cap layer (thin) and one with a regrown GaSb cap layer (thick).

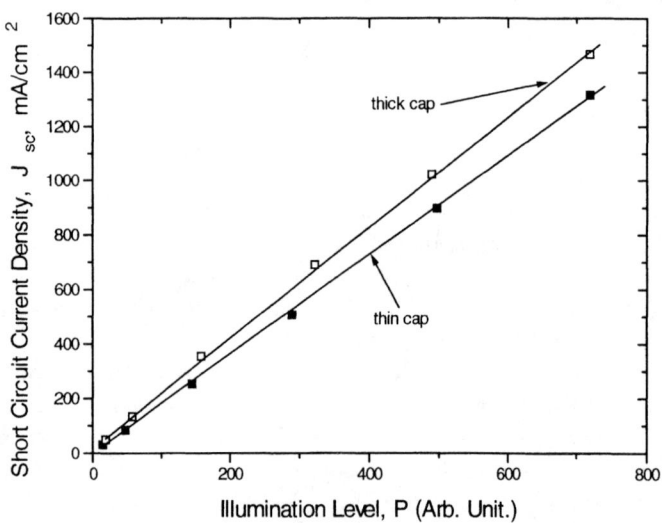

FIGURE 3. Short-circuit current density photoresponse for two heterojunction TPV cells: one with the original GaSb etched cap layer (thin) and one with a regrown GaSb cap layer (thick).

J_{sc} is proportional to the quantum efficiency. The curves are linear, and at the highest excitation level, the difference in J_{sc} for the two cells is 12 percent.

The thick GaSb cap layer reduces the cap-layer/emitter recombination velocity, S. For a given excitation level, J_{sc} and V_{oc} increase in cells with a thick cap layer. J_0 decreases. To calculate these effects, we use the simple one-dimensional diffusion model to describe minority-carrier dynamics in the emitter (4). The cells have high quantum efficiencies, implying that the electron diffusion length, L, is much longer than the emitter width, W; and that S is not much different from the natural bulk recombination velocity, D/L, in which D is the electron diffusion constant. Also, we assume that the absorption length of the exciting radiation, α^{-1}, is much shorter than L.

Eliminating all extraneous prefactors, the dependence of J_{sc} on S is given by

$$J_{sc} \propto \frac{\alpha L + S/(D/L)}{\alpha L - 1} \frac{1}{1 + S/(D/W)}$$

and with our assumptions that

$$\alpha L \gg 1 \text{ and } S/(D/L) \sim 1$$

$$J_{sc} \cong \frac{1}{1 + S/(D/W)} \equiv \frac{1}{1 + C_{sc}}$$

C_{sc} defines the factor $S/(D/W)$, through which J_{sc} depends on S.

In a similar way, the dependence of V_{oc} on S can be obtained. V_{oc} is a logarithmic function of N_0, the photogenerated electron concentration in the emitter:

$$V_{oc} = kT \ln(N_0 / n_p)$$

where n_p is the background electron concentration in the p-type emitter. The dependence of N_0 on S is given by

$$N_0 \propto \frac{\alpha L + S/(D/L)}{\alpha L - 1} \frac{1}{1 + SL^2/DW} \cong \frac{1}{1 + SL^2/DW} \equiv \frac{1}{1 + C_{oc}}$$

and the factor $C_{oc} = SL^2/(DW)$ has been defined. The ratio of C_{sc} to C_{oc} is much less than unity, showing that N_0 is more sensitive to S than is J_{sc}:

$$\frac{C_{sc}}{C_{oc}} = \frac{W^2}{L^2} \ll 1$$

Another way to view this effect is to compare the time constants associated with J_{sc} and N_0. The time constant associated with J_{sc} is the diffusion time of electrons out of the emitter, τ_D, which is given by

$$\tau_D = W^2/D.$$

The time constant associated with N_0 is the bulk recombination time τ, given by

$$\tau = L^2/D.$$

Thus, the ratio of these two time constants equals C_{sc}/C_{oc} and is much less than unity:

$$\tau_D/\tau = C_{sc}/C_{oc} = W^2/L^2 \ll 1$$

J_{sc} is less sensitive to S than N_0 because under short-circuit conditions, photogenerated electrons quickly diffuse out of the emitter. On the other hand, photogenerated electrons under open-circuit conditions linger in the emitter for the much longer bulk-recombination time, thereby increasing their chance to recombine at the cap-layer/emitter interface.

Figure 4 shows the calculated relative decrease of J_{sc} with increasing S, and Fig. 5 shows the decrease in V_{oc} with increasing S.

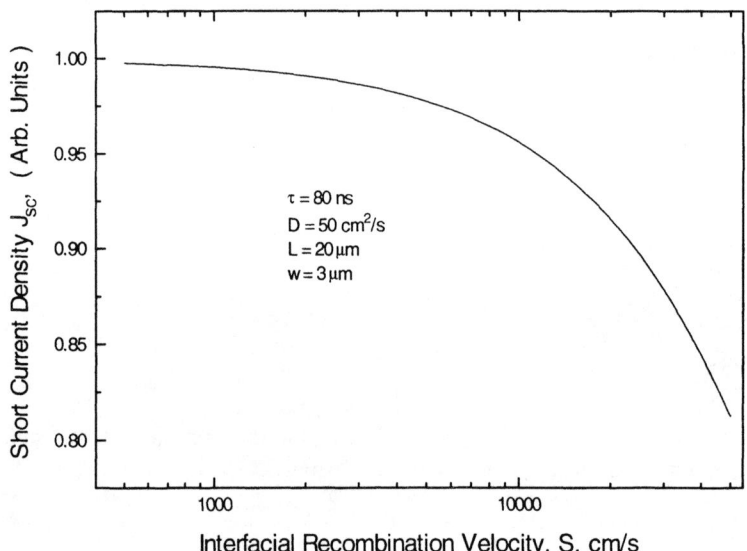

FIGURE 4. The calculated relative decrease in the short-circuit current with increasing cap-layer/emitter interfacial recombination velocity.

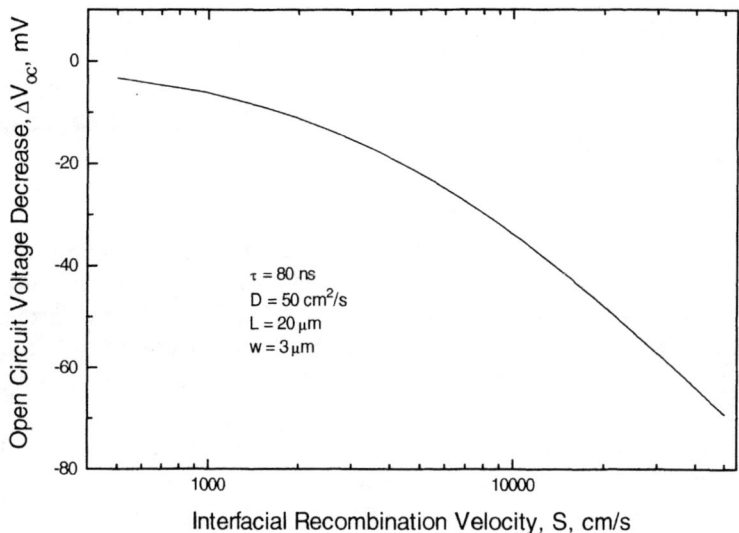

FIGURE 5. The calculated decrease in open-circuit voltage with increasing cap-layer/emitter interfacial recombination velocity.

These curves were calculated using the one-dimensional diffusion model. The parameter values were $W = 3$ µm, $\tau = 80$ ns, and $D = 50$ cm^2/s, giving $L = 20$ µm, and $\tau_D = 1.8$ ns. $D/L = 2.5 \times 10^4$ cm/s. The difference in V_{oc} is shown because V_{oc} depends logarithmically on N_0, and it is the ratio of N_0 values that is compared.

Increasing S from 5×10^2 to 5×10^4 cm/s produces a 20-percent decrease in J_{sc} (Fig. 4) and a 70-meV drop in V_{oc} (Fig. 5).

For our range of parameters, the simplified diffusion model also gives the expression for the dependence of the dark current density, J_0, on S:

$$J_0 = A\frac{S/(D/L)+W/L}{S/(D/W)+1} \cong A\frac{S}{(D/L)}$$

J_0 is directly proportional to S, and A is a constant of proportionality.

Returning to the fitted curves in Fig. 2, the value of J_0 for the cell with the thick cap layer is 10 µA/cm², and if we assume $S = 2.5 \times 10^4$ cm/s (*i. e.*, S/(D/L) = 1), then $A = 10$ µA/cm². For the cell with the thin cap layer the value of J_0 is 22 µA/cm², which, since A is a constant, implies that $S = 5.05 \times 10^4$ cm/s. As can be seen from Fig. 4, the corresponding calculated decrease in J_{sc} is about 10 percent, which is in reasonable agreement with the 12-percent decrease for the highest data points shown in Fig. 3. Likewise, from Fig. 5 the calculated decrease in V_{oc} is 20 mV, again in reasonable agreement with the 25 mV difference in the highest data points shown in Fig. 2. For both cells the values of S/(D/L) are close to unity. We see experimentally that as S increases by a factor of 2.2, J_{sc} decreases by 12 percent, V_{oc} decreases by 25 mV, and J_0 increases by a factor of two.

Similar results are obtained for homojunction TPV cells. Figure 6 shows the J_{sc} - V_{oc} characteristics of two 4 x 4 mm^2 homojunction cells.

Cell #1 has a thin cap layer, and Cell #2 has a thick cap layer containing a built-in field. At short-circuit current densities approaching 3 A/cm^2, Cell #2 produces a 15-mV larger open-circuit voltage than does Cell #1. We interpret this increase as a decrease in S. The lower value of S reduces J_0 by about a factor of two from 32 to 16 µA/cm^2. As was the case for heterojunction cells, a thick GaSb cap layer reduces S by a factor of two.

At excitation levels below 2000 mA/cm^2, depletion-layer recombination in Cell #2 is evidenced by the branch in its characteristic with an ideality factor of n = 1.8. Note that this is larger than that of the heterojunction cells, in which n = 1.2 (Fig. 2), indicating the stronger depletion-region recombination in the homojunction cells.

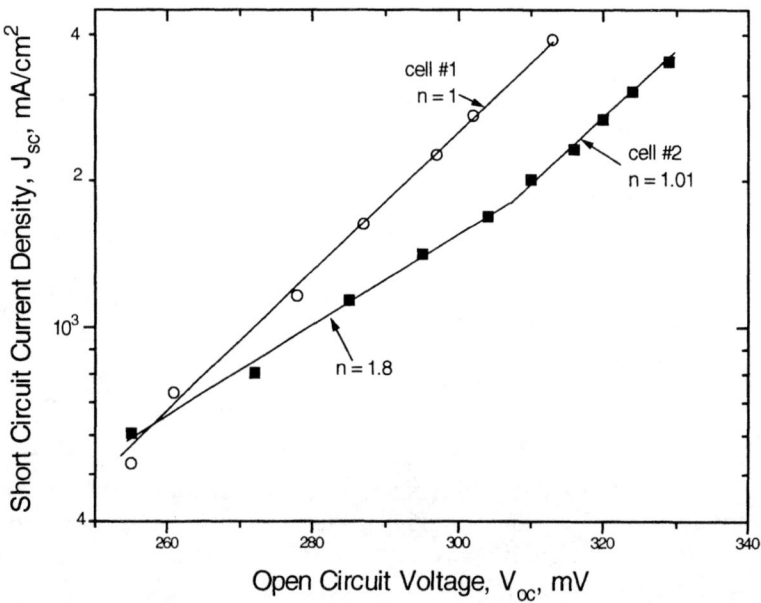

FIGURE 6. J_{sc} - V_{oc} characteristics of two homojunction TPV cells: one with a thin cap layer, Cell #1, and one with a thick cap layer, Cell #2.

SUMMARY AND CONCLUSIONS

Efficient p-on-n homo- (InGaAsSb/GaSb) and hetero-junction (InGaAsSb/AlGaAsSb/GaSb) thermophotovoltaic (TPV) cells were studied with respect to the recombination velocity at the cap-layer/emitter interface, S. In both cell types, the open-circuit voltage, V_{oc}, and the short-circuit current, J_{sc}, each have about the same sensitivity to S. The dark current, J_0, is much more sensitive to S. An examination of the essential factors in the one-dimensional minority-carrier diffusion model shows that under short-circuit conditions, photogenerated electrons diffuse rapidly away from the interface to the junction. Under open-circuit conditions, the electrons remain in the emitter for a much longer bulk-recombination time, and therefore they are more likely to recombine at the interface. J_0 is proportional to S. An increase in S from 2.5 to 5.05×10^4 cm/s produces a 25 mV decrease in V_{oc} and a 12-percent decrease in J_{sc} or quantum efficiency. J_0 increases by a factor of two. This work points out the critical importance of interfacial recombination even in efficient TPV cells.

REFERENCES

1. Martinelli, R. U., Garbuzov, D. Z., Lee, H., Morris, N., Odubanjo, T., Taylor, G. C., and Connolly, J. C., *"The Quantum Efficiency of InGaAsSb Thermophotovoltaic Diodes"*, in Space Technology and Applications International Forum - 1998, edited by M. S. El-Genk, AIP Conf. Proc. 420, pp 1394-1399 (1998).
2. Garbuzov, D. Z., Martinelli, R. U., Khalfin, V. B., Lee, H., Morris, N., Odubanjo, T., Taylor, G. C., Connolly, J. C., Charache, G. W., and Dupoy, D. M., *"A Novel Approach for the Improvement of Open Circuit voltage and Fill Factor of InGaAsSb/GaSb Thermophotovoltaic Cells"*, in Space Technology and Applications International Forum - 1998, edited by M. S. El-Genk, AIP Conf. Proc. 420, pp 1400 - 1409 (1998).
3. Charache, G. W., Baldasaro, P. F., Danielson, L. R., Dupoy, D. M., Freeman, M. J., Wang, C. A., Choi, H. K., Garbuzov, D. Z., Martinelli, R. U., Khalfin, V. B., Saroop, S., Borrego, J. M., and Gutmann, R. J., *"InGaAsSb Thermophotovoltaic Diode Physics Evaluation"*, presented at the Fortieth Electron Materials Conference, June 24 - 26, 1998, Charlottesville, VA. Also submitted to the J. Appl. Phys. for publication.
4. Sze, S. M., *"Physics of Semiconductor Devices"* 2nd ed., John Wiley and Sons, New York, 1981, 802-805.

Extending the Cutoff Wavelength of Lattice-Matched GaInAsSb/GaSb Thermophotovoltaic Devices

C.A. Wang*, H.K. Choi*, D.C. Oakley*, and G.W. Charache[+]

*Lincoln Laboratory, Massachusetts Institute of Technology, Lexington, MA 02420-9108
[+]Lockheed Martin, Inc., Schenectady, NY 12301

Abstract. This paper reports the growth, materials characterization, and device performance of lattice-matched GaInAsSb/GaSb thermophotovoltaic (TPV) devices with cutoff wavelength extended from 2.3 to 2.5 µm. GaInAsSb epilayers were grown lattice matched to GaSb substrates by organometallic vapor phase epitaxy using all organometallic precursors including triethylgallium, trimethylindium, tertiarybutylarsine, and trimethylantimony with diethyltellurium and dimethylzinc as the n- and p-type dopants, respectively. The growth temperature was 525°C. Although these alloys are metastable, a mirror-like surface morphology and room-temperature photoluminescence (PL) are obtained for alloys with PL peak emission at room temperature as long as 2.5 µm. Lattice-matched GaInAsSb/GaSb TPV devices exhibit internal quantum efficiency as high as 90% for devices with a cutoff wavelength of 2.5 µm. The open circuit voltage for extended wavelength devices is 239 mV at 3.6 A/cm^2.

INTRODUCTION

$Ga_{1-x}In_xAs_ySb_{1-y}$ alloys are of interest for lattice-matched thermophotovoltaic (TPV) devices because of the high performance attainable at 2.3 µm [1]. Extension of the TPV device cutoff wavelength λ_c to beyond this wavelength is especially desirable since the emissive power of the radiator is significant at these longer wavelengths, and higher power density can be attained [2]. However, the $Ga_{1-x}In_xAs_ySb_{1-y}$ quaternary alloy system exhibits a miscibility gap [3] in the wavelength range of interest. Thus, to increase λ_c requires the growth of high quality alloys that penetrate further into the miscibility gap. The growth has been difficult because these alloy compositions are metastable, and consequently, no devices with $\lambda_c > 2.3$ µm have been demonstrated previously.

In this paper, the successful preparation of $Ga_{1-x}In_xAs_ySb_{1-y}$ alloys for TPV devices with extended λ_c is reported. Epitaxial $Ga_{1-x}In_xAs_ySb_{1-y}$ layers of various composition, and thus λ_c, were grown lattice matched to GaSb substrates by organometallic vapor phase epitaxy (OMVPE). The materials properties are presented for epilayers with increasing x and y values, and thus λ_c varying from 2 to 2.5 µm. High optical quality is achieved for alloys with room-temperature photoluminescence (PL) peak emission at 2.5 µm, and the first demonstration of TPV devices with λ_c out to 2.5 µm is reported.

EPITAXIAL GROWTH AND CHARACTERIZATION

$Ga_{1-x}In_xAs_ySb_{1-y}$ epilayers were grown in a vertical rotating-disk reactor with H_2 carrier gas at a flow rate of 10 slpm, reactor pressure of 150 Torr, and a typical rotation rate of 100 rpm [4]. Solution trimethylindium, triethylgallium, tertiarybutylarsine, and trimethylantimony were used as organometallic precursors. Diethyltellurium (10 ppm in H_2) and dimethylzinc (1000 ppm in H_2) were used as n- and p-type doping sources, respectively. For lattice-matched epilayers, $Ga_{1-x}In_xAs_ySb_{1-y}$ was grown on (001) Te-doped GaSb substrates misoriented 6° toward (111)B. The growth temperature was 525°C.

The surface morphology was examined using Nomarski contrast microscopy and atomic force microscopy (AFM) in the tapping mode. High-resolution x-ray diffraction (HRXRD) was used to measure the degree of lattice mismatch to GaSb substrates. PL was measured at 4 and 300 K using a PbS detector. The In and As contents of $Ga_{1-x}In_xAs_ySb_{1-y}$ epilayers were determined from the combination of 1) HRXRD splitting of ω-2Θ scans; 2) peak emission in 300 K PL spectra; and 3) energy gap dependence on composition, $E(x,y) = 0.726 - 0.961x - 0.501y + 0.08xy + 0.415x^2 + 1.2y^2 + 0.021x^2y - 0.62xy^2$, where $y = 0.867x/(1 - 0.048x)$ for alloys lattice matched to GaSb substrates. For electrical characterization, GaInAsSb was grown on semi-insulating (001) GaAs substrates misoriented 6° toward (111)B. Because of the lattice mismatch between the epilayer and the substrate, a 0.4-µm-thick GaSb buffer layer was grown [4]. Carrier concentration and mobility of GaInAsSb epilayers, which were grown about 3 µm thick, were obtained from Hall measurements based on the van der Pauw method.

GROWTH RESULTS

$Ga_{1-x}In_xAs_ySb_{1-y}$ epilayers of various x and y values were grown lattice matched to GaSb substrates. Figure 1 shows the surface topography of four layers, each about 2 µm in thickness, as imaged by AFM. The In and As concentrations were varied with $0.09 < x < 0.23$ and $0.08 < y < 0.20$. These compositions correspond to values that are predicted to be thermodynamically metastable at a growth temperature of 525°C [3], with higher x and y values more metastable. The AFM images of Figs. 1a – 1c reveal periodic surface features that are oriented perpendicular to the substrate misorientation direction, which is suggestive of bunched supersteps. On the other hand, the irregular features observed for the sample in Fig. 1d imply a three-dimensional growth mode, which is unfavorable.

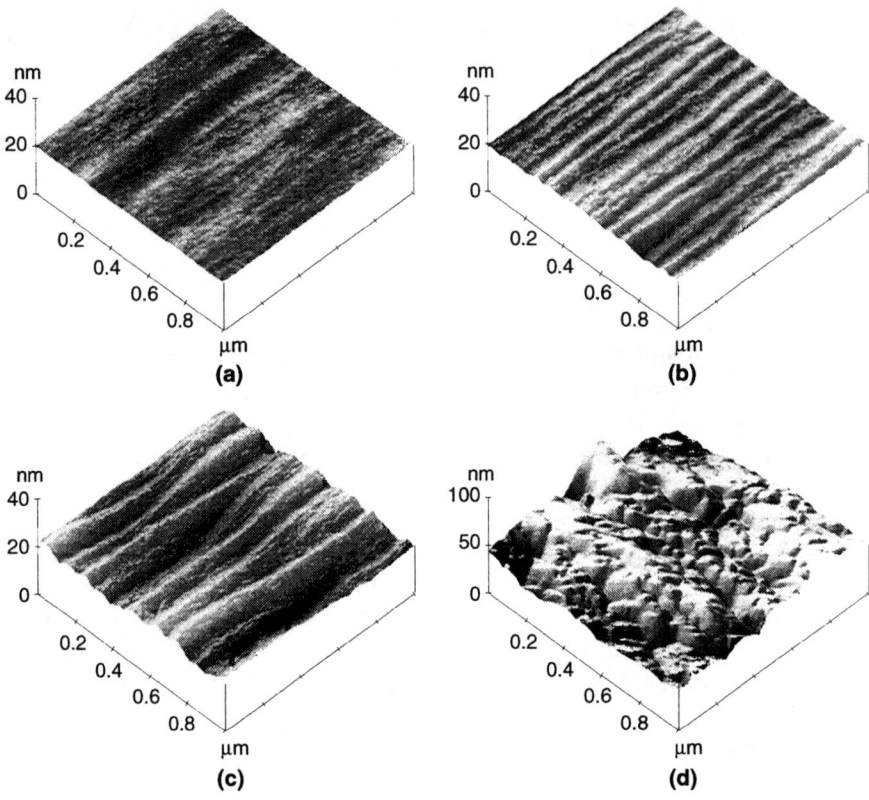

FIGURE 1. Atomic force microscopy images showing surface topography of nominally lattice-matched $Ga_{1-x}In_xAs_ySb_{1-y}$ epilayers grown at 525°C on (001) 6° toward (111)B GaSb substrates. The composition of the layers is: (a) $x = 0.09$, $y = 0.08$; (b) $x = 0.16$, $y = 0.15$; (c) $x = 0.20$, $y = 0.18$; and (d) $x \sim 0.23$, $y \sim 0.21$. The values for the layer shown in (d) are estimated from the In distribution coefficient [4] since no room temperature photoluminescence (see text) was observed from this epilayer. Note the change in vertical scale for (d). Scan area is 1 μm x 1 μm.

The root-mean-square roughness values for samples shown in Figs. 1a – 1d are 0.2, 0.4, 1.4, and 8 nm, respectively. The layers with a periodic structure (Figs. 1a – 1c) have a mirror-like appearance to the eye, and exhibit a slight 'wavy' texture under Nomarski contrast microscopy. Conversely, the layer shown in Fig. 1d is hazy to the eye. The breakdown in the growth mode from supersteps to a three-dimensional mechanism may be related to the higher x- and y-values of this sample, since they correspond to a region well within the miscibility gap [3].

Figure 2 shows HRXRD ω-2Θ scans of the $Ga_{1-x}In_xAs_ySb_{1-y}$ epilayers shown in Figs. 1a – 1c. The scans are plotted on a log scale. All layers are nominally lattice matched to the GaSb substrate, with full width at half-maximum values (FWHM) that are on the same order (23-32 arc s) as the GaSb substrate (18 arc s). Only the layer with the highest x- and y- values (Fig. 2c) exhibits some broadening, of relatively low intensity, which corresponds to a maximum lattice mismatch of only 2 x 10^{-3}.

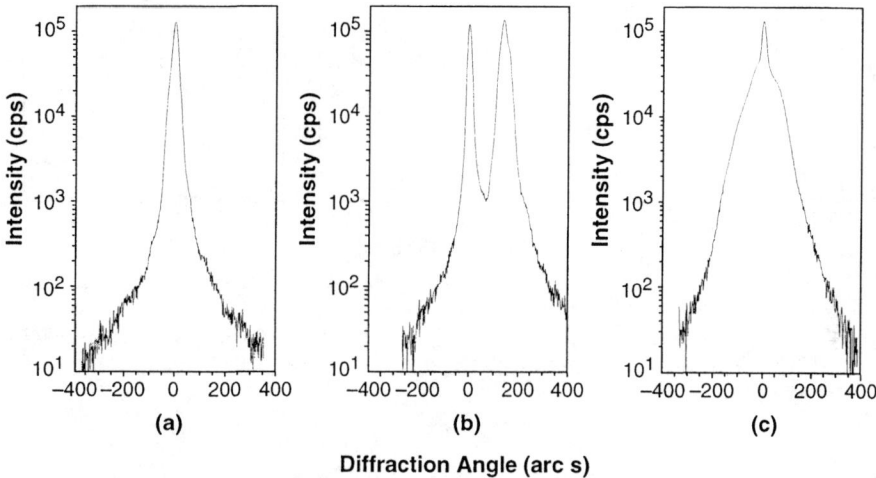

FIGURE 2. High resolution x-ray diffraction of $Ga_{1-x}In_xAs_ySb_{1-y}$ epilayers grown at 525°C on (001) 6° toward (111)B GaSb substrates with compositions: (a) x = 0.09, y = 0.08; (b) x = 0.16, y = 0.15; (c) x = 0.20, y = 0.18.

Figure 3 shows the 4 and 300 K PL spectra for $Ga_{1-x}In_xAs_ySb_{1-y}$ epilayers discussed in Figs. 2a – c. The peak emission for the sample with x = 0.09, y = 0.08 (Fig. 3a) is 1818 and 2035 nm at 4 and 300 K, respectively. The 4 K FWHM is 5.3 meV. With increasing x and y values, the 4 and 300 K emission increases to 2080 and 2320 nm, respectively, for x = 0.16, y = 0.15 (Fig. 3b); and to 2225 and 2505 nm, respectively, for x = 0.2, y = 0.18 (Fig. 3c). The 4 K FWHM also increases to 7.5 and 25 meV, respectively. Although lattice-matched $Ga_{1-x}In_xAs_ySb_{1-y}$ of higher In and As composition (see Fig. 1d) was epitaxially grown, this layer did not exhibit PL at 4 or 300 K. The longest 300 K PL emission that was observed in this study is 2525 nm, which is longer than any other values previously reported for this alloy system grown by OMVPE [3,6,7].

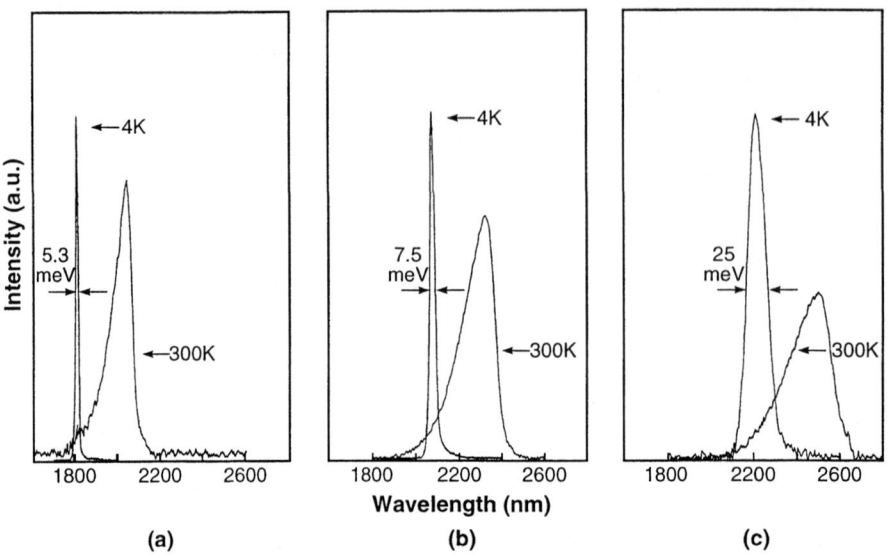

FIGURE 3. Photoluminescence spectra measured at 4 and 300 K of $Ga_{1-x}In_xAs_ySb_{1-y}$ grown on (001) 6° toward (111)B GaSb substrates. Layers were grown at 525°C: (a) x = 0.09, y = 0.08; (b) x = 0.16, y = 0.15; (c) x = 0.20, y = 0.18.

The 300 K electrical properties of p- and n-doped $Ga_{1-x}In_xAs_ySb_{1-y}$ are summarized in Figs. 4a and 4b, respectively. The plots include data for layers grown at 525 and 550°C [4] on (001) 2° toward (110) and (001) 6° toward (111)B substrates. The majority of the data, which are shown as filled circles, corresponds to layers with 300 K PL emission at ~2.25 μm. The hole concentration ranges from 4.4×10^{15} to 1.9×10^{18} cm^{-3} with mobility values between 560 and 180 cm^2/V-s, respectively. The two higher mobility values measured for p-GaInAsSb at the lower concentration ~ 8×10^{15} cm^{-3} are data for samples with a 0.8-μm-thick GaSb buffer layer.

The data in Fig. 4a indicate that the p-type electrical characteristics depend on the $Ga_{1-x}In_xAs_ySb_{1-y}$ alloy composition. As the x and y values increase, the mobility values decrease for comparable hole concentration. For $p \sim 1 \times 10^{17}$ cm^{-3}, the hole mobility is 441 cm^2/V-s for 2.05-μm material, compared to 239 cm^2/V-s for the 2.5-μm material. On the other hand, the data shown in Fig. 4b for n-type GaInAsSb suggest that the electrical characteristics are independent of alloy composition. The electron concentration ranges from 2.2×10^{17} to 3.2×10^{18} cm^{-3}, with corresponding mobility values between 5208 and 2084 cm^2/V-s, respectively.

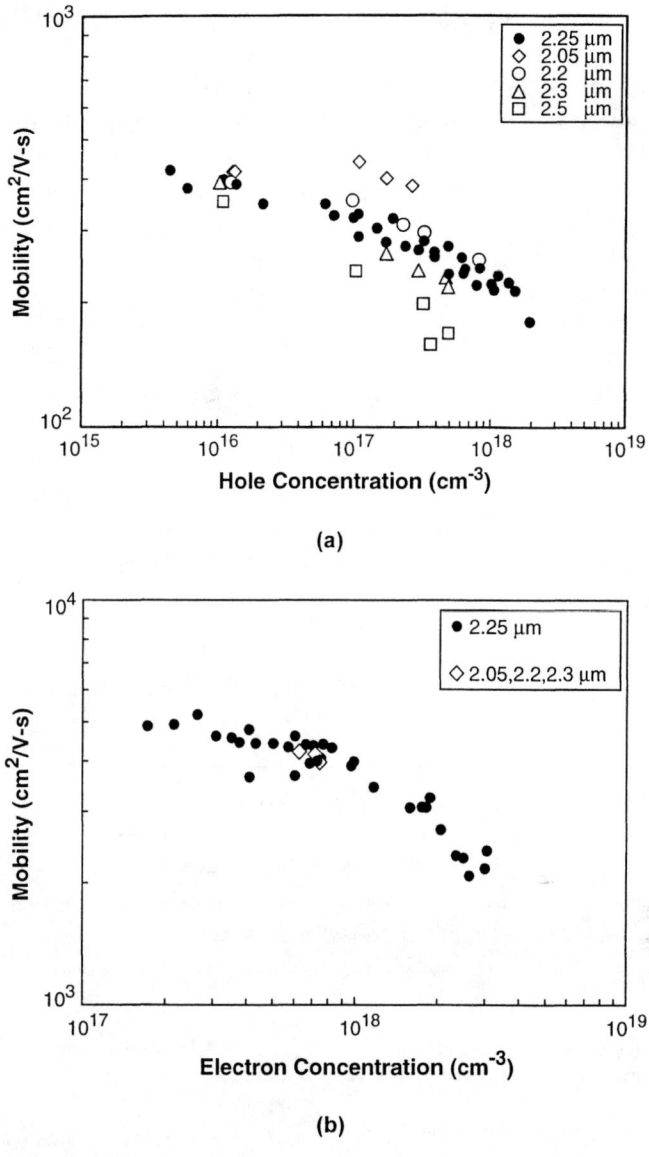

FIGURE 4. Electrical properties measured at 300 K of (a) p-GaInAsSb and (b) n-GaInAsSb.

THERMOPHOTOVOLTAIC DEVICES

GaInAsSb TPV devices were grown on (001) 6° toward (111)B GaSb substrates with a p-on-n configuration (Fig. 5). The GaInAsSb base layer is doped n-type to ~5 x 10^{17} cm^{-3} and is 1 µm in thickness, while the GaInAsSb emitter layer is doped p-type to ~2 x 10^{17} cm^{-3} and is 3 or 4 µm in thickness. The upper 0.05-µm-thick p-GaSb layer (p-doped to ~2 x 10^{18} cm^{-3}) serves as the window/contact layer. Although an AlGaAsSb window layer was incorporated previously to reduce surface recombination [1,8], a GaSb window layer was used in this study to simplify the layer structure. Thick emitter layers were incorporated to take advantage of the longer minority carrier diffusion lengths in the p-type layer [9]. Three structures with different $Ga_{1-x}In_xAs_ySb_{1-y}$ composition were grown to obtain various λ_c.

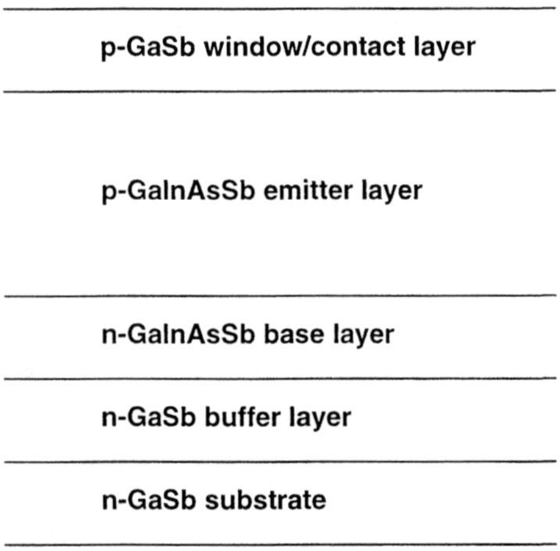

p-GaSb window/contact layer
p-GaInAsSb emitter layer
n-GaInAsSb base layer
n-GaSb buffer layer
n-GaSb substrate

FIGURE 5. Schematic structure of GaInAsSb/GaSb TPV device. The GaInAsSb alloy composition was varied to evaluate device performance at different λ_c.

Large-area (1 cm^2) TPV cells were fabricated using a conventional photolithographic process. A single 1-mm-wide central busbar connected to 10-µm-wide grid lines spaced 100 µm apart was used to make electrical contact to the front surface. Ohmic contacts to p- and n-GaSb were formed by depositing Ti/Pt/Au and Au/Sn/Ti/Pt/Au, respectively, and alloying at 300°C. Mesas were formed by wet chemical etching to a depth of ~5 µm. No antireflection coatings were deposited on these test devices.

The external quantum efficiency QE as a function of wavelength is plotted in Fig. 6 for three TPV devices with a 4-μm-thick emitter. The λ_c ranges from 2.3 to 2.5 μm, which is the longest wavelength that has been reported for TPV devices in any materials system. A maximum QE is measured at ~1.6 μm, and is 59, 57, and 58% for devices with λ_c of 2.3, 2.4, and 2.5 μm, respectively. The QE decreases below 1.6 μm because of absorption in the GaSb window layer. The spectral response is nearly flat between 1.6 and 2.2 μm for all three devices. For the device with λ_c = 2.5 μm, the QE is as high as 41% at 2.5 μm. At 1.6 μm, the internal QE of the three devices is 91, 88, and 90%, respectively, assuming a surface reflection of 34% and absorption of 2% in the GaSb window layer. These values of QE are consistent with estimates of minority carrier diffusion lengths of ~20 μm and surface recombination velocity of ~4000 cm/s reported for lattice-matched GaInAsSb TPV devices with λ_c = 2.3 μm [10].

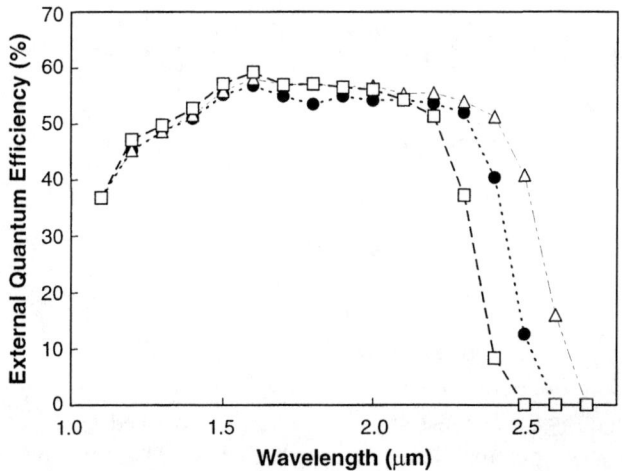

FIGURE 6. External quantum efficiency of GaInAsSb/GaSb TPV devices as a function of wavelength. The emitter thickness is 4 μm. No antireflection coatings were deposited.

Measurements of short circuit current vs open circuit voltage V_{oc} were performed at room temperature. Table 1 summarizes data for three TPV devices with various λ_c. For the 2.3-μm device, V_{oc} is 313 mV at a short circuit current density J_{sc} of 3.4 A/cm^2. This value is comparable to the value reported for TPV structures with the AlGaAsSb window layer [1,8]. With increasing λ_c, V_{oc} decreases to 269 and 239 mV for the 2.4- and 2.5-μm devices, respectively. This reduction may result from increases in dark current, Auger recombination, surface recombination velocity, or carrier trapping at the p-GaInAsSb/p-GaSb interface due to the valence band offset.

The best V_{oc} values for lattice-mismatched InGaAs/InP TPV devices are similar even though those devices had a shorter λ_c of 2.2 μm [11], thus showing the advantage of lattice-matched GaInAsSb/GaSb TPV structures. Fill factors for the GaInAsSb/GaSb devices range from 58 to 66% at current densities of ~3 A/cm².

TABLE 1.

DATA FOR GaInAsSb TPV CELLS UNDER ILLUMINATION

λ_c (μm)	V_{oc} (mV)	J_{sc} (A/cm²)	FF (%)
2.3	313	3.4	66
2.4	269	3.5	62
2.5	239	3.6	58

SUMMARY

$Ga_{1-x}In_xAs_ySb_{1-y}$ epilayers were grown lattice matched to GaSb substrates by OMVPE. Room-temperature PL emission at wavelengths as long as 2.5 μm was achieved, which is the longest 300 K wavelength obtained for this alloy grown by OMVPE. Low-temperature PL spectra exhibit FWHM as narrow as ~5 meV, the smallest value reported for GaInAsSb alloys grown by OMVPE. This high optical quality of metastable GaInAsSb epilayers was achieved by using a low growth temperature of 525°C. The first lattice-matched GaInAsSb/GaSb TPV devices with extended λ_c to 2.5 μm have been demonstrated. The internal QE of these TPV devices is 90% at 1.6 μm, and as high as 62% at 2.5 μm. At J_{sc} of 3.6 A/cm², V_{oc} is 239 mV, with a fill factor of 58%. The high performance of these GaInAsSb/GaSb TPV devices should be especially attractive for TPV systems utilizing 1100 K radiator temperatures.

ACKNOWLEDGMENTS

The authors gratefully acknowledge D.R. Calawa for x-ray diffraction, J.W. Chludzinski for photoluminescence, K.J. Challberg for manuscript editing, and D.L. Spears for continued support and encouragement. This work was sponsored by the Department of Energy under AF Contract No. F19628-95-C-0002.

REFERENCES

1. Choi, H.K, Wang, C.A., Turner, G.W., Manfra, M.J., Spears, D.L., Charache, G.W., Danielson, L.R., Depoy, D.M., Appl. Phys. Lett. **71**, 3758 (1997).

2. Baldasaro, P.F., Brown, E.J., Depoy, D.M., Campbell, B.C., Parrington, J.R., AIP Conference Proceedings **321**, 29 (1995).

3. Cherng, M.J., Jen, H.R., Larsen, C.A., Stringfellow, G.B., Lundt, H., Taylor, P.C., J. Cryst. Growth **77**, 408 (1986).

4. Wang, C.A., J. Cryst. Growth **191**, 631 (1998).

5. Singh, J. and Bajaj, K.K., Appl. Phys. Lett. **44**, 1075 (1984).

6. Shin, J., Hsu, T.C., Hsu, Y., Stringfellow, G.B., J. Cryst. Growth **179**, 1 (1997).

7. Sopanen, M., Koljonen, T., Lipsanen, H., Tuomi, T., J. Cryst. Growth **145**, 492 (1994).

8. Wang, C.A., Choi, H.K., Turner, G.W., Spears, D.L., Manfra, M.J., Charache, G.W., AIP Conference Proceedings **401**, 75 (1997).

9. Hitchcock, C.W., Gutmann, R.J., Ehsani, H., Bhat, I.B., Wang, C.A., Freeman, M.J., Charache, G.W., presented at the 9th International Conference on Metalorganic Vapor Phase Epitaxy, La Jolla, CA, June 1998.

10. Charache, G.W., Baldasaro, P.F., Danielson, L.R., DePoy, D.M., Freeman, M.J., Wang, C.A., Choi, H.K., Garbuzov, D.Z., Martinelli, R.U., Khalfin, V., Saroop, S., Borrego, J.M., Gutmann, R.J., presented at the Electronic Materials Conference, Charlottesville, VA, June 1998.

11. Charache, G.W., Egley, J.L., Danielson, L.R., DePoy, D.M., Baldasaro, P.F., Campbell, B.C., Hui, S., Frass, L.M., Wojtczuk, S., Proc. 25th IEEE Photovoltaic Specialist Conf., p. 137 (1996).

SESSION 5: CHARACTERIZATION

FTIR and FT-PL Spectroscopic Analysis of TPV Materials and Devices

J. D. Webb, L. M. Gedvilas, M. R. Olson, X. Wu, A. Duda,
M. W. Wanlass, K. M. Jones

National Renewable Energy Laboratory, 1617 Cole Blvd., Golden, CO 80401

Abstract. Fourier transform (FT) spectroscopic techniques are useful in determining properties of thermophotovoltaic (TPV) materials and devices. The III-V TPV absorber materials have energy bandgaps that can be optimized for conversion of the near-infrared radiation emitted by thermal sources in the $1000°–1200°C$ temperature range. The bandgaps of these materials can be measured at room temperature using FT-photoluminescence spectroscopy, which can be done with a modified FT-Raman spectrophotometer operating in the near-infrared spectral region. The intensities and bandwidths of the FT-PL spectra also provide information on the extent of non-radiative recombination and the compositional uniformity of the materials. To achieve adequate operating efficiencies, TPV converters must return sub-bandgap radiation to the thermal source. The percent reflectance of the device in the mid-infrared spectral region is therefore an important operating parameter that can be accurately measured using FT-infrared (FTIR) spectroscopy with total reflectance optical accessories. In this paper, we discuss applications of these techniques to TPV materials and devices, and variations on these approaches, such as scanning micro-FT-PL spectroscopy, that enable microanalysis of TPV device structures at the 1–100-μm scale.

INTRODUCTION

Infrared reflectance and transmittance are important properties of thermophotovoltaic (TPV) devices. In order to convert infrared radiation supplied by conventional energy sources, i.e., fossil fuel, to electricity with high efficiencies, TPV devices must absorb band-gap radiation while reflecting most of the sub-bandgap radiation back to the thermal radiative source (1). Fourier transform infrared (FTIR) spectroscopy is the dominant technique for measuring the infrared spectra of materials at wavelengths ranging from 1.3 to 25 μm (2). The use of FTIR specular reflectance spectroscopy to assess the performance of TPV front surface (3) and back surface (4) reflectors is well-established.

Photoluminescence (PL) spectroscopy is used extensively to measure properties of semiconductor thin films used in photovoltaic (PV), electronic, and optoelectronic devices. For example, near-infrared (NIR) PL spectroscopy has been used to detect light- or heat- induced defects ("dangling bonds") in hydrogenated amorphous silicon (5) and has been correlated with the efficiency of PV devices produced from copper indium diselenide ($CuInSe_2$, or CIS) thin films (6). Although some of this work has been done at room temperature, detection of weak NIR PL emission beyond the wavelength range of Si and photomultiplier detectors when using dispersive spectrophotometers has typically required cryogenic cooling of semiconductor samples (5, 6).

FTIR spectroscopy offers substantial improvements in the analysis of weak signals over dispersive infrared spectroscopy through Jacquinot's throughput advantage (2). A variety of PL measurements, with FTIR detection that exploits this advantage, have been reported on NIR PL emitters. Recent work has involved the use of FT-Raman spectrophotometers, rather than laser-excitation sources, coupled with FTIR spectrophotometers, to obtain room-temperature and cryogenic PL and electroluminescence spectra from a variety of bulk and thin-film semiconductor materials having PV applications (7–10). Use of an FT-Raman spectrophotometer for FT-luminescence measurements is advantageous because the luminescence processes occur with quantum efficiencies many orders of magnitude higher than that for the Raman scattering process, which for NIR excitation is of the order of 10^{-6} (11). Thus, the FT-Raman instrumentation is highly sensitive to NIR PL even at room temperature; in fact, excitation power must be reduced to sub-milliwatt levels in some cases to avoid band-filling artifacts and localized sample heating (10).

Silicon, germanium, and the III-V and II-VI compound semiconductors all emit luminescence bands within the range of the FT-Raman instrumentation, with some modifications that will be discussed. We have used FT-PL spectroscopy with other analytical techniques to solve problems such as junction formation and surface passivation in CIS PV devices (10), determination of optical gap and strain-induced effects on recombination rates in III-V TPV materials and devices (12), and band-gap analysis of graded-composition TPV devices (13). In the present work, we will focus on the results of FTIR reflectance and FT-PL analyses of TPV materials and devices.

EXPERIMENTAL APPROACH

Cadmium stannate films were grown as described previously (3). TPV device structures and undoped device analog structures (12) incorporating $Ga_xIn_{1-x}As$ or $InAs_yP_{1-y}$ absorber and graded composition (graded x) layers were grown using metal-organic vapor-phase epitaxy (MOVPE) on InP substrates (1). The detailed structure of the multilayer $Ga_xIn_{1-x}As$ samples is shown in the figures accompanying the results and discussion. Continuously graded layers (CGL) and step-graded layers (GL) are used in lattice-mismatched (LMM) devices and analog structures to enable epitaxial growth while controlling bandgap by varying x or y, respectively, in the $Ga_xIn_{1-x}As$ or $InAs_yP_{1-y}$ films (1).

We used a SpectraTech FT-30 specular reflectance (SR) accessory operated at a fixed-incidence angle of 30° from the sample normal to obtain large-area SR spectra of the TPV samples. SR measurements were also made using a Spectra-Tech Model 500 variable-angle accessory with incidence-angle variable between 20° and 80°. We used a Labsphere Model RSA-NI-550D integrating sphere with fixed-incidence angle of 10° and integral deuterated triglycins sulfate (DTGS) detector to obtain large-area total reflectance (TR) data on the TPV samples. The latter instrument was fitted with a ½"-diameter circular aluminum aperture at the ¾"-diameter sample port to allow analysis of smaller-sized samples, plus a light trap placed below the samples to eliminate spurious reflection from the sample mount. Attempts to incorporate smaller apertures reduced the accuracy of the TR measurements. All of these attachments could be fitted with a Perkin-Elmer AgBr wire-grid IR polarizer. We also submitted some samples to Surface Optics Corporation (San Diego, CA 92131) for TR analysis using their SOC-100 hemispherical reflectometer accessory, which has an integral DTGS detector and incidence-angle variable between 10° and 80°. All of these accessories were used in conjunction with a Nicolet Magna 550 FTIR spectrophotometer with DTGS detector operated at 16 cm^{-1} resolution.

A novel FT-PL system, based on the Nicolet 960 dedicated FT-Raman spectrophotometer (operated at 8 cm^{-1} resolution), was used for the measurements on TPV devices and materials. The instrument includes a diode-pumped, Nd:YVO$_4$ excitation laser operating at 1064 nm, an LN$_2$-cooled, Zn-doped Ge detector, 180° reflective sampling optics in the main sample compartment, and a scanning, auto-focusing Raman microscope with 1-μm positioning precision and video capture for visible light microphotography of the samples analyzed. A Spectra-Tech 32x Reflachromat reflective objective and an Olympus 100x MS Plan 100 objective enabled microscope-excitation spot diameters of 10.0 μm and 1.2 μm, respectively.

Several features were added to the 960 instrument to extend its capabilities for FT-PL spectroscopy. Its long-wavelength range was extended from 1750 nm (the Ge detector limit) to 2700 nm by adding a. cold-filtered InSb detector (Graseby Infrared). A Melles-Griot Model 06 DLL 807 GaAs diode laser operated at 847 nm by a Melles-Griot Model 06 DLD 103 controller was installed in the interlocked instrument housing with an optical isolator. The laser delivers 30 mW of excitation power to the main sampling compartment. Together with a Kaiser 850-nm holographic notch filter in the emission-beam path, the GaAs laser extended the short-wavelength range of the spectrophotometer from 1075 nm to 870 nm for PL measurements. Mounts for neutral density filters in the main sampling compartment and microscope excitation-beam paths enabled low-excitation power levels, which are especially important to avoid band-filling and sample overheating effects in FT-PL micro-spectroscopy. The FT-PL data were taken at 300 K, although sample cooling to 30 K is possible (10, 13).

RESULTS AND DISCUSSION

FTIR Reflectance Measurements

The influence of sub-bandgap reflectance on the efficiency of TPV systems has been described (1, 14). The strong correlation between these variables means that even variations of a few percent in the return of sub-bandgap (mid-IR) photons to the thermal radiator will have a profound influence on TPV system performance. For this reason, accurate measurements of TPV device or device filter reflectance are needed. To assess accuracy, we compared the mid-IR reflectance data measured for a series of cadmium stannate plasma filters using unpolarized SR (30° incidence) with unpolarized TR data taken using the integrating sphere (10° incidence) and the absolute reflectometer (30° incidence). We found that for rough samples (diffuse reflectors), the two TR measurements were comparable (Fig. 1), while the SR measurement gave lower values, as expected to result from Rayleigh scattering. For smooth samples, reflectance was higher for the accessories having a higher incidence angle, also as expected. The agreement between the results obtained using the two different TR accessories is encouraging. The SR measurements are within three percent of the TR measurements, a numerically small but significant difference.

Reflectance measurements on TPV devices are more challenging because of their fragility, complexity, and small size. In Fig. 2 we present the effects on the TR spectrum of successive application of a magnesium fluoride/titanium/gold back-surface reflector (BSR) and a magnesium fluoride/zinc selenide anti-reflective (AR) coating onto a TPV device with SiO_2 backing. The spectra were measured using the FTIR integrating sphere. The effect of the BSR, which appeared highly specular, in increasing device reflectance is very evident. However, application of the AR coating had little or no effect on the sub-bandgap reflectance of the device. The reflectance minima at 10 and 15 microns in Fig. 2 correspond to lattice absorption by the InP substrate and absorption by Fe centers in the InP substrate, respectively.

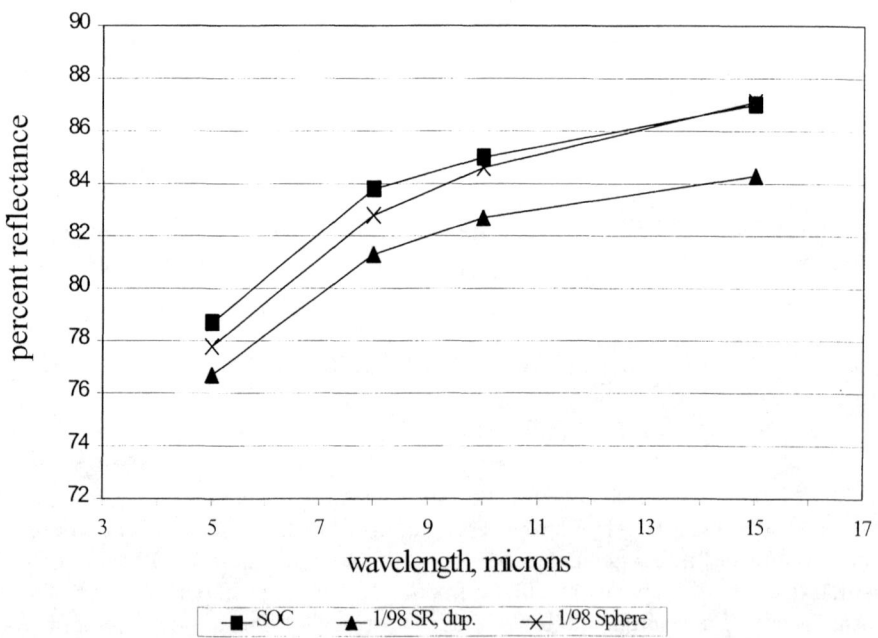

FIGURE 1. FTIR reflectance measurements of a diffuse CdSnO4/glass film.

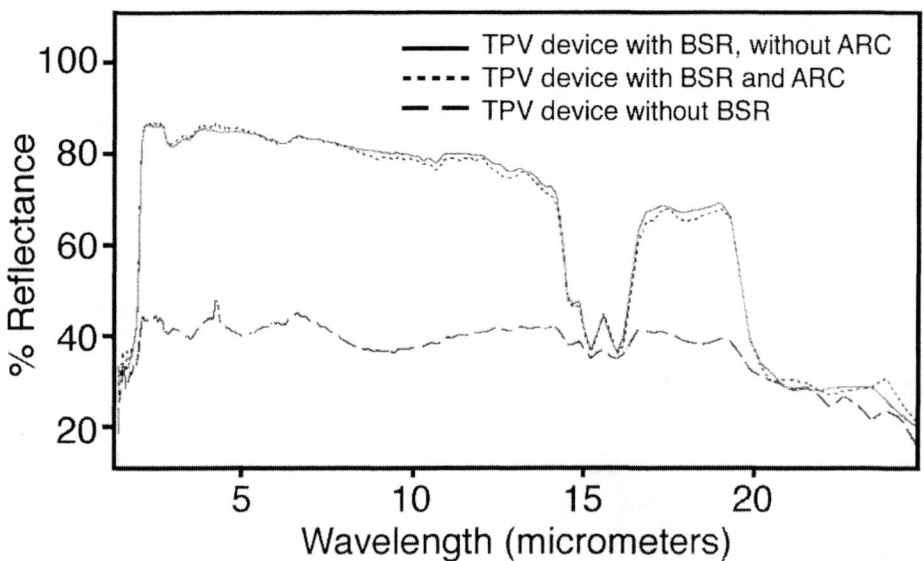

FIGURE 2. Effect of back-surface reflector and antireflective coating on total reflectance of TPV Device 1-236b.

We verified the effect of orientation of the gridlines on monolithic interconnected minimodule (MIM) TPV device structures with respect to the plane of incidence of IR radiation in the integrating sphere reported by Ballinger (15). At 2.5 µm, the TR of a typical MIM structure without a BSR is about 2% lower with the device trenching and gridlines perpendicular to the incidence plane in the integrating sphere, and 5% lower for the same device with a gold BSR, regardless of polarization of the incident beam. By performing polarized transmittance measurements on the MIM structure without a BSR, we found that the device transmittance is sensitive to polarization. The device acts similarly to a wire-grid polarizer in that it polarizes light in the direction perpendicular to the conductive grid pattern. The device-polarization ratio (Fig. 3) exhibits a minimum near 4 µm and a continuing decrease at longer wavelengths corresponding with orientation of the device gridlines parallel to the direction of polarization, with some fine structure corresponding to lattice absorption bands. The long-wavelength decrease correlates with the grid spacing (100 µm between trenches and top contacts). However, the light-trapping effect dominates the polarization effect in the reflectance measurements. We made variable-angle SR measurements (not shown) of MIM TPV Device 1-223b with a mechanically contacted gold BSR, which showed that the device reflectance decreased strongly with increasing incidence angle from 20° to 45° when the device gridlines were perpendicular to the incidence plane. The SR measurements may overestimate this decrease because of Rayleigh scattering from the device structures, however, no such decrease in SR was noted when incidence angle was increased with the device gridlines oriented in the plane of incidence. More accurate measurements of the angle-dependent TR of the MIM structures could be made using the SOC-100 variable-angle hemispherical reflectometer, but the light-trapping character of the device structure for non-normal incident radiation is evident from our current results.

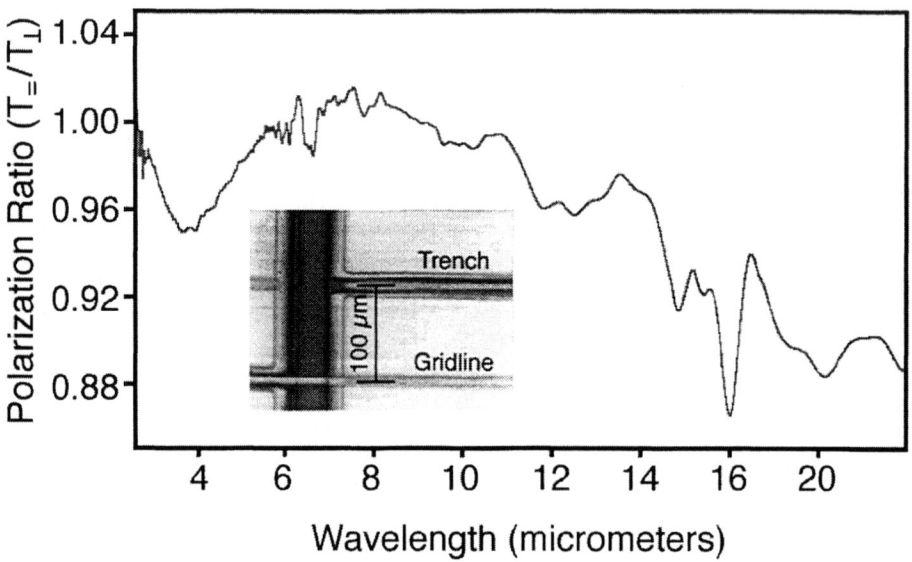

FIGURE 3. Polarization ratio for MIM device structure #1-223b (no BSR), ratio of T with [gridlines parallel]/[gridlines perpendicular] to direction of polarization.

FT-Photoluminescence Measurements

Figure 4 shows the relationship between the TR and the FT-PL spectra of a $Ga_{0.47}In_{0.53}As$ MIM TPV device with a gold BSR. The correspondence between the low-energy TR and PL band edges is close for this device, which was lattice-matched (LM) to InP and did not incorporate a graded layer (GL). The PL maximum is higher in energy than the reflectance band edge by less than kT, and it may be possible to use TR spectra collected from non-graded BSR films and devices to validate an algorithm for calculating the bandgap from FT-PL spectra.

In (12) we showed that the intensity of PL collected from undoped, lattice-mismatched (LMM) $Ga_xIn_{1-x}As$ samples is more than an order of magnitude lower than that collected from a LM sample. The compositional red shift of the PL band maxima as x is decreased from the InP-LM value of 0.47 is within kT of the predicted bandgap for this material (1, 12, 16).

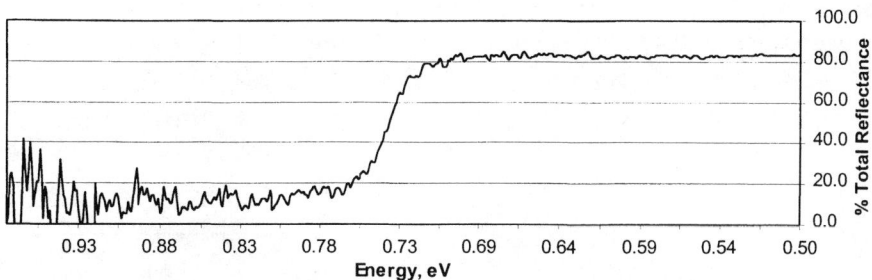
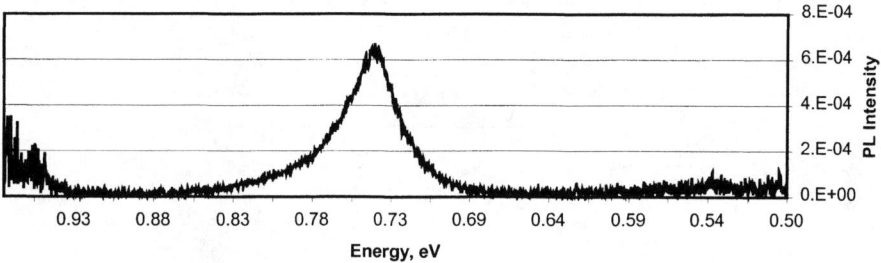

FIGURE 4. Total reflectance and FT-photoluminescence spectra of TPV device #1-153.

In (13) the results of high spatial resolution (~1 µm) FT-PL spectroscopy of TPV device 1-213 were given. The spectra were taken on the cleaved edge of the device, with one spectrum collected in the $Ga_xIn_{1-x}As$ p/n absorber layer (x=0.32), and another in the GL, close to the absorber layer. The device structure is given in Fig. 5. The technique has sufficient resolution to distinguish the two thicker $Ga_xIn_{1-x}As$ layers. A lower PL intensity, implying a relative increase in the rate of non-radiative recombination because of LMM in the GL, is evident in the spectrum of the absorber layer relative to that of the GL.

Fig. 5 shows the locations of an 18-step line map superimposed on a micrograph of angle-lapped TPV device 1-213. Angle-lapping was done at an angle of $2.5°$ or less to the device plane using a chemical/mechanical polishing medium consisting of 0.05 µm non-crystallizing colloidal silica in basic solution. This procedure exposed lapped surfaces of 180 to 400 microns, enabling use of the 32x microscope objective (10 µm excitation spot diameter) to resolve device layers as thin as [10 µm *sin(2.5°)] = 0.4 µm or less. The scan began on the contact cap layer surface, and progressed through the lapped absorber and GL to the InP substrate. A thin vertical line is visible just to the left of the center of the lapped device surface, which corresponds to the position of the lower confinement layer (1) in this device. Note that the "device layer" scale on Figure 5 and in subsequent figures is approximate in that it assumes planarity of the angle-lapped surface and is given only for reference. The scale ignores the thinner device layers, which were not detectable by FT-PL microspectroscopy. The PL intensity decreases as the lapped region is traversed during the linear scan.

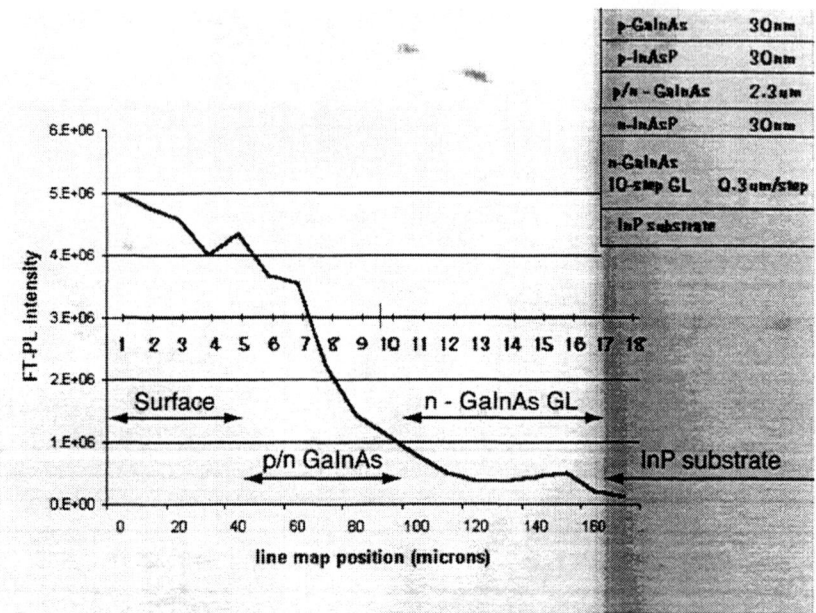

FIGURE 5. FT-PL peak intensity (InSb detector) vs. line map position for TPV device #1-213, superimposed on a micrograph of the device. The number line shows where the FT-PL spectra were collected on the angle-lapped device.

Figure 6 is a three-dimensional plot of the corrected FT-PL spectra obtained from the line map of the angle-lapped device, versus scan position in microns. A strong, sharp PL spectrum was obtained from the device surface, apparently from the absorber layer through the thin cap and upper confinement layers. The PL intensity dropped to some extent as the absorber layer was traversed during the scan, but stayed relatively high until the GL was approached. PL intensity dropped rapidly as the GL was traversed, implying a relative increase in the rate of non-radiative recombination due to LMM, and the energy and bandwidth of the PL spectrum both increased. No PL was collected from the final scan point, in the InP substrate. During this scan, only PL emission characteristic of $Ga_xIn_{1-x}As$, i.e., having energies between 0.74 and 0.60 eV, was observed. Even with angle-lapping, the 32x objective has insufficient spatial resolution to collect PL from the cap and confinement layers, and the confinement layer and substrate bandgaps (1 eV and 1.34 eV, respectively) are beyond the upper limit of the InSb detector response range. The detector cannot be switched to Ge during a scan with the current scanning software, and a higher-energy laser source would be required to excite PL from the substrate. Attempts to resolve PL from the confinement layers in the angle-lapped device using the 100x objective with 1064-nm excitation and Ge detector were not successful. Apparently the limited optical throughput of the 100x objective, the shallow absorption depth of the confinement layer, or damage to the confinement layer during lapping reduced the PL signal below detection limits.

The energies of the corrected PL band maxima are plotted as a function of scan position, with the predicted $E_g(x)$ of the $Ga_xIn_{1-x}As$ device structure, in Fig. 7. The agreement between the PL peak energies and the predicted direct bandgap of $Ga_xIn_{1-x}As$ at 300 K (16),

$$E_g(x) = (0.555 \, x^2 + 0.505 \, x + 0.356), \tag{1}$$

is within kT (0.026 eV), with the maxima of the PL peaks at energies above the predicted bandgap. The value of x is constant at 0.32 in the absorber layer, and varies from 0.32 to 0.47 (InP-LM) in the GL. The difference between the PL maxima and

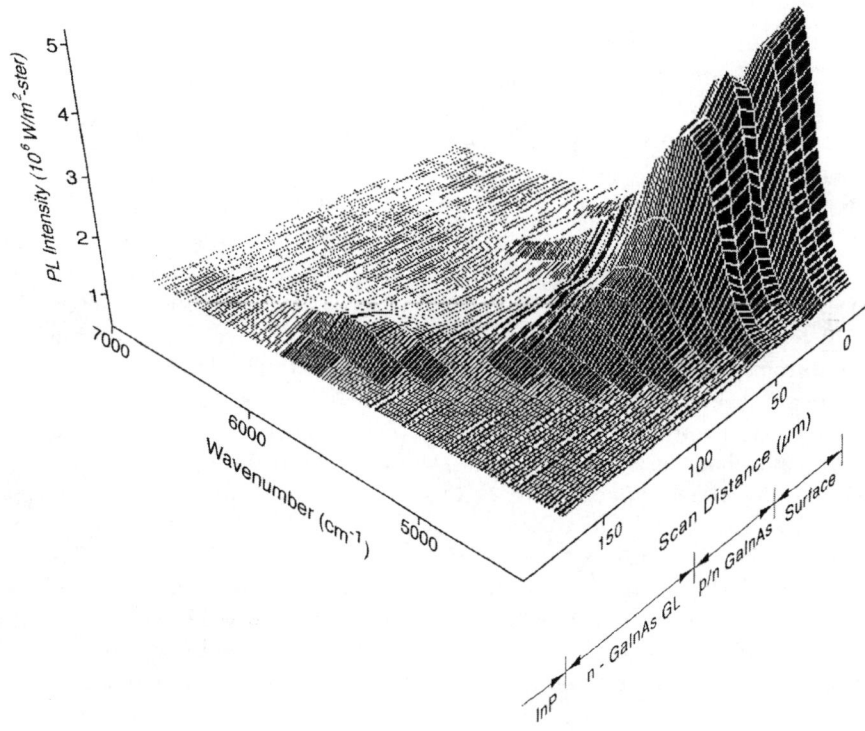

FIGURE 6. FT-PL spectra vs. line map position on angle-lapped TPV device #1-213.

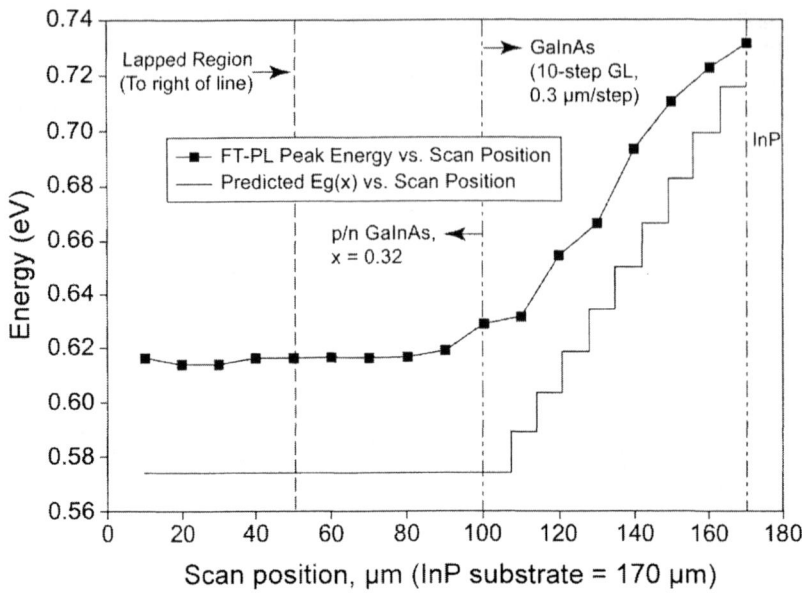

FIGURE 7. FT-PL peak energies vs. line map position on angle-lapped TPV device #1-213, compared to the predicted bandgap of the device.

$E_g(x)$ is expected to arise from thermal band filling, which is significant at room temperature, and could be more pronounced in doped material (1). Allowing for these effects, the PL band maxima track the composition of the material closely. There is some "anticipation" of the increase in $E_g(x)$ in the GL at 100 μm in the line scan, indicating that the excitation penetration depth may correspond to about a single scan interval, or that the effective excitation depth is approximately [10 μm *sin(2.5°)], or about 0.4 μm. A similar variation in PL peak energies with scan position was obtained from a micro-FT-PL line map of angle-lapped TPV device 1-216, which incorporated a continuously graded layer (CGL) of $Ga_xIn_{1-x}As$. However, the PL peak intensities measured during the linear scan of Device 1-216 (not shown) were much more variable than those measured for Device 1-213 (Figure 5), perhaps in part because the angle-lapped portion of Device 1-216 was visibly rougher than the lapped portion of Device 1-213.

In Fig. 8, the PL intensities obtained from two successive micro-FT-PL line scans of angle-lapped TPV device 1-217 are shown, together with the device structure. Two scans in identical locations with different detectors were required because the $Ga_xIn_{1-x}As$ and $InAs_yP_{1-y}$ layers emit PL within the response ranges of the InSb and Ge detectors, respectively, and the detectors cannot collect data simultaneously.

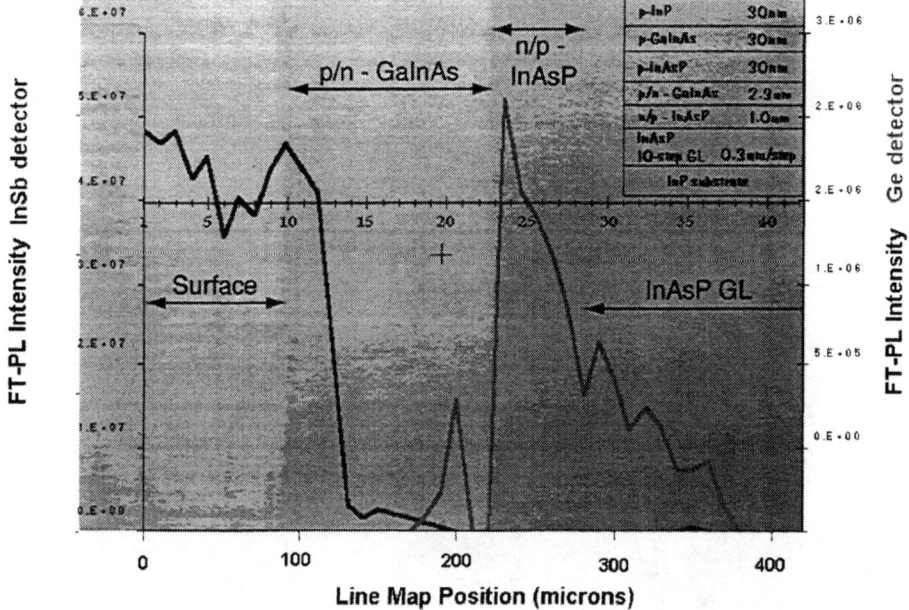

FIGURE 8. FT-PL peak intensity (InSb and Ge detectors) vs. line map position for TPV device #1-217, superimposed on a micrograph of the device. The number line shows where the FT-PL spectra were collected on the angle-lapped device.

The maximum PL intensities apparently differ by about an order of magnitude between the two sets of data in Fig. 8, but the detectors are uncalibrated, although an instrument spectral response correction has been applied to the FT-PL data. An interesting feature in Fig. 8 is the overlap in PL emission characteristic of $Ga_{0.32}In_{0.68}As$ and $InAs_{0.32}P_{0.68}$ in the line scan region between 175 and 200 μm, which corresponds to the lower portion of the p/n $Ga_{0.32}In_{0.68}As$ layer. This is apparently related to penetration of the lower few tenths of a micron of this layer by the excitation radiation, as described previously. The non-luminescent area at the n/p $InAs_{0.32}P_{0.68}$/ $InAs_yP_{1-y}$ step-GL interface is also of interest. The abrupt drop in PL intensity midway through the line scan of the p/n $Ga_{0.32}In_{0.68}As$ layer is especially evident in Fig. 9, which is a superposition of the line scan micro-FT-PL spectra collected using the InSb and Ge detectors. The apparently high rate of non-radiative recombination in this region may coincide with defects in the n-layer, although replicate scans of this device would be needed to verify this hypothesis.

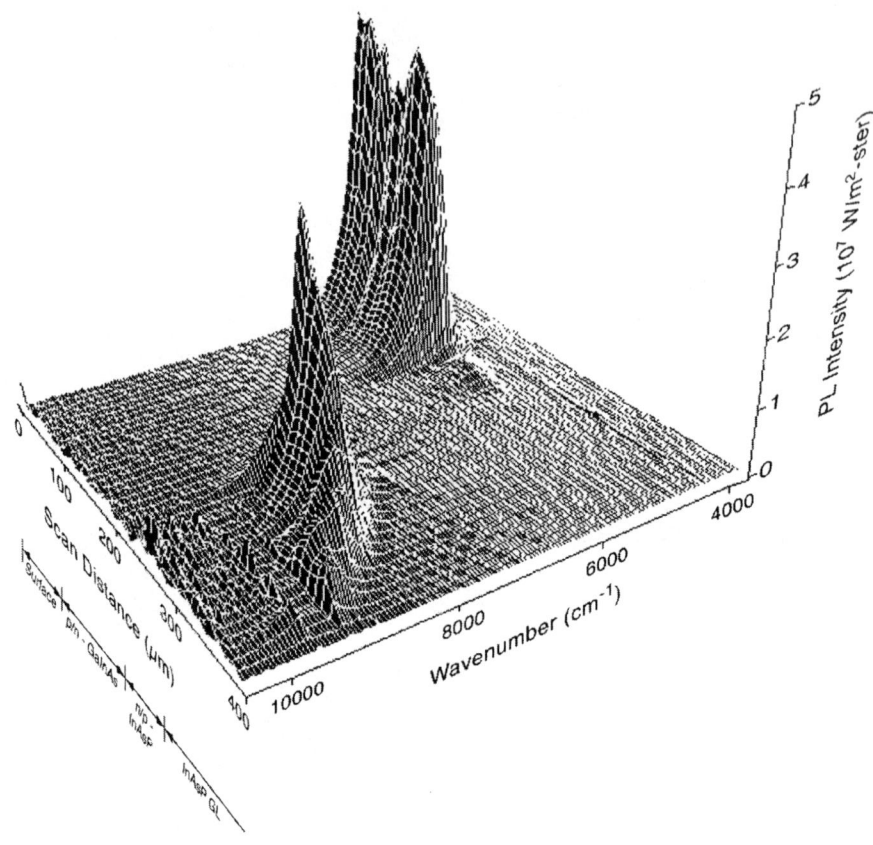

FIGURE 9. FT-PL spectra vs. line map position on angle-lapped TPV device #1-217.

CONCLUSIONS

Fourier transform spectroscopic techniques for photoluminescence and infrared reflectance offer powerful tools for TPV device characterization. We are integrating these techniques into a regular protocol with inputs to TPV device characterization, modeling, quality control, and troubleshooting. Algorithms for extracting bandgap estimates from the FT-PL spectra of $Ga_xIn_{1-x}As$ and $InAs_yP_{1-y}$ are also being developed.

ACKNOWLEDGMENT

We acknowledge the contribution of the late Paul McKenna of Surface Optics Corporation, San Diego, CA 92131 in measuring the reflectance of the cadmium stannate samples using the SOC-100 absolute reflectometer.

REFERENCES

1. Wanlass, M. W., *Solar Energy Materials and Solar Cells*, 41/42 (1996), p. 405.
2. Griffiths, P.R., and De Haseth, J.A., *Fourier Transform Infrared Spectroscopy*, John Wiley & Sons, New York, 1986.
3. Wu, X., Mulligan, W. P., Webb, J. D., and Coutts, T. J., "TPV Plasma Filters Based on Cadmium Stannate," presented at the 2nd NREL Conference on TPV Generation of Electricity, *AIP Conf. Proc. 358*, 1995, pp. 329–338.
4. Charache, G. W., DePoy, D. M., Baldasaro, P. F., and Campbell, B. C., "Thermophotovoltaic Devices Utilizing a Back Surface Reflector for Spectral Control," presented at the 2nd NREL Conference on TPV Generation of Electricity, *AIP Conf. Proc. 358*, 1995, pp. 339–345.
5. Gu,S.Q., Chen, D. Viner, J.M., Raikh, M.E., and Taylor, P.C., in Noufi, R., (Ed.), AIP Conference Proc., Vol. 268, American Institute of Physics, New York, 1992, p. 363.
6. A. Yamada, M. Tanda, S. Manaka, H. Sano, M. Konegai, and K. Takahashi, in L. Guimaraes (Ed.), *Proc. 11th European Photovoltaic Solar Energy Conference*, Harwood, Switzerland, 1992, p. 128.
7. J. D. Webb, D. J. Dunlavy, T. Ciszek, R. K. Ahrenkiel, M. W. Wanlass, R. Noufi, and S. M. Vernon, *Appl. Spectrosc.*, 47, no. 11 (1993), 1814.
8. J. D. Webb, D. J. Dunlavy, T. Ciszek, R. K. Ahrenkiel, M. W. Wanlass, R. Noufi, and S. M. Vernon, in O. J. Glembocki (Ed.), *Proc. Mat. Res. Soc. Symp.*, Vol. 324, Materials Research Society, Pittsburgh, 1994, p. 233.
9. J. D. Webb, M. Contreras, and R. Noufi, in D. J. Flood (Chair), *Proc. 24st IEEE Photovoltaic Specialists Conference*, Institute of Electrical and Electronic Engineers, New York (1994), p. 275.
10. J. D. Webb, B. M. Keyes, K. Ramanathan, P. Dippo, D. W. Niles, and R. Noufi, in C. E. Witt (Ed.), *Proc. AIP*, Vol. 314, American Institute of Physics, Woodbury, New York, 1996, p. 573.
11. B. Schrader, in D. B. Chase and J. F. Rabolt (Eds.), *Practical Fourier Transform Infrared Spectroscopy*, Academic Press, New York, 1990, Chap. 4.
12. T. L. Wangensteen, M. W. Wanlass, J. J. Carapella, H. R. Moutinho, A.R. Mason, J. D. Webb, and F. A. Abulfotuh, in *Proc. 26th IEEE Photovoltaic Specialists Conference*, Institute of Electrical and Electronic Engineers, New York (1997), p.967.
13. Webb, J. D., Keyes, B. M., Ahrenkiel, R. K., Wanlass, M. W., Ramanathan, K., and Jones, K. M., "Fourier Transform Luminescence Spectroscopy of Semiconductor Thin Films and Devices," *Vibrational Spectroscopy*, in press (1998).
14. Broman, L., Jarefors, K., Marks, J., and Wanlass, M., "Efficiency Measurements of TPV Cells," presented at the 3rd NREL Conference on TPV Generation of Electricity, *AIP Conf. Proc. 401*, 1995, pp. 463–469.
15. Ballinger, C., *et al.*, "Computational Results for a MIMS Device with Back Surface Reflector (BSR)," to be presented at the 4th NREL Conference on TPV Generation of Electricity, *AIP Conf. Proc. XXX*, 1998, in press.
16. P. Bhattacharya (Ed.), *Properties of Lattice-Matched and Strained Indium Gallium Arsenide*, Short Run Press, Ltd., Exeter, UK, 1993, p. 7.

Recombination Lifetime of $In_xGa_{1-x}As$ Alloys Used in Thermophotovoltaic Converters

R.K. Ahrenkiel, R. Ellingson, S. Johnston, J. Webb, J. Carapella, and M. Wanlass

National Renewable Energy Laboratory
1617 Cole Blvd, Golden, CO 80401
Golden, CO 80401

Abstract. The family of ternary compounds of composition $In_xGa_{1-x}As$ are of considerable interest for thermophotovoltaic energy converters. The recombination lifetimes of the various compositions are critical to the successful application of these materials as efficient converters. Here we will describe experimental results on the composition, $In_{0.53}Ga_{0.47}$, that is lattice-matched to InP. We will also describe lifetime results on the compositions $In_{0.68}Ga_{0.32}As$, with a bandgap of 0.60 eV to compositions $In_{0.78}Ga_{0.22}As$ with a bandgap of 0.50 eV. Double heterostructure confinement devices have been made over a range of both n- and p-type doping. These results are preliminary, but the goal is to obtain the radiative and Auger recombination coefficients for the alloys in this composition range.

INTRODUCTION

InGaAs is a ternary III-V semiconductor that is of current interest for thermophotovoltaic (TPV) devices. Of particular interest are the compositions that produce bandgaps of 0.5 eV to 0.6 eV, because these can be used with lower-temperature thermal radiators. The minority-carrier and recombination (high-injection) lifetimes are intimately related to photovoltaic efficiency. As the minority-carrier diffusion length is directly proportional to the square root of the lifetime, the short-circuit current J_{sc} varies accordingly. In addition, the reverse-saturation current J_o increases as the inverse square root of the lifetime. Obtaining acceptable lifetimes and even maximizing the lifetime is a crucial component of material development.

For the both lattice-matched and small-bandgap ternaries, limited lifetime data have been available (1). Henry and coworkers (2) measured the minority-carrier lifetime in n- and p-type lattice-matched $Ga_{0.47}In_{0.53}As$ grown on InP substrates. They measured the lifetime over a doping range from 5×10^{17} cm^{-3} to 1×10^{19} cm^{-3} in both n- and p-type $Ga_{0.47}In_{0.53}As$. They found that the lifetime decreased with doping density over these doping levels. The lifetime first deceased with the inverse of the doping level, which is indicative of radiative recombination. Finally, they found relatively higher lifetimes in n-type than in p-type $Ga_{0.47}In_{0.53}As$. Higher doping levels produced a lifetime that decreases as the inverse square, which is indicative of Auger recombination. In the low doping range, Gallant and Zemel (3) measured a photoconductive lifetime of 18.5 μs in an undoped $InP/Ga_{0.47}In_{0.53}As/InP$ double heterostructure (DH). Recent work by the NREL group has measured the lifetime of the lattice-matched composition $In_{0.53}Ga_{0.47}$ over a doping range from 2×10^{14} cm^{-3} to 2×10^{19} cm^{-3} (4). As a result of these measurements, the radiative (B-coefficient) and Auger (C-coefficient) parameters were obtained for this composition from an analysis of the data. Some more recent results with lattice-matched compositions will be described here.

Very little information exists in the literature on the properties of the compositions of $In_xGa_{1-x}As$ with $x > 0.47$ and $E_g < 0.73$ eV. The author and coworkers (5) described some early results on these compositions that were both undoped and doped. We measured the lifetime of an undoped film of composition $In_{0.78}Ga_{0.22}As$ with a bandgap of 0.5 eV. The lifetime decreased to 52 ns as compared with lifetimes of several microseconds in a similar lattice-matched film. The decrease was attributed to recombination at the threading dislocations arising from mismatch. Owing to large improvements in the growth and materials technology, the results to be described here are markedly improved.

EXPERIMENTAL RESULTS

MOVPE Film Growth

Our samples were grown by metalorganic vapor phase epitaxy (MOPVE), as described in another paper presented at this conference (6). The $In_xGa_{1-x}As$ DH was grown on step-graded layers from a InP substrate. The graded structure consists of 10 steps that are each 0.3 μm thick. The composition of these layers is $InAs_xP_{1-x}$ (as x varies from 0 to the final composition $InAs_yP_{1-y}$). Here the lattice constant of $InAs_yP_{1-y}$ matches that of the active layer $In_xGa_{1-x}As$. After the step-graded layer is complete, a 1.0-μm window layer of the final composition $InAs_yP_{1-y}$ is grown. Next, an active layer of $In_xGa_{1-x}As$ is grown that varies between 0.25 and 2.0 μm in thickness. Finally, a lattice-matched top window layer of $InAs_yP_{1-y}$ is grown that is 300 Å thick. These passivating window layers are critical improvements over earlier lattice-mismatched devices (5).

Lifetime Measurement Techniques

Recombination lifetime measurements were made by two complementary techniques. A radio-frequency photoconductive decay (RFPCD) technique was used (7) to measure the samples in the lower doping range. The RFPCD technique can resolve lifetimes greater than about 20 ns. A second RFPCD system provides variable sample temperature from about 80 to 300 K. Using temperature as a variable provides additional information for identifying the recombination mechanism, as well as providing data for low-temperature device applications. The pulsed-light sources were used here included light-emitting diodes (LED), an attenuated pulsed YAG laser, and a tunable optical parametric oscillator (OPO) driven by a tripled YAG laser. The latter could be tuned to a wavelength of about 400 nm to 2.2 μm. The long wavelength range was used to find the InGaAs bandgap by the onset of photoconductivity.

Samples in the higher doping ranges were measured by photon-counting technique that involved up-conversion of photoluminescence. As the intrinsic photoluminescence for InGaAs is in the infrared (about 1.7 to 2.4 μm), the sum-frequency up-conversion technique allows the use of photon counting (8). The samples were pumped by 90 femtosecond (fs) pulses at 82 MHz repetition rate. The pump laser was a Spectra-Physics mode-locked titanium-sapphire (Ti:S) laser tuned to 762 nm wavelength output. The collected luminescence was mixed with the Ti:S laser pulse for sum-frequency generation (SFG) in a 1-mm lithium iodate ($LiIO_3$) crystal. The SFG signal was dispersed by a SPEX 270 meter monochromator and detected by a Hamamatsu R464 photomultiplier tube. The system time resolution is approximately 110 fs.

The bandgaps of the alloy films were determined by either Fourier Transform infrared photoluminescence or Raman scattering. The bandgaps were also confirmed by using the tunable OPO in the infrared range and measuring an excitation spectrum for photoconductivity.

Recombination Mechanisms

The low-injection lifetime in dislocation-free films of InGaAs can be written as:

$$\frac{1}{\tau_B} = BN + CN^2 + \frac{1}{\tau_{SRH}}, \qquad 1)$$

where B is the radiative B-coefficient, C is the Auger coefficient, and N is the majority-carrier density. The bulk Shockley-Read-Hall recombination lifetime due to point defects is given by τ_{SRH}. When two surfaces are included with recombination velocity S_1 and S_2, the total lifetime becomes:

$$\frac{1}{\tau_B} = BN + CN^2 + \frac{1}{\tau_{SRH}} + \frac{S_1 + S_2}{d}, \qquad 2)$$

Lifetime of In(0.53)Ga(0.47) as a Function of Doping Level

Figure 1 shows the minority-carrier lifetime of a series of $In_{0.53}Ga_{0.47}$ double heterostructures of both n- and p-type over five orders of magnitude of doping concentration. The data could be fit with Eq. 2 where the B- and C-coefficients were found to be:

$$B = 1.43 \times 10^{-10} \text{ cm}^{-3}\text{s}^{-1}$$

$$C_{n,p} = 8.1 \times 10^{-29} \text{ cm}^{-6} \text{ s}^{-1},$$

These values compare very favorably with earlier values found by Sermage and coworkers (9). The lifetime at the lowest carrier concentrations is defect dominated as predicted by the Shockley-Read-Hall theory (10,11). In this sample set, the largest lifetimes in

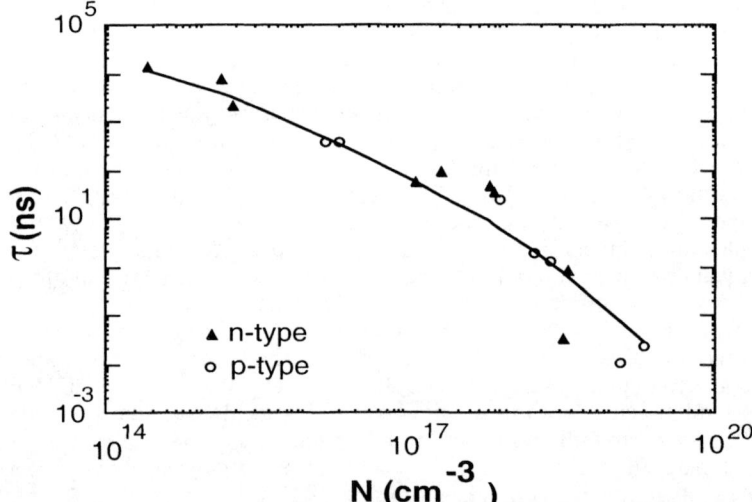

FIGURE 1. The minority-carrier lifetime of lattice-matched $In_{0.53}Ga_{0.47}$ versus free-carrier concentration. The solid line is a fit of Eq. 1. to these data.

undoped films were about 20 μs. These data on lattice-matched samples form a basis for comparison of the new work on lattice-mismatched alloys. For undoped DH structures, one hopes to get lifetimes in the microsecond range, indicating that dislocation-induced recombination has been reduced. For the heavily doped alloys, the B- and C-coefficients for lattice-matched material serve as starting points and estimates for finding these constants in mismatched material.

Lattice-mismatched Films

Undoped n-type In(0.68) Ga(0.32)As DH Structure

A series of lattice-mismatched DHs of composition In_xGa_{1-x} with x varying between 0.68 and 0.77 were grown on InP substrates. A grading layer was grown between the InP substrate and DH structure. The indium composition was increased from 0.53 until the active layer composition x was reached. The InGaAs film thicknesses are about 2 μm for each sample. Fig. 2 shows RFPCD data on two undoped DH structures with bandgaps of 0.58 eV (Curve A) and 0.55 eV (Curves B and C), respectively. The excitation source for Curves A and B is the doubled-YAG wavelength of 532 nm. For sample A, the capacitance-voltage (C-V) measurement show the conductivity to be n-type with n = 1.43 x 10^{15} cm^{-3}. The low-injection lifetime of this sample is 5.5 μs, as seen in the figure. Sample B is n-type with an electron concentration of 2.6 x 10^{15} cm^{-3} as measured by C-V. The lifetimes here, as shown by curves B and C, are 3.3 μs. Curve C was obtained by exciting the sample with the optical parametric amplifier tuned to about 2.3 μm wavelength. This wavelength corresponds to a photon energy of 0.54 eV, which is slightly less than the bandgap determined by photoluminescence. The response died for slightly longer wavelengths. As the samples show lifetimes of several microseconds, the results indicate high-quality epitaxial films with low dislocation densities. These lifetimes are slightly smaller than found in the best lattice-matched films of prior work (12). These data also indicate that dislocations must be of fairly low density, contrasted with the results of earlier work (5).

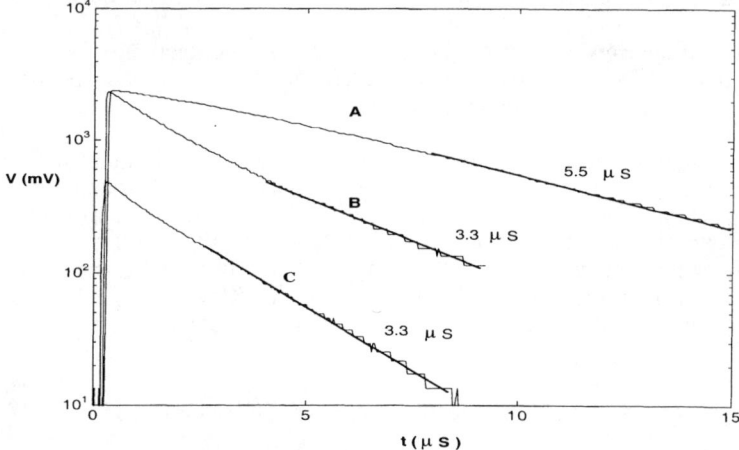

FIGURE 2. RFPCD data on two undoped, n-type DH structures with bandgaps A: 0.58 eV and B,C: 0.55 eV. The laser excitation wavelengths are: A: 532 nm, B: 532 nm, C: 2.3 μm. The donor concentrations are: A: 1.43 x 10^{15} cm^{-3}; B, C:2.6 x 10^{15} cm^{-3}.

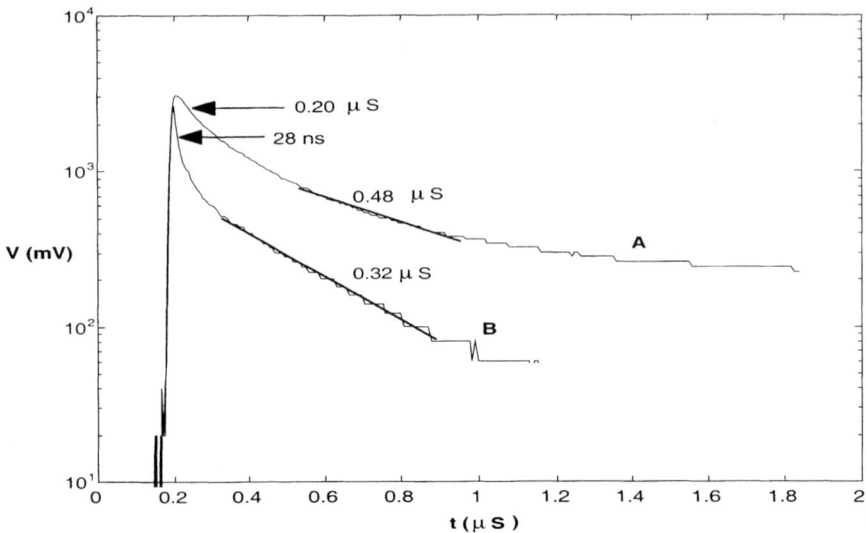

FIGURE. 3. RFPCD data of a p-type InGaAs DH with Eg = 0.52 eV and $N_A = 1.7 \times 10^{16}$ cm^{-3}. The data of Curve A was measured with a excitation wavelength of 1064 nm. The data of Curve B used a wavelength of 532 nm.

P-type In(0.78) Ga(0.22) As DH Structure

This sample was grown with the above target composition and was Zn-doped for p-type conductivity. C-V measurements showed p-type conductivity with a hole concentration of 1.7×10^{16} cm^{-3} at the front window layer. However, the carrier type appears to reverse deeper into the structure. The RFPCD lifetime is shown in Fig. 3 using two excitation wavelengths. Curve A is drawn from data with the laser wavelength set to 1064 nm, and Curve B from data with the wavelength set to 532 nm. The light pulse is absorbed very near the front surface for the 532 excitation. The steep initial slope with a 28-ns decay time is indicative of a high surface recombination velocity between the active InGaAs region and the InAsP window layer (13). In both cases, the excess carrier decay has positive curvature. Such behavior can be indicative of either bimolecular (high-injection) radiative recombination or a shallow SRH-defect dominated recombination process (14). Using the B-coefficient obtained from the analysis of lattice-matched InGaAs, the calculated radiative lifetime is about 0.39 µs. The portion of the decay between 0.4 µs and 1.0 µs is in the range of the predicted radiative lifetime. The dominant recombination mechanism can be clarified with further studies such as the temperature dependence of the lifetime (15). The general conclusion however, is that the lifetimes are much improved compared to the earlier devices reported by the author and coworkers (5).

Heavily doped n-type In(0.68) Ga(0.32) As

The RFPCD technique is only useful for lifetime measurements when the lifetime exceeds about 10 ns. For shorter lifetimes, we have developed that photon up-conversion technique that has been described above. With up-conversion, we can resolve lifetime values to less than 1 ps. Figure 4. shows up-conversion data on two samples that are heavily doped with sulfur to produce n-type conductivity. Sample A has a doping-density profile that varies from 2.8×10^{18} cm^{-3} to 8.7×10^{18} cm^{-3}. The lifetime measured here is 2.2 ns measured at the peak of the photoluminescence spectrum. Optical measurements indicate that the bandgap of this sample is 0.63 eV. Sample B has a lifetime of 925 ps and the electron concentration is 1.0×10^{19} cm^{-3} as measured by CV. Using the B-coefficient from the lattice-matched material, the calculated radiative lifetime of sample A is about 1.6 ns compared with the 2.23 ns measured here. We see from Fig. 1 that the lifetimes in this doping range is in the 10-to-100-ps range and are controlled by Auger recombination. The C-coefficient predicts an Auger lifetime of 367 ps for sample A.

For sample B, we calculate a radiative lifetime of 700 ps and an Auger lifetime of 11 ps, compared with a measured lifetime of 930 ps. The measured lifetime for the mismatched films are much larger than one would extrapolate from the lattice-matched case. These preliminary data may indicate that the lifetimes are consistent with predicted radiative lifetimes. The Auger coefficients appear to be *much smaller* then predicted from the lattice-matched C-coefficient. The reduction of the Auger effect is a very puzzling feature of these data.

These are preliminary experiments and further work will be required to establish reliable recombination parameters. The Auger coefficient is a very important parameter for devices that are operating in a high-injection mode. A suspected weak Auger coefficient awaits confirmation by characterizing more samples as a function of carrier concentration.

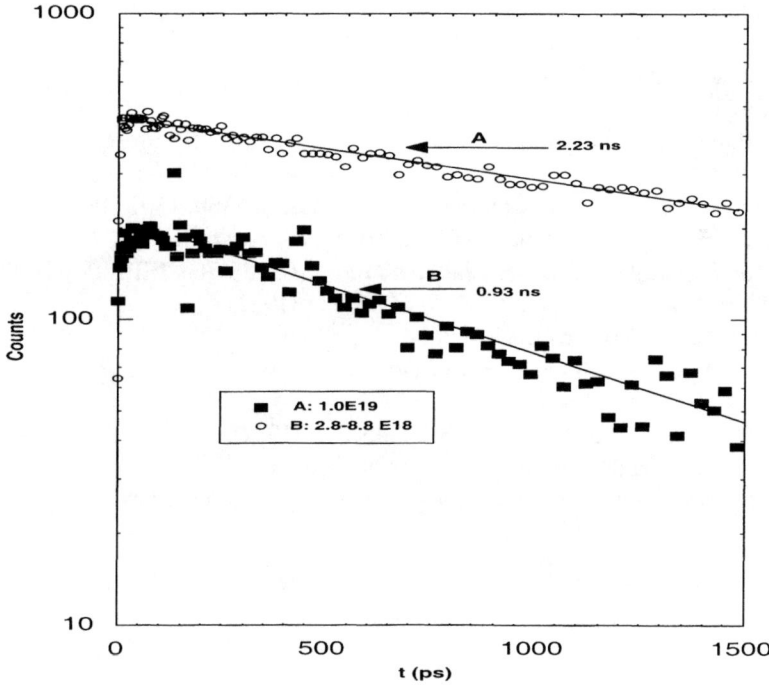

FIGURE 4. The minority-carrier lifetime measured by the upconversion technique described in the text. The DH samples are n-type with the doping levels shown. The bandgaps of these samples are about 0.63 eV.

CONCLUSIONS

Measurements on lattice-matched InGaAs were made over a wide doping range to produce values of the radiative and Auger recombination coefficients. These measurements provide a basis or starting point for comparing the radiative and Auger coefficients for mismatched ternaries. Measurements on lattice-mismatched films with bandgaps ranging from 0.5 to 0.6 eV has also shown much improved electronic properties. Lifetimes of about 2 to 5 μs were measured in undoped DH film structures. Measurements were made on n-type samples doped in the range of 5×10^{18} cm^{-3} to 1×10^{19} cm^{-3} by the new up-conversion technique. These preliminary lifetimes were much larger than found in lattice-matched material of comparable doping.

REFERENCES

1. *Properties of Indium Gallium Arsenide*, INSPEC, The Institution of Electrical Engineers, London and New York (1993).
2. C.H. Henry, R.A. Logan, F.R. Merrit, and C.G. Bethea, Electron. Lett . (UK), **20**, 358 (1984).
3. M. Gallant and Z. Zemel, *Appl. Phys. Lett.*, **52**, 1686 (1988).
4. R.K. Ahrenkiel, R. Ellingson, S. Johnson, and M. Wanlass, *Appl. Phys. Lett.* **72**, 3470 (1998).
5. R.K. Ahrenkiel, T. Wangensteen, M.M. Al-Jassim, M. Wanlass, and T. Coutts, *AIP Conference Proceedings* **321**, 412 (1995).
6. M. Wanlass et al. (presented at this conference).
7. R.K. Ahrenkiel, *AIP Conference Proceedings,* **353,** 161, AIP Press, New York (1996).
8. J. Shah, T.C. Damen, and B. Deveaud, *Appl. Phys. Lett.* **50**, 1307 (1987).
9. B. Sermage, J.L. Benchimol, and G. M. Cohen, *Proceedings of the 10th Intern. Conf. on Indium Phosphide and Related Materials*, 758-760, IEEE Press (1998).
10. W. Shockley and W.T. Read, *Phys. Rev.*, **86**, 335 (1952).
11. R.N. Hall, *Phys. Rev.*, **87**, 387 (1952).
12. R.K. Ahrenkiel, R. Ellingson, S. Johnston, and M. Wanlass, *Appl. Phys. Lett.*, **72**, 3470 (1998).
13. R.K. Ahrenkiel, *Solid State Electronics* **35,** 239 (1992).
14. R.K. Ahrenkiel and S. Johnston (to be published).
15. G.W. t' Hooft and C. van Opdorp, *Appl. Phys. Lett..*, **42**, 813 (1983).

RF Photoreflectance Characterization of Binary and Quasi-Binary Substrates and Antimonide-Based TPV Devices

Sudesh Saroop[†], Jose M. Borrego[†], Ronald J. Gutmann[†],
Partha S. Dutta[†], Aleksander G. Ostrogorsky[†], and Greg W. Charache[*]

[†]*Center for Integrated Electronics and Electronics Manufacturing,
Rensselaer Polytechnic Institute, Troy, New York 12180*

[*]*Lockheed-Martin Inc., Schenectady, New York 12301*

Both starting substrates and complete TPV device structures have been characterized using a radio-frequency (RF) photoreflectance technique, in which a Nd-YAG pulsed laser is used to excite excess carriers, and the short-pulse response and photoconductivity decay are monitored with an inductively-coupled non-contacting RF probe. The initial exponential transient decay, indicative of bulk recombination and surface recombination mechanisms as demonstrated previously for doubly-capped sample structures, is approximately 30-40 ns for GaSb substrates, with the decay constant increasing with increasing optical excitation (similar to Shockley-Read-Hall (SRH) high injection behavior). In the InGaAsSb quasi-binary substrates two distinct decays are observed, an initial decay transient of 15-20 ns which is independent of optical intensity and a subsequent decay of 30-60 ns which decreases with increasing optical intensity. This latter dependence on optical intensity was observed with doubly-capped epitaxial layers and is indicative of radiative recombination.

Similar measurements on quaternary device structures indicate that both the pulse amplitude and initial decay are reduced significantly without a front-surface capping layer that reduces surface recombination velocity. With reduction of the front surface recombination velocity, initial decays of 20-25 ns were obtained under open-circuit conditions.

These results indicate that the RF photoreflectance technique can be useful in characterizing and qualifying starting substrates and can be used to qualify epitaxial structures as well, particularly when doubly-capped standards are available for initial understanding of recombination processes in the material systems being investigated.

INTRODUCTION

The optimization of minority carrier devices, including photovoltaic devices, depends heavily on the various recombination processes which exist to return an excited system to its equilibrium state. These processes include recombination at surfaces and interfaces, Shockley-Read-Hall (SRH) recombination, Auger recombination, and, especially in III-V compounds, radiative recombination. An understanding of these processes is critical to the improvement of materials and the design of antimonide-based thermophotovoltaic (TPV) devices[1,2].

CP460, *Thermophotovoltaic Generation of Electricity: Fourth NREL Conference*
edited by T. J. Coutts, J. P. Benner, and C. S. Allman
© 1999 The American Institute of Physics 1-56396-828-2/99/$15.00

A radio-frequency (RF) photoreflectance technique, similar to that reported by Yablonovitch and Gmitter[3] and Ahrenkiel, *et al.*[4], is employed to measure recombination in unprocessed substrates and epitaxial layers using double heterostructure confinement or capping layers to isolate the bulk lifetime from the interface recombination. Variation of the laser pulse energy was used to differentiate between the recombination mechanisms in these capped structures, as well as in TPV device structures and substrates (binary and quasi-binary).

EXPERIMENTAL TECHNIQUE

Measurements of the bulk and surface lifetimes were made using a contactless inductively-coupled RF reflectance system[5]. The system senses changes in the sample conductivity as carriers recombine following excitation by a Laser Photonics YQL-102 Q-switched Nd:YAG laser operating at 1.06 μm with a nominal pulse width and decay time of 13.5 ns and 5 ns, respectively.

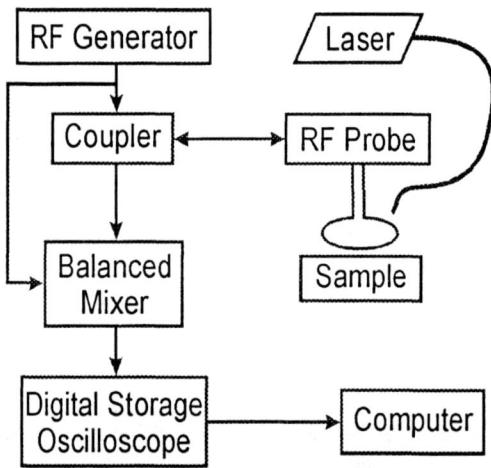

FIGURE 1. RF reflectance system diagram

A schematic of the RF system as shown in Figure 1 has been described elsewhere[5]. The system is designed to detect the sample conductivity using the probe coil (inductor) to sense currents induced in the sample by the continuous-wave (CW) RF signal from the generator. When the sample is illuminated by the laser, excess carriers are generated causing an increase in conductivity. This produces a transient signal whose height is proportional to the change in sample conductivity, assuming the change in sample resistivity is small compared to the system impedance. Plotting the transient response on a log scale allows the decay time of this signal to be readily extracted. Typically, more than one decay time is observed, the origin of which can be understood via modeling of the recombination processes.

THEORETICAL ANALYSIS

Analysis of the photoreflectance decay transient in thick boule samples was accomplished by solving the transient diffusion equation for a uniform distribution of bulk recombination centers with the assumption that the RF signal penetrates the entire sample and that the optical excitation is an impulse. The latter assumption is valid for times greater than the laser pulse decay time. The boundary conditions for the solution follow a Beer's law distribution with N_{ph} incident photons/cm² for the excess carriers at $t=0$, a surface recombination velocity (SRV) of S_R at the front surface, and a vanishing excess carrier concentration at the back surface for a semi-infinite solid. For p-type material with bulk lifetime τ_B and electron diffusion constant D_n, the excess electron concentration obeys the relations:

$$\frac{\partial n}{\partial t} = -\frac{n}{\tau_B} + D_n \frac{\partial^2 n}{\partial x^2}, \quad n(x,0) = \alpha N_{ph} e^{-\alpha x} \tag{1}$$

$$\text{and} \quad D_n \frac{\partial n}{\partial x}\bigg|_{x=0} = S_R n, \quad n(\infty, t) = 0 \;,$$

where α is the absorption constant of the monochromatic light. The preceding equations are simplified by substituting

$$n(x,t) = f(x,t) \cdot e^{-t/\tau_B} \tag{2}$$

in the continuity equation, thus giving

$$D_n \frac{\partial^2 f}{\partial x^2} = \frac{\partial f}{\partial t} \;. \tag{3}$$

Solution of this equation (subject to the boundary conditions given above) is accomplished through the Laplace transform in the time domain as follows:

$$\mathcal{L}\left[\frac{\partial f(x,t)}{\partial t}\right] = sF(x,s) - f(x,0+) = sF(x,s) - \alpha N_{ph} e^{-\alpha x}$$

$$D_n \frac{d^2 F(x,s)}{dx^2} - sF(x,s) = -\alpha N_{ph} e^{-\alpha x} \tag{4}$$

$$\Rightarrow F(x,s) = Ae^{-\lambda x} + Be^{\lambda x} + \frac{\alpha N_{ph}}{s - \alpha^2 D_n} e^{-\alpha x}, \quad \lambda = \sqrt{\frac{s}{D_n}} \;.$$

The value of the constant B must be 0 for the solution to remain finite throughout the semi-infinite slab and the value of the constant A can be obtained by applying the boundary condition on the front surface, giving

$$F(x,s) = \frac{\alpha N_{ph}}{s - \alpha^2 D_n} \left(e^{-\alpha x} - \frac{S_R + \alpha D_n}{S_R + \lambda D_n} e^{-\lambda x} \right) . \quad (5)$$

Integrating $F(x,s)$ over the thickness to obtain the total free carrier concentration in the Laplace domain, i.e. $G(s)$, yields

$$G(s) = \int_0^\infty F(x,s) dx = \frac{N_{ph}}{s - \alpha^2 D_n} - \frac{N_{ph} \alpha (S_R + \alpha D_n)}{\sqrt{s} \left(\sqrt{s} + \frac{S_R}{\sqrt{D_n}} \right) (s - \alpha^2 D_n)} , \quad (6)$$

which has poles at $s = 0$, $s = \alpha^2 D_n$, and $\sqrt{s} = -S_R/\sqrt{D_n}$. This expression can be rewritten using partial fraction expansion as

$$G(s) = \frac{A}{\sqrt{s}} + \frac{B}{\sqrt{s} + \frac{S_R}{\sqrt{D_n}}} \quad (7)$$

$$A = \frac{N_{ph}(S_R + \alpha D_n)}{\alpha S_R \sqrt{D_n}} , \quad B = \frac{N_{ph} \alpha D_n^{3/2}}{S_R (S_R - \alpha D_n)} .$$

Taking the inverse Laplace transform, the integrated carrier concentration is

$$N(t) = \left[\frac{S_R}{\alpha (S_R - \alpha D_n) \sqrt{\pi D_n t}} - \frac{\alpha D_n}{S_R - \alpha D_n} e^{\frac{S_R^2 t}{D_n}} \text{erfc} \, S_R \sqrt{\frac{t}{D_n}} \right] N_{ph} e^{-\frac{t}{\tau_B}} . \quad (8)$$

For low and high values of SRV, the above expression simplifies to the following

$$N(t) = \begin{cases} N_{ph} e^{-\frac{t}{\tau_B}} & , S_R = 0 \\ \frac{N_{ph}}{\alpha \sqrt{\pi D_n t}} e^{-\frac{t}{\tau_B}} & , S_R \to \infty \end{cases} . \quad (9)$$

For intermediate values of SRV, the solution can be simplified for $t >> D_n/S_R^2$ using the approximation

$$\text{erfc} \, x = \frac{e^{-x^2}}{x \sqrt{\pi}} \quad \text{for } x >> 1 \quad (10)$$

to give

$$N(t) = \frac{N_{ph}(S_R + \alpha D_n)}{\alpha S_R} \frac{e^{-\frac{t}{\tau_B}}}{\sqrt{\pi D_n t}} . \qquad (11)$$

By plotting $\ln[\sqrt{t}N(t)]$, a linear plot is obtained with a slope of $-1/\tau_B$ and, extrapolating back to the y-axis, a y-intercept of

$$\ln\left|\frac{N_{ph}(S_R + \alpha D_n)}{\alpha S_R \sqrt{\pi D_n}}\right| . \qquad (12)$$

Thus, it is possible to extract the bulk lifetime and SRV from the decay transient through this analysis.

For passivated samples (i.e. SRV $<10^4$ cm/s) the expressions in Eqs. (11) and (12) are applicable to the long decay of the transient. In this case, the analysis of the initial decay is achieved by first integrating the transient diffusion equation, Eq. (1), over the thickness of the sample to give

$$D_n \frac{\partial n(x,t)}{\partial x}\bigg|_0^\infty - \frac{N(t)}{\tau_B} = \frac{dN(t)}{dt} . \qquad (13)$$

Substituting the boundary conditions of Eq. (1) for the first term results in

$$-S_R n(0,t) - \frac{N(t)}{\tau_B} = \frac{dN(t)}{dt} \qquad (14)$$

from which the instantaneous exponential decay time may be obtained by dividing by $N(t)$. Evaluating the result at $t=0$ gives the following expression for the initial decay:

$$\frac{-1}{N(t)} \frac{dN(t)}{dt}\bigg|_{t=0} = \frac{S_R n(0,0)}{N(0)} + \frac{1}{\tau_B} . \qquad (15)$$

From the initial condition in Eq. (1) and the fact that $N(0)=N_{ph}$, the initial decay time can be expressed as

$$\frac{1}{\tau_{eff}} = \alpha S_R + \frac{1}{\tau_B} . \qquad (16)$$

Thus, for passivated samples the initial decay time will depend on the bulk lifetime, the SRV, and the absorption constant, i.e. wavelength of light, whereas the long decay time will depend on the bulk lifetime; the effect of bulk traps have not been considered here.

The bulk lifetime, τ_B, depends on SRH, radiative, and Auger recombination. Separation of the more conventional SRH recombination from radiative and Auger recombination is important to guide material research and to design optimum TPV device structures. For SRH recombination, the lifetime is given by τ_n under low-level injection, and $\tau_n+\tau_p$ under high-level injection levels, where $\tau_{n,p}=\sigma_{n,p}v_{th}N_t$, $\sigma_{n,p}$ is the electron (hole) capture cross-section, v_{th} is the thermal velocity, and N_t is the defect density. The full intensity of the 1.06 µm laser is 1 mJ, which results in an initial excess carrier concentration of roughly 10^{20} cm^{-3}. This high level of injection (almost 3 orders of magnitude above the active layer doping of 2×10^{17} cm^{-3}) necessitates the inclusion of high level injection effects on radiative and SRH recombination.

If radiative recombination is the dominant bulk recombination process, the following equation describes the distribution of excess carriers:

$$\frac{\partial n}{\partial t} = -\frac{G_{th}(n_0+p_0+n)n}{n_i^2} + D_n\frac{\partial^2 n}{\partial x^2}, \qquad (17)$$

where G_{th} is the thermal generation rate; n_0 and p_0 are the equilibrium electron and hole concentrations, respectively; and n_i is the intrinsic carrier concentration. A closed form solution of the above partial differential equation is not possible in general; but if the carrier concentration is assumed to be uniform, as is the case when both surfaces are good (low SRV) on a thin epitaxial layer, the resulting ordinary differential equation can be solved[6] to give:

$$n(t) = \frac{(n_0+p_0)n(0)}{(n_0+p_0+n(0))e^{t/\tau_{rad}} - n(0)}, \quad \tau_{rad} = \frac{n_i^2}{G_{th}(n_0+p_0)}, \qquad (18)$$

where $n(0)$ is the initial concentration of excess carriers following the photo excitation. A similar analysis for the thick samples requires a numerical solution as the spatial dependence is significant.

Auger recombination involves three carriers and is therefore a relatively weak effect except in heavily doped semiconductors or in the case of high level injection. Under high level injection, the Auger lifetime is approximated[6] by $\tau_A = 2\tau_i n_i^2/n(0)^2$ which has an inverse dependence on the excitation level squared. Assuming $n_i=3\times10^{13}$ cm^{-3} in 0.55 eV TPV devices and a conservative estimate for τ_i of 39 ms, extremely short values for τ_A of 10^{-16} to 10^{-8} s are obtained for similar injection levels as for the radiative recombination analysis. Thus, the effects of Auger recombination will not impact the measured decay times with the present setup.

The bulk recombination processes may be combined into a bulk lifetime by $1/\tau_B=1/\tau_{SRH}+1/\tau_{rad}$. Since these components exhibit a different dependence on excitation level, varying the laser pulse energy can differentiate between SRH and radiative recombination.

This paper reports on the experimental results with uncapped GaSb and InGaAsSb boule samples. A comparison is made with capped and uncapped organometallic vapor phase epitaxy (OMVPE) grown InGaAsSb devices for which electrical measurements have been taken.

RESULTS WITH CAPPED AND UNCAPPED QUATERNARY OMVPE DEVICE STRUCTURES

Three p/n junctions were grown on Te-doped GaSb substrates with an emitter (p-layer) thickness of 3 µm. One sample had a bare (uncapped) InGaAsSb surface, while the other two samples had a cap of either p-GaSb or p-AlGaAsSb to provide a retarding field for minority carriers at the surface. The samples were later subjected to metal deposition and patterning to form a TPV cell. These samples were subsequently characterized electrically using current-voltage (I-V) and quantum efficiency (QE) measurements[2]. The photoreflectance data for these samples at full intensity is shown in Table 1.

A typical waveform of the photoreflectance decay transient is shown in Figure 2 along with a graphical depiction of the parameters measured. Analysis of thin active layer samples presented previously[7] indicates that the initial decay depends only on surface and bulk recombination. As expected from the electrical data[2], the uncapped device had the shortest initial decay due to the high SRV at the front surface (estimated to be on the order of 10^7 cm/s from electrical and optical characterization). There is no significant difference in the initial decay times of the capped devices, but a higher pulse height was obtained with the GaSb/AlGaAsSb cap. It has been reported[8] that in the presence of a junction, the effective lifetime reduces to the diffusion lifetime $\tau_D = 4W^2/\pi^2 D_n$, which in this case ($W=3$ µm) is about 1 ns. However, the data in Table 1 were measured with the device open-circuited, allowing the separated charges to build up in each layer within a few nanoseconds, creating a retarding field a short time after the optical pulse.

TABLE 1. p/n InGaAsSb Junction Photoreflectance Data

Sample	Capping Layer	Pulse Height (mV)	Initial Decay (ns)	Long Decay (ns)	Long Decay Amp (%)
97-463	uncapped	54	9	179	7
97-570	p-GaSb	77	23	71	19
97-548	p-AlGaAsSb	88	22	83	13

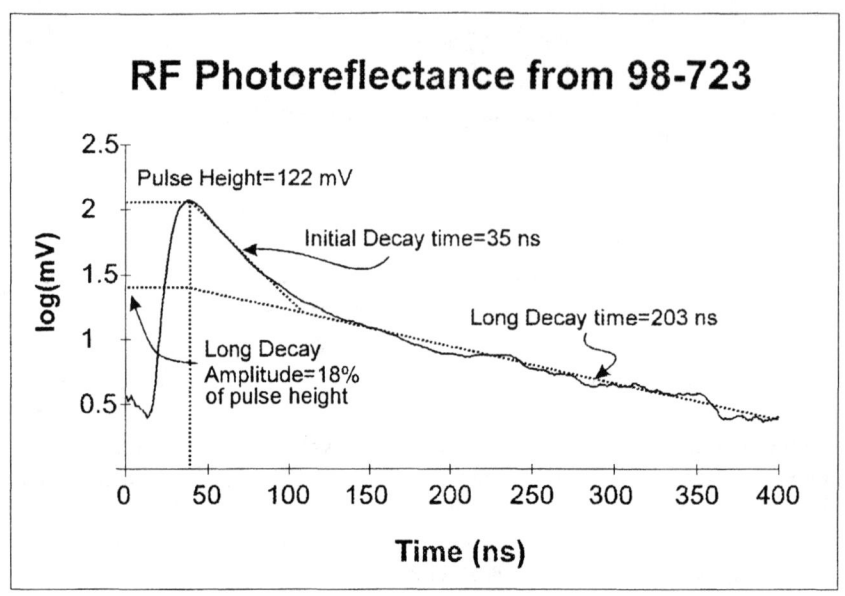

FIGURE 2. Graphical depiction of measured parameters

The decay times measured as a function of the laser intensity using neutral density filters to reduce the optical intensity are plotted in Figure 3. Due to the low response, a decay time could not be extracted from sample 97-463 at the lowest intensity. Note that the decay times exhibit an inverse dependence on excitation level. This behavior was also obtained in doubly capped OMVPE InGaAsSb structures[5] which showed a good fit to the radiative recombination model yielding radiative lifetimes on the order of 95 ns. This similarity in dependence upon optical intensity provides further evidence of radiative recombination being dominant in these materials.

EXPERIMENTAL RESULTS: BINARY AND QUASI-BINARY SUBSTRATE MATERIALS

The initial decay time from a commercial bulk p-GaSb (compensated) wafer from Firebird® (FB1), and Rensselaer grown bulk compensated p-GaSb (GaSb15) and uncompensated p-GaSb (GaSb17) were measured as a function of excitation level and is presented in Figure 4. The dependence on intensity is opposite to that of the OMPVE InGaAsSb material in each case; that is, the decay times increase with intensity and resembles classic SRH high level injection behavior, even though the decay times are of comparable value.

Three InGaAsSb quasi-binary samples (IGAS1, IGAS2, IGAS3) were grown by

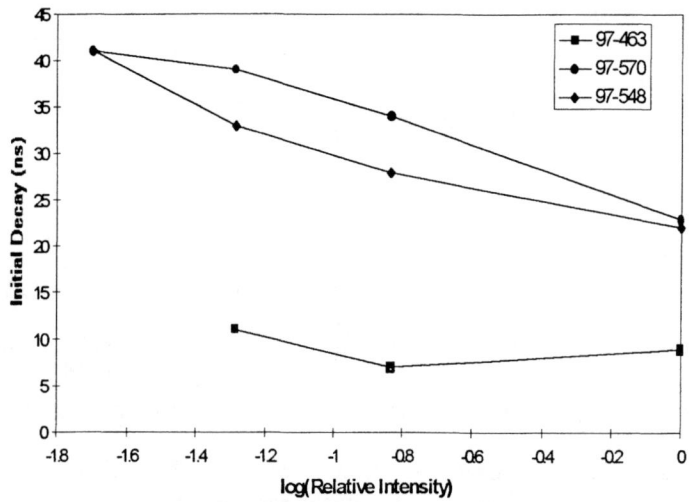

FIGURE 3. Photoreflectance Decay with InGaAsSb Junctions vs. Optical Intensity

the vertical Bridgman method from GaSb and InAs sources and polished at Rensselaer[9,10] in a manner similar to that of the GaSb15 and GaSb17 samples. Comparing the transients of the quasi-binaries (Figure 6) with those of the binary samples (Figure 5) it is evident that these samples have at least two distinct decays (plotted in Figures 7a and 7b) preceding a long decay (latter presumably due to bulk traps). No trend is observed in the initial decay, whereas the second decay time decreases with increasing intensity, as did the OMVPE InGaAsSb material.

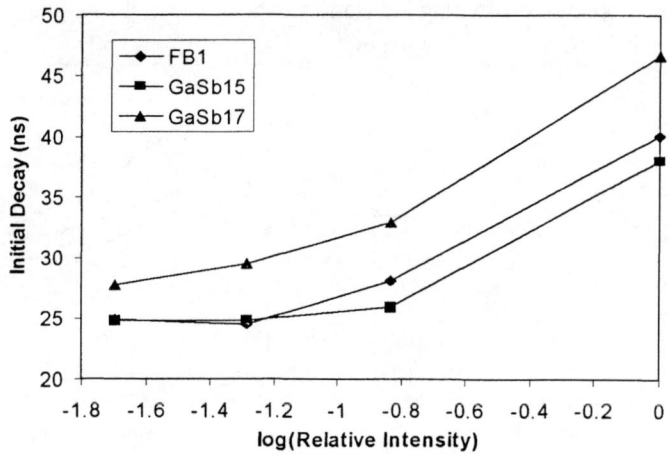

FIGURE 4. Photoreflectance Decay with GaSb Substrates vs. Optical Intensity

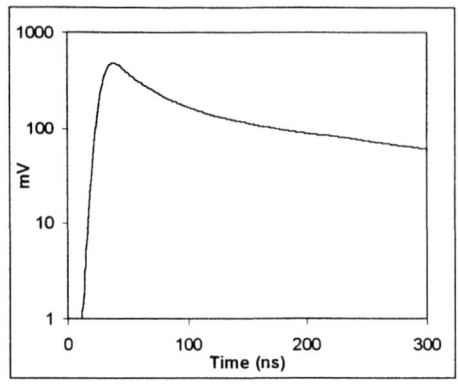
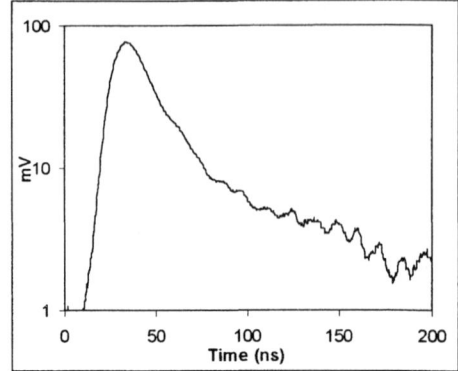

FIGURE 5. Photoreflectance from bulk GaSb

FIGURE 6. Photoreflectance from IGAS1

Such behavior can be characteristic of the presence of ordering and the separation of carriers at the type II heterointerfaces of the resulting domains. These effects were observed[11] in InGaAs materials lattice matched to InP, where transmission electron diffraction images show evidence of ordering on {111} planes (CuPt-type ordering). However, such ordering does not appear to be present in these antimonides based upon recent X-ray diffraction images.

Three additional quasi-binary samples were prepared with differing amounts of mixing prior to crystal growth. As before, two decays were observed preceding a long exponential tail. The initial and second decay times are shown in Figure 8a and 8b. For the relatively unmixed sample, the initial decay does not show a significant dependence on laser intensity while the second decay shows a strong inverse dependence similar to the previous quasi-binary samples. As the mixing time is increased, the initial decay shows increasing dependence on intensity. One hypothesis is that well mixed samples should be homogenous and exhibit no ordering. While this data supports the assumption, further

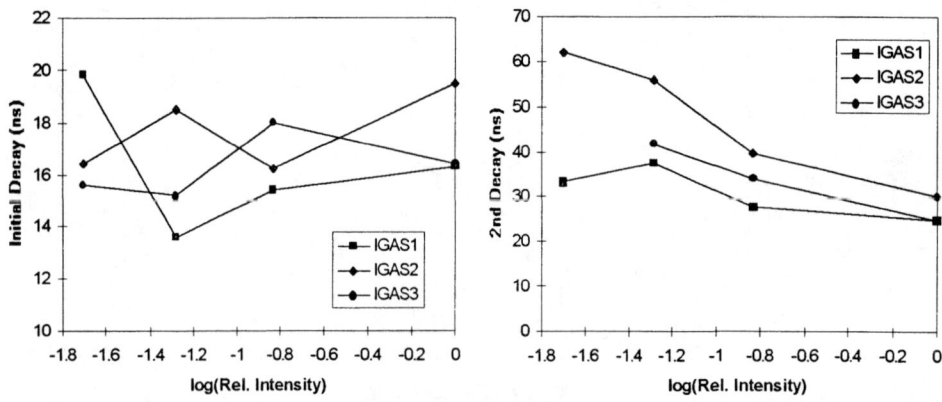

FIGURE 7. Initial (a) and second (b) decays from bulk quasi-binary samples

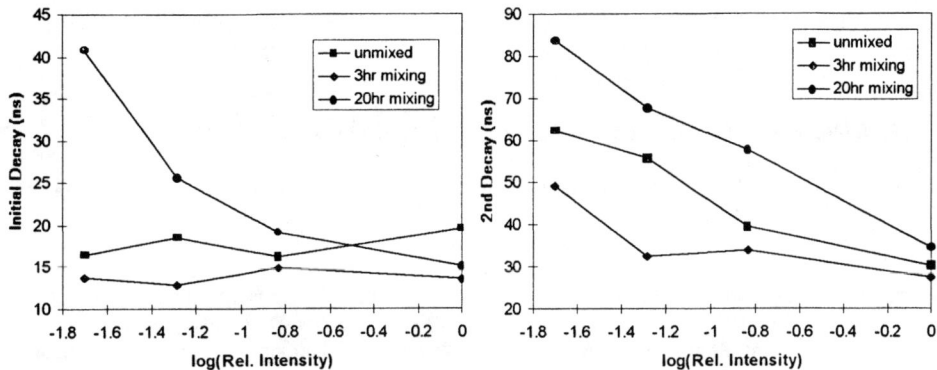

FIGURE 8. Initial (a) and second (b) decays from quasi-binary samples with different mixing times

material characterization is required for corroboration.

CONCLUSION

An RF photoreflectance system was used as a non-contacting, non-destructive tool for the characterization of recombination processes in GaSb and InGaAsSb samples. The system was applied to both TPV device structures and starting material. The strong difference in pulse height and decay time between capped and uncapped devices demonstrates the system's ability to discriminate surface passivation. In both the InGaAsSb device structures and the InGaAsSb quasi-binary material a dependence on optical intensity was observed which is indicative of radiative recombination being dominant. In contrast, in the GaSb boule material the optical intensity exhibited SRH recombination behavior.

ACKNOWLEDGMENTS

The authors would like to thank Dr. Christine Wang of Massachusetts Institute of Technology's Lincoln Laboratory for her work in the growth of the OMVPE doubly-capped and device structures which have been instrumental in the development of the RF photoreflectance system and the interpretation of data.

REFERENCES

1. C. Hitchcock, R. Gutmann, J. Borrego, H. Ehsani, I. Bhat, M. Freeman, and G. Charache, "GaInSb and GaInAsSb Thermophotovoltaic Device Fabrication and Characterization", *Thermophotovoltaic Generation of Electricity, 3rd NREL Conference*, 89 (1997).

2. C. W. Hitchcock, R. J. Gutmann, H. Ehsani, I. B. Bhat, C. A. Wang, M. J. Freeman, G. W. Charache, "Ternary and Quaternary Antimonide Devices for Thermophotovoltaic Applications", accepted for publication, *J. Crystal Growth*.

3. E. Yablonovitch and T. J. Gmitter, *Solid-State Electronics* **35** (3), 261 (1992).

4. R. K. Ahrenkiel, T. Wangensteen, M. M. Al-Jassim, M. Wanlass, and T. J. Coutts, "Recombination Lifetime of $In_xGa_{1-x}As$ Ternary Alloys", *AIP Conference Proceedings 321* (AIP. New York. 1994) p. 412.

5. S. Saroop, J. M. Borrego, R. J. Gutmann, H. Ehsani, I. Bhat, S. Dakshina Murthy, A. Ostrogorsky, P. Dutta, M. Freeman, G. Charache, "RF/Microwave Non-Destructive Measurements of Electrical Properties of Semiconductor Wafers for Thermophotovoltaic Applications", *Thermophotovoltaic Generation of Electricity, 3rd NREL Conference*, 139 (1997).

6. J. S. Blakemore, *Semiconductor Statistics*, (Pergamon Press, New York, 1962).

7. S. Saroop, J. M. Borrego, R. J. Gutmann, G. Charache, and C. Wang, "Characterization of Recombination Processes in Doubly-Capped Quaternary Thin Films for Antimonide-Based Thermophotovoltaic Devices", (Submitted to *J. Appl. Phys.*)

8. R. K. Ahrenkiel, *J. Appl. Phys.* **62** (7), 2937 (1987).

9. P. S. Dutta and A. G. Ostrogorsky, "Melt Growth of GaInAsSb Crystals", (Accepted for publication in *J. Crystal Growth*).

10. P. S. Dutta and A. G. Ostrogorsky, "Band Gap Narrowing Phenomenon in GaInAsSb", (Submitted to *J. Crystal Growth*).

11. R. K. Ahrenkiel, S. P. Ahrenkiel, and D. J. Arent, "Recombination Lifetime in Ordered and Disordered InGaAs", *Thermophotovoltaic Generation of Electricity, 2nd NREL Conference*, 434 (1996).

Thermophotovoltaic Cell Temperature Measurement Issues

T. Moriarty and K. Emery

NREL, 1617 Cole Blvd., Golden, CO, 80401

Abstract. The power produced by photovoltaic devices changes with temperature, ranging from 0.1% to nearly 1% per degrees Celsius depending on the structure. The temperature across the surface of TPV cells will vary depending on the amount of absorbed power. Thus the temperature over a region of a wafer where there is no cell will be different from a region of the wafer containing a cell with an antireflection coating and back surface reflector. Vacuum hold-downs or back surface probes may result in local hot spots. Bonding a cell to a heat sink may not be practical in a research environment, and a temperature gradient between the heat sink and space-charge region will still exist. Procedures for determining the current *versus* voltage (*I-V*) characteristics at a given temperature are discussed. For continuous illumination measurement systems, the temperature of the heat sink or backside of the device can be directly measured. The temperature can also be inferred by placing the sample at a known temperature in the dark, and monitoring the open-circuit voltage (V_{oc}) as a high-speed shutter is opened. The maximum V_{oc} from this method corresponds to the temperature in the dark and the plate temperature can then be lowered until this maximum V_{oc} is reached. The temperature can also be indirectly determined from the dark *I-V* characteristics, assuming negligible series resistance in the ideal case that the voltage in the dark at a given current and temperature corresponds to the V_{oc} and short circuit voltage (I_{sc}) at that temperature. A high-intensity flash simulator will produce negligible cell heating during the flash and therefore the cell temperature may be easily set before the flash.

INTRODUCTION

Many thermophotovoltaic (TPV) device performance parameters, and hence overall power output, vary with device space-charge region temperature. As with any photovoltaic device, it is important to quantify these variations in the form of temperature coefficients for the essential current versus voltage (*I-V*) parameters. It is even more important for TPV devices because they are likely to be tested and operated in harsh conditions at elevated temperatures and with large temperature gradients. This paper discusses some difficulties in measuring TPV device temperatures under testing conditions, consequences of inaccurate temperature measurements, methods for mitigating the difficulties, and temperature coefficients for the open-circuit voltage (V_{oc}), maximum power (P_{max}), fill factor (*FF*), and the short-circuit current (I_{sc}) for a InGaAs TPV device fabricated at NREL with an energy gap of 0.6-eV.

CP460, *Thermophotovoltaic Generation of Electricity: Fourth NREL Conference*
edited by T. J. Coutts, J. P. Benner, and C. S. Allman
© 1999 The American Institute of Physics 1-56396-828-2/99/$15.00

CAUSES OF TEMPERATURE GRADIENTS

Vacuum holes, back surface voltage or temperature probes or voids in bonding solder may all cause localized hot spots in a cell. Figure 1 shows the temperature increase of a 3 mm x 3 mm cell when placed in various positions over a back surface voltage probe in a 1-mm-diameter hole. The short-circuit current density (J_{sc}) was the same in each case, about 1.5 A/cm^2, and the cell temperature in the dark was 25°C. The temperature was determined using the "fast V" method explained below. Although the probe occupies only about 9% of the back surface area of the cell, its presence at the center of the cell results in a 34°C temperature rise when light is applied. However, when the hole is 2 millimeters beyond the perimeter of the cell, but still under the wafer, the temperature rise is only 14°C.

Top surface voltage and current probes may cause temperature gradients in a cell if they are exposed to light and subsequently heated. The probes may be shielded from the light, but shielding increases the probe size and may reduce light uniformity. Probes may also conduct heat to or away from the sample and will shade the portion of the cell that they touch.

The quality of the temperature plate surface finish may have a strong effect on the difference between the illuminated and nonilluminated temperatures. Figure 2 shows the cooling-plate temperature adjustment required to return a GaInAs cell to its dark temperature after it has been illuminated on two different cooling plates. One is thermoelectrically cooled nickel-plated aluminum, and after 10 years of use in the laboratory has a visibly poor surface quality (machining marks, scratches, and dents). The other is a new, commercially available, water-cooled plate with a high-quality surface finish and high thermal conductivity. The same source and optics were used in each case

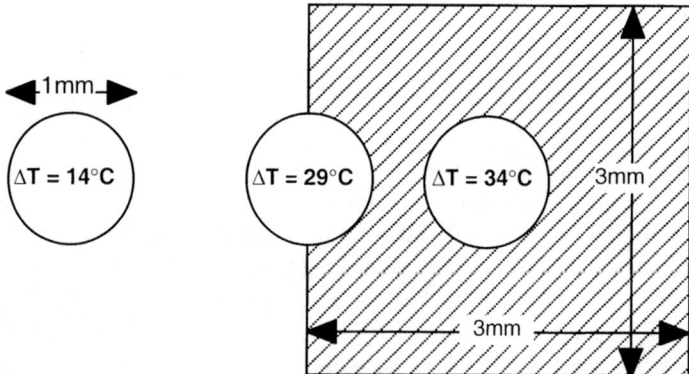

FIGURE 1. The effect of a 1-mm-diameter hole in the cooling plate below a wafer containing a 9-mm^2 GaInAs cell. The cell temperature rise for three different positions is shown. In each case the plate temperature was 25°C, the light intensity was the same, and current density was about 1.5 A/cm^2.

and the fast V method was used to determine the temperature change.

The nonuniform temperature distribution that results from holes will yield reduce power, and for a given current density V_{oc}, FF, and P_{max} will all suffer [1]. The drop in V_{oc} caused by temperature gradients across a cell becomes greater as illumination increases, and calculations of the ideality factor, n, based on V_{oc} versus $\log J_{sc}$ may yield a diode quality factor of less than one [2].

Temperature gradients caused by nonuniform illumination become more severe as the cell size increases. Of course, measurements of the I-V characteristics of series connected cells with nonuniform light are subject to other errors since voltage contributions will vary among cells and current will be limited by the cell exposed to the least amount of light.

METHODS FOR MEASURING TEMPERATURE

Direct Contact

Direct contact measurement is the easiest method conceptually and to implement but may yield very misleading results. A temperature probe on the top surface is exposed to intense IR from one side; it casts a cooling shadow on the spot that it is contacting and may work as a heat sink on the other [3]. It is usually necessary to hold the temperature plate at a level much lower than the desired junction temperature (see Figure 2), so if a bottom contact probe is in thermal contact with the cooling plate it will report a temperature that is lower than the junction temperature. If a bottom contact probe is thermally isolated from the plate, then it is, in effect, a hole under the wafer and may report a temperature much higher than adjacent areas on the wafer (as in Figure 1).

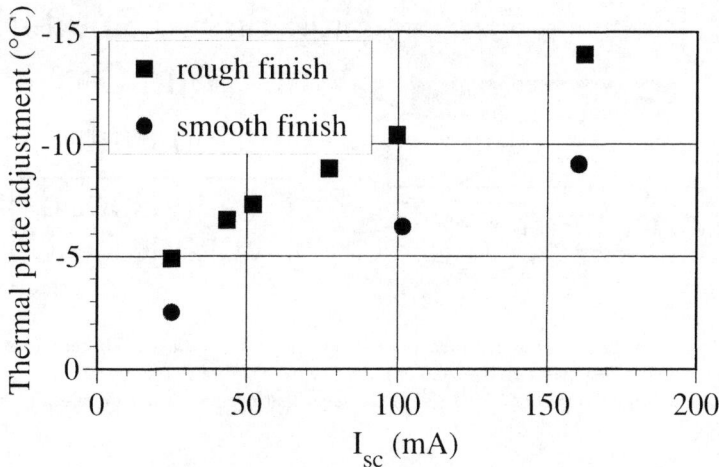

FIGURE 2. The cooling-plate temperature adjustment necessary to return a cell to 25°C after continuous illumination depends in part on the light level and surface quality of the plate.

The Fast V Method

When a cell in the dark is suddenly exposed to light, there is a rapid increase in cell temperature. After its initial increase, V_{oc} drops asymptotically to a level that reflects its increased steady-state temperature. If the cooling-plate temperature is adjusted until V_{oc} is the same as it was immediately after exposure to the light, then it can be assumed that the cell temperature has returned to nearly its initial value as well. This procedure assumes that V_{oc} can respond in less than 1 ms to light being turned on, and the temperature rise during the first ms that the shutter is opening is negligible. The effect of V_{oc} response time can be mitigated by letting the sample sit at V_{oc} at a light level that causes negligible heating prior to opening the shutter. The fast V method for setting cell temperature takes advantage of this effect. This method uses a closed high-speed shutter to block the light from the cell while the cell temperature comes to equilibrium with the cooling plate. Then, the shutter is opened and V_{oc} measured every few milliseconds (see Figures 3 and 4). The change in cell temperature, ΔT, between the voltage measurements immediately

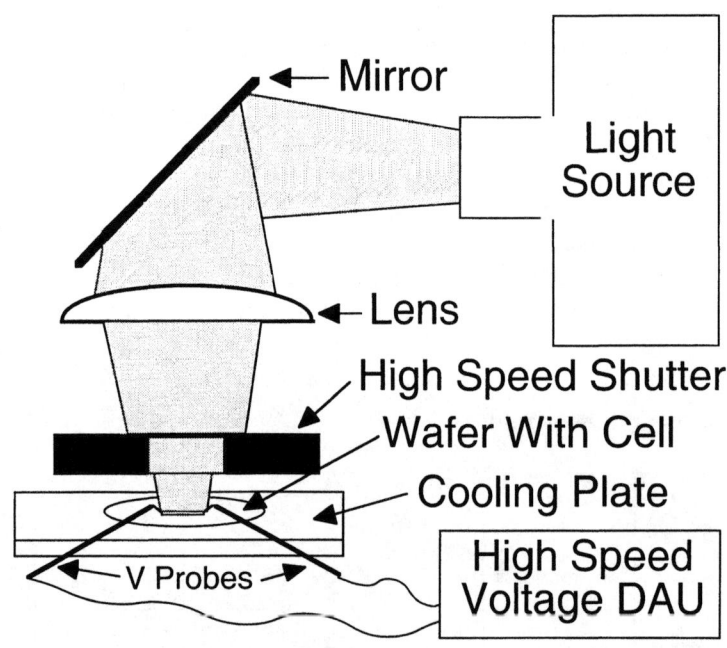

FIGURE 3. Apparatus for fast V measurement of the drop in V_{oc} as device heats. The data acquisition unit (DAU) collects voltage data every millisecond or longer starting before shutter opens. The cooling plate reduces the device temperature until V_{oc} rises to the level that occurred immediately after the shutter opened.

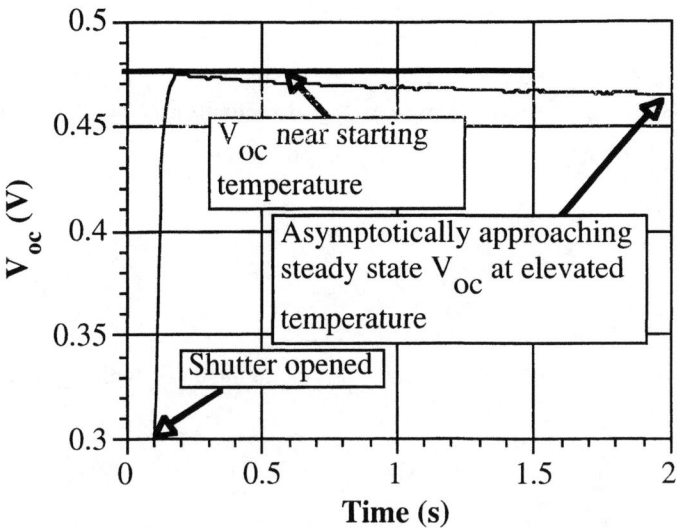

FIGURE 4. V_{oc} rises rapidly after the shutter is opened, but immediately starts to drop as the cell heats in a fast V temperature measurement. The cell may be brought back to its dark temperature by adjusting the cooling plate temperature until V_{oc} rises back to the value that occurred immediately after the shutter was opened.

before and after the shutter opens represents the inherent error in this method and may be estimated [4] by

$$\Delta T = \frac{Qt^{1/2}}{\sqrt{\pi \rho c k}} \qquad (1)$$

where Q is the heat flux or total irradiance, t is the time that the sample is exposed to light, ρ is the density, c is the specific heat, and k is the thermal conductivity. For an InP substrate $\rho = 4.8$ g/cm^3, $c = 0.322$ J g^{-1} °C^{-1} and $k = 0.68$ W cm^{-1} °C^{-1} [5]. Assuming a worst-case scenario of all incident light being absorbed, a 0% efficient cell, and a total

$$\Delta T = 27.5 t^{1/2} . \qquad (2)$$

power density Q of 50 W/cm^2, then the expected temperature rise is
If the time between voltage measurements is 5 ms, $\Delta T = 2.1$ °C.

The fast V method is limited by the size of the shutter, the V_{oc} response time to a transient light pulse, and the shutter speed. Its advantage is that it allows the space-charge-region temperature to be set to within the uncertainty discussed above and given in Equation (2).

Dark I-V

The *I-V* characteristic of a diode with low series resistance, large shunt resistance, diode quality factor (*n*), dark current (I_0), and illumination (I_L) can be written as

$$I = I_0 \left\{ \exp\left[\frac{qV}{nkT}\right] - 1 \right\} - I_L \tag{3}$$

Then at V_{oc}

$$I_L = I_0 \left\{ \exp\left[\frac{qVoc}{nkT}\right] - 1 \right\} \tag{4}$$

Equation (4) is of the same form as the characteristic of the diode with negligible series resistance in the dark, or

$$I = I_0 \left\{ \exp\left[\frac{qV}{nkT}\right] - 1 \right\} \tag{5}$$

Under normal conditions, where the short-circuit current is not affected by series resistance, I_L and I_{sc} are identical. So, voltage-current pairs extracted from a dark *I-V* curve should correspond to V_{oc}-I_{sc} pairs for the same device under illumination at the same temperature, assuming that the I_0 and *n* are independent of I_L. When a voltage-current pair from the dark *I-V* is "dialed in" at the *I-V* test bed (as V_{oc} and I_{sc}) by adjusting light intensity and cooling plate temperature, then it can be assumed that the junction temperature is the same as it was during the dark *I-V* measurement. An advantage of the dark *I-V* method of setting temperature over the fast V method is that it can be used on modules whose performance is sensitive to the spatial uniformity of the light source. Figure 5 compares the voltage-current pairs of a dark *I-V* with the V_{oc}-I_{sc} pairs extracted from a set of *I-V* curves under various illumination levels for a low series-resistance cell. The dark *I-V* approach becomes susceptible to large errors at high currents for even relatively small series resistances. If series resistance is included in Equation (5) then,

$$I = I_0 \left\{ \exp\left[\frac{q(V - IR_s)}{nkT}\right] - 1 \right\} \tag{6}$$

So,

$$\frac{dT}{dR_s} = \frac{I(q/nk)}{\ln(1 + J/J_0)} \tag{7}$$

Figure 6 shows the dependence of the temperature uncertainty on I_L and I_0. Notice that for a cell with $I_L = 0.5$ A and $I_0 = 10$ µA the uncertainty is more than 0.5 °C/mΩ and increases to 1.0 °C/mΩ at a current of 0.9 A. Because it may be difficult to measure a cell's series resistances to within a few milli-ohms, uncertainties can easily reach several degrees.

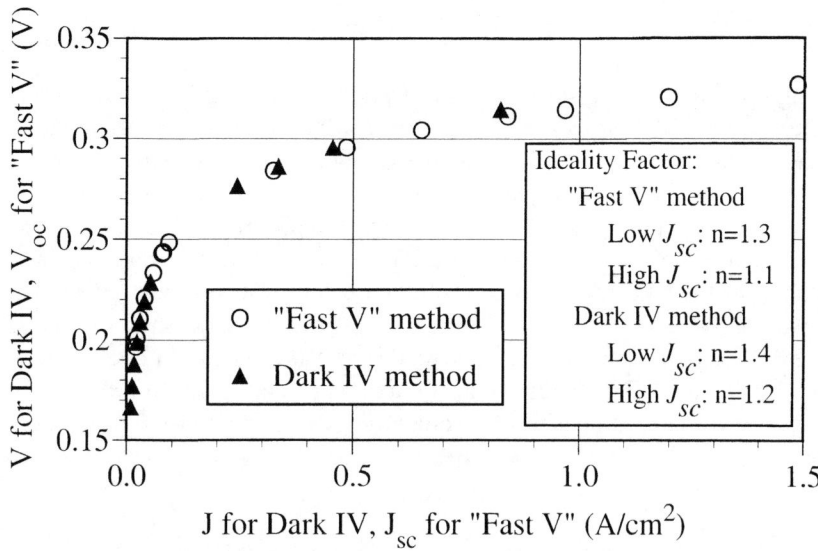

FIGURE 5. Dark I-V and V_{oc}-J_{sc} pairs for a 0.6 eV GaInAs cell. Note the difference between V_{oc} from fast V and V from dark I-V at high current density. This corresponds to about a 2.5°C difference in temperature that may be due to heating in the first few milliseconds after the shutter opens during fast V measurement.

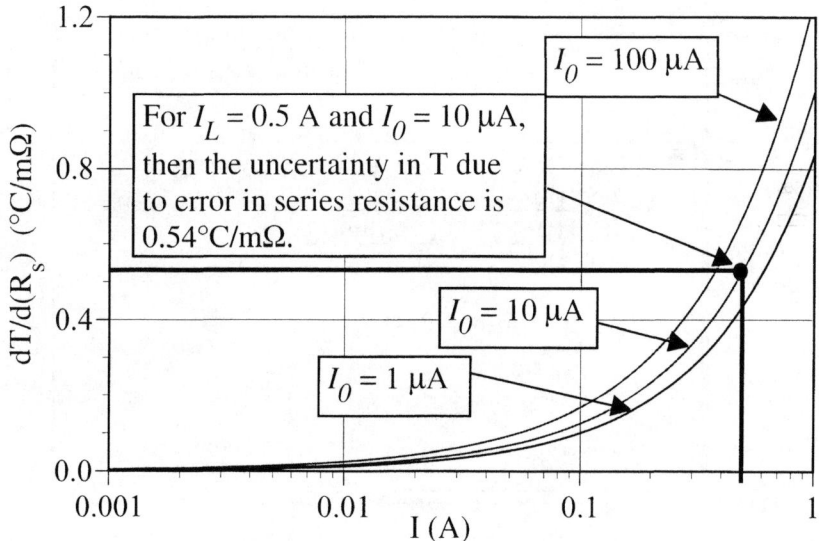

FIGURE 6. Temperature error caused by series resistance when using current-voltage pairs from a dark I-V to set V_{oc} and I_{sc} under illumination. It is assumed here that the ideality factor, n, is equal to 1.0. If the series resistance in not well-known to within a few milli-ohms, then temperature errors may be large.

Flash Simulator

A high-intensity pulsed solar simulator (HIPSS) is frequently used to measure I-V characteristics of TPV devices at NREL. The HIPSS is capable of delivering up to a 200-W/cm^2 (2000 suns) pulse of light with a spatial uniformity within 2% over a 10 cm by 10 cm area for about 1 ms. Typically, only about one-half of the 1-ms pulse is used; therefore, from Equation (2) at 50 W/cm^2, the temperature error is approximately 0.6 °C. Flash simulators have their own set of difficulties [6], but by incorporating correction algorithms for light fluctuations during the pulse and other transient effects we have measured the I-V characteristics of TPV devices with current densities as high as 8 A/cm^2 at 25°C. It would be very difficult to do this measurement under continuous illumination while maintaining junction temperature. The validity of HIPSS data has been verified at lower current densities by comparing it to data collected using the fast V method to set the temperature (see Figure 7).

TEMPERATURE COEFFICIENTS

The measurement of the PV performance at a given temperature requires the accurate determination of the space-charge-region temperature and minimization of temperature-related artifacts discussed in the preceding sections. The data used to generate

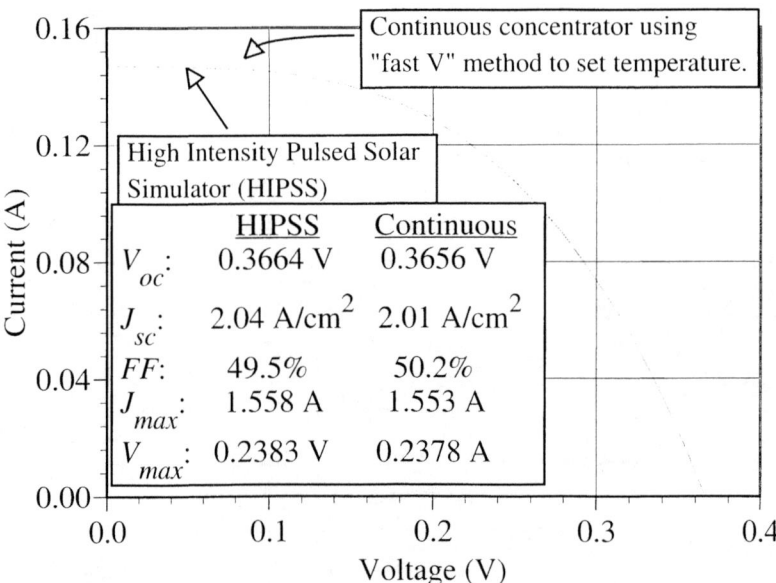

FIGURE 7. I-V data generated under continuous illumination using the "fast V" method to set the temperature to 25°C compared to data generated by HIPSS for a 0.073 cm^2, 0.6 eV GaInAs cell [8].

the temperature coefficients in Figure 8 were collected with the HIPSS for a 0.6-eV energy gap GaInAs cell [8]. The temperature coefficient, TC, for a given parameter, Z, may be written [7] as:

$$TC(\text{part}/°C) = \frac{1}{Z}\frac{\partial Z}{\partial T}\bigg|_{T_n=25°C} . \qquad (8)$$

The temperature coefficient is by convention evaluated at a normalization temperature, T_n, of 25°C. The parameters V_{oc}, I_{sc}, FF, and P_{max} for these devices were linear with temperature. Figure 9 illustrates how the quantum efficiency near the bandgap shifts with temperature. Depending on the spectral content of the light source, this shift may or may not be captured, causing potentially large errors in the temperature coefficient of parameters dependent on the photocurrent (i.e. I_{sc} and P_{max}). Ideally, the relative spectral irradiance of the light source that the I-V parameters are measured under should be the same as the intended application's light source.

SUMMARY

Problems that occur in determining temperature while taking I-V data of TPV devices have been noted. These problems are related to poor thermal contact to a temperature-controlled surface, temperature gradients between the front surface and back sur-

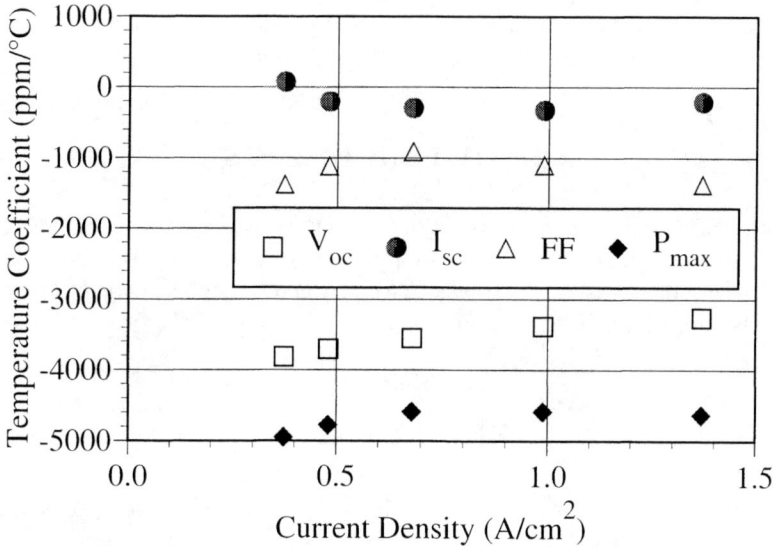

FIGURE 8. The temperature coefficients as a function of current density for a 0.6-eV GaInAs TPV cell.

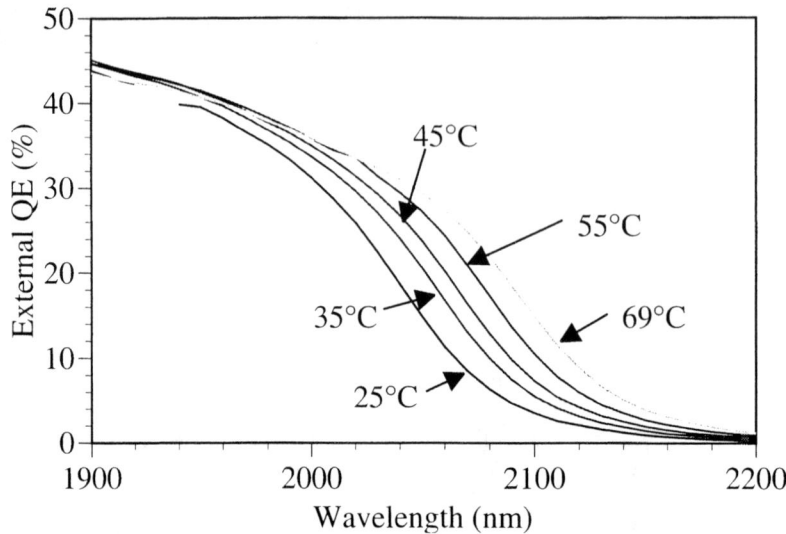

FIGURE 9. Variation in the external quantum efficiency with temperature for a 0.6-eV GaInAs cell [8] showing the temperature dependence of the energy gap.

face, and the junction temperature. Temperature-related issues of determining the dark current and diode quality factor from V_{oc} versus J_{sc} data compared with the limitations of dark I-V measurements were discussed. Several strategies for mitigating temperature-related problems and their potential errors have been presented. Temperature coefficients for several cell parameters, and temperature dependence of the quantum efficiency near the energy gap for 0.6 eV GaInAs TPV devices, have been plotted.

ACKNOWLEDGEMENTS

The authors wish to thank M. Wanlass, S. Ward, A. Duda and J. Carapella of the National Renewable Energy Laboratory for providing the TPV devices that were evaluated for this paper. The authors wish to also thank C. Murray and B. Wernsman of Bettis Atomic Physics Laboratory for their helpful suggestions.

REFERENCES

1. Sanderson, R. W., O'Donnell, D. T., and Backus, C. E., "The Effects of Nonuniform Illumination and Temperature profiles on Silicon Solar Cells Under Concentrated Sunlight," *Proc. 14th IEEE Photovoltaic Specialists Conference*, San Diego, CA, Jan 6-10, 1980, pp. 431-436.

2. Nasby, R. D., and Sanderson, R. W., "Performance Measurement Techniques for Concentrator Photovoltaic Cells," *Solar Cells*, **6**, pp. 39-47, (1982).
3. Emery, K. A., and Osterwald, C. R., *Efficiency Measurements and Other Performance Rating Methods, Current Topics in Photovoltaics*, **3**, pp. 301-349, T.J. Coutts editor, Academic Press, 1988.
4. Borden, P. G., LaRue, R. A., Gregory, P. E., and Boettcher, R., "The Flash Testing of AlGaAs/GaAs Concentrator Solar Cells," *Proc. 15th IEEE Photovoltaic Specialists Conference*, Kissimmee, FL, May 12-15, 1981, pp. 193-196.
5. Adachi, S., *Physical Properties of III-V Semiconductor Compounds InP, InAs, GaAs, GaP, InGaAs and InGaAsP*, New York: John Wiley & Sons, 1992.
6. Winter, S., and Metzdorf, J., "Correction Procedures for the Flasher Calibration of PV Devices Resulting in Reduced Restrictions and Uncertainties," *Proc. 2nd World PV Conference and Exhibition*, Vienna, Austria, July 6-11, 1998.
7. Emery, K., Burdick, J., Caiyem, D., Dunlavy, D., Field, H., Kroposki, B., Moriarty, T., Ottoson, L., Rummel, S., Strand, T., and Wanlass, M. W., "Temperature Dependence of Photovoltaic Cells, Modules, and Systems," *Proc. 25th IEEE Photovoltaic Specialists Conference*, Washington D.C. May 13-17, 1996, pp. 1275-1278.
8. Wanlass, M. W., Carapella, J. J., Duda, A., Emery, K., Gedvilas, L., Moriarty, T., Ward, S., Webb, J., Wu, X., and Murray, C. S., "High Performance 0.6 eV, $Ga_{0.32}In_{0.68}As/InAs_{0.32}P_{0.68}$ Thermophotovoltaic Converters and Monolithic Minimodules," *Proc. 4th NREL Thermophotovoltaic Generation of Electricity Conference*, Denver, CO, October 11-14, 1998.

A Single TPV Cell Power Density and Efficiency Measurement Technique

Lewis Fraas, Mitch Groeneveld,
Galen Magendanz, and Paul Custard

JX Crystals Inc., Issaquah, WA 98027

Abstract. In order to design improved TPV generators, it is important to characterize the performances of the key components. While it is possible to project high cell electric power densities and conversion efficiencies using measured cell quantum efficiency, filter spectra, and emitter spectra, it is important to confirm these projections with actual measurements of cell electric power and conversion efficiency. It is possible to both measure the spectral properties of TPV spectral control components and to document the radiation to electric conversion efficiency for single cell assemblies by using the "FlatPack" test unit described here. For a SiC emitter temperature of 1450° C, a cell electric power density of 2.5 W/cm^2 was measured using the FlatPack test unit. The measured radiation heat load was 15.2 W/cm^2 resulting in a radiation conversion efficiency of 16.4%.

INTRODUCTION

In the TPV community, it has been customary to calculate TPV cell electric power density and radiation to electric (spectral) efficiency by starting with cell quantum efficiency and the separately measured spectral properties of emitters and filters. While this procedure should be scientifically correct, a more direct measurement of cell power density and spectral efficiency is desirable. JX Crystals has developed a "FlatPack" unit in order to make these direct measurements, both in-house and at Creare. A separate paper by Creare is being presented at this conference.

In the FlatPack spectral control test station, the four key elements are a heat source, the IR emitter, the IR filter, and a TPV cell. These four components along with insulation are supported in four parallel plains in a FlatPack. This unit can be used to characterize IR emitters, IR filters, as well as cell spectral control efficiency. With the ability to duplicate distances, operating temperatures, and aperture configurations combined with the small 2"x 2" sample sizes, this unit is ideal for laboratory research. The second generation FlatPack design is shown in Figure 1.

The idea of the FlatPack stemmed from the need at JX Crystals to test cells, filters, and emissive materials in the simplest form possible. The idea became a design and through several in-house experiments grew into the unit pictured in Figure 2. The success of meeting the in-house needs at JX Crystals suggests the potential use of the FlatPack throughout the TPV community and a move toward a much-needed standard.

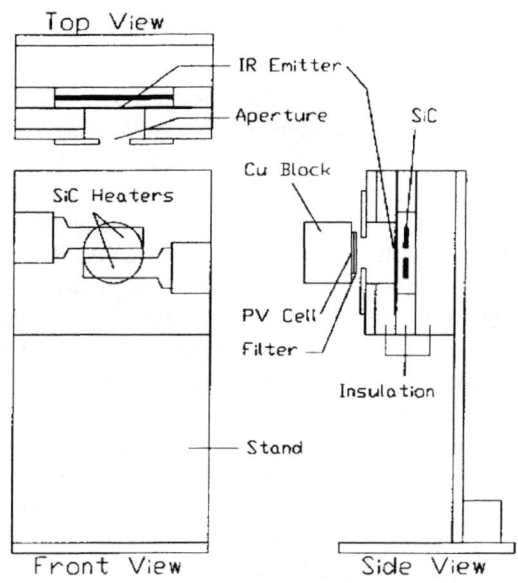

FIGURE 1. FlatPack design.

FIGURE 2. FlatPack test station used to characterize IR emitters, IR filters, and cell spectral control efficiency.

DESCRIPTION OF OPERATION

Using the FlatPack test unit allows one to measure both the spectral properties of thermophotovoltaic (TPV) spectral control components and to document the radiation to electric conversion efficiency for single cell assemblies.

To characterize an IR emitter coupon, a 2" x 2" emitter coupon is inserted into the slot at the top of the FlatPack. A pair of electrically heated SiC source elements heats the flat emitter coupon. When possible, a thermocouple is mounted inside the emitter coupon so that its temperature can be measured. Alternatively, if the emittance of the emitter is known at a given pyrometer wavelength, a pyrometer can be used to measure the emitter temperature. The IR emitter spectra exiting through a ½ inch diameter aperture can be measured with a spectrometer and the total radiant power exiting through the aperture can be measured with a power meter.

Next, to characterize an IR filter, the appropriate filter is placed in front of the aperture and, using a spectrometer, the transmitted spectra can be measured. Transmission measurements can be made conventionally by dividing the spectra with the filter in place by the IR source spectra without a filter in place.

Finally, in order to characterize the cell spectral conversion efficiency, an appropriate filter is bonded to a TPV cell and the cell is bonded to a brass block. A thermocouple inserted into the block is used to monitor block temperature. After heating the IR emitter to the desired temperature, the Cell block is positioned in front of the aperture. The cell electric output and the block temperature rise as a function of time are recorded. From the mass of the block and its specific heat capacity and from the rate of temperature rise, the total heat input into the block can be calculated. The spectral control efficiency will then be simply the cell electrical output divided by the total radiant heat input.

A SIMPLE EFFICIENCY MEASUREMENT

As the previous discussion indicates, the FlatPack offers a lot of opportunities for characterizing TPV components. However, it is desirable to start with a simple case and describe some of the measurement details. We shall describe the case where the IR emitter is a SiC near gray body emitter and the goal is to measure the radiation to electric efficiency for a GaSb TPV cell with a simple five layer dielectric filter deposited on its surface. The intention is to measure the electric power produced by the cell and the radiation heat absorbed by the cell as a function of temperature.

In order to accomplish the above objective, there are three subtle points that require discussion. First in order to limit the heat transfer to the cell block to primarily radiation, the aperture plate must be water cooled. If it is not cooled, its temperature will rise well above ambient and heat will transfer from it to the cell block through air conduction. Second, heat will still flow from the emitter through the aperture to the cell by conduction through the air. Since the interest is in the radiation heat flow, a gold coated reference block is provided with which the heat transfer through the air can be measured and subtracted away from the heat transfer to the cell block to yield the radiation heat load.

Third, note that for an emitter temperature in the 1400°C range, the cell current will be in the 5 to 9 Amp range. This is a very high current. While it would be desirable to measure the cell electric power by measuring a complete current vs. voltage curve, most people do not have the equipment to perform this measurement. As the emitter temperature varies, the cell short circuit current will vary but the cell fill factor and voltage will not change significantly. Therefore, JX Crystals provides a measured cell I vs. V curve from which the fill factor and open circuit voltage can be obtained. The problem is then reduced to a measurement of the cell short circuit current. Since the cell only produces approximately 0.46 V and a short circuit current measurement is desired, JX Crystals provides the GaSb cell on a block with a calibrated small shunt resistor (resistance R_{shunt}) across the cell terminals. A measurement of the voltage across this shunt resistor (V_{shunt}) translates to a short circuit current measurement (I_{sc}).

Provided here are the equations used in the calculation of cell output power and spectral efficiency.

Relevant Equations:

$$q = mc \frac{dT}{dt} \quad (watts)$$

q – heat flux (watts)
m – mass of the brass block (g)
c – specific heat capacity of brass (.3777 J/g K)
dT/dt – change in temperature over time (measured K/sec)

$$I_{sc} = \frac{V_{shunt}}{R_{shunt}}$$

I_{sc} – short circuit current
R_{shunt} – shunt resistance (supplied by JX Crystals)
V_{shunt} – measured shunt voltage

$$P_{cell} = V_{oc} * FF * (V_{shunt}/R_{shunt})$$

P_{cell} – calculated cell output power
V_{oc} – Cell open circuit voltage = 0.46 V
FF – Cell fill factor = 0.75

$$spectral\ efficiency = \frac{P_{cell}}{q_{cell} - q_{block}}$$

q_{cell} – calculated using Cell block
q_{block} – calculated using Gold reference block

In order to provide a more standardized emissivity, a Dow Corning SiC based material, Sylramic S200, has been used to obtain the data shown in Table 1. The same notation indicated previously in the relevant equations section is used. The loss ratio refers to that percent of the calorimetric power subtracted to account for convective heat transfer losses versus the measured cell powers.

TABLE 1. Cell Power & Spectral Efficiency

Temp (°C)	P_{cell} (W/cm^2)	q_{cell} (W/cm^2)	q_{block} (W/cm^2)	Efficiency (%)	Loss ratio
1450	2.5	18.5	3.3	16.4	0.18

FUTURE WORK

Jx Crystals is currently fabricating a temperature controller for the FlatPack in order to eliminate some margin of error inherent to the use of an optical pyrometer. A higher purity SiC source is also sought due to the graybody nature of Sylramic S200 since a blackbody material could also narrow the potential margin of error.

JX Crystals does plan to continue improving the FlatPack as an in-house standard of comparison. While the TPV community may not readily embrace this test device as a standard, the need for one is unquestionable. As in the past, the intention is to meet the challenges set forth by this technology and urge others to do the same.

CONCLUSIONS

- The FlatPack provides a simple means of measuring material emissivity, filter effectiveness, cell efficiencies and cell power densities utilizing small samples.

- The FlatPack allows one to quickly compare performances of multiple materials in an actual TPV system.

- In order to advance the field of TPV, some standard must be introduced and the FlatPack has definite potential to accommodate these advancements.

Measurement Techniques for Single Junction Thermophotovoltaic Cells

L.R. Danielson, J.R. Parrington, G.W. Charache, G. J. Nichols, and D.M. Depoy

Bin 103, Lockheed Martin, Inc., 2401 River Road, Niskayuna, NY 12309

Abstract. Several measurement systems and techniques for the electrical and thermal characterization of thermophotovoltaic(TPV) cells are discussed. One computer controlled system measures the quantum efficiency of cells from 0.8 to 2.6 microns. A probe resistor is used to account for cells with low shunt resistances. In the second system, a production-style robot provides automated measurements of I-V characteristics under dark, blackbody, and flashed illumination conditions. The system measures the length and width of each cell, and calculates the open circuit voltage, short circuit current, fill factor, and maximum power for each cell. The mean and standard deviation of the measured parameters are also computed. The third system measures the overall cell efficiency by a calorimetric technique. Heat losses due to radiation, conduction, and convection are factored into the analysis method.

INTRODUCTION

Accurate measurements of electrical and thermal properties of thermophotovoltaic(TPV) cells are important to improve such cells, and to provide a basis for arranging the cells in a generator configuration. Experimental test systems to measure quantum efficiency vs. wavelength as well as dark and light I-V response curves have been developed. Also, the overall efficiency test for a single cell (1) has been modified and is described.

QUANTUM EFFICIENCY MEASUREMENTS

The external quantum efficiency (QE) of a TPV cell is defined as the probability that an incident photon will create an electron-hole pair that contributes to the current produced by the cell. The internal quantum efficiency of the cell is the external quantum efficiency divided by one minus the wavelength dependent reflection. The generated current and overall efficiency of a cell are proportional to the photon-weighted average internal quantum efficiency (2).

An experimental system for measuring the external quantum efficiency has been developed. A monochrometer illuminator (63W) provides the incident radiation, which passes through a 167 Hz chopper before entering the two-grating monochrometer. Three filters are cycled in as the wavelength varies to remove higher order reflections. The cell contacts are 25 spring loaded pins on the back, and one spring loaded pin on the top central busbar. The multi-pin back contact was found to be crucial for repeatable measurements. The current, typically about 7µA at 1600nm for InGaAs cells, is amplified by a current-to-voltage amplifier (10^4 A/V), and the resulting signal is measured by a Lock-in amplifier. The light intensity is measured by a calibrated (NIST traceable) pyroelectric detector. The system is completely computer controlled, with both data acquisition and analysis performed with LabVIEW software.

The repeatability of the measurements improved by making electrical contact resistances as low as possible. All electrical connections from the cell to the Lock-in amplifier are either gold-to-gold or soldered.

A probe resistor is required to account for the finite shunt resistances of the cells. The probe resistor is particularly critical for some of the recently developed InGaAsSb cells with shunt resistances as low as 1-2 ohms. For the probe resistor switched out of the circuit (Figure 1), the total cell current, I_t, is

$$I_t = I\left(1 + \frac{R_M}{R_S}\right), \tag{1}$$

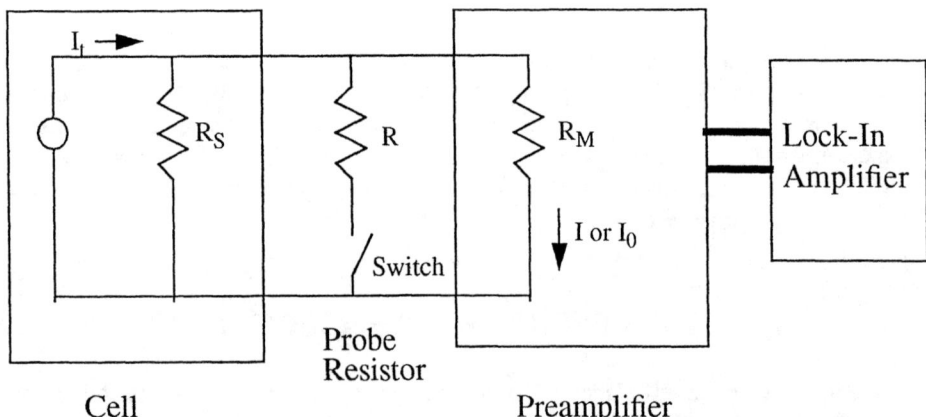

Figure 1. Electrical diagram of QE system showing the location of the probe resistor.

where I is the current through the preamplifier with the switch open, R_M is the input impedance of the preamplifier, and R_S is the shunt resistance of the TPV cell. For the probe resistor switched in the circuit (Figure 1), the total cell current, I_t, is

$$I_t = I_0 \left(1 + \frac{R_M}{R_S} + \frac{R_M}{R}\right), \qquad (2)$$

where I_0 is the current through the preamplifier with the switch closed, and R is the resistance of the probe resistor. The basic working equation is obtained by combining Eq. 1 and Eq. 2 to eliminate R_S:

$$I_t = \frac{R_M I I_0}{R(I - I_0)}. \qquad (3)$$

The procedure is to first place a cell with a known high shunt resistance, R_{S1}, on the fixture and calculate R_M/R at one wavelength (e.g., 1600nm) according to

$$\frac{R_M}{R} = \frac{1}{\dfrac{I_0}{I - I_0} - \dfrac{R}{R_{S1}}}. \qquad (4)$$

The value of R is chosen to be approximately equal to the value of the input impedance of the preamplifier in order for $I - I_0$ to be less than half of I. Note that the calculation of R_M/R depends only weakly on R and R_{S1}, if R is much smaller than R_{S1}. This is easily arranged by selecting R to be 1 ohm and R_{S1} to be several hundred ohms.

The second part of the procedure is to place the sample cell on the fixture and calculate $I_0/(I - I_0)$ at a fixed wavelength, in this case 1600 nm. The value of $I_0/(I - I_0)$ was determined to be independent of wavelength. The final step is to run the QE measurement with the probe resistor out of the circuit and calculate I_t according to Eq. 3.

The external QE's for cells with shunt resistances of 60.5 ohms and 1.42 ohms with and without the use of a probe resistor are shown in Figure 2. The proper accounting for the shunt resistance increases the QE at 1600nm by 4.5% for the 60.5 ohm cell and by 20.8% for the 1.42 ohm cell.

As a check on the QE measurement system, the photon weighted integral of the external QE is compared to the measured short circuit current from a blackbody source at 1200 C (see Figure 3). The measured and calculated currents are proportional, but not equal. This is because of view factor uncertainty and light absorption in the light pipe attached to the blackbody source. Any points which markedly deviate from the

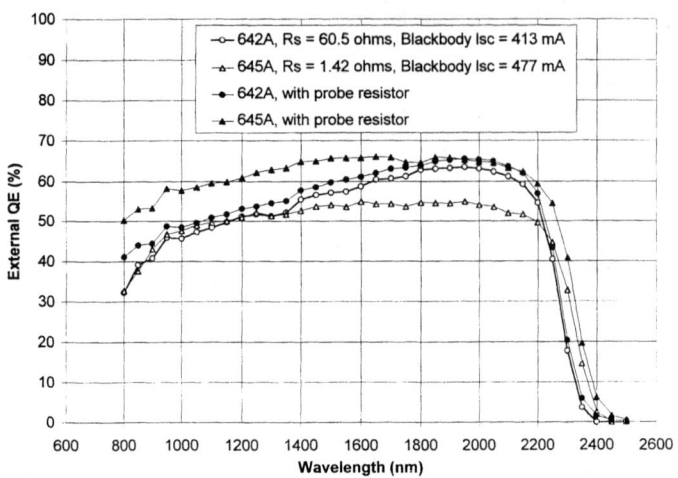

Figure 2. QE vs. wavelength showing the necessity of the probe resistor, especially for cells with low shunt resistances.

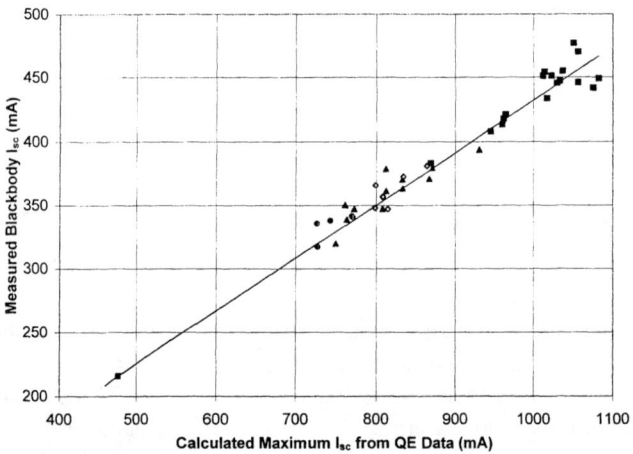

Figure 3. Plot showing a linear relationship between integrated QE vs. wavelength data and I_{sc} measurements. Different symbols indicate different batches of cells.

straight line in Figure 3 act as a flag that there may be a measurement problem.

Wavelength dependent and average internal quantum efficiencies are calculated from the reflection data obtained with an Fourier Transform Infrared (FTIR) spectrometer (2). The internal QE is modeled with a program which inputs fundamental parameters including Auger and radiative recombination coefficients, Schockly-Reed-Hall hole and electron lifetimes, front and back surface recombination velocities, and absorption of cap layers.

CURRENT-VOLTAGE MEASUREMENTS WITH A ROBOT

A robot was designed and constructed to measure in less than a minute the cell dimensions and I-V curves under dark, blackbody, and flashlamp conditions (Figure 4). The computer-controlled system consists of a three axis linear track robot with a vacuum cell pickup, a cell width and length measurement stage, a moveable cell stage which translates linearly along a fourth axis linear track, a blackbody source, a flashlamp with a fast shutter, a curve tracer, a control box, and a computer. The data are displayed on the computer monitor and are saved as ASCII-type data files.

The sequence of a cell measurement is as follows. The vacuum pickup lifts a cell from a box of 25 and transports it to the X-Y measurement station. Two air-actuated probes position the sample against supports and measure the length and width of the cell.

The cell is next transported to the moveable stage. The electrical contacts to the cell consist of 25 spring loaded pins (24 for current, 1 for voltage) on the bottom of the cell, and 7 spring loaded pins (6 for current, 1 for voltage) on the top central busbar. As with the QE measurement system, these multi-pin probes are crucial to minimize errors due to back and front contact errors and to enhance repeatability. The spring constants of the pins are selected so that the 80 psi of air pressure applied to the top pins is sufficient to push the cell against a baseplate, which is maintained at 25C.

Figure 4. Photograph of automated robot for measurement of dark, blackbody, and illuminated I-V curves and cell dimensions.

All I-V data are obtained with a Tektronix 370A curve tracer set to 2 sweeps with 512 data points per sweep. After a dark I-V measurement is performed, the cell is moved along the track and blackbody cell data is acquired. The blackbody source is held at 1200C, and the light is concentrated with a gold-coated air-cooled light pipe inserted into the bottom of the blackbody source. The short circuit current(I_{sc}), open circuit voltage(V_{oc}), fill factor(FF), and maximum power(P_{max}) are calculated, along with the mean and standard deviation of these parameters for a group of cells.

The cell is moved further along the track to the flashlamp illumination position. The high intensity flashlamp is energized, the high speed (5 ms to open and stabilize) shutter is opened, and the curve tracer collects the data. The configuration of the curve tracer requires that the data acquisition times are ≥200msec. The values of I_{sc}, V_{oc}, FF, and P_{max} and statistical parameters are calculated as for the blackbody data. The cell is then transferred back to the box, another cell is picked up, and the process is repeated.

The maximum temperature rise of the cell during the flash test was estimated by measuring V_{oc} as a function of time for lattice mismatched InGaAs and lattice matched InGaAsSb (both about 0.55eV) cells. The decrease in V_{oc} after opening the shutter is approximately linear with time with a slope of 69mV/s for InGaAsSb and 70mV/s for InGaAsSb. The dependence of V_{oc} on temperature is approximately 1.5mV/C for lattice mismatched InGaAs and 1.45mV/C for lattice matched InGaAsSb (3). The total time for the flash measurement is 200msec, so the maximum temperature rise during the measurement is estimated to be 9C. A comparison of V_{oc} obtained by the robot flash test and by a steady state test with the cell soldered down show a difference of 4mV. This indicates a temperature rise during the flash test of approximately 3C.

An internally generated standard cell is run before and after each group of cells to verify the day to day measurement repeatability. The mean, standard deviation, and fractional percent deviation of the dimensional and electrical parameters for this standard cell are shown Table 1. For all of the measurements, one standard deviation is less than 3%.

A separate non-linear least squares fitting routine takes the dark I-V input data and performs a fit to the following equation for the I-V characteristics:

$$I = -I_L + I_S \left[e^{\frac{q(V-IR_S)}{nkT}} - 1 - e^{\frac{-q(V-IR_S+V_b)}{bkT}} \right] + \frac{V-IR_S}{R_{SH}} \quad (5)$$

where I_L = light generated current, I_S = saturation (dark) current, V = voltage across the cell, I = current through the cell, R_S = cell series resistance, n = ideality factor, V_b = breakdown voltage, b = breakdown parameter, and R_{SH} = cell shunt resistance. This equation modifies the ideal diode equation by accounting for Zener breakdown and for cell shunt resistance. The terms I_L, I_S, R_S, n, V_b, b, and R_{SH} are possible adjustable parameters. The program performs the fit by minimizing the sum squared of the errors, by the Gauss-Newton method and singular value decomposition (SVD) for the

Table 1. Measurement uncertainty for standard sample S2 measured 30 times over 107 days.

Measurement	I-V Type	Mean(M)	Standard Deviation (σ)	Fractional Percent ($100*\sigma/M$)
V_{oc}(mV)	Blackbody	235.7	2.1	0.9
I_{sc}(mA)	Blackbody	340.4	2.4	0.7
FF(%)	Blackbody	64.9	1.0	1.6
P_{max}(mW)	Blackbody	52.1	1.3	2.5
V_{oc}(mV)	Flash	305.2	2.0	0.7
I_{sc}(mA)	Flash	2859.1	55.3	1.9
FF(%)	Flash	64.7	0.6	0.9
P_{max}(mW)	Flash	564.5	16.5	2.9
X(mm)	-	10.40	0.03	0.3
Y(mm)	-	10.32	0.03	0.3

resulting system of equations (4). The parameters resulting from the data fit are input to a commercial electrical network modeling program (SABRE) to determine both the I-V characteristics and the electrical output power of a multicell TPV generator.

OVERALL EFFICIENCY MEASUREMENTS

The overall cell efficiency is measured by a calorimetric technique. Light from a calibrated blackbody source illuminates a cell soldered to a copper block. The block is fitted with a thermocouple, and is surrounded by gold shields on the top and by thermal insulation on the sides and bottom. The system is instrumented with current and voltage probes to measure the electrical characteristics of the cell. The rise in temperature after a shutter is pulled from the exit port is monitored by an automated data acquisition system.

If no heat losses occur, then the slope of the temperature vs. time curve is constant and can be used to calculate the heat absorbed by the cell. However, in real systems, there are losses due to conduction, convection, and radiation which cause a decrease in the slope of the temperature-time curve. The conduction and convection losses are proportional to the temperature difference (ΔT) between the cell and the ambient. For the small temperature differences we measured ($\Delta T \leq 20C$), the radiant temperature loss is also well approximated by a linear dependence on ΔT. With these assumptions, the rate of temperature rise of the copper block for relatively small excursions above ambient can be written as

$$dT/dt = P_{abs}/(mC_p) - B\Delta T, \tag{6}$$

where T is the cell temperature, t is the time since the shutter was removed, P_{abs} is the heat absorbed by the cell plus the block, m is the mass of the cell plus block, C_p is the specific heat of the cell plus block, B is a heat loss parameter, and ΔT is the temperature rise of the cell above ambient.

A plot of dT/dt vs. ΔT should give a straight line with the y-intercept equal to

$$(dT/dt)_0 = P_{abs}/(mC_p). \tag{7}$$

A typical temperature vs. time curve is shown in Figure 5. A plot of dT/dt vs. ΔT (Figure 6) shows the expected linear behavior.

The system is calibrated by measuring the current (I) and the voltage (V) with the cell in reverse bias, so

$$dT/dt = IV/(mC_p) - B\Delta T \tag{8}$$

and the y-intercept of the calibration run is

$$(dT/dt)_{0,cal} = IV/(mC_p). \tag{9}$$

Dividing Eq. 7 by Eq. 9 yields

$$P_{abs} = IV[(dT/dt)_0/(dT/dt)_{0,cal}]. \tag{10}$$

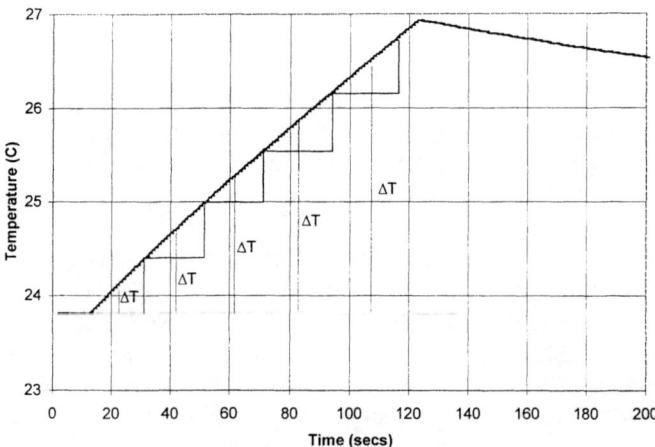

Figure 5. Temperature vs. time for overall efficiency test showing the multiple slope technique.

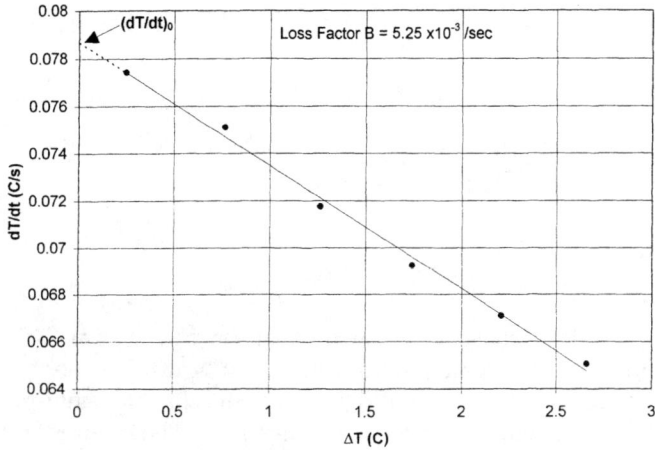

Figure 6. Rate of temperature rise of the copper block showing the "no heat loss" point at the y-intercept.

There are several advantages to this multiple slope method for measurement of overall efficiency as outlined above. The absorbed power is independent of specific heat and mass, since the system is calibrated by supplying a known power into the cell. This method accounts for losses due to conduction, convection, and radiation. Losses are extracted from a single normal run, so that no additional background runs are necessary. Approximately 100 data points are used to calculate the y-intercept. Finally, the method provides a quantitative value of the thermal losses in the system by providing the loss factor B. This loss factor can help characterize the heat losses for various configurations of the copper block set-up.

The efficiencies measured directly in the copper block experiment are for relatively small currents, e.g., 0.5A. To calculate the efficiency expected at larger currents of 3A, for example, use the efficiency equations:

$$\eta_1 = (qV_{oc}/E_G)_1 (FF)_1 (QE)_1 (F_u)_1 (F_o)_1 \qquad (11)$$

$$\eta_2 = (qV_{oc}/E_G)_2 (FF)_2 (QE)_2 (F_u)_2 (F_o)_2, \qquad (12)$$

where η is the efficiency, q is the electronic charge, V_{oc} is the open circuit voltage, E_G is the band gap, FF is the fill factor, QE is the average internal quantum efficiency, F_u is the spectral control factor, and F_o is the "overexcitation" factor. The subscript "1" refers to the low current case directly measured by the copper block test, and the subscript "2" refers to the high current (3A) case.

Since F_u, F_o, and E_G do not depend on current, and we assume that the average QE does not change from 0.5A to 3A, dividing Eq. 12 by Eq. 11 yields

$$\eta_2 = \eta_1[(V_{oc})_2*(FF)_2/(V_{oc})_1*(FF)_1]. \qquad (13)$$

The values of $(V_{oc})_2$ and $(FF)_2$ are obtained from a steady state measurement with a cell soldered down or from a flash test.

CONCLUSIONS

The systems described above are used routinely to characterize TPV cells. The inclusion of a probe resistor in the quantum efficiency system is important to account for the low shunt resistances of some of the InGaAsSb cells. The multi-pin probes for both the quantum efficiency and the I-V measurements allow more repeatable and accurate measurements to be made.

The automated data acquisition of I-V curves by a robot has greatly facilitated the rapid quantitative evaluation of cells. The advantage of collecting two illuminated I-V curves is that the blackbody source provides a prototypical current for the cell at low illumination levels (about 0.4A), while the flashlamp provides prototypical open circuit voltages and fill factors at higher current levels (about 3A).

The calorimetric evaluation of overall cell efficiency is an accurate and convenient method. The calibration of the heat absorbed by applying a known power to the cell eliminates the need to know the specific heats of the copper block and the cell. The determination of the parasitic heat loss is a useful addition to the system.

Since no NIST standard TPV cell exists, round-robin measurements of standard laboratory cells among a variety of laboratories are recommended to identify and understand more fully the measurement uncertainty for these systems.

REFERENCES

1. J.P. Benner, T.J. Coutts, D.S. Ginley, eds.,*The Second NREL Conference on the Thermophotovoltaic Generation of Electricity*, AIP Conference Proceedings #358, AIP Press, Woodbury, NY, 1995, p351.
2. J.L. Egley, D.M. Depoy, L.R. Danielson, M.J. Freeman, R.J. Dziendziel, J.F. Moynihan, P.F. Baldasaro, B.C. Campbell, C.A. Wang, H.K. Choi, G.W. Turner, S.J. Wojtczuk, P. Colter, P. Sharps, M. Timmons, R.E. Fahey, and K. Zhang, J. Elect. Mat., Vol 27, No. 9 (1998), 1138.
3. G.W. Charache, J.L. Egley, L.R. Danielson, D.M. Depoy, P.F. Baldasaro, and B.C. Campbell, Current Status of Low-Temperature Radiator Thermophotvoltaic Devices, presented at the 25th Photovoltaic Specialists Conference, Washington, D.C. 1996.
4. Numerical Recipes in C, Press, Teukolsky, Vetterling and Flannery, Cambridge University Press, 1994.

Optical Properties of Thin Semiconductor Device Structures with Reflective Back-Surface Layers

M. B. Clevenger*, C. S. Murray*, S. A. Ringel[†],
R. N. Sacks[†], L. Qin[†], G. W. Charache[§] and
D. M. Depoy[§]

*Bettis Atomic Power Laboratory, West Mifflin, PA 15122

[†]Department of Electrical Engineering
Ohio State University, Columbus, OH 43210-1272

[§]Lockheed Martin, Inc., Schenectady, NY 12301-1072

Abstract. Ultrathin semiconductor device structures incorporating reflective internal or back surface layers have been investigated recently as a means of improving photon recuperation, eliminating losses associated with free carrier absorption in conductive substrates and increasing the above bandgap optical thickness of thermophotovoltaic device structures. However, optical losses in the form of resonance absorptions in these ultrathin devices have been observed. This behavior in cells incorporating epitaxially grown FeAl layers and in devices that lack a substrate but have a back-surface reflector (BSR) at the rear of the active layers has been studied experimentally and modeled effectively. For thermophotovoltaic devices, these resonances represent a significant loss mechanism since the wavelengths at which they occur are defined by the active TPV cell thickness of ~2-5 microns and are in a spectral range of significant energy content for thermal radiators. This study demonstrates that ultrathin semiconductor structures that are clad by such highly reflective layers or by films with largely different indices of refraction display resonance absorptions that can only be overcome through the implementation of some external spectral control strategy. Effective broadband, below-bandgap spectral control using a back-surface reflector is only achievable using a large separation between the TPV active layers and the back-surface reflector.

Introduction

Previously, photovoltaic and thermophotovoltaic (TPV) cells have been shown to benefit significantly from the introduction of highly reflective layers at the back surfaces of devices[1-2]. In these applications, the back surface reflector (BSR) serves a number of important functions. The BSR increases the above-bandgap optical thickness of device structures by providing multiple internal reflections of incident light, aids in the recuperation of below-bandgap light (important in TPV) and, when incorporated into the device above the substrate, eliminates losses associated with free carrier absorption in conducting substrates and heavily doped layers used for lateral current conduction. In this work, two strategies for the use of BSR's were investigated in order to improve the performance of TPV devices.

First, the fabrication of UltraThin TPV (UTTPV) cells was investigated wherein the p-InP substrates of 0.74 eV and 0.55 eV singe junction TPV devices were replaced with unalloyed gold back contacts. This work was completed using a peeled epitaxial lift-off process or a chemical etching process for InP removal followed by thermal deposition of the gold back surface reflector/contact. This strategy left an n-on-p InGaAs TPV cell backed by a gold BSR/contact layer.

The second strategy employed was the replacement of the lateral conduction layer (LCL) in a Monolithic Interconnected Module (MIM) with a buried, epitaxial layer of FeAl. This technique of incorporating a highly reflective layer into the TPV cell architecture, above the substrate, offered the above-stated advantages afforded by the use of BSR's while also minimizing the series resistance normally associated with the semiconducting LCL. This work was completed through the epitaxial thermal evaporation of iron and aluminum to yield a $Fe_{0.5}Al_{0.5}$ alloy that is lattice matched to InP, followed by the growth of a p-on-n InGaAs TPV cell structure.

These types of TPV structures exhibit similar optical responses. Notably, destructive interference patterns are established as a result of the effective TPV cell thickness being on the order of incident wavelengths. Similar behaviors have been observed previously in photovoltaic structures where light trapping for increased optical pathlength was an objective[2]. However, in TPV structures, such resonance absorption bands decrease significantly the devices' spectral utilization factors and overall efficiencies. Described herein are efforts to model and measure this optical behavior in UTTPV cells and those incorporating buried metal layers.

Optical Modeling Method

The optical response of ultrathin TPV cells was modeled using the commercially-available code TFCalc.™ In order to accurately predict the optical response of a multilayer structure, TFCalc™ requires that the wavelength dispersion of the refractive index (n) and the extinction coefficient (k) be entered in tabular form or be specified by identifying accurately the coefficients for a dispersion equation (e.g. the Drude formula). For the TPV cell structures modeled in this work, n and k data tables were generated using two methods. For undoped InGaAs, n-InGaAs, some p-InGaAs and undoped InP layers, n and k data was calculated from front surface reflectance measurements[3]. For other p-InGaAs and doped InP layers, n and k data acquired via spectroscopic ellipsometry was used[4]. In either analysis method, the production of accurate, comprehensive data tables over a range of wavelengths can be difficult, and iterative comparison to experimentally obtained data is essential. The structure employed for the ultra-thin TPV cell study is shown in Figure 1.

| 500Å 3x10^{18} n-InP |
| 500Å 3x10^{18} n-InGaAs |
| 2.1μm 2x10^{17} p-InGaAs |
| 500Å 5x10^{18} p-InP |
| 150Å 5x10^{18} p-InGaAs |
| 30Å Palladium |
| 1600Å Gold |

Figure 1. Ultra-thin TPV cell structure used in modeling studies and reflectance measurements

The behavior of TPV cells incorporating epitaxially grown metal layers was also modeled. Here, the device structure shown in Figure 2 was used to model a TPV device with a Fe_xAl_{1-x} lateral conduction layer. The wavelength dispersions of n and k for the semiconductor layers were acquired as described above for the UT cells. The

| 1000Å 1x10^{19} p-InGaAs Emitter |
| 3μm 0.74eV n-InGaAs Base |
| 500Å 3x10^{18} n-InGaAs Back Surface Field |
| Buried Metal Layer |
| 1200Å Undoped InGaAs |
| 350μm InP Substrate |

Figure 2. Device structure modeled using TFCalc.

dispersions of the refractive index and coefficient of absorption for Fe_xAl_{1-x} were obtained from calculations of the self-consistent band structure of FeAl [5] and Kramers-Kronig analysis of near-normal reflectance from $FeAl_{1-x}Cu_x$[6]. The $FeAl_{1-x}Cu_x$ alloy possesses a dielectric constant and joint density-of-states that are in good agreement with the values obtained from band structure calculations for FeAl and it has been shown that the substitution of copper for aluminum in FeAl does not change the band structure significantly[6]. As a result, both methods produced very similar results for the wavelength dispersion of n and k.

Results and Discussion

Modeled and measured reflectances for the ultra-thin TPV cell structure are provided in Figure 3. Here, a pattern of destructive interference results in resonance losses across the infrared. This is a direct result of making the effective TPV cell thickness on the order of the wavelengths of incident light. Thus, it is important to note that photonic devices should have thicknesses substantially different from the wavelength of light in the region in which they are operating. This can be seen from the analysis below.

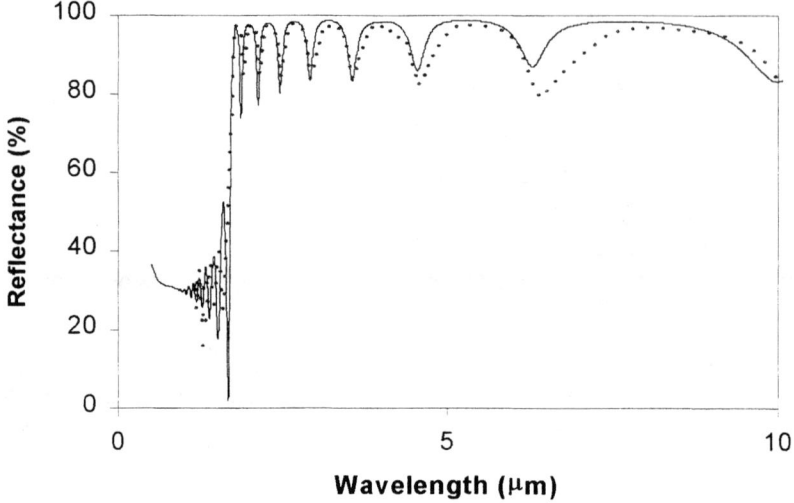

Figure 3. (—)Modeled and (•••)measured reflectance of an ultra-thin TPV cell.

The resonance absorption bands displayed by a sample exhibit a destructive interference pattern that can be described by the principle of superposition of waves[7-8]. Accordingly, for waves in phase that are incident on a TPV cell, a reflected phase difference, δ, arises from a difference in the pathlength traveled by each wave:

$$\delta = \frac{2\pi(x_1 - x_2)}{\lambda} \qquad \text{Eq. 1}$$

where x_1 and x_2 are the distances from the source of the two waves to the point of observation and λ is the wavelength in the medium (cell). The pathlength differences for individual waves propagating through a TPV structure arise from partial reflectance of the waves at each interface in the structure. The reflectance at each interface can be evaluated using Equation 2:

$$R = \frac{(n_2 - n_1)^2 + k^2}{(n_2 + n_1)^2 + k^2} \qquad \text{Eq. 2}$$

where n_1 and n_2 are the refractive indices of the incident medium and the layer being encountered at the interface, respectively, and k is the coefficient of absorption of the encountered layer. The total distance traveled by a wave through a structure will determine its phase difference with other waves within the structure. For example, depicted in Figure 4 are three of the possible paths through a UTTPV cell structure for incident light (I_0) with a wavelength of 1.1 µm. Because the reflected wave R_2 has traveled 4.4 µm farther than R_1 (2.2 µm into the device and 2.2 µm out), Equation 1 states that the two waves will be in phase. Arriving at the front surface in phase with R_1 means that R_2 will constructively interfere with R_1, and, assuming no absorptive losses, the intensity of the interference wave ($R_1 + R_2$) will match that of I_0. However, in the case of wave R_3, it has traveled a total of 4.536 µm farther than R_1. This will result in R_3 being ~45° out of phase with R_1. The resulting partial destructive interference will produce a decrease in net reflected intensity from the device at 1.1 µm.

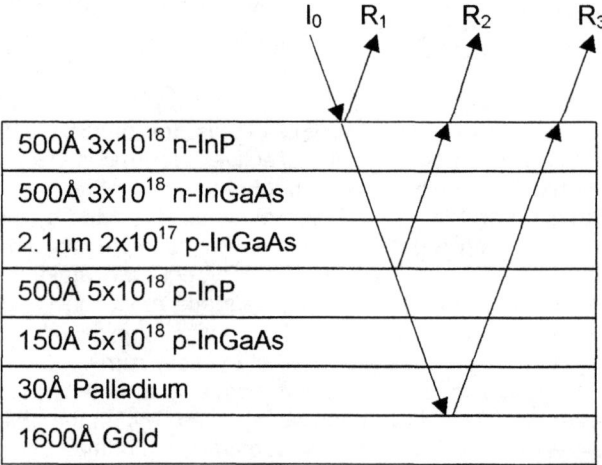

Figure 4. Schematic depicting different optical paths for an incident wave that will produce varying degrees of constructive and destructive interference.

In high index materials such as InP and InGaAs, it is useful to relate Equation 1 to the media through which the waves are propagating. Thus, given that the refractive index of each layer of the structure is $n=\lambda_0/\lambda$, where λ_0 is the wavelength of incident light in vacuum, the phase difference between interfering waves at each interface is evaluated using Equation 3:

$$\delta = \frac{2\pi n(x_1 - x_2)}{\lambda_0} \quad \text{Eq. 3}$$

For the thermophotovoltaic device structures under consideration here, two simplifications can be made in calculating phase differences and resultant reflected intensities. First, InP and InGaAs layers possess approximately equal indices of refraction in the infrared. Second, the incorporation of a reflective back surface or internal layer dictates that the principle reflection from within the device takes place at the interface to that layer. The ramification of these two facts is that calculations of the phase difference of waves exiting the TPV structure can be completed accurately by assuming that the multilayer TPV structure possesses a single dispersion of the refractive index (i.e. That it behaves as one thick layer) and that the quantity (x_1-x_2) is simply twice the cell thickness, t. Equation 3 is then simplified to:

$$\delta = \frac{2\pi n(2t)}{\lambda_o} \qquad \text{Eq. 4}$$

which can be applied effectively to the entire device without serious and tedious consideration being given to reflections at every interface. From Equation 4, it can be seen that complete destructive interference will always occurs when $\delta = m\pi$ or

$$t = \frac{(2m+1)\lambda_0}{4n} \qquad \text{Eq. 5}$$

where m is an integer (m=1, 2, 3,..).

As stated previously, in thermophotovoltaic cells, the principle reflection is from the back-surface reflector or from a highly reflective layer that is incorporated into the device. When this highly reflective layer is behind the active layers of the device, the effective cell thickness is on the order of the wavelengths emitted by thermal radiators and the destructive interference patterns are established in the infrared. However, if a BSR is placed behind a thick substrate in a TPV cell, the major effective optical pathlength difference between any internal or surface reflections and that from the BSR will be on the order of 350μm. This shifts significantly the destructive interference patterns out of the infrared region of the spectrum.

By a similar approach to that described above, the reflectance of a TPV cell incorporating a reflective internal layer can be modeled. When the optical behavior of a FeAl layer grown epitaxially above the cell substrate is modeled using the results of band structure calculations and Kramers-Kronig analyses, the results converge to yield a reflectance typical of that shown in Figure 5.[6] As in the case of the ultra-thin TPV cells, a destructive interference pattern is produced by the superposition of waves reflected from each of the layers within the device, but principally from the reflective FeAl layer.

However, in the case of buried metal layers, there are more losses associated with the destructive interference. These losses can be attributed primarily to the use of Group III-transition metal alloys as buried metal layers. Alloys of this kind are needed to achieve an effective lattice match to InP and GaAs substrates. However, the transition metal component, in this case iron, absorbs significantly in the infrared, particularly at wavelengths below 4μm. Thus, the reflectance of buried metal layer structures is observed to decrease with wavelength.

Applying the analysis of destructive interference patterns detailed above, it is possible to predict accurately the location of resonance absorption bands for ultrathin structures and determine the reflection order, m, that is resulting in the loss. Assuming that the semiconductor layers within both the UTTPV and buried metal layer structures all possess dispersions of the refractive index that can be described by the Drude formula (i.e. they act as a single n-InGaAs layer), the locations of resonance absorption bands resulting from destructive interference were calculated using Equation 5. The results of these calculations are provided with the TFCalc™-predicted and experimentally determined values in Table 1.

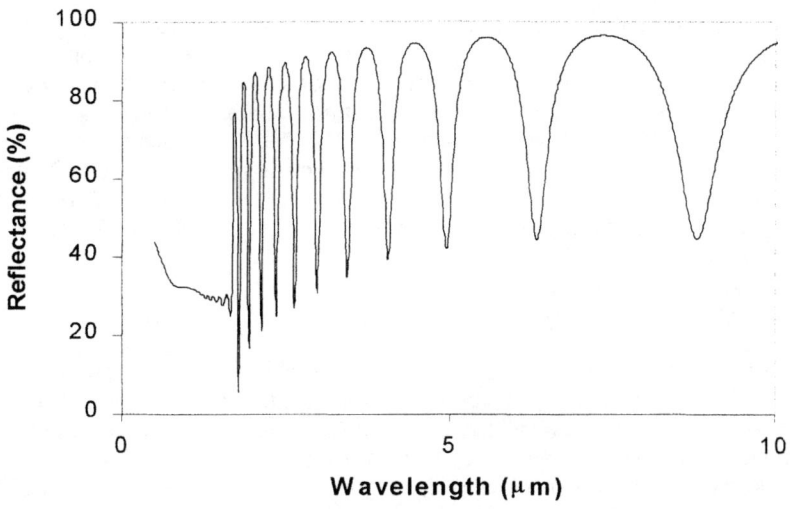

Figure 5. Modeled reflectance of a TPV cell incorporating a buried FeAl layer.

From the results presented in Table 1, it is clear that destructive interference patterns in thin TPV structures may be predicted accurately. Optical models require knowledge of individual layer thicknesses, compositions and dopant densities and measurements require the growth and characterization of a sample in order to determine the location of resonance absorption bands that result from destructive interference. However, use of the equations that result from the principle of superposition allows the same determination to be made accurately from knowledge only of the total structure thickness. This approach should benefit significantly future efforts to either avoid or utilize light trapping structures in thermophotovoltaics and photovoltaics.

Table 1. Calculated, Modeled and Measured Wavelengths at Which Destructive Interference Occurred in Thin TPV Devices with Reflective Layers.

Reflection Order, m	Ultrathin TPV Cells			Buried Metal Layer Cells	
	Calculated (μm)	TFCalc (μm)	Measured (μm)	Calculated (μm)	TFCalc (μm)
1	9.95	9.98	10.00	12.95	-
2	6.25	6.28	6.39	8.45	8.79
3	4.55	4.55	4.58	6.20	6.33
4	3.55	3.57	3.57	4.90	4.94
5	2.90	2.92	2.94	4.05	4.05
6	2.50	2.48	2.50	3.40	3.43
7	2.15	2.15	2.17	3.00	2.98
8	1.85	1.90	1.86	2.65	2.63
9	-*	-	-	2.35	2.36
10	-	-	-	2.15	2.13
11	-	-	-	1.95	1.95
12	-	-	-	1.80	1.79

*Values below the bandedge of 0.74 eV cells.

Importantly, for either of these strategies for the use of back-surface reflectors to be practical in thermophotovoltaics, the resonance losses associated with the destructive interference patterns must be attenuated. This can most effectively be achieved using some front-surface spectral control element. In such a strategy, it is possible to utilize a discrete interference or tandem filter approach to reflecting above bandgap light. This would prevent the resonance losses by preventing the long-wavelength, useless light from entering the cell. If device spectral utilization and efficiency are to be maintained with an ultrathin TPV structure, front-surface spectral control is essential.

Summary

The incorporation of reflective internal or back-surface layers into thermophotovoltaic devices offers a variety of advantages, including: an improvement of photon recuperation, an elimination of losses associated with free carrier absorption in conductive substrates and an increase in the above bandgap optical thickness of device structures. However, placing the highly reflective layers directly behind the active layers of a cell renders the effective device thickness comparable to the wavelengths emitted by thermal radiators. Consequently, a pattern of destructive interference is established that is governed by the principle of superposition. The multiple, absorptive layers that comprise a TPV device augment this problem. The resonance losses that result occur principally in the spectral range of significant energy content for thermal radiators and a device structure with a low spectral utilization results. In order to combat this problem effectively, a large separation between the cell active layers and the back-surface reflector must be maintained, or a broadband, front-surface spectral control strategy must be employed.

[1] P. A. Ilies and C. L. Chu, *Proceedings of the 25th IEEE Photovoltaics Specialists Conference*, 1996, pp.109-112.
[2] M. G. Mauk, P.A. Burch, S. W. Johnson, T. A. Goodwin and A. M. Barnett, *Proceedings of the 25th IEEE Photovoltaics Specialists Conference*, 1996, pp. 147-150.
[3] M. B. Clevenger, C. S. Murray and D. R. Riley, *Proceedings of the 1997 Fall Meeting of the Materials Research Society*, 1997, in press.
[4] S. Adachi, *Physical Properties of III-V Semiconductor Compounds: InP, InAs, GaAs, GaP, InGaAs, and InGaP*, John Wiley and Sons, New York, 1992.
[5] C. Koenig and M. A. Khan, *Phys. Rev. B* **27**, 6129 (1983).
[6] A. S. Saleh and M. Y. Alaqra, *Solid State Comm.* **66**, 521 (1988).
[7] J. D. Ingle, Jr. and S. R. Crouch, *Spectrochemical Analysis*, Prentice Hall, New Jersey, 1988.
[8] J. I. Pankove, *Optical Processes in Semiconductors*, Dover Publications, New York, 1971.

Lessons Learned on Closed Cavity TPV System Efficiency Measurements

C. K. Gethers, C. T. Ballinger, and D.M. DePoy

Lockheed Martin Corporation, Schenectady, NY

Abstract: Previous efficiency measurements [1] have highlighted that to accurately measure and predict thermophotovoltaic (TPV) integrated cell or array efficiencies, a thorough understanding of the system is required. This includes knowledge of intrinsic diode and filter characteristics, radiative surface properties of all materials used within the cavity, and an intimate knowledge of the radiator / photon source. As a result of these and other lessons learned, the cavity test fixture used in earlier experiments was redesigned. To reduce radiator temperature gradients, the radiator was oversized and thickened, cavity walls were eliminated, the diode heat sink and shielding material were separated, and the cold side was redesigned to incorporate a steady state heat absorbed measurement technique. This redesigned test fixture provides an isothermal radiator and significantly enhances calorimetry capabilities. This newly designed cavity test fixture, in conjunction with the Monte Carlo Photon Transport code RACER-X, was used to improve and demonstrate the understanding of "in-cavity" TPV diode / module system efficiency testing. A single TPV diode was tested in this new fixture and yielded good agreement between measurements and predictions.

INTRODUCTION

Years of thermophotovoltaic (TPV) system experience have shown that TPV efficiency tests in prototypical environments often have results that are contrary to intuition. Small gaps, thin busbars and other macroscopically small features can have an unexpected impact on efficiency. TPV efficiency, η, is measured by dividing the diode / module maximum output power (P_{out}) by its total absorbed heat (P_{abs}), illustrated by equation 1.

$$\eta = \frac{P_{OUT}}{P_{ABS}} = \frac{V_{OC}I_{SC}ff}{\dot{m}C_P\Delta T} \qquad (1)$$

where,
P_{OUT} = maximum power output by the array,
P_{ABS} = the array's total absorbed radiative heat,
V_{oc} = diode / module open circuit voltage,

I_{sc} = diode / module short circuit current,
ff = diode / module fill factor,
$\dot{m}$ = mass flow rate of the coolant,
C_p = specific heat of the coolant, and
ΔT = coolant temperature rise between inlet and outlet of the array.

TPV diode / module efficiency is primarily dependent upon first order effects such as diode quality and the type and quality of spectral control used. However, engineering factors also affect efficiency, including:

- Geometry (i.e., gaps and grids lines)
- Radiator temperature, profile and emissivity
- Cavity spectral properties
- Networking losses (i.e., cell mismatching and series resistance)
- Fabrication flaws (i.e., solder and flux residue)

These factors, as well as others, need to be thoroughly understood, if predictable TPV efficiency measurements are to be made. Once a demonstrated understanding of TPV cavity system efficiency measurements is made (via agreement between measurements and predictions), then prototypical and efficient TPV systems can be engineered.

Previous TPV efficiency measurements agreed well with predictions to within 10%. However, the efficiency predictions for the individual components (P_{out} and P_{abs}) were approximately 20% lower than measurements. Possible reasons for this large discrepancy were initially attributed to uncertainties in radiator temperature, profile and emissivity, spectral quantum efficiency, filter reflectivity, and various other modelling assumptions. Based upon these results, a concerted effort was made to more thoroughly understand the fundamentals of TPV diode system efficiency measurements in a closed cavity geometry. This was accomplished by simplifying the test articles used, redesigning the experiment and refining predictive capabilities.

Two computer codes were developed and refined to model in-cavity TPV system efficiency measurements. The first, TPVCalc, assumes an infinite parallel plate arrangement, subdivides the infrared spectrum into several bands, and then numerically integrates over the spectrum to provide diode performance predictions [2]. However in some cases, a one dimensional (1-D) code such as this, may not be able to model the desired effect; but rather account for it through the application of an engineering factor applied to the final calculated result. The second model, a 3-D Monte Carlo Photon Transport code called RACER-X, tracks photons from birth at the radiation source until they either escape or are absorbed; and is more geometrically flexible than TPVCalc [1]. A 3-D analysis capability such as RACER-X, could possibly provide more reliable predictive analysis.

This paper discusses the lessons learned from this testing effort and the steps taken to improve our understanding of in-cavity TPV system efficiency measurements.

ORIGINAL EXPERIMENTAL SETUP

Reference 1 briefly describes the test fixture and procedure used to obtain the original TPV in-cavity system efficiency measurements (Figure 1). The test rig consisted of an aluminum heater cavity and diode / module heat sink. The heater cavity was simply a small hollowed out aluminum block designed to provide a blackbody photon source for diode / module illumination. A window was machined in the base of the block to provide a path for the radiant heat to the target positioned beneath it. A small electrically resistive heater was mounted to the lid of the heater cavity and used to radiatively heat the blackbody radiator. The polished internal walls of the cavity reduced the power required to radiatively heat the radiator material. And, the polished window edge helped the test fixture to simulate an infinitely large radiator to the target positioned centrally beneath it.

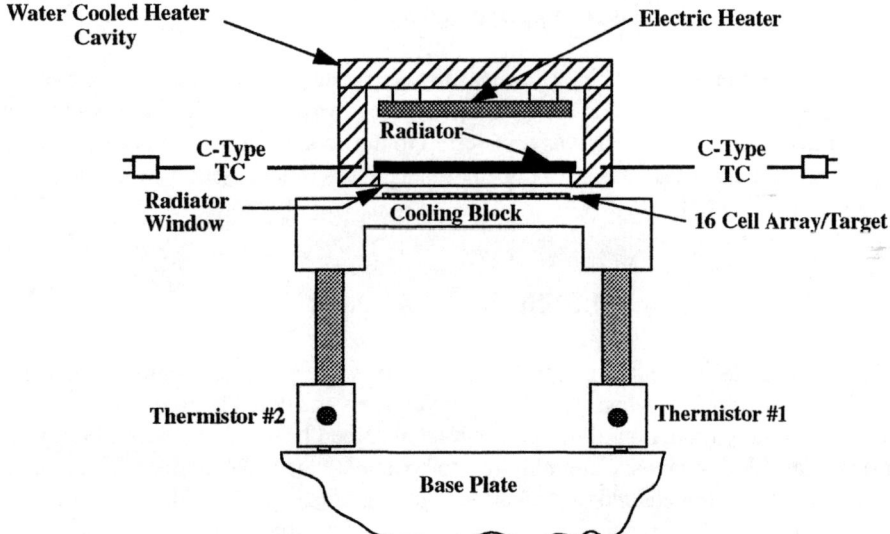

Figure 1. Schematic representation of the previous experimental setup

The 50.80 mm^2 poco graphite radiator was approximately 2.032 mm thick and rested atop the window. The radiator was thermally isolated from the aluminum structure via small diameter ceramic pins. This arrangement locates the radiator approximately 7.62 mm above the surface of the cold side heat sink and reduces conductive gradients caused by direct contact with the water cooled structure. Two - 1.02 mm diameter thermocouple holes were drilled 6.35 mm deep on two sides of the radiator, so that the average radiator temperature could be measured during testing.

The diode / module heat sink used to assist in measuring the absorbed heat was a simple "U" shaped aluminum cooling block. The cooling block was fabricated with internal flow channels to ensure even cooling across the face of the block. Diodes or targets to be tested were attached to the cooling block via a two part silicon based epoxy material. To address parasitic absorption concerns, the perimeter of the target was typically shielded with gold foil. The cooling block and target could then be centered beneath the heater cavity's radiator window for testing.

All experiments were conducted under steady state, high vacuum conditions. Once the fixture was assembled and properly aligned in the vacuum chamber, testing was initiated. This consisted of raising radiator temperature to a desired level and allowing the system to equilibrate. At steady state, the electrical performance of the diode / module was measured with a Tektronix 370A Curve Tracer. The complete diode / module IV curve was collected and the values of interest (i.e., I_{sc}, V_{oc}, ff and P_{out}) were extracted from the curve via an in-house computer program.

The diode / module absorbed heat was measured by monitoring the temperature rise and coolant flow rate. Absorbed heat measurements were always taken with the module open circuited to ensure that the heat was transferred to the coolant, and was not lost to joule heating of an external load. Once obtained, this data was ratioed with the electrical data to determine efficiency.

LESSONS LEARNED

Initial efficiency tests resulted in good agreement between previous efficiency measurements and predictions. However, a significant variation was noted between the measured and predicted maximum output power and total absorbed heat (~20%). Since both the numerator and denominator were high by approximately the same amount, the comparison of measured and predicted efficiency was not significantly affected. To more thoroughly understand these results, the test apparatus and computer models used were critically evaluated. This evaluation was performed by simplifying the experiment, modifying the computer codes to more accurately model the experiment, and by conducting various sensitivity predictions using RACER-X. In this set of experiments, small square silicon wafers were used as cold side targets. Since silicon has a relatively flat spectral profile and no surface features, this type of test represents the simplest case possible. Analysis of the test was conducted using RACER-X. This analysis reinforced the importance of understanding geometry effects, explicitly knowing the radiator's temperature, profile and the spectral properties of all in-cavity materials.

For instance, one analysis showed how the combination of the gap between the heater cavity and cold side heat sink and the imperfectly reflecting cavity window walls results in an illumination profile on the target. Even with a flat temperature profile on the radiator, a nonuniform photon flux distribution at the target level still exist. This effect is illustrated in Figure 2. This Figure shows the normalized distribution of above bandgap energy deposition on a sixteen cell module; notice the nonuniformity. If an array were fabricated of perfectly matched cells, this illumination profile would degrade the performance of the array and lower the efficiency. To circumvent this effect, the cavity window walls must be either eliminated or be perfectly reflecting. This nonuniform illumination effect is further amplified by the known parabolic profile of the radiator used in this experiment. This uneven temperature distribution across the radiator, therefore, makes this cavity test fixture ill-suited for efficiency testing of multi-cell arrays.

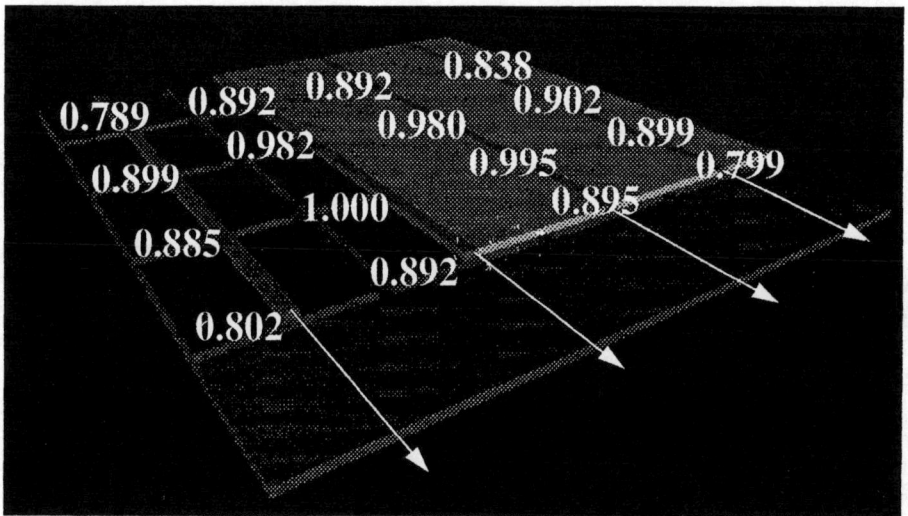

Figure 2. Computer model illustrating the normalized above bandgap energy deposition for a module exposed to an isothermal radiator.

Figure 3 shows good agreement between the computational results for the old cavity test at two different temperatures (using the silicon targets) and measurements. This agreement came, in part, after carefully modeling the "as-built" geometry of the cavity, the uneven temperature profile across the radiator, and the actual room temperature radiator emissivity.

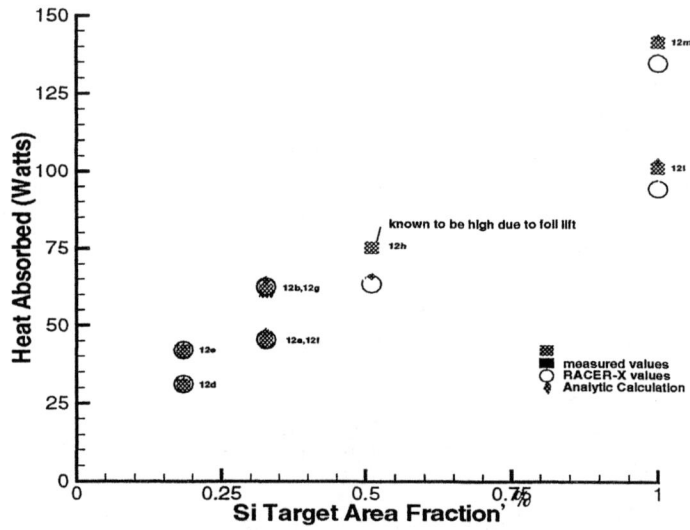

Figure 3. Comparison of Silicon experimental data with predictions.

Even though the computer model was carefully constructed, final adjustments to the results were made by varying the reflectivity of the gold foil used to shield the heat sink from the previously modelled 98% to 95%; both of which are reasonable for this material. As this data suggests, small variations in the reflectivities of in-cavity materials can significantly change the predicted results. Accurate verification of the gold foil's reflectivity after testing is needed but currently not possible. This is due to the complexity of the measurement. The gold foil surface is largely specular, however, a purely specular reflectivity measurement cannot be made due to the wrinkled condition of the foil after testing. A measurement of this type would only give an idea as to what the material's reflectivity is. Thus, a better shielding procedure or explicit knowledge of the shielding material is required to ensure more accurate predictions.

However, it should also be noted that accurate modelling of the radiator's temperature profile is difficult at best. Unmodelled variations in the profile and / or inaccuracies in the measurement of radiator temperature or other factors could also lead to similar agreement between measurements and predictions. Other noteworthy spectral effects that cannot presently be measured include: (1) radiator spectral emissivity at temperature, and (2) the effect of directly depositing filter material over grid lines, and (3) fabrication induced anomalies. These effects may have a significant impact on the model's predictions. However, they are extremely difficult, if not impossible, to quantify. These effects are also not explicitly accounted for in the computer models.

Many of the lessons learned mentioned seem to be apparent observations to the skilled experimentalist. Small geometries and / or other minor effects are often dismissed as being trivial and having no significant impact on the measured or predicted results. However, careful consideration should be given to them, as they may cause greater than expected deviations.

To summarize the findings of this testing effort, the careful evaluation of the test fixture and computer model sensitivity studies revealed the following lessons learned:

- Cavity geometry effects must be understood and modeled appropriately
- The radiator's temperature, profile, and emissivity must be explicitly known
- Reflectivity characteristics of all cavity components must be explicitly modeled
- Angle of incidence effects have to be modeled accurately
- Minor fabrication flaws must be accounted for as they can impact measurements and predictions (i.e., take nothing for granted)

It is also worthy to note that a 1-D type of analysis could not perform most of the geometric sensitivity studies conducted in this experimental evaluation. A 3-D analysis technique such as RACER-X, however, can handle them easily.

IMPROVED EXPERIMENTAL SETUP

Based upon previous testing experience, the agreement achieved in this simplified testing effort and RACER-X sensitivity analysis results, the test fixture used in this experiment was redesigned. It now contains a water cooled heater cavity and a diode / module heat sink and shield (Figure 4a). The water cooled heater cavity was designed to provide a blackbody photon source.
The heater cavity is a water cooled aluminum block with a 63.5 mm^2 window machined

in its bottom. This window acts as a view port for a 60.325 mm^2 poco graphite radiator. The radiator is 31.75 mm thick and, within the cavity, rests atop 12.70 mm thick alumina insulation. The insulation also has a window machined in it and is thermally decoupled from the aluminum cavity via ceramic spacers. Ceramic locator pins allow for the central location of the radiator between the two windows. Alumina insulation is attached to the cavity's lid along with two electrically resistive heaters. The heaters are used to heat the poco graphite radiator, and can be independently controlled to reduce or induce thermal gradients at the radiator surface.

Sixteen - 1.02 mm diameter thermocouple holes are located approximately 0.508 mm from the surface (four on a side) and are at various depths. This type of thermocouple arrangement enables a mapping of the radiator's thermal profile which can be provided as input into the photon transport code RACER-X. Using this fixture, the diode heat sink can be located as close as 0.127 mm away from the radiator surface, thus ensuring a 1:1 view factor.

Radiator temperature is typically measured with C and K type 1.02 mm diameter thermocouples. Occasionally, to calibrate these thermocouples, a calibrated high-temperature single-color sapphire pyrometer is used. To accommodate the sapphire pyrometer in the radiator, one of the thermocouple holes was bored out to 1.65 mm in diameter. Since all pyrometers suffer from measurement inaccuracies due to emissivity uncertainties at temperature, the thermocouple hole into which the pyrometer was inserted was made to emulate a blackbody cavity. As detailed in reference 3, a cavities emissivity can be altered by varying its length to diameter ratio. For this experiment the ratio of these two values was approximately 8. The pyrometer highlighted a delta of 40F and 20F between the C and K type thermocouples, respectively.

The fixture's array / diode heat sink and shield are two separate water cooled structures (Figure 4b). These two fixtures were separated to reduce measurement uncertainties associated with the posited change in the optical properties of the shielding material after application to the array heat sink. The diode / array heat sink consists of a round water cooled copper block that has a square copper pedestal extending from its top surface. The pedestal is 12 cm long and has three evenly spaced (placed 5 cm apart) 1.02 mm diameter thermocouple holes along its center. Thermocouples of equivalent diameter were soldered into the holes to improve thermal contact between them. The pedestal's cross sectional area was dependent upon the cross sectional area of the target. The cold side was mounted onto a three axis positioner to accurately position the target beneath the thermal shield.

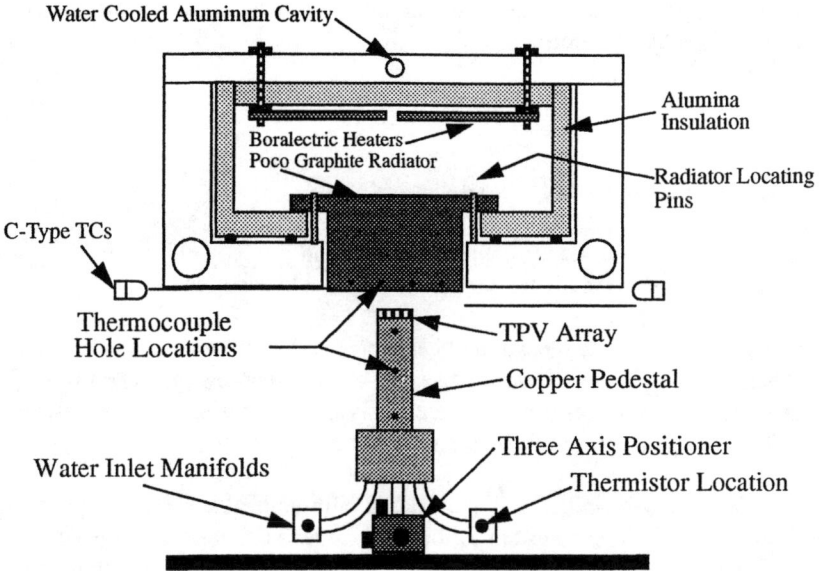

Figure 4a. Schematic of the redesigned cavity test fixture (thermal shield not shown).

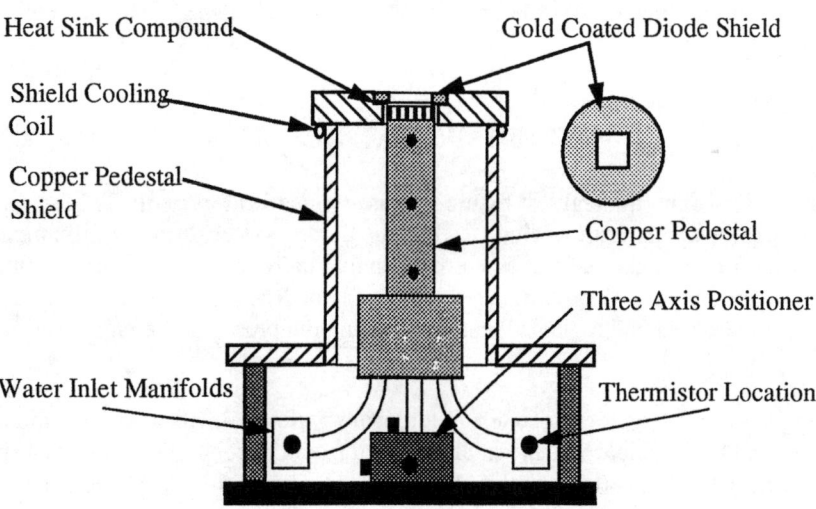

Figure 4b. Schematic of the diode / target thermal shield.

This unique fixture allows for two independent thermal measurements. Since the structure is water cooled and testing is conducted under steady state conditions, the diode or target's absorbed heat can be measured by monitoring the temperature rise and coolant flow rate illustrated by equation 2.

$$P_{abs} = Q = KA\frac{\Delta T}{\Delta x} \qquad (2)$$

where,
  K = the thermal conductivity of the copper pedestal,
  A = the cross sectional area of the copper pedestal,
  ΔT = the temperature rise of the copper pedestal, and
  Δx = the distance between the temperature difference measurements.

This technique has proven to be extremely effective and accurately measures the absorbed heat of small cross sectional areas. Un-shuttered in-cavity single diode TPV system efficiency measurements can now be easily made. These types of measurements were previously extremely complicated due to cell cooling issues.

Another benefit of this technique is that the diode or module attached to the pedestal can also be used to perform in situ system calibrations. Due to the need for large temperature rise measurements, this type of calibration technique was not possible with the old test fixture. The copper pedestal technique also significantly enhances absorbed heat absorbtion measuring capabilities. Whereas, based upon calibration data, the accuracy of this measurement was limited to ~±5%, the improved copper pedestal technique renders accuracies of ~±1%.

EXPERIMENTAL RESULTS

The redesigned experimental test fixture was used to perform in-cavity TPV system efficiency measurements. Prior to diode testing, the fixture was evaluated for illumination uniformity. Three diodes were characterized, individually wired, and mounted onto a special rotating fixture ~2.54 mm beneath the radiator. The short circuit current of each diode was measured and indicated that the illumination profile of the radiator was uniform.

Scoping tests with the redesigned test fixture were performed with a two part diode / module shield. The shield was thermally isolated from the copper pedestal via a small perimeter gap that was ~0.356 mm wide on average (Figure 5). A RACER-X analysis was conducted of this configuration and highlighted the fact that a significant amount of heat strikes the edges of the target and copper pedestal; manifesting itself as parasitic absorbtion. Figure 6 shows the predicted sensitivity of the heat absorption to the gap spacing around the target. Notice that small increases in the gap, substantially increase the heat absorbed. This situation was corrected by redesigning the shield as a single unit slightly smaller than the target.

Following these results, a single TPV diode efficiency test was conducted. Typically test temperatures ranged from ~870C-1200C. A poor-performing Indium Gallium Arsenide (InGaAs) 0.55eV diode with a direct deposited interference filter was used. A comparison of the experimental results and predictions are presented in Table 1. The data in this Table show good agreement with both RACER-X and TPVCalc predictions.

Table 1: Comparison of Single Diode Measured Data With Predictions.

	Measured Results	RACER-X Results	Difference (%)	TPVCalc Results	Difference (%)
Absorbed Heat Flux, Watts/cm^2	5.90	6.0	-2.3	5.28	10.5
V_{oc}, V	0.296	N/A	N/A	0.283	4.4
I_{sc}, A	2.191	2.41	-9.9	2.228	-4.0
ff, %	57.38	N/A	N/A	60.2	-4.9
P_{out}, W	0.372	N/A	N/A	0.389	-4.5
η, %	6.30	N/A	N/A	7.36	-16.8

The diode's absorbed heat and electrical properties (i.e., I_{sc}, P_{out}, etc.) agree to within ~10%. However, the TPVCalc predicted efficiency is ~16% higher than the measured value. This is due to the under prediction of absorbed heat and the over prediction of short circuit current. It should be noted that both codes over predict the diode's current. Typically, modelling predictions are lower than measurements.

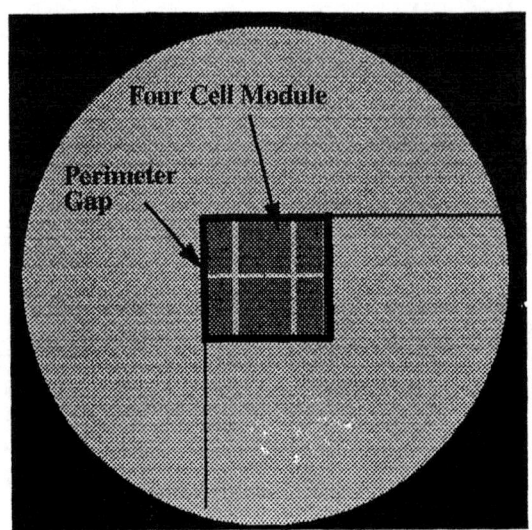

Figure 5. Computer model of the top view of the two part shield design.

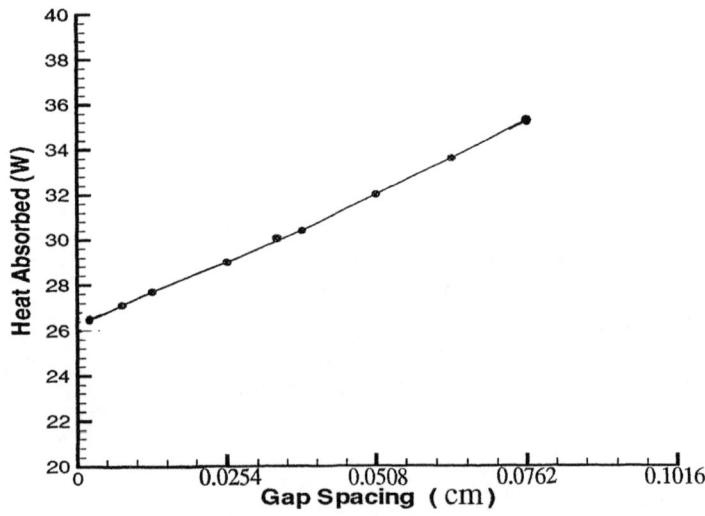

Figure 6. RACER-X perimeter gap sensitivity study results.

A possible cause for this discrepancy may be related to the size of the shield. The redesigned one piece shield has a ~1cm^2 opening machined in it (Figure 4b). The diode tested in this experiment was ~1.08 cm^2. Therefore, accurate positioning of the diode beneath the shield was critical. The diode was positioned beneath the shield via a three axis positioner (Figure 4b).

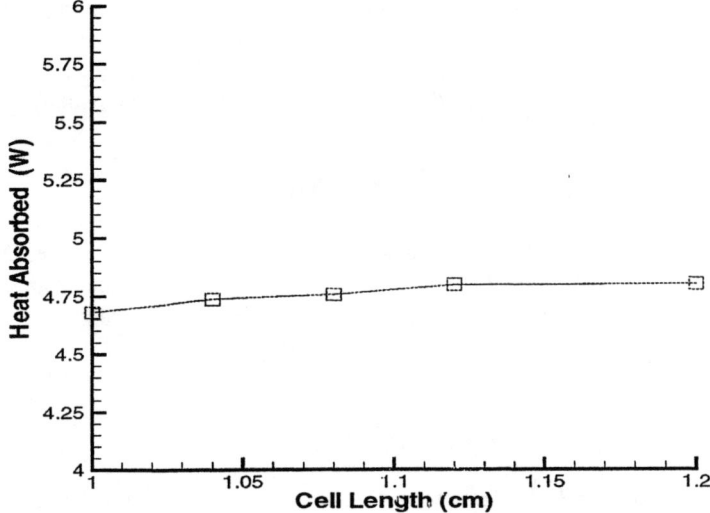

Figure 7. Heat absorbed as a function of target size.

However, accurate positioning of the target beneath the shield relies heavily on the operator. Therefore, it is possible that the active area was slightly shaded by the shield causing a reduction in current, and an increase in heat absorbtion. After analysis of these results, a RACER-X sensitivity study was performed to determine the effects of reducing the size of the thermal shield (Figure 7). This study showed that having a shield smaller than the active area causes a negligible rise in absorbed heat.

CONCLUSIONS

In conclusion, extensive research experience has shown that obtaining predictable in-cavity system efficiency measurements are not intuitive. Even though TPV efficiency is greatly dependent upon first order effects such as diode and spectral control quality, many other engineering factors can significantly impact expected results. This research effort has shown that test fixture design plays a key role in obtaining predictable results. The geometry of the fixture (i.e., cavity walls and gaps) may create an undesirable thermal profile within the cavity, thus directly impacting predictive capabilities and the general performance of the system.

Other analysis has shown that the spectral properties of all in-cavity materials can also significantly impact efficiency. Spectral properties, as difficult as they are to measure, must be accurately known and modeled as they can account for a significant portion of parasitic heat absorbtion. Also, small gaps, thin busbars and other macroscopic features can also impact efficiency. All of the factors mentioned need to be explicitly understood if predictable in-cavity TPV system efficiency measurements are to be made.

ACKNOWLEDGEMENTS

We would like to acknowledge the insight and assistance of Dr. Greg Charache, Dr. Paul Baldasaro, Dr. Josef Parrington and Michael Postlethwait for testing support and analysis. We would also like to thank Dr. Lee Danielson, Steve Berger, Joe Nehrbauer and Nick Brooks for their valuable insight, suggestions and help.

REFERENCES

1. Gethers, C.K., Ballinger, C.T., Postlethwait, M.A., DePoy, D.M., and Baldasaro, P.B., "TPV Efficiency Predictions and Measurements for a Closed Cavity Geometry", The Third NREL Thermophotovoltaic Conference, AIP Conference Proceeding, 471, 1997.

2. Charache, G. W. et. al., "Measurement of Conversion Efficiency of Thermophotovoltaic Devices", Proc. of the 2nd NREL Conference on TPV Generation of Electricity, 351, 1995.

3. Modest, M. F., <u>Radiative Heat Transfer</u>, McGraw-Hill, New York, 1993.

SESSION 6:
SYSTEMS DESIGN AND EXPERIENCE

Design Of A Thermophotovoltaic Battery Substitute

Edward F. Doyle, Frederick E. Becker, and Kailash C. Shukla

Thermo Power Corporation, Tecogen Division
45 First Avenue, P.O. Box 9046
Waltham, Massachusetts 02454-9046

Lewis M. Fraas

JX Crystals Inc.
1105 12th Avenue NW, Suite A2
Issaquah, WA 98027-8994

Abstract. Many military platforms that currently use the BA-5590 primary battery or the BB-390A/U rechargeable battery are limited in performance by low storage capacity and long recharge times. Thermo Power Corporation, with team members JX Crystals and Essential Research Inc. is developing an advanced thermophotovoltaic (TPV) battery substitute that will provide higher storage capacity, lower weight, and instantaneous recharging (by refueling). The TPV battery substitute incorporates several advanced design features including: an evacuated and sealed enclosure for the emitter and PV cells to minimize unwanted convection heat transfer from the emitter to PV cells; selective tungsten emitter with a well matched gallium antimonide PV cell receiver; optical filter to recycle nonconvertible radiant energy; and a silicon carbide thermal recuperator to recover thermal energy from exhaust gases.

INTRODUCTION

The Army purchases large quantities of batteries to meet the needs of the Dismounted Soldier for both military missions and for training. For actual missions, the lower weight and higher stored energy of primary (non-rechargeable) batteries make them the clear choice. For training, rechargeable batteries are being used more often because they can be recharged hundreds of times thereby making their life-cycle costs much lower. The higher weight and lower stored energy of rechargeable batteries requires more batteries and a higher weight for the soldier to train for the same mission. The time to recharge the batteries is also an issue.

At a Thermophotovoltaic (TPV) workshop conducted for the Army Research Office in July 1996 (reference 1), several potential applications of TPV for the Army were identified. One of these was as a substitute for batteries such as the BA-5590. This battery and its rechargeable alternative, the BB-390A/U, are useable in many military platforms, and represent a significant fraction of the cost for the batteries purchased by the Army.

The overall goal of the program described in this paper is to develop a TPV power source for the Dismounted Soldier which provides:
- higher stored energy than electrochemical batteries
- environmental benefits compared to batteries related to disposal

Specifically, this project aims to develop a propane-fueled TPV power source suitable for replacement of:
- BA-5590 primary batteries used in military missions
- BB-390A/U rechargeable batteries used in military training

The performance characteristics of these batteries are presented in Table 1. While these batteries are identical in size and configuration, the weight of the rechargeable BB-390A/U battery is 65% higher than the primary BA-5590 battery. The capacity of the rechargeable BB-390A/U is also only 47% of the BA-5590 primary battery. It is basically because of these advantages that primary batteries are clearly preferable for military missions.

Specifications	BA-5590 Battery	BB-390 A/U Battery
Dimensions (l×w×h), in	4.4×2.45×5.0	Same
Volume, in^3	53.9	Same
Weight, lb	2.33	3.85
Max. Current, amps	4/2	7.2/3.6
Nominal Voltage, volts	12/24	Same
Max. Power, watts	48	86
Storage Capacity, w-h	163	77

TABLE 1. Battery Performance Specifications

DESCRIPTION OF TPV BATTERY SUBSTITUTE SYSTEM

The TPV battery substitute system under development will be comprised of a 20 watt TPV power source used in combination with a BB-390A/U rechargeable battery. It will use replaceable propane fuel cartridges to provide an immediate recharging capability. This combination of a TPV power source with a rechargeable battery has the capability to provide both high instantaneous power and high stored energy and to substitute for BA-5590 and BB-390A/U batteries in many military missions and for military training.

The TPV battery substitute system will have the capability to provide:

- 106 watts maximum power output
- 20 watts recharge rate

The electric stored energy capacity of the TPV battery substitute system energy is time dependent since the stored energy for the TPV power source is in the form of fuel energy and must be converted to electric energy by the TPV power source. The time dependent capacity of the TPV battery substitute system is presented in Figure 1.

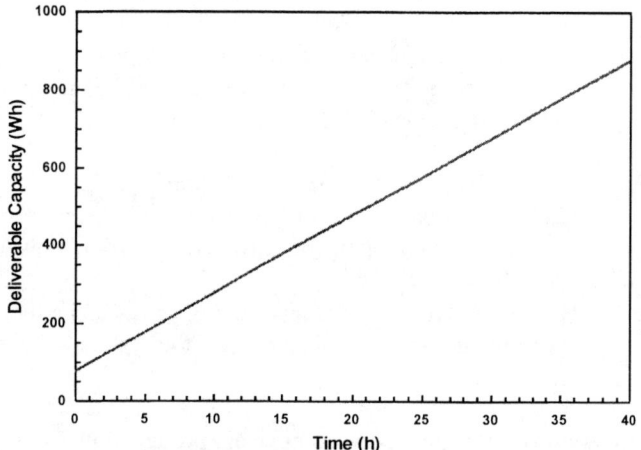

FIGURE 1. TPV Battery Substitute System Capacity

In designing the TPV battery substitute system, the following constraints have been placed on the design:

- **Environmental**
  - can be exposed to and submerged under water
  - combustion and cooling air may have contaminants that can damage the emitter, optical filter, and PV cells
- **Orientation**
  - must be capable of operation in any orientation

The approach taken to meet these design constraints is to use a hermetically sealed power conversion module. With this approach, the emitting surface, optical filter and PV cells are contained within a hermetically sealed enclosure to prevent exposure to water and ambient air. This approach excludes the use of flow-through emitters where combustion gases directly contact the emitting surface. This design approach is also suitable for future conversion to liquid fuels that may have contaminants potentially damaging to components.

The key features of the TPV power source design are:

- **Hermetically Sealed Power Conversion Module**
  - allows enclosed space to be evacuated or charged with inert gas, thereby reducing thermal convection losses
  - provides close packaging for high view factor
  - permits use of tungsten, which would oxidize in air, as the wavelength-selective (for gallium antimonide cells) emitting surface
- **Fuel Pressure Powered Combustion System**
  - fuel pressure used in a jet pump to inspirate combustion air, thereby eliminating the need for a combustion blower
- **Orientation Independence**
  - Forced convection cooling of the PV array and a fuel pressure powered combustion system allow the system to be orientation independent

The design specifications for the TPV power source are presented in Table 2.

PV Cell Type	GaSb
Emitter Type	Tungsten
Emitter Temperature	1700 K
In-band PV Cell Efficiency	30%
Off-band Optical Recovery	75%
Thermal Recuperator Effectiveness	70%
Air Preheat Temperature	1350 K
Power Output	20 W, Net
System Efficiency	7%, Net

TABLE 2. Design Specifications

The burner-emitter-recuperator (BER) assembly for the TPV power source uses a single ended coaxial radiant tube/burner/recuperator configuration. It will be made of silicon carbide with a tungsten layer as the emitting surface. It will be fabricated using techniques developed at Thermo Power and our parent company, Thermo Electron, for thermionic diodes (reference 2). These include chemical vapor deposition (CVD) and chemical vapor composite deposition (CVC). With the hermetically sealed and evacuated power conversion module, a highly effective thermal insulation, Thermo Electron MULTI-FOIL, can be used to minimize radiant heat losses. Convective heat losses are eliminated with a sealed and evacuated chamber separating the emitter from the PV cells.

The gallium antimonide (GaSb) PV array will be provided by JX Crystals. With the tungsten emitter, it will have a 30% in-band conversion efficiency. The array will use 40 GaSb cells, connected in series, to produce 26 watts of gross electric power at 12 volts. The PV array and heat sink will be configured as an eight-sided octagon with its axis concentric to the emitter. The array will be comprised of eight individual circuits with five cells mounted and connected on each circuit. A cover glass, with an integral dielectric filter for spectral control, is bonded to each five cell circuit. A gold reflecting surface on one edge of the cover glass reflects the radiant energy falling in the space between the circuits back to the emitter.

The power conversion module, which is comprised of the BER, PV array, and air-cooled heat sink is illustrated schematically in Figure 2. The direction of flow of the combustion air and exhaust gas is shown on the figure. The way in which the module is hermetically sealed is also illustrated. The hermetically sealed module has two basic subassemblies; the heat sink with the array, and the BER. These two subassemblies will be joined together by a flanged connection and sealed with a metallic o-ring for the initial prototypes. Ultimately this would be a welded seal.

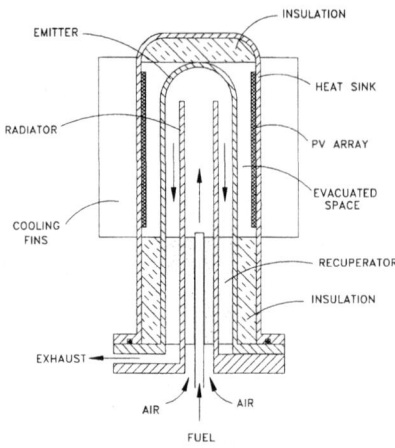

FIGURE 2. Power Conversion Module Schematic

Two package configurations are being developed. These are a rectangular package and a cylindrical package. Except for the cooling fins on the heat sink, the components used in each package configuration are identical.

A schematic for the rectangular package is shown in Figure 3. With this configuration, cross-flow fins are used on the heat sink. This allows the cooling fan and motor to be located to the side of the power conversion module with the fuel controls and piezoelectric igniter located below the fan and motor. This is the more compact packaging approach. This approach, however, has more difficult cooling air flow distribution problems to solve, through the use of air baffles, and could result in greater non-uniformity with respect to the temperature of individual PV cells.

Figure 4 shows the schematic for the cylindrical package. Axial-flow fins are used on the heat sink with this configuration. This allows the cooling fan and motor to be located directly above and on the same axis as the power conversion module. The fuel controls and igniter are located below the power conversion module. This approach produces a somewhat less compact package, but results in very uniform cooling air flow distribution and greater uniformity with respect to the temperature of individual PV cells.

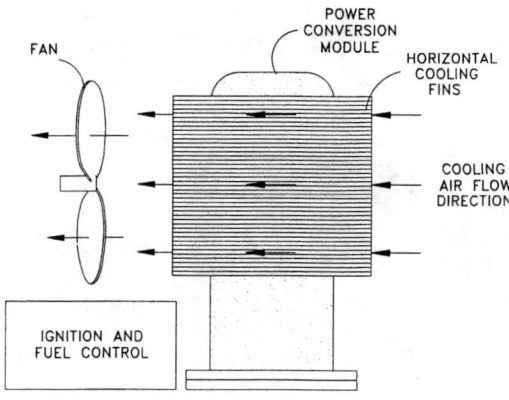

FIGURE 3. Rectangular Package Schematic

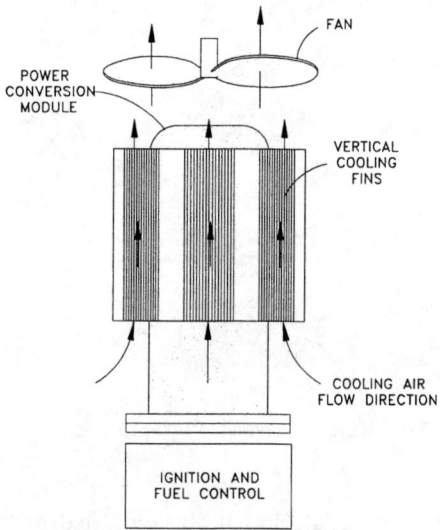

FIGURE 4. Cylindrical Package Schematic

An exploded view of the rectangular package design is shown in Figure 5, and an exploded view of the cylindrical package design is shown in Figure 6. The various components of the TPV power source are labelled on each of the drawing.

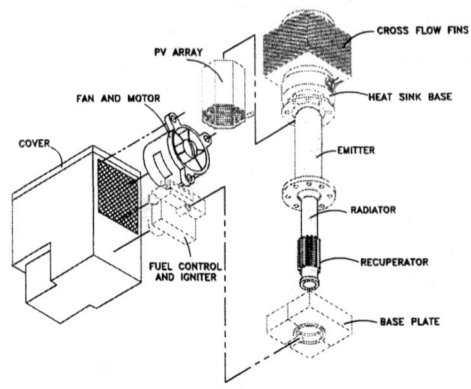

FIGURE 5. Rectangular Package Design

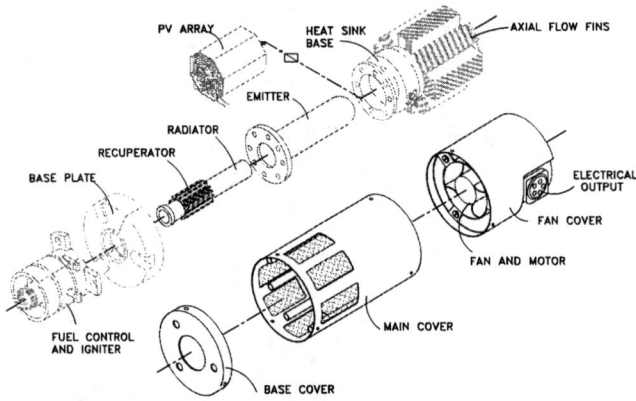

FIGURE 6 - Cylindrical Package Design

The design of the TPV power module has been completed and the procurement and fabrication of some of the key components is underway. Figure 7 shown the first set of heat sink components for the power conversion module. These are the copper heat sink base and the stainless steel heat sink base and cover. The first set of eight PV array circuits has been fabricated and is shown Figure 8. The five GaSb cells on each circuit are hidden by the cover glass. The cover glass with its integral filter and gold reflective edge is visible on each circuit assembly. A picture of the power converter assembly is presented in Figure 9.

FIGURE 7. Power Converter Components

FIGURE 8. PV Array Circuits

FIGURE 9. Power Converter Assembly

PERFORMANCE CHARACTERISTICS

The projected performance characteristics for the TPV battery substitute system are presented in Figures 10 and 11. Figure 10 shows the system weight as a function of capacity for the TPV battery substitute system, for BA-5590 primary batteries and for BB-390A/U rechargeable batteries. The number of each type of battery needed to obtain a given capacity are shown on the figure. For military training or military missions using rechargeable BB-390A/U batteries, the TPV battery substitute system provides a lower weight system if more than one rechargeable battery is required. For military missions using BA-5590 primary batteries, the TPV battery substitute system provides a lower weight system if more than three primary batteries are needed to perform the mission. Figure 11 shows the system volume as a function of capacity for the TPV battery substitute system, for BA-5590 primary batteries and for BB-390A/U rechargeable batteries. The conclusions with respect to system volume are the same as for system weight.

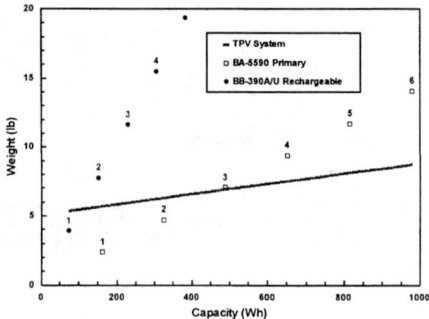

FIGURE 10. Weight Comparison

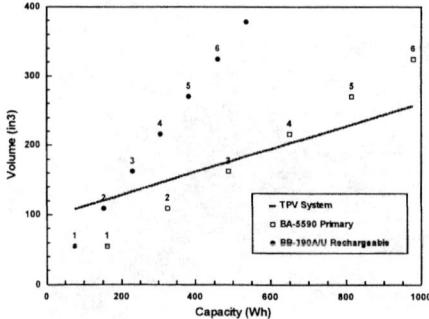

FIGURE 11. Volume Comparison

CONCLUSIONS

A TPV battery substitute system is being developed which is comprised of a 20 watt TPV power source used in combination with a BB-390A/U rechargeable battery. It will use replaceable propane fuel cartridges to provide an immediate recharging capability. This combination of a TPV power source with a rechargeable battery has the capability to provide both high instantaneous power and high stored energy and to substitute for BA-5590 and BB-390A/U batteries in many military missions. The projected performance characteristics indicate that the TPV battery substitute system will be lower in weight and volume for military training or military missions requiring more than one BB-390A/U rechargeable battery or more than three BA-5590 primary batteries.

ACKNOWLEDGEMENTS

This work has been supported by the U.S. Department of Defense through DARPA and NASA-Lewis. The authors acknowledge the guidance provided by Dr. Robert Rosenfeld of DARPA and Mr. David Wilt of NASA-Lewis.

REFERENCES

1. Rose, M. Frank, editor, "Prospector VIII: Thermophotovoltaics -- An Update on DOD, Academic, and Commercial Research", sponsored by Space Power Institute, Auburn University, Alabama, and Army Research Office Research, Triangle Park, North Carolina July 14-17, 1996, pp. 1-6.

2. Reagan, P., Lieb, D., Miskolczy, G., Goodale D., and Huffman, F., "CVD Fabrication of Thermionic Converter and Heat Pipe," Ceram. Eng. Sci. Proc., 4, 520, 1983.

Component Development for 500 Watt Diesel Fueled Portable Thermophotovoltaic (TPV) Power Supply

Authors: Crispin L. DeBellis, Mark V. Scotto*,
Lewis Fraas, John Samaras[†], Ron C. Watson,
Stephen W. Scoles*

McDermott Technology, Inc., Alliance, Ohio
[†]*JX Crystals, Issaquah, Washington*
BWX Technologies Inc., Lynchburg, Virginia

Abstract. McDermott Technology, Inc. (MTI) and JX Crystals have developed an innovative design for a compact, 500-watt, thermophotovoltaic (TPV) power supply using diesel fuel. Under a contract with the Defense Advanced Research Projects Agency (DARPA) and managed by the U.S. Army Communications-Electronics Command (CECOM), this design is being reduced to hardware. Prototypes of the two main subsystems, the power converter assembly (PCA) and the burner / emitter / recuperator (BER), have been designed, fabricated, and tested. The PCA uses low-band-gap gallium antimonide (GaSb) photovoltaic (PV) cells for high efficiency and power density. The prototype PCA will be air cooled for system simplicity and portability. However, initial testing was performed on a water-cooled PCA. The BER uses a thermal vaporizer to produce a stable, high-intensity, low-emissions combustion zone inside an impervious emitter. A thermally integrated recuperator is utilized to boost system efficiency by transferring the unused energy in the exhaust stream to the incoming fuel and combustion air. This paper describes the design, testing and performance of the first-generation PCA and BER along with model predictions used for design and evaluation.

INTRODUCTION

Thermo means heat or energy. *Photo* means light or electromagnetic radiation. *Voltaic* means voltage or electricity. Thermophotovoltaics (or TPV), therefore, is the conversion of energy to radiation to electricity. The heart of a TPV system is the photovoltaic (PV) cell. PV cells convert electromagnetic radiation produced by a heat source — like the sun or a flame — directly into DC power without moving parts.

The components of a portable TPV power system are simple, as shown in Figure 1. The core of the system is the heat source, which for this system is a diesel-fired burner. The purpose of the burner is to convert the chemical energy in the fuel into heat and transfer that heat to the emitter. The emitter is a high-temperature ceramic material that is heated to 1600 to 1700 K (2420 to 2600°F). Because of its temperature, the emitter emits electromagnetic radiation in the infrared region. The radiation is focused on the PV cells. Most systems include a filter, because the PV cells only convert a narrow band of the infrared spectrum into power. The filter reflects the unusable part of the spectrum back into the system to increase efficiency. The PV cells convert the radiation into DC power. The portion of the energy that is not converted to power must be removed as waste heat. A recuperator also boosts system efficiency by transferring the waste heat in the combustion exhaust stream to the incoming combustion air. Finally, an air / fuel delivery system is required; this may include air fans, a fuel pump, and controls.

TPV power systems have many advantages when compared with existing technologies. In the 500-watt range, TPV systems have the highest power conversion efficiency of any static device that uses diesel fuel directly. The expected conversion efficiency of the system described in this paper is 8% to 10%. In the long term, systems are being designed with efficiencies of 15% to 20%. There are few moving parts in the system, significantly increasing the mean time between equipment failures and simplifying repairs. Steady-state combustion will produce significantly less noise than an internal combustion engine. With proper design, the noise level should be inaudible at close distances. Because the system is designed for diesel fuel, it can easily burn other liquid or gaseous fuels.

As a portable power system, TPV can be used for battery charging or powering computers and communications equipment in military or remote applications. In cogeneration applications, TPV can be used in self-powered appliances such as furnaces and water heaters. TPV can also be combined with solar energy to produce a hybrid system that produces power regardless of the time or weather. In the long term, TPV may become a major power producer as a distributed power source. Distributed power refers to the move away from central power stations to local and even on-site power production in homes and businesses.

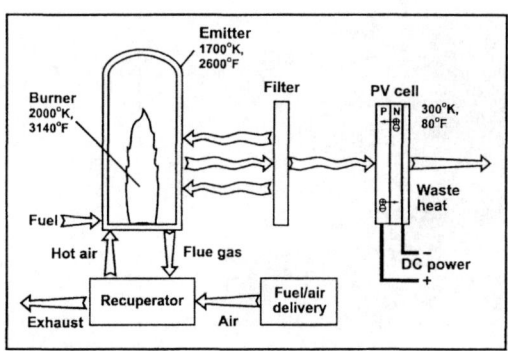

FIGURE 1. TPV System Components.

SYSTEM DESIGN

The design of the 500-watt portable TPV power system is shown in Figure 2 and is described by DeBellis (1). This arrangement utilizes a cylindrical power converter assembly (PCA) surrounding the burner / emitter / recuperator (BER). A horizontal air fan and cylindrical airflow ducting for cooling the PV cells complete the basic TPV portion of the system. Balance-of-plant components (not shown) include housing and ducting, air fans, fuel tank, fuel pump, lines and valves, control system, and power conditioner if required.

A cylindrical geometry was selected for the PCA and the BER. The PCA is composed of multiple interlocking circuits with heat transfer fins on the outside and the photocells on the inside. Each circuit contains a string of GaSb PV cells. The PV cells are covered with an integral dielectric filter, which is deposited directly on the cells. The filter reflects out-of-band radiation back into the system, reducing the cooling load on the PV array and increasing system efficiency. An axial flow fan at the top of the PCA provides cooling for the PV cells. At the bottom of the PCA, hot exhaust from the BER is vented to mix with the cooling air to produce a low-temperature exhaust.

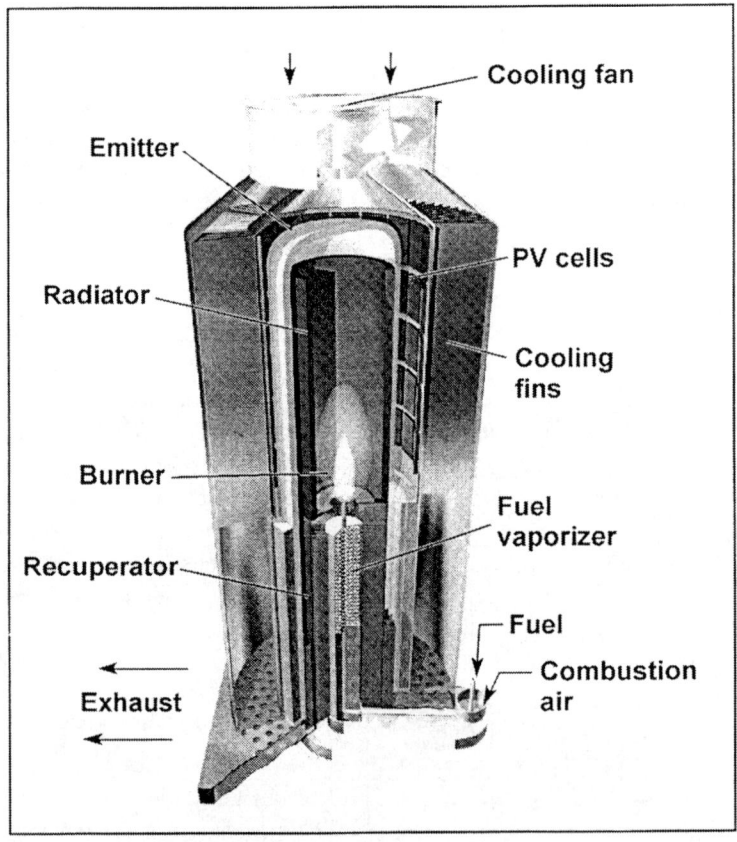

FIGURE 2. Schematic of portable TPV generator.

The BER incorporates a fuel feed system and vaporizer in the center of an annular recuperator located in the bottom half of the unit. The upper half of the unit contains the combustion zone with an internal radiator and external emitter. The radiator contains the combustion flame and channels the hot flue gas to the top of the emitter. The flue gas then flows down the annulus between the radiator and emitter and exhausts into the recuperator. This geometry is simple and provides a means to control the emitter temperature profile by way of radiation from the radiator and convection from the flue gas. The emitter is impervious to gas flow and can be either a gray-body emitter like SiC or a matched emitter as described in Ferguson (2). A matched emitter is being developed for the system; however, a SiC emitter was used for the work described in this paper.

PCA DEVELOPMENT

For development purposes, a water-cooled PCA was used for first-generation testing. The fabrication of the PCA was a joint effort by JX Crystals and Western Washington University. These two organizations had previously developed the PCA with funding from the Army Research Office under a separate program. The cross section of the PCA is shown in Figure 3. The PCA contains 20, 19-cell circuits. Mirrors between the circuits are used to cover the cell interconnects and reflect radiant energy back into the system. Details on the design can be found in Fraas (3). The PCA was partially populated with three electrically live circuits. The remaining 17 circuits had the same cells and filters but no interconnects. Therefore, the thermal and optical performance of the PCA should be the same as a fully populated one. The GaSb cells are covered with an integral dielectric filter, which is deposited directly on the cells.

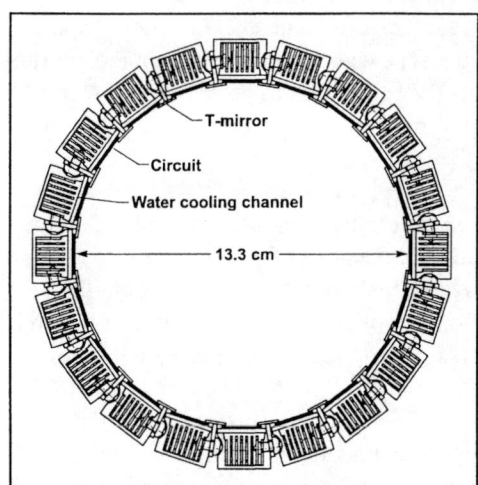

FIGURE 3. Cross section of water-cooled PCA.

PCA testing was performed on an electrically heated glowbar test station, also described in Fraas (3) for a SiC emitter with a diameter of 7 cm (2.75 inches). An 11.4 cm (4.5-inch)-diameter quartz shield was placed between the emitter and the PCA. The cell plane diameter was 13.3 cm (5.24 inches). A data acquisition system was used to record emitter temperature, circuit current / voltage characteristics, and energy absorbed by the PCA water. Testing was performed by turning on the glowbar and recording data as the system heated up over a 2 to 3-hour period.

Test results are shown in Figure 4. PCA efficiency is defined as the total circuit power divided by the heat absorbed by the water. The total power is calculated by multiplying the individual circuit power by 20 because only one circuit was measured. This assumes that the other 19 circuits will act the same. Based on fully populated cylinder data taken by JX Crystals, this is a fair assumption. At an emitter temperature of 1700 K (2600°F), which is the design point for this system, the PCA produced 665 watts of power at an efficiency of 9%. The circuit fill factor at these conditions was 70%. Measured power densities as high as 2 W/cm² active cell area were achieved.

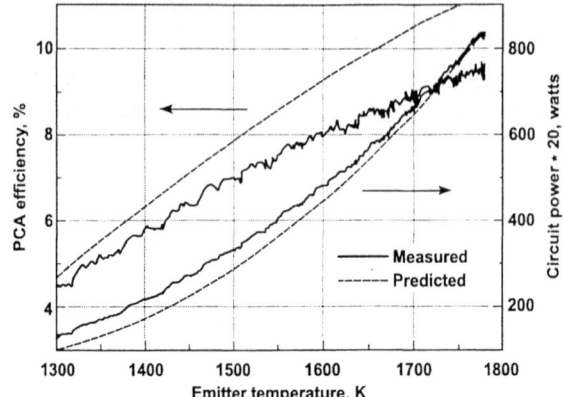

FIGURE 4. Water-cooled PCA test results.

PCA MODELING AND VALIDATION

To understand the PCA performance and develop tools to design future applications, a computer model of the PCA is being developed. The approach taken was to simplify the problem by modeling it as quasi-one dimensional. The solution of the optical part of the problem is essentially that of Schock (4). The optical performance of the TPV cavity is determined by doing a radiant energy balance (including absorption) at each surface over finely discretized wavelength (spectral) intervals. Three-dimensional radiation shape factors are used to account for the cavity geometry. The wavelength-dependent radiant flux incident on the PV cells is used in a standard PV cell model to determine the resulting power output from the PV cell array. Integration over the wavelength intervals gives overall performance.

The optical analysis provides overall radiation heat transfer input for use in the thermal model. The convection, conduction, and radiation heat transfer in the TPV cavity are determined by a thermal energy balance. The temperatures determined from the heat transfer analysis are compared to those used for the optical analysis, and if the problem is not converged, the optical and thermal analyses are repeated iteratively until convergence is achieved.

The PCA model geometry consists of concentric cylinders, including the emitter, multiple windows (or shields), and the circuit array. The circuit array includes cells with filter, grids, mirrors, and gaps between cells and mirrors. All critical dimensions are user specified. The top and bottom of the PCA are modeled with one-dimensional conduction calculations with user-supplied material definitions. The PV circuit heat sink has user-defined finned passages with air or water cooling. The spectral radiation properties of the emitter, shields, mirrors, filters, and cells can be accessed through a built-in database or alternatively supplied by the user. Figure 5 shows a typical set of data used in the model. The emitter and filter properties were measured on MTI's spectral radiation properties facility.

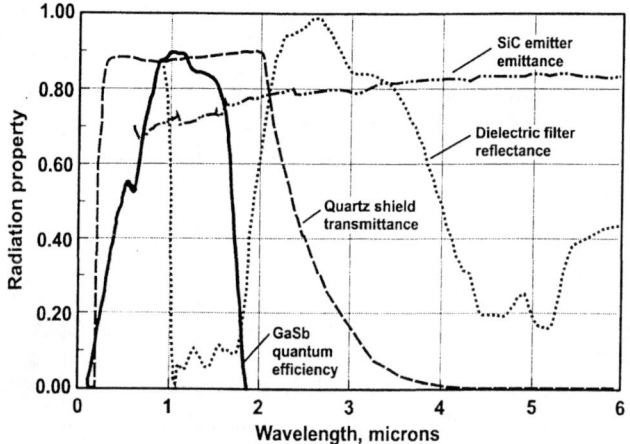

FIGURE 5. Spectral properties of PCA components.

Predicted results are shown in Figure 4 as dashed lines. The model predicts the power output quite well. However, model predictions can vary considerably, depending on how certain physical details are handled. One of these is how well emitter temperature and temperature uniformity are known. As can be seen in the figure, a small change in emitter temperature makes a large difference in power output. Another issue is associated with filters and cell grids. The filter reflectance or transmittance is usually measured off the cell. In the real system, however, the filter is on top of grids, which add to the equivalent reflectance. This effect is significant when the grids cover 20% or more of the cell. For the predictions shown, the grid coverage is accounted for in the model with an area-weighted adjustment of the filter reflectance and transmittance.

Efficiency is overpredicted by the model by 1% to 2%. There are at least two reasons for this:

- The standard conduction / convection correlations used for the air gaps between the emitter, shield, and array were developed with much smaller temperature gradients. Therefore, the model may be underpredicting this loss term.
- Cracks between cells and mirrors (along with ridges in the geometry) may act like black bodies and absorb all the incident radiation on them. Therefore, 5% crack and ridge area can significantly reduce PCA efficiency. This effect is only handled on an area-weighted basis in the model and may also be underpredicted.

BER DEVELOPMENT

A BER test stand shown in Figure 6 was built to allow for interchangeable components. The test stand is fully instrumented to characterize flow, temperature, pressure, and gas species.

The objective of the BER design was to use a simple geometry because of the high-temperature materials involved to produce a uniform emitter temperature. It is difficult to produce a uniform emitter temperature because convective heat transfer from the flue gas as it flows down the annulus between the radiator and emitter decreases rapidly. The key is to offset this with the radiant heat transfer from the radiator by making it hottest at the bottom. This is accomplished by having a very short flame. MTI's numerical heat transfer, flow, and combustion model (known as COMO) was used to design the burner. COMO results are shown in Figure 7. The

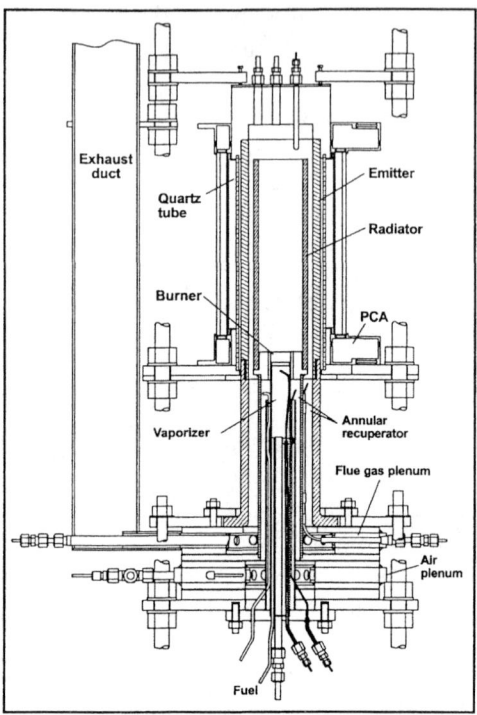

FIGURE 6. BER test stand.

right side of the figure shows gas temperature contours and streamlines. The left side shows emitter and radiator temperature profiles.

A number of fuel-feed systems were investigated. Two are currently showing promise. One uses a low-pressure atomization technique. The second uses a sonic orifice approach. Fuel is supplied by a low-pressure (<20-psig) gear pump. Both approaches produce good mixing, flame shape control, and low emissions, as shown in Figure 8. CO emission limits for small diesel fueled systems are in the 400- to 500-ppm range.

Emitter temperature and uniformity were evaluated by simulating the heat loss to the PCA. A small amount of insulation was ap-

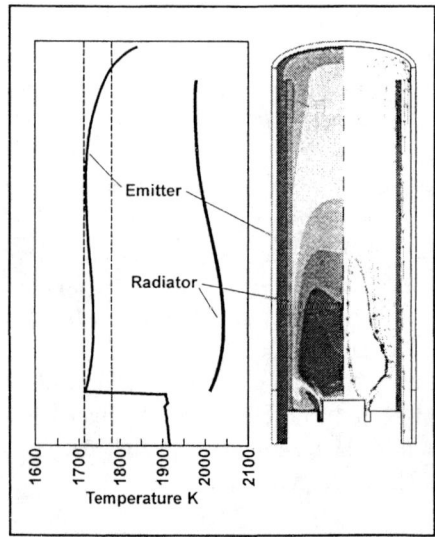

FIGURE 7. Burner design results.

plied on the outside of the emitter. The thickness was sized to give the same heat loss as the PCA. Figure 9 shows the vertical temperature profile on the emitter for different fuel flow rates. At high fire, the emitter reached an average temperature of 1600 K (2420°F), with an axial gradient of 80 K (144°F). The measured combustion efficiency of the system was approximately 70%.

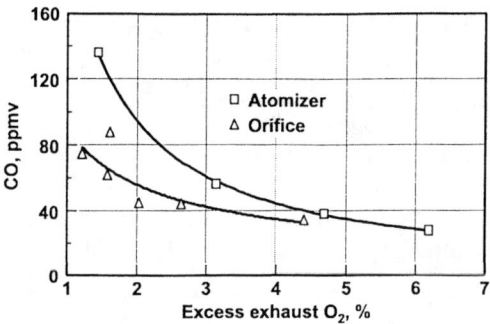

FIGURE 8. Burner CO emissions.

CONCLUSIONS

A 500-watt, portable, diesel-fueled TPV generator has been designed, and the first-generation components developed and tested. The PCA consisted of 380-GaSb cells arranged in a cylindrical array of 20 circuits and included integral dielectric filters deposited directly on the cells. The PCA demonstrated power output that exceeded the goal of 600 watts gross, with an emitter temperature of 1700 K (2600°F). Measured power densities as high as 2 W/cm^2 active area were achieved. PCA modeling successfully predicted power output but overpredicted efficiency. The BER consisted of a SiC-impervious emitter coupled with a high-intensity diesel burner and annular recuperator. Two diesel-fuel feed systems were demonstrated (atomizer and orifice) with good combustion, low emissions, and flame shape control. Emitter temperatures of 1600 K (2420°F) were achieved, with a low temperature gradient of 80 K (144°F). Combustion modeling was useful in designing the BER, and the predicted emitter temperature profile qualitatively agreed with data.

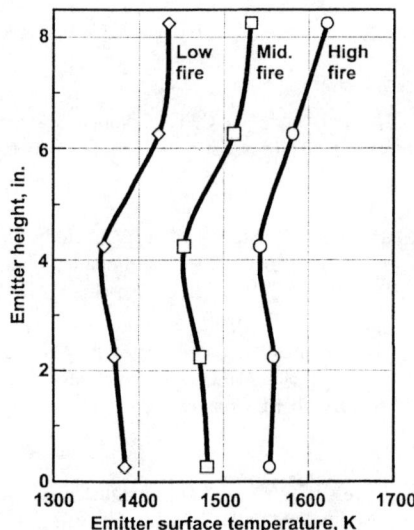

FIGURE 9. Emitter temperature profile.

FUTURE ACTIVITIES

A fully integrated breadboard system is being designed and will be fabricated and tested. Two key issues will be addressed in the breadboard system:

- The first is air cooling of the PCA. Based on measured and predicted PCA performance, the air cooling system will be at the limits of conventional designs.

- The second issue is PCA efficiency. Although measured efficiencies of 9% to 10% may be the best result to date for a PCA, higher efficiencies are needed to reduce the load on the cooling and combustion components.

Because of these factors, JX Crystals has developed a new circuit design known as the "shingle" circuit and described in Fraas (5). This circuit design will be used in the breadboard with air cooling. PCA efficiencies of 11% to 14% are expected. With combustion-side efficiency of 65% expected, breadboard system efficiency should be between 7% and 9%.

ACKNOWLEDGEMENTS

This work is funded by the Defense Advanced Research Projects Agency (DARPA) and McDermott International, Inc. The contract is managed by the U.S. Army Communications-Electronics Command (CECOM). The authors would like to acknowledge the guidance provided by Dr. Robert Nowak of DARPA and Selma Nawrocki of CECOM. The project is also indebted to JX Crystals and Western Washington University for the development of the water-cooled PCA. This work was funded by the Army Research Office under a separate program.

REFERENCES

1. DeBellis, C.L., Scotto, M.V., Scoles, S.W., Fraas, L.M.; "Conceptual Design of 500-Watt Portable Thermophotovoltaic Power Supply Using JP-8 Fuel," Colorado Springs, Colo., May 1997, Thermophotovoltaic Generation of Electricity: Third NREL Conference, AIP Conference Proceedings 401, Woodbury, N.Y.

2. Ferguson, L., Fraas, L.M.; "Matched Infrared Emitters for Use with GaSb TPV Cells", Colorado Springs, Colo., May 1997, Thermophotovoltaic Generation of Electricity: Third NREL Conference, AIP Conference Proceedings 401, Woodbury, N.Y.

3. Fraas, L., Samaras, W., Haung, H., Seal, M., West E.; "Development Status on a TPV Cylinder for Combined Heat and Electric Power for the Home," Denver, Colo., October 1998, Thermophotovoltaic Generation of Electricity: Fourth NREL Conference, AIP Conference Proceedings, Woodbury, N.Y.

4. Schock, A., Mukunda, M., Or, C., Summers, G.; "Analysis, Optimization, and Assessment of Radio-isotope Thermophotovoltaic System Design for an Illustrative Space Mission," Copper Mountain, Colo., 1994, Thermophotovoltaic Generation of Electricity: First NREL Conference, AIP Conference Proceedings 312, Woodbury, N.Y.

5. Fraas, L., Ballantyne, R., Hui, S., Ye, S.Z., Gregory, S., Keyes, J., Avery, J., Lamson, D. Daniels, B.; "Commercial GaSb Cell and Circuit Development for the Midnight Sun® TPV Stove," Denver, Colo., October 1998, Thermophotovoltaic Generation of Electricity: Fourth NREL Conference, AIP Conference Proceedings, Woodbury, N.Y.

Development Status on a TPV Cylinder for Combined Heat and Electric Power for the Home

Lewis Fraas, John Samaras, Han-Xiang Huang
Michael Seal* and Edward West*

JX Crystals Inc., Issaquah, WA 98027
*Western Washington University, Bellingham, WA 98225

Abstract. Several first-generation water-cooled TPV cylinders have been built and tested. The existing units contain 380 GaSb cells mounted on 20 circuits; the design and test results on these photovoltaic converter arrays are presented here. Tested with a 1600° C glowbar, one of these cylinders generated 990 Watts from a cell active area of 396 cm^2, which is an electric power density of 2.5 Watts per cm^2. A second-generation design is presented, using a new shingled circuit assembly. These shingled circuits allow for a slightly larger cylinder design with nearly double the cell active area. Using a SiC emitter operating at 1425° C, this second-generation cylinder should produce over 1.5 kW of power with improved efficiency.

INTRODUCTION

GaSb infrared sensitive photovoltaic cells responding to wavelengths out to 1.8 microns make it practical to generate electricity using ceramic infrared emitters operating at 1400° C. JX Crystals is now selling combined heat and electric power generating thermophotovoltaic (TPV) systems using GaSb cells for the off-grid market. In the longer term, TPV systems can be used to generate heat and electric power for homes connected to the electric grid.

TPV holds the promise of tremendous fuel-efficiency by cogenerating electricity in otherwise traditional furnaces at a reasonable cost. Heating fuel is converted to heat energy at 80% - 90% efficiency in homes, and TPV units can convert fuel to heat and electricity combined at levels at least as high. By comparison, electricity generated at a central power plant and delivered to homes by transmission and distribution lines will often have a dismally low 30% fuel to delivered-electric efficiency. These future TPV cogeneration systems for the home should be capable of generating at least 1 kW of electric power. In order to provide systems for this future home generator market, JX Crystals is developing a water-cooled TPV cylinder with a target electric power output exceeding 1 kW.

A completed TPV generator will consist of two subassemblies, the photovoltaic converter array (PCA) and the hydrocarbon burner and heat recuperator. In this paper, the focus is on the development of the PCA. First-generation PCAs have already been built and tested. First-generation TPV cylinders like the one shown below in figure 1 have been operated up to 990 Watts and have demonstrated emitter to electric conversion efficiencies as high as 9.4%. In the next two sections of this paper, this first-generation cylinder design is described and detailed test results are presented. Analysis of these test results then suggests a second-generation PCA design. This second-generation design is then described in a subsequent section along with performance projections for that design. While this second-generation cylinder is only 19% larger than the first, it has a projected power output of 1.5 kW with an emitter to PCA electric efficiency of over 12%.

FIGURE 1. First-generation water-cooled TPV PCA, tested at 990 Watts.

FIRST-GENERATION PCA

Several first-generation water-cooled TPV cylinders have already been built and tested. These cylinders are 12.1" long and contain 380 GaSb cells arrayed on the inside surface of the cylinder at a diameter of 5.3". The cells are mounted on twenty circuits with 19 cells per circuit. Each circuit has a length of 8.5". There are two 2" tall water galleries at each end of the cylinder. The twenty circuits are wired in four parallel groups with each group containing five series-connected circuits. Figure 1 (previous page) shows a photograph of one of these cylinders along with two circuits mounted on water cooling channels. A cross-section drawing through this cylinder is also shown in figure 2.

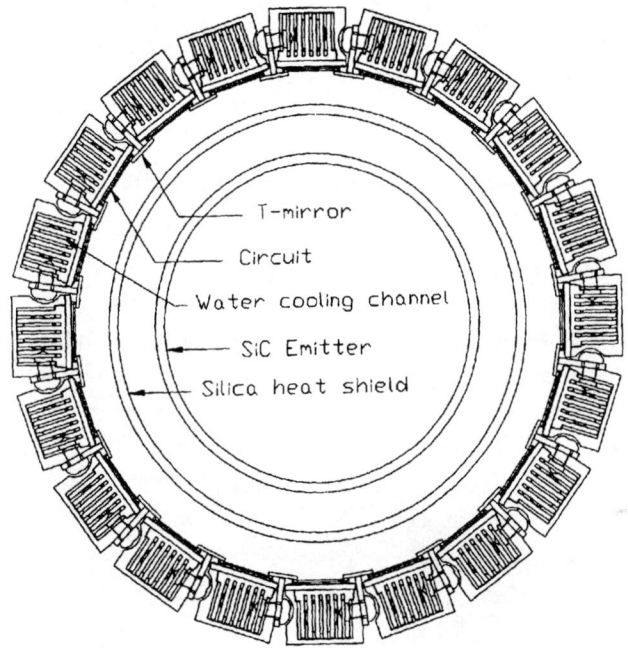

FIGURE 2. First-generation water-cooled TPV PCA cross-section drawing.

Two sample first-generation circuits are shown at the bottom of the photograph in figure 1. In this design, GaSb TPV cells are soldered down onto metal pads on a copper substrate where a thin insulating layer isolates the metal pads from the copper substrate. The GaSb cell has two buses on opposite edges for current extraction. Flexible interconnects are bonded from these pads on the top of the cell over to pads at the sides of the circuit. In the first–generation cylinder, mirrors are mounted over the lead bond areas. Since the leads have to carry large currents in TPV circuits, the lead area is fairly large and, consequently, the mirrors can cover typically 40% of the PCA circuit area.

A simple 5-layer dielectric filter covers the cells with three low index layers separated by 2 high index layers. These 5 layers are deposited in sequence directly on the cells at the wafer level prior to wafer dicing. Figure 3 shows the transmission function for this filter.

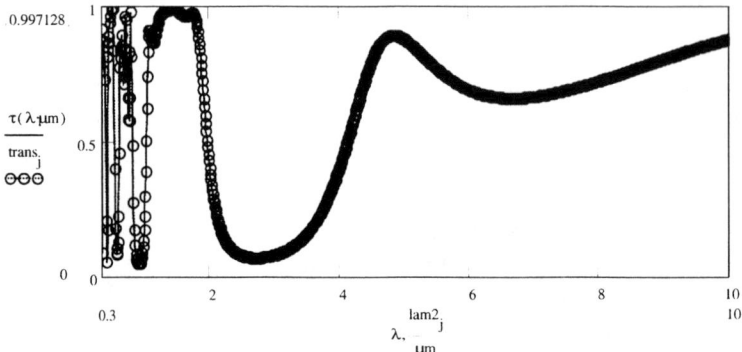

FIGURE 3. Five layer dielectric filter transmission function, deposited directly on cells.

TEST STATION AND TEST RESULTS

The first-generation TPV cylinders have been tested at both JX Crystals and at Western Washington University, using the glowbar test stations located at both of these sites. Western Washington University has a combustion-fired test facility as well. Photographs of the glowbar test system at JX Crystals are shown in figure 4.

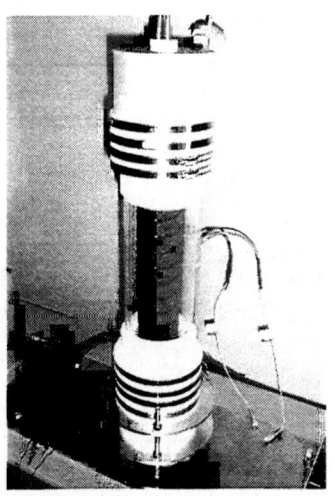

FIGURE 4. Electric glowbar test system; PCA mounts around central glowbar.

Four relevant tests have been run on the first-generation cylinders to date. The configuration tested in every case included a heated SiC emitter surrounded by the cell array with a 4.5" diameter fused silica heat shield tube located between the emitter and the cell array. In the first test, a cylinder was measured at Western Washington University using a 2.1" diameter electrically heated SiC glowbar. The cylinder produced a steady state output of 400 W. An open circuit voltage of 44.2 V and a short circuit current of 13.2 A were recorded. The complete array fill factor was a remarkable 69%.

This cylinder was then tested with a combustion heated 3.3" diameter SiC infrared emitter. In this second test, the cylinder produced a higher power of 603 W. The cylinder used for these first two tests used filters deposited on glass and then glued onto the cells. These first two experiments were done jointly by JX Crystals and Western Washington University and funded by Army and DOE contracts.

Subsequent to the above two experiments, it was decided that the cylinder performance could be improved if a dielectric filter were deposited directly on the cells. Absorption loses in the glass filters could then be eliminated. Under Army contract funding, JX Crystals then fabricated a new cylinder using integral dielectric filters directly on the cells. This cylinder was then tested at JX Crystals with a 2.75" diameter electrically heated SiC glowbar. The PCA output reached 776 W as shown in figure 5.

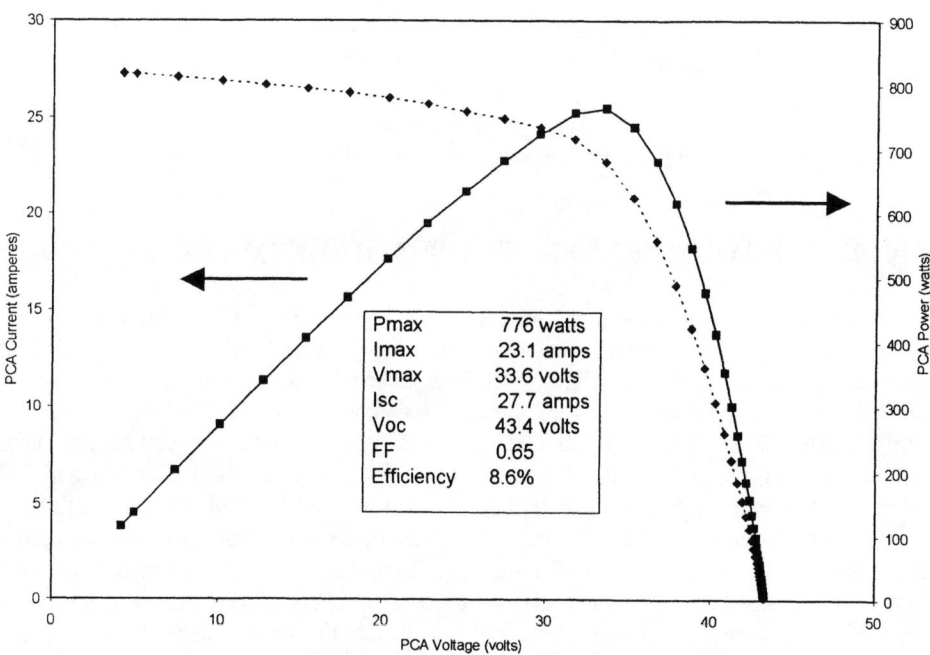

FIGURE 5. Current and Power vs. Voltage using a 2.75" electrically-heated glowbar.

The heat load in the cooling water was also measured during the glowbar test. The ratio of the array electrical output to the water heat load in this last experiment was 9.4%. Figure 6 shows efficiency and power curves as a function of emitter temperature for this third experiment. Subsequent to this third experiment, Western Washington University updated its glowbar test system, using a 3.2" diameter glowbar operating at 1600° C. With this configuration, the University was able to produce 990 W with a first-generation TPV cylinder. Since the active cell area for one of these cylinders is 396 cm^2, a power output of 990 W represents a cell electrical power density of 2.5 W/cm^2.

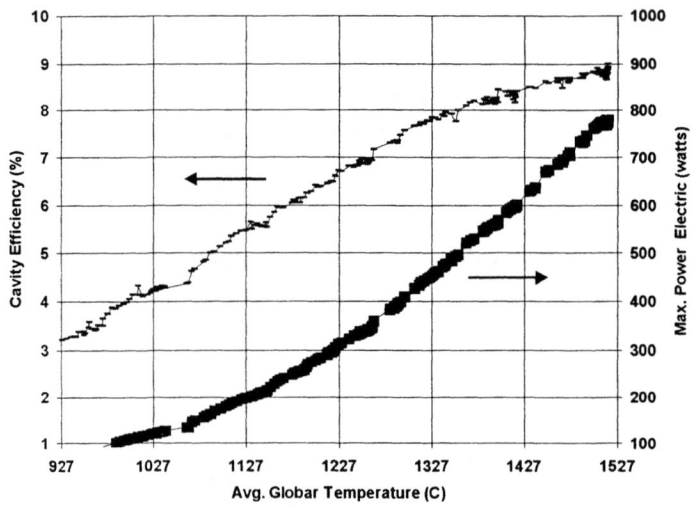

FIGURE 6. Efficiency and Power vs. Temperature from an electrically-heated glowbar.

CYLINDER WITH "PERFECT" SPECTRAL CONTROL

While the results reported here set new records for TPV, higher emitter to electric conversion efficiencies were expected. A very simple heuristic calculation of PCA efficiency follows. Initially, conductive and convective heat transfer through the air will be ignored (and reexamined later). First note that at 1400° C for a SiC gray body emitter, 25% of the radiant energy falls in the cell useable wavelength band below 1.8 microns, 50% of the radiant energy falls in the band between 1.8 and 3.6 microns where the dielectric filter reflectivity is high, and 25% of the energy falls at wavelengths longer than 3.6 microns. Note that fused silica transmits energy below 3.6 microns and absorbs energy at longer wavelengths and, therefore, it can act as a heat shield. A silica heat shield should then return half of the longer wavelength energy to the emitter. Finally, note that a cell should convert 33% of the radiant energy in the useful band to electric power.

Given the above, the emitter to electric conversion efficiency should be: (0.33*25%) / (25% + 0.2*50% + 0.5*25%) = 17.5%. This was inconsistent with the test results, so it appeared that the conductive and convective heat transfer terms might not be negligible. The first-generation cylinder test produced 668 W at an emitter temperature of 1425° C. The measured heat input at this point, after correcting for end gallery heat losses, was 7.5 kW. Conduction heat transfer was calculated at 824 W, but this is not large enough to explain the problem.

In order to check the calculation of conductive and convective heat transfer assuming "perfect" spectral control, a water-cooled cylinder with gold-coated reflectors instead of cells was fabricated. Then it was possible to measure the heat transferred to this cylinder as a function of emitter temperature. The results are shown in figure 7.

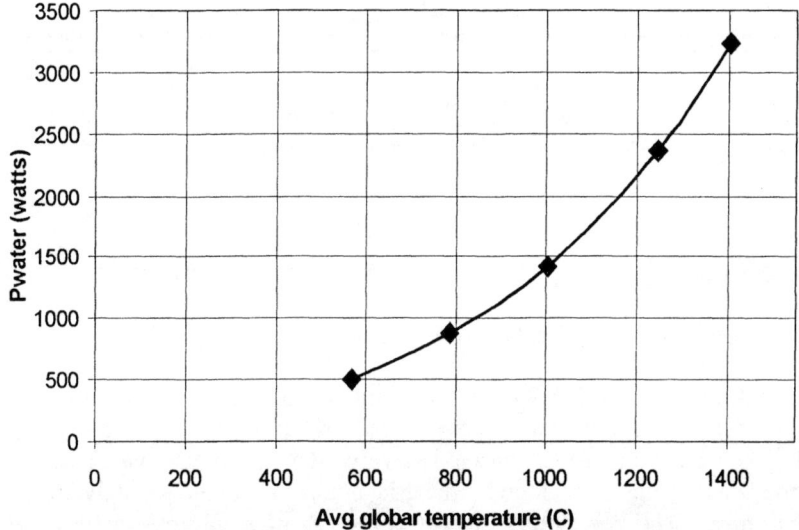

FIGURE 7. Convective heat transfer: power into water vs. average glowbar temp.

Correcting for end gallery losses as before, heat transfer was measured at 2 kW for a 1425° C glowbar temperature. This result is still inconsistent with predictions. Because of this inconsistency, the reflectivity of one of the gold-coated mirrors was measured. The reflectivity was indeed in excess of 98%. The conclusion is that it is possible that the convective theory is based on smaller temperature differences of e.g. 200° C rather than the larger than 1000° C temperature difference involved in this experiment. Convective heat transfer could be non-linear as is seen in the data in figure 7.

The simple heuristic calculation of PCA efficiency now breaks down. It had been assumed that mirrors are benign but mirror area is parasitic via conduction and convection.

SECOND-GENERATION PCA

The above discussion leads to the conclusion that one should increase the percent of active cell area by decreasing the mirror area as much as possible. The necessity of eliminating edge mirrors in the second-generation design led to the invention of the shingle circuit shown in figure 8. The cell used in this new circuit design has a single busbar, which is covered by the next cell in the shingled circuit.

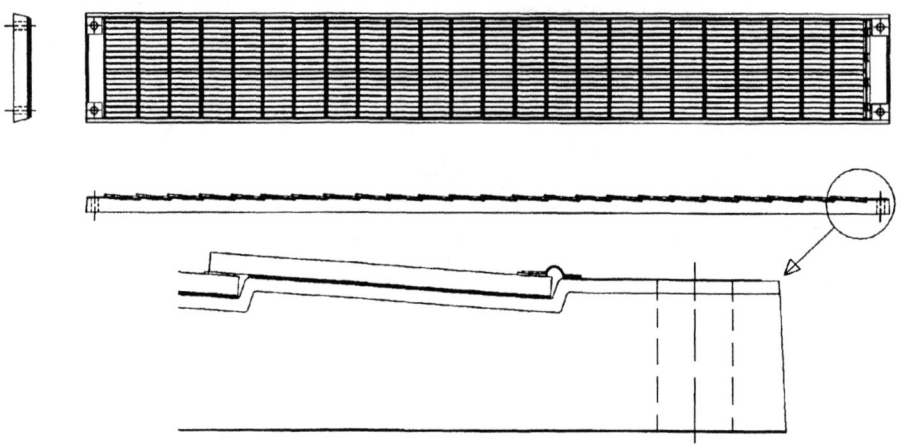

FIGURE 8. Shingled circuit design used in second-generation PCA.

In this new circuit, GaSb cells are laid down like shingles on a terraced substrate with the bottom of each cell overlapped on top of the front metal bus of the next cell. The advantage of this concept is a very high circuit active area where only the cell grids obscure the IR radiation. The grid area is only approximately 20% of the total circuit area. The key to this concept is the use of a substrate material with a coefficient of thermal expansion very well matched to that of GaSb. Having discovered such a substrate material, this shingle circuit concept now has the advantage of a low-cost simple circuit assembly process. Front to back interconnect is now made in a single solder-flow process step.

This circuit utilizes a terraced AlSiC substrate. An insulating film is deposited over the terraces and then copper pads are deposited on the terrace top faces. GaSb TPV cells are then bonded to the copper pads and connected in series in a shingle pattern. AlSiC is a packaging material developed for GaAs devices. AlSiC composites consist of SiC particles in an aluminum matrix. This material can be cast in various shapes and has a thermal expansion coefficient of $7.5 \times 10^{-6} / °C$. GaSb TPV cells have a thermal expansion coefficient of $7 \times 10^{-6} / °C$. In order to prove this shingle circuit concept, GaSb cells on terraced AlSiC substrates were fabricated and tested. These tests included 30 thermal cycles with no loss in performance, at which point the testing was ended.

Given the shingled circuits of figure 8, the second-generation cylinder design is shown in figure 9. Twelve shingled circuits are arrayed to form a TPV cylinder with a 5.2" inner diameter and a 10.3" active length. Table 1 shows a comparison of the specifications for our first and second-generation water-cooled PCAs.

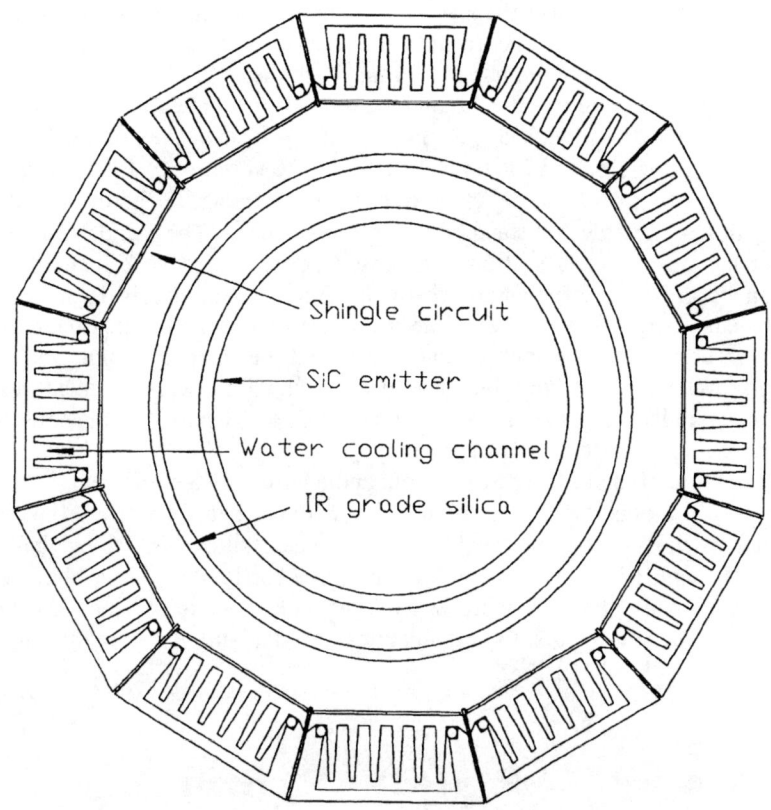

FIGURE 9. Second-generation PCA design using 12 circuits, each two cells wide.

TABLE 1. Old & new PCA specifications.

	1st Generation	2nd Generation
ID	5.3"	5.2"
Active length	8.5"	10.3"
# Circuits	20	12
# Cells	380	576
Active Area	396 cm^2	786 cm^2
Total Area	907 cm^2	1078 cm^2

Comparing the first and second-generation PCA designs, the major benefit in the second-generation design is in the elimination of the edge mirrors which occupied 40% of the area in the first design. Reducing the mirror area really has two benefits. The first obvious benefit is that the conductive and convective heat transfer losses directly to the mirrors are eliminated. The second benefit is not as obvious; eliminating the mirrors reduces the amount of radiant heat returned from the mirrors to the fused silica heat shield. This reduces the heat shield temperature and makes the heat shield more effective at reducing heat transfer to the cells.

Attempts to quantify the second benefit above have led to more careful monitoring of the fused silica heat shield temperature. This has led to the discovery of a second flaw in the simple heuristic calculation of the emitter to PCA efficiency. It had been assumed that fused silica is transparent to radiation below 3.6 microns. However, this is not true for standard grade fused silica. The standard grade fused silica that has been used has strong OH absorption centered at 2.7 microns. This absorption occurs at a wavelength where the filter reflectivity is high and where emitter radiant energy is very high. The result is that the fused silica temperature is being driven up to near the emitter temperature and the "shield" is not working as a heat shield. It is possible to buy IR grade fused silica with low OH content. IR grade fused silica is indeed transparent out to 4 microns, and using IR grade fused silica should improve the emitter to PCA efficiency.

In order to predict PCA power output and emitter to PCA efficiency, the simple one-dimensional model depicted in figure 10 was developed. There are three objects in this model: the IR emitter, the PCA, and a fused silica heat shield halfway in between. The emitted spectrum is broken into three bands: the cell usable band (CB) with wavelengths less than 1.8 microns, the filter high reflectivity band (RB) between 1.8 and 4 microns, and the quartz heat shield band (QB) with wavelengths longer than 4 microns.

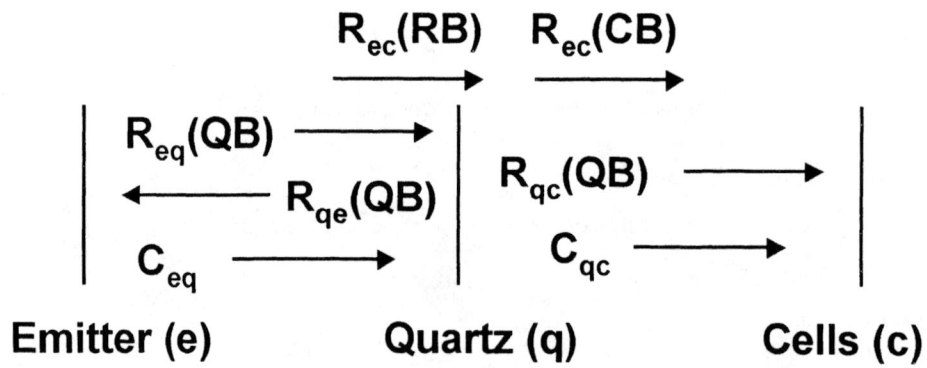

FIGURE 10. Model of emitter to PCA efficiency, used to predict PCA power output.

$R_{ec}(CB)$ is defined as the net radiation density in the cell usable band leaving the emitter and being absorbed by the cell, and $R_{ec}(RB)$ is the net radiation density in the mirror reflection band leaving the emitter and being absorbed by the cell. $R_{eq}(QB)$ and $R_{qc}(QB)$ represent the radiation density absorbed by the quartz from the emitter and passed on from the quartz to the cell array, respectively, in the long wavelength band. Finally, C_{eq} and C_{qc} represent conduction/convection heat transfer from the emitter to the quartz and from the quartz to the cell array, respectively. All of these densities are referred to the emitter surface area. If η represents the cell energy conversion efficiency for cell useful radiation, then the cell electric power density is given by:

$$P_e(c) = \eta R_{ec}(CB)$$

And the PCA heat load is given by:

$$Q_c = R_{ec}(CB) + R_{ec}(RB) + R_{qc}(QB) + C_{qc}$$

Finally, the emitter to PCA efficiency is given by:

$$\text{Effic.} = P_e(c) / Q_c$$

The model shown in figure 10 and described above is used to predict the performance characteristics of the second-generation PCA. Table 2 summarizes the model predictions for a SiC emitter operating at 1325° C or 1425° C. In the following, the assumptions and calculation procedures for the 1425° C case are described in detail. It is assumed that SiC has a flat emittance of 0.8 and that there is 20% grid coverage. The filter function is that shown in figure 3.

TABLE 2. Second-generation PCA performance projections.

Emitter Temperature	1325° C	1425° C	
$R_{ec}(CB)$	4.7	7	(W/cm2)
$R_{ec}(RB)$	3.1	4	(W/cm2)
$R_{eq}(QB)$	6.2	7	(W/cm2)
$R_{qc}(QB)$	2.1	2.3	(W/cm2)
C_{eq} & C_{qc}	1.5	1.6	(W/cm2)
$P_e(c)$	1.5	2.3	(W/cm2)
Q_c	11.4	14.9	(W/cm2)
Efficiency	13.1%	15.4%	
PCA Electric Power	1.1 kW	1.7 kW	
PCA Heat Load	8.6 kW	11.2 kW	

Starting with $R_{ec}(CB)$, one integrates the blackbody function with the filter transmission function times the 0.8 emittance to obtain a power density of 8.7 W/cm². Then note that given grid reflection, $R_{ec}(CB)$ will be 80% of this, or 7 W/cm². Next, calculate $R_{ec}(RB)$, which is the integral of the blackbody function times the filter function times 0.8 or 4 W/cm².

The long wavelength band is more complex. Begin by calculating the energy radiated from the emitter to the quartz, $R_{eq}(QB)$. This is simply an integral of the blackbody spectrum beyond 4 microns times 0.8, or 7 W/cm². To obtain $R_{qc}(QB)$, note that the quartz temperature will equilibrate at a value such that there is a zero net energy flow to it. It is assumed that the average filter reflectivity beyond 4 microns is 30% and that the grids reflect 20% so that 50% of the energy radiated by the quartz to the cell plane is returned to the quartz. Since the quartz radiates equally in both directions but receives half the energy it radiates to the cell plane back again, the energy it receives form the emitter is divided into three equal parts with two thirds being returned to the emitter and one third being passed on to the cells. $R_{qc}(QB)$ is then one third of 7 W/cm², or 2.3 W/cm².

The conduction/convection terms, C_{eq} and C_{qc} are again uncertain. It is known they are equal when referenced to the emitter area. Also, they should be smaller than in the first-generation design because the quartz will be running cooler, appreciably smaller because the convection term is nonlinear. A value is assumed for C_{qc} of 1.6 W/cm².

Assuming a value for η of 33%, the cell electric power density is then 2.3 W/cm². The heat load sums to 14.9 W/cm². Finally, the predicted emitter to PCA efficiency is then 15.2%. Given a 3.5" diameter 10.5" long emitter with an area of 750 cm², the PCA array power output should be 2.3*750 = 1.7 kW and the PCA heat load should be 14.9*750 = 11.2 kW.

In the interest of simplicity, certain assumptions have been made in the above arguments. As a first example, the fused silica reflections have been neglected. This will reduce the electric power produced by about 8% and reduce the PCA efficiency slightly. As a second example, it is assumed that there is a wavelength independent emittance of 0.8. The emittance for SiC can vary slightly from 0.75 to 0.85. Because of these small corrections and others, power and PCA efficiency are stated in projected ranges. The second-generation PCA power output is projected to be between 1.5 kW and 1.7 kW and PCA efficiency is anticipated to be between 12% and 15%.

SUMMARY

Future TPV home cogeneration systems should be capable of generating at least 1 kW of electric power. In order to provide systems for this market, JX Crystals is developing a water-cooled TPV cylinder with a electric power output exceeding 1 kW. Several first-generation TPV cylinders have already been built and tested. These cylinders contain 380 GaSb cells mounted on twenty circuits arrayed on the inside surface of the cylinder at a diameter of 5.3". The mirrors used to cover the cell-to-cell interconnects on these circuits occupy 40% of the cylinder area. A cylinder of this first-generation design has been tested at 600 Watts with a combustion heated SiC emitter and at 990 Watts with an electrically heated SiC glowbar. Emitter to array efficiency increases with emitter temperature. The highest full array efficiency measured to date is 9.4%.

Based on test results on the first-generation TPV cylinders, a second-generation array has been designed. In this new array, edge mirrors are eliminated by using a new circuit design. In this new circuit, GaSb cells are laid down like shingles on a terraced-substrate with the bottom of each cell overlapped on top of the front metal bus of the next cell. This allows a dramatically improved circuit active area, because only the cell grids obscure the IR radiation. Twelve double-wide shingled circuits are arrayed in a cylinder that is only 19% larger than the first-generation array, with projected power output of 1.7 kW for an emitter temperature of 1425° C. It is also projected that by removing the mirrors and using a fused silica heat shield with low OH content, the emitter to PCA electric efficiency can be increased to over 12%.

In this paper, the focus has been on near term PCA configurations using currently available materials and components. In the longer term, better filters and emitters can be developed. Improvements in the GaSb cells are also possible. These improved components will lead to improved PCA designs. Meanwhile, however, this paper has not focused on burner and recuperator development. Work on these components for TPV systems has barely begun.

ACKNOWLEDGEMENTS

The work described here on PCA development has been funded largely by a Multidisciplinary University Research Initiative from the Army Research Office. The authors are grateful to Dr. Richard Paur and Dr. John Kruger for their valuable guidance on this contract.

Portable TPV Generator Based on Metallic Emitter and 1.5-Amp GaSb Cells

V.D.Rumyantsev, V.P.Khvostikov, S.V. Sorokina,
V.I. Vasil'ev, V.M.Andreev

*Ioffe Physico-Technical Institute, 26 Polytechnicheskaya, St.Petersburg, 194021, Russia
fax:+7(812)2471017, e-mail: rumyan@pcl.ioffe.rssi.ru*

Abstract: We present the results of development of a TPV generator which may be considered as a prototype for a portable 5-20 watt TPV power supply using propane or butane fuel. The system of a cylindrical geometry consists of the following parts: burner; "gray" emitter at the temperature of about 1200^0C fabricated with high-temperature resistant metallic alloy; quartz window; PV array consisting of 15 series-connected GaSb cells surrounding the window. The cells of 2×1 cm^2 in area are mounted on copper prisms having passively air-cooled aluminum fins. Generated photocurrent in the circuit of the TPV system is about 1.5A.

Advanced cells on the base of liquid phase epitaxial GaSb material and lattice-matched InGaAsSb solid alloys (Eg = 0.72-0.55 eV) are under development. External quantum yield as high as 90-96% was measured in such cells. Also, two-terminal monolithic tandem GaSb(top)/InGaAsSb(bottom) TPV devices with a built-in tunnel junction have been designed and fabricated. External quantum yields of 80-85% in the top cell and 75-80% in the bottom cell have been measured.

INTRODUCTION

Thermophotovoltaics in Russia has to be a very promising field of activity. Such a situation occurs through a number of natural features. Extended territories are characterized here by a cold climate. Many families are living away from the electric power grid. Their need is a need for heat and electricity. At the same time, Russia possesses the powerful sources and deposits of the natural gaseous and liquid fuel. That is why the thermophotovoltaic (TPV) concept as well as a concept of heat and electricity cogeneration is of a great interest for Russia.

Photovoltaic Converters Laboratory of the Ioffe Physico-Technical Institute has a many years experience in the field of III-V based devices, especially concentrator solar cells [1]. Last time this experience was applied to TPV research and development, including growth, treatment and packaging the narrow-gap cells (GaSb-, or InGaAsSb-based), as well as design and manufacturing the demonstration systems. In this paper we present the results of development of the TPV cells prepared by Zn-diffusion into "bulk" GaSb wafers, as well as into epitaxial GaSb and InGaAsSb layers grown by liquid phase epitaxy. Also, two-terminal monolithic tandem GaSb (top) /InGaAsSb (bottom) TPV device is described. After this, practical design and results on the TPV generator, which may be considered as a prototype for a portable 5-20 watt propane/butane fueled power supply, are given.

"BULK" GaSb CELLS

Our main technological approach to producing the GaSb TPV cells is the use of "bulk" wafers (wafers prepared with bulk ingots grown by a Czochralski technique, Te-doped) as a base material. At the first stage, GaSb wafers were exposed for the first zinc diffusion procedure in a "pseudo-closed box" technique to form a shallow p-n junction in photoactive area of a cell. Anodic oxidation and selective etching were employed for precise thinning the diffused p-GaSb layer. At the second stage, a deep p-n junction (1-1.5 μm) was formed by additional spatially selective diffusion process to avoid the current leakages under contact grid fingers.

A blanket Au (Ge) ohmic contact was deposited on the back surface by vacuum evaporation and was alloyed at 300^0C. Cr-Au-Ni-Au contact was deposited on the front surface and alloyed at 250^0C. Front contact grid was thickened by electroplated Au up to 3 μm. A two-layer ZnS/MgF$_2$ antireflection coating was applied to minimize the reflection loses in a spectral range of 700-2200 nm. The cell area is 1x1cm^2 or 2x1cm^2. Contact grid spacing is 0.2 mm. Scheme of a "bulk" GaSb cell is given in Figure 1.

Figure 2 shows the spectral response of the cells based on "bulk" GaSb with different p-layer thicknesses. As it is seen from Figure 2, the maximum photocurrent can be obtained from a cell with p-region thickness in the range 0.1-0.3 μm. That is why the second selective diffusion process is necessary.

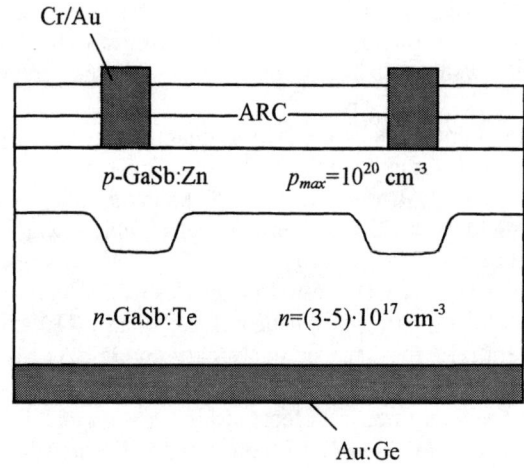

FIGURE 1: Cross section of a "bulk" GaSb cell.

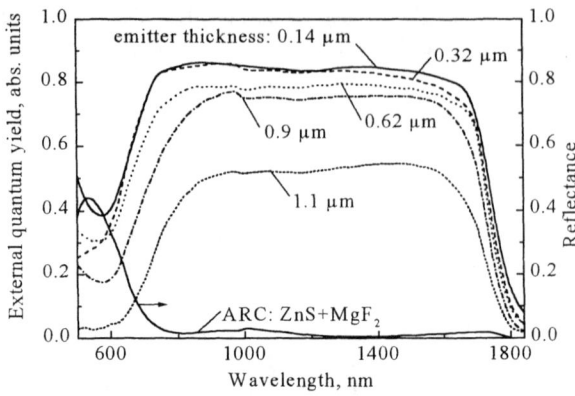

FIGURE 2: Spectra of photoresponse (on the area including contact grid) for "bulk" GaSb cells with different p-emitter thicknesses. Spectrum of reflectivity in the case of ZnS+MgF$_2$ antireflection coating is shown as well.

"EPITAXIAL" GaSb AND InGaAsSb CELLS

Epitaxial growth of Te-doped GaSb layers was carried out by liquid phase epitaxy (LPE) in attempts to prepare material of a higher crystal quality. Ga-, Sb- and Pb-rich melts were used in the LPE processes.

In the case of LPE growth from Ga-rich melt we employed a "piston" boat technique, which enabled us to grow epilayers on 50 substrates of 7 cm^2 in area for one run. Epilayers from Ga-rich melts were grown at temperature of 450-400^{0}C with the cooling rate of 0.5-1.0 degree/min.

Epilayers from Sb- (and Pb-) rich melts were grown in a graphite slider boat. The growth was carried out from Sb-Ga-Te liquid phase: $x^l_{Ga}=0.12$-0.13, $x^l_{Te}=(3$-$7) \cdot 10^{-5}$ with the supercooling values of 3-7 ^{0}C by a step - cooling technique at 610-597 ^{0}C during 20-30 sec.

Pb-rich melt was employed as an electrically inactive solvent. Epilayers were grown from Pb-Sb-Ga-Te liquid phase: $x^l_{Pb}=0.47$, $x^l_{Ga}=0.127$, $x^l_{Te}=3.85 \cdot 10^{-5}$ at 608^{0}C.

Figure 3 shows the spectral responses of the "epitaxial" TPV cells. A maximum of external quantum yield was obtained in the cells, which were prepared using the epilayers grown from Ga-rich melt.

Fabricated GaSb TPV cells were tested under flash illumination. Figure 4 (curve 1) shows the current-voltage characteristic of a 1 cm^2 GaSb TPV cell at output current density of 0.67 A/cm^2. In the case of quaternary In$_x$Ga$_{1-x}$As$_y$Sb$_{1-y}$ epilayers Sb-rich melts were used for LPE process. The use of Sb-rich melts made it possible to decrease the concentration of stoichiometric defects, which are typical for GaSb-related solid solutions. The effect of a negative factor connected with interphase nonequilibrium in GaSb (solid) - Ga-In-As-Sb (liquid) was also reduced. Crystal perfection of the grown solid solutions was comparable with that of GaSb epilayers. Heterostructures for TPV

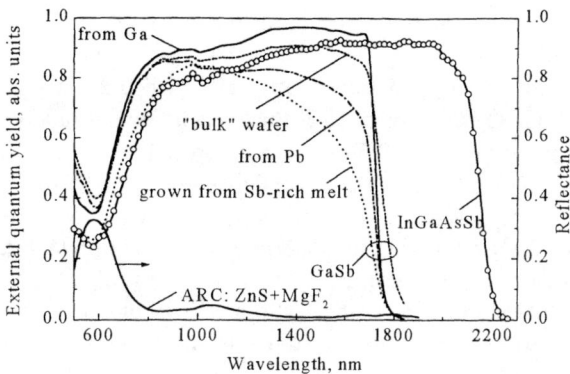

FIGURE 3: Spectral responses (on the photoactive area) of the "epitaxial" TPV cells. Reflectance from a cell with ZnS+MgF$_2$ antireflection coating is shown as well.

application were grown on GaSb:Te (100) substrates. LPE growth was carried out at 600^{0}C from supercooled liquid phases by the step-cooling method. The value of supercooling was varied in the 4-10^{0}C range. *P-n* junction was formed both by epitaxial growth and by additional gas-phase Zn-diffusion treatment. Corresponding curve in Figure 3 represents spectral distribution of photoresponse for one of the InGaAsSb-based (E_g=0.55 eV) TPV cells. Figure 4 (curve 2) demonstrates the current-voltage characteristic of a 1cm^2 InGaAsSb TPV cell with Zn-diffused emitter.

TANDEM GaSb/InGaAsSb TPV CELLS

Tandem approach has been tried out by us recently in respect to TPV application [2]. Multilayer GaSb/InGaAsSb structures with built-in tunnel *p-n* junction were fabricated

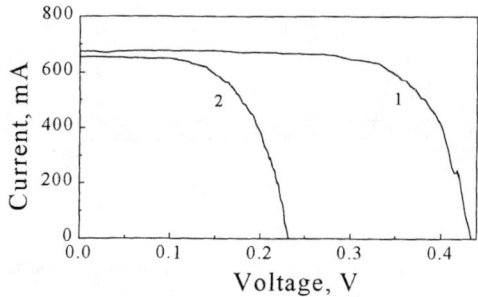

FIGURE 4: Flash illuminated current-voltage characteristics of a 1cm^2 GaSb TPV cell (1) and 1cm^2 InGaAsSb (E_g=0.55 eV) TPV cell (2).

using two-stage LPE growth and two Zn-diffusion processes. The heterostructure consists of the following layers (see, also, Figure 5): n-GaSb (substrate); n-p In$_x$Ga$_{1-x}$As$_y$Sb$_{1-y}$ (E_g=0.56 eV, 1-3 μm thick n-layer and 0.2-0.5 μm thick p-layer) - bottom cell; p^{++}-n^{++}GaSb (0.8 μm of total thickness) - tunnel junction; n-p GaSb (n-layer 3-5 μm thick and p-layer 0.2-0.5 μm thick) - top cell.

The first step included LPE growth of n-InGaAsSb layer and cap GaSb layer from Sb-rich melt on a GaSb substrate. Zn-diffusion was carried out after the first LPE growth to form both the p-n junction in quaternary solid solution and the first layer (p^{++} GaSb) of the tunnel junction in GaSb.

Te-doped n^{++} GaSb-nGaSb layers (n^{++} GaSb as the n-region of the tunnel junction) were grown from Ga-rich melts. Finally, Zn-diffusion was carried out to form the p-n junction in GaSb top cell.

Concentration of Te measured by SIMS was more than $5 \cdot 10^{19}$cm^{-3} and concentration of Zn was about 10^{19}cm^{-3} in the p^{++}-n^{++} junction region. As a test, after annealing at 600°C during 20 min, the tunnel diodes demonstrated ohmic characteristics under voltage biases of 10-20 mV corresponding to the current density of about 10 A/cm^2. V_{oc}=0.61 V and FF=0.75 at 0.7 A/cm^2 were achieved in such a tandem. Spectral response of the monolithic tandem cell is shown in Figure 6.

Cr/Au:Zn/Au			
ZnS+MgF$_2$			
p -GaSb	(Zn)	0.2-0.5 μm	top cell
n -GaSb	(Te)	5.0 μm	
n^{++}-GaSb	(Te)	0.5 μm	tunnel diode
p^{++}-GaSb	(Zn)	0.3 μm	
p-InGaAsSb	(Zn)	0.2-0.5 μm	bottom cell
n-InGaAsSb	(Te)	1.0-3.0 μm	
n-GaSb (Te) substrate			
Au:Ge/Au			

FIGURE 5: Cross section of the monolithic two-junction two-terminal TPV cell.

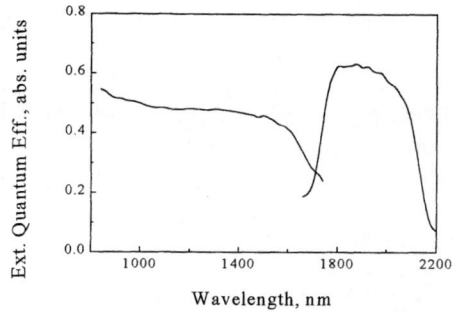

FIGURE 6: Spectral response of the monolithic GaSb/InGaAsSb tandem TPV cell.

TPV GENERATOR

A small TPV demonstration system was realized in our laboratory. Possible applications of such a system are stand-alone generators of heat and electricity able to charge an electrochemical battery. Simplicity of this device is one of the basic principles. The overall system efficiency depends on how three key components – PV cells, radiant emitter and heat source - operate as a system.

PV cells

As may be seen from above text concerned with description of the developed TPV cells, "bulk" GaSb cells are characterized by quite good parameters in respect to

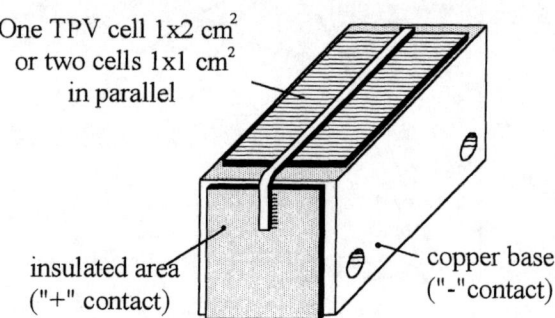

FIGURE 7: TPV cell design.

external quantum efficiency of photoresponse, open circuit voltage and fill factor of the illuminated I-V curve. On the other hand, fabrication process for them is relatively simple. This is the reason why we have employed such cells in an experimental TPV generator. Passive air cooling of the cells has been planned in the installation. For this, each cell of 1x2 cm^2 in area (or two cells of 1x1 cm^2 in a parallel connection) was soldered to a copper prism (see Figure 7) equipped with two aluminum fins. Some of cells had been mounted on a special holder giving a possibility to change the method of cooling - by water, or by air (with blower, or passively).

Radiant emitter

There are different approaches to design of an efficient TPV emitter. One of them is the use of high temperature-resistant metals as the emitter material [3, 4]. It has been found by us, that metallic alloy Fe+Cr+Al+Co+C in view of a wire (about 1 mm in diameter) is a proper material for emitter operating up to temperature of 1250^{0}C. Attractive features of such a material are the following: spectral selectivity of emission; resistance against thermoshock; possibility to arrange developed emitter surface - "transparent" for exhaust gas, but "solid" in respect to radiation. The latter feature may be realized in practice, if emitter of a cylindrical geometry has a view of a "basket". Figure 8 shows measured spectral distribution of emissive power of the basket-type metallic emitter heated by flame ut to 1200-1250^{0}C. In this figure, calculated curve for blackbody at T=1523 K is given as well. One can see a reduced long-wavelength tail of spectrum for "gray" metallic emitter in comparison with that for blackbody. Probably, such a behaviour of metallic emitter is caused by the same reason as in the case of tungsten lamps. It is well know, that in lamps used for calibration in spectroscopic

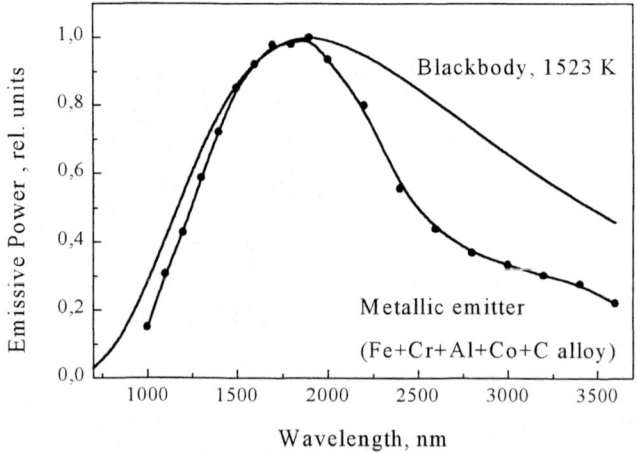

FIGURE 8: Emissive power spectra for the basket-type metallic emitter (experimental) and for blackbody at T=1523 K (calculated).

measurements a drop in emissivity takes place in a long-wavelength part of spectrum due to corresponding rise of reflectivity from tungsten surface [5]. Of course, presence of oxides on the emitter surface can deform emissive power curve as well.

Heat source and general design

In Figure 9, a schematic of the designed TPV generator is shown. A burner with gas inlet of 0.15 mm in diameter is connected (directly) with butane tank, or (through a pressure reducing unit) with propane tank. Gas stream force and natural convection process are employed for gas-air premixing and transport to combustion zone. Premixing zone is separated from the combustion zone by a wire screen.

Basket-type metallic emitter has external diameter of 3.6 cm being 1.8 cm in height. Combustion process is carried out directly at its surface. There is a quartz window of 5 cm in diameter arranging a stream of the exhaust gases. Fifteen GaSb cells of 1x2 cm^2 each are situated around the emitter and window. Heat sink of each cell consists of copper prism and two aluminum fins.

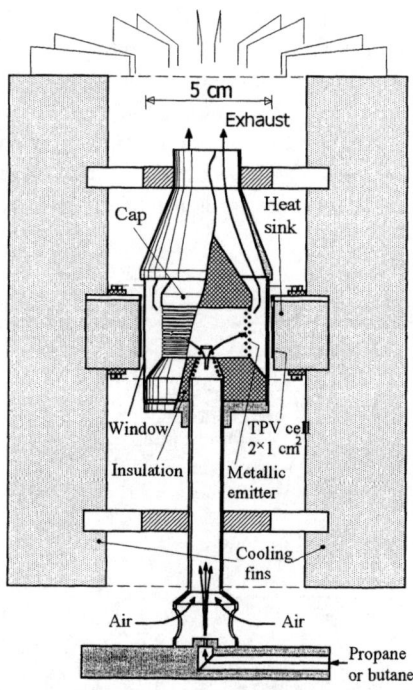

FIGURE 9: Schematic of portable TPV generator based on 15 GaSb cells.

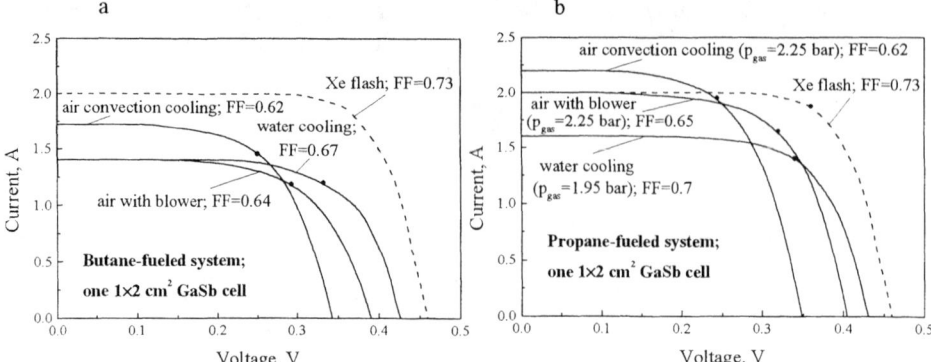

FIGURE 10: I-V curves for one of the GaSb cells under illumination. Solid lines-illumination by metallic emitter. Different cooling conditions are indicated for: a – butane-fueled system; b – propane-fueled system. Dotted lines – I-V curve under Xe-lamp flash illumination.

Experimental results

Figure 10, a-b represents experimental I-V curves for one of the mounted 1x2 cm² GaSb cells under gas-fueled metallic emitter illumination. Different cooling conditions (by water, or by air) are indicated. Dotted line show I-V curve of the same sample under Xe-lamp flash illumination (no heating the cell). In the case of butane-fueled system it was possible to measure gas consumption by weighing, that gave the chemical power input of about 550-600 watt. In the case of propane-fueled system, the same conditions took place at input gas pressure of about 2 bars. In both cases one can

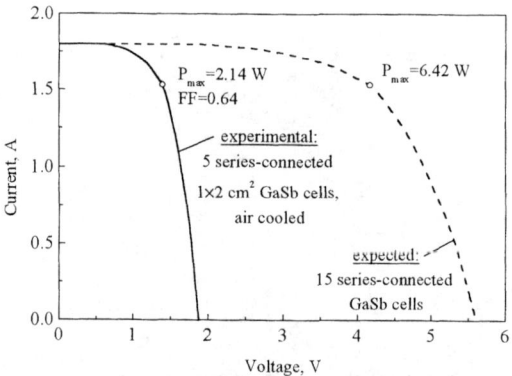

FIGURE 11: Solid line - illuminated I-V curve of a fragment of the TPV array: 5 series-connected 1x2 cm² GaSb cells; propane-fueled (p_{gas}=2.1 bar) TPV system. Dotted line - expected I-V curve of the whole array (15 cells)

see a strong effect of cell heating under continuous illumination by emitter.

Final result to date on illuminated I-V curve for a fragment of the TPV array (5 series-connected 1x2 cm^2 GaSb cells, propane-fueled TPV system) is given in Figure 11 by solid line. Operation current lies in the vicinity of 1.5 A. Dotted line in this Figure represent an expected I-V curve for the whole TPV array (15 series-connected GaSb cells), if all cells are characterized by the identical parameters similar to that for 5-cells fragment under test. Expected maximum power is about 6.4 W. An increase in output power may be reached. Our experiments with larger gas inlets and emitters have shown that a more powerful operation of the TPV generator up to about 20 W is possible. For this, emitter has to be at temperature not more than 1250^{0}C. Increase in emitter area and corresponding increase in quantity of cells is implied.

CONCLUSION

Research and development of the narrow-gap GaSb and InGaAsSb TPV cells has been carried out. A portable TPV generator based on metallic basket-type emitter and GaSb PV array has been designed and fabricated. Test operation of the 1/3-fragment of the array with output electric power of 2.14 W has been demonstrated.

ACKNOWLEDGEMENTS

This work was supported by Russian Fundamental Research Foundation, Grant N97-02-18112, and by INTAS under Grant 96-1887.

REFERENCES

1. Andreev, V.M., Grilikhes, V.A., Rumyantsev, V.D., *Photovoltaic Conversion of Concentrated Sunligh*, Chichester: John Wiley & Sons Ltd, 1997, 294 p.
2. Andreev, V.M., Khvostikov, V.P., Larionov, V.R., Rumyantsev, V.D., Sorokina, S.V., Shvarts, M.Z., Vasil'ev, V.I., Vlasov, A.S., "Tandem GaSb/InGaAsSb Thermophotovoltaic Cells", presented at the 26 th Photovoltaic Specialists Conference, Anaheim, CA, September 1997.
3. Schubnell, M., Gabler, H., Broman, L., "Overview of European Activities in Thermophotovoltaics", presented at the Third NREL Conference on Thermophotovoltaic Generation of Electricity, Colorado Springs, CO, 1997 (pp. 12-13).
4. Gabler, H., Hein, M., Xenker, M., "A Propane-fueled Thermophotovoltaic Energy Converter Using low-bangap Photovoltaic Cells", presented at the 2-nd World Conference on Photovoltaic Solar Energy Conversion, Vienna, July 1998.
5. Malyshev, V.I., *Introduction into Experimental Spectroscopy* (in Russian), Moscow: Nauka, 1979, ch. 4, p.p. 352-354.

Operating Experience Of A Portable Thermophotovoltaic Power Supply

Frederick E. Becker, Edward F. Doyle, and Kailash Shukla

Thermo Power Corporation, Tecogen Division
45 First Avenue, P.O. Box 9046
Waltham, Massachusetts 02454-9046

Abstract. Two configurations of man-portable thermophotovoltaic (TPV) power supplies based on Thermo Power's supported continuous fiber emitter have been designed, built, and are being tested. The systems use narrow-band, fibrous, ytterbia emitters radiating to bandgap matched silicon photovoltaic arrays with dielectric stack filters for optical energy recovery and recuperators for thermal energy recovery. The systems have been designed for operation with propane and with combustion air preheat temperatures of up to 1250 K. To operate at air preheat temperatures above the auto-ignition temperature of the fuel, a unique fuel delivery system was devised which results in the micromixing and rapid combustion of the fuel and air right in the emitter fibers. This allows the ytterbia emitter fibers to run much hotter (~2000 K) than any of the surrounding structure.

INTRODUCTION

Thermophotovoltaic (TPV) energy conversion, which uses photovoltaic (PV) cells to directly convert radiant thermal energy into electric power, has a number of important advantages for portable power generation in military applications. Since TPV is a direct energy conversion technology with no moving parts in the energy conversion system, it has the potential to provide quiet, reliable, and maintenance-free electric power. Thermo Power Corporation, with funding support from DARPA/NASA-Lewis, is developing portable, propane-fueled TPV power supplies in the range of 100 to 125 Watts to demonstrate the integrated performance of the major components of a TPV system – emitter, optical filter, PV array, and thermal recuperator.

TECHNICAL APPROACH

The emitter system chosen for this program is based on Thermo Power's patented supported continuous fiber radiant structure that can operate up to temperatures of 2200 K with good thermal shock resistance and rapid response time. The emitter, shown in Figure 1, consists of continuous fibers woven into a porous ceramic base or substrate. As such, the emitter preserves the advantages of gas light mantles that have traditionally operated at high temperatures for long times. Unlike gas mantles, the Thermo Power emitter can be made in planar, 15 cm x 15 cm or larger tiles without becoming fragile. A cellulose support process that is based on textile precursors is employed to fabricate the fibrous emitter structure. Continuous filament rayon yarn impregnated with aqueous metal salt solution is tufted into the porous ceramic support much like a rug making process that yields an uncut looped pile. After tufting, the treated rayon is converted into a ceramic by means of a controlled heat treatment process.

The technical approach taken in this project is to use Thermo Power's rare-earth oxide, wavelength selective, flow-through emitter radiating to a bandgap matched photovoltaic array to minimize thermal and optical recuperation requirements. Two combinations of emitter material and matching photocells, ytterbia/silicon and erbia/gallium antimonide, were evaluated relative to system efficiency, project cost and delivery time. By using a simplified systems analysis model, the system efficiency of the ytterbia/silicon combination was calculated to be about 10% higher than the erbia/gallium antimonide combination, for an emitter temperature of 2000 K, thermal recuperator effectiveness of 80%, and off-band optical recovery of 75%. More importantly, the cost and delivery for the silicon arrays were more favorable (at the time of the decision), since the supplier, SunPower Corporation, was producing the planar water-cooled arrays for another application. The ytterbia/silicon combination was therefore selected for the initial prototypes.

FIGURE 1. Supported Continuous Fiber Emitter

There are two combustion issues related to air-preheat, auto-ignition and thermal decomposition, which affect the system design. If the air preheat temperature is below the auto-ignition temperature of the fuel (about 700 K for propane), fuel and preheated air can be premixed prior to the burner/emitter system. For our supported fiber emitter, the air/fuel premix would flow through the open pores in the substrate and burn above the substrate, with the flame located in the fibers. If the combustion air preheat temperature is above the auto-ignition temperature of the fuel, air and fuel will have to be delivered separately to the point of combustion at the emitter fibers. These two alternative combustion approaches, premixed and micromixed, are illustrated with reference to our emitter in Figure 2.

The thermal decomposition temperature for propane is 848 K, above which propane breaks down into carbon and other hydrocarbons. For the low air-preheat premixed case, thermal decomposition in not an issue since the preheat temperature is lower than the decomposition temperature. For the high air-preheat micromixed case, however, the thermal breakdown needs to be controlled to prevent the products of decomposition from depositing in the fuel tube and plugging it over time.

PROTOTYPE DEVELOPMENT

Based on these two combustion approaches, two configurations of TPV prototype systems were designed, fabricated, and are being tested. In designing these systems, we established the prototype size based on the commercial availability of critical component parts like ceramic emitter substrate, ceramic core for the recuperator, and the water-cooled silicon array. With these design criteria, the active emitter area and photovoltaic array area was determined to be 12 cm x 12 cm, and the gross power output is projected to be in the range of 100 to 150 Watts.

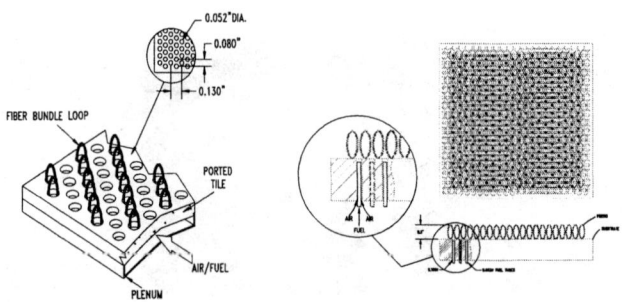

FIGURE 2. Alternative Combustion Approaches

The schematic drawings of the two prototype systems are shown in Figure 3. In the low air-preheat premixed system, the two quartz windows separating the PV array from the emitter are used as a recuperator. The quartz windows (with optical filter) are used to protect the PV array from direct contact with hot combustion gases, to reduce the heat load on the photocells and to recycle a portion of the off-band radiation and exhaust gas heat. In order to maintain the windows at temperatures consistent with the capability of the optical coatings, the air preheat temperature is kept below 550 K. The photographs of the low air-preheat prototype are shown in Figure 4.

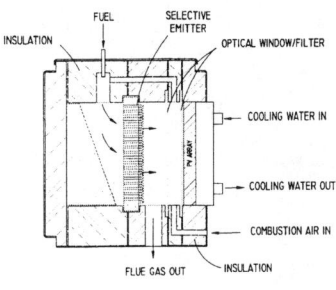
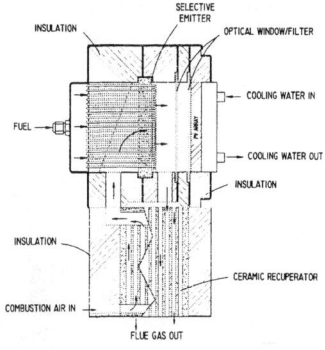

Low Air-Preheat Premixed SystemHigh Air-Preheat Micromixed System

FIGURE 3. System Schematic Drawings

FIGURE 4. Low Air-Preheat TPV Prototype

In the high air-preheat micromixed system, a cordierite ceramic recuperator, shown in Figure 5, is used to preheat the combustion air to 1250 K. The preheated air then flows through the back of the porous ceramic substrate that supports the ytterbia emitter in a filament form. The fuel is delivered from the fuel plenum through fuel tubes to very near the surface of the substrate and micromixed with preheated air. A sufficient number of fuel tubes are used to provide uniformity of temperature distribution, to achieve rapid air-fuel mixing, and to complete combustion within the fiber bed for good fuel efficiency. In addition, a small portion (10-15%) of the unheated combustion air is premixed with propane upstream of the fuel plenum to control the thermal breakdown of propane and avoid build-up of deposits in the fuel tubes. The photographs of the high air-preheat prototype are shown in Figure 6.

FIGURE 5. Cordierite Ceramic Counterflow Recuperator

FIGURE 6. High Air-Preheat Prototype

The PV cells used in the systems are supplied by SunPower Corporation. The spectral response of these cells is very well matched to the ytterbia emitter characteristics. Figure 7 shows the spectral efficiency of the SunPower cells and the spectral exitance of the ytterbia emitter as a function of wavelength. The peak monochromatic efficiency for these cells is 40% and occurs near the ytterbia peak of 980 nm. The overall efficiency of these cells in converting radiation below 1180 nm is 32% when matched with the ytterbia emitter. The open circuit voltage, short circuit current, fill factor, peak power, responsivity and overall efficiency for these cells are listed in Table 1. A water-cooled heat sink coupled to an air-cooled fan coil was used for cooling the PV arrays. The photograph of the PV array is shown in Figure 8.

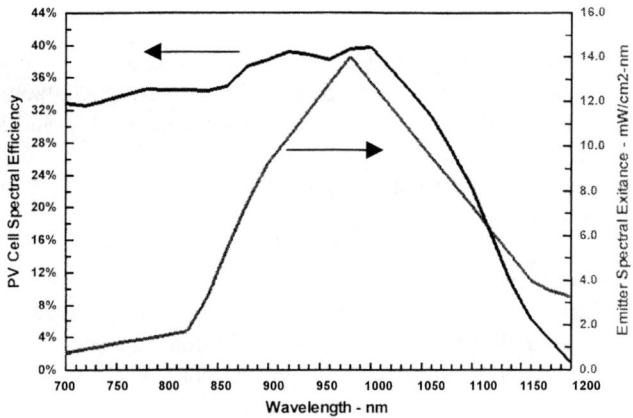

FIGURE 7. PV Cell Efficiency and Emitter Spectral Exitance vs Wavelength

Voc – Volts	0.78
Isc – A/cm^2	1.75
FF	0.79
Pm – A/cm^2	1.08
Responsivity – A/W	0.52
Efficiency %	32

TABLE 1. PV Cell Characteristics

FIGURE 8. PV Array

Three types of filters were investigated for optical performance and mechanical stability by Essential Research Inc. These were a multilayer dielectric stack filter, a thin "transparent" conductive coating also known as a solar control film, and a tandem filter consisting of both dielectric and solar control film. The dielectric stack filter was chosen for use during this phase of the project, because of its higher in-band transmission, lower absorption and high reflectivity in the out-of-band region.

Figure 9 shows the measured power output for the high air-preheat system as a function of convertible or in-band radiation at the array. The power output measured at a convertible exitance of 2.7 W/cm^2 was 90 Watts. Additional tests are planned with higher air preheat levels to give higher exitance and power output.

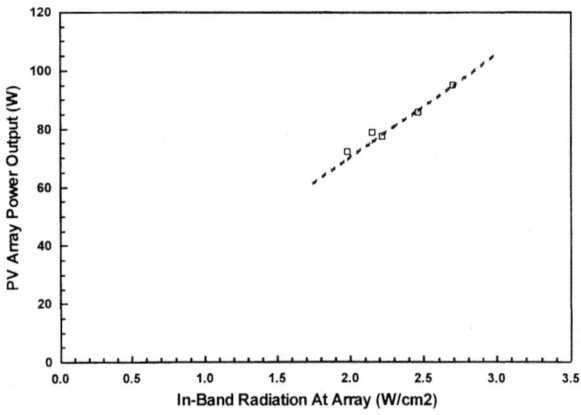

FIGURE 9. Power Output vs In-Band Radiation at PV Array

The design specifications for the two current prototypes, a low air-preheat system and a high air-preheat system, as well as for two future advanced prototypes, are presented in Table 2. The first advanced prototype will be designed to operate at an emitter temperature of 2070 K and an air preheat temperature of 1350 K. This prototype will require the use of a higher temperature substrate with a low emissivity surface, and will produce a gross power of 150 Watts at a gross efficiency of 7%. The net power and efficiency are 115 Watts and 5.4% respectively, after allowing power for ancillary equipment. For the second advanced prototype, the emitter temperature will be increased to 2100 K and the air preheat temperature to 1650 K. This will require the use of a silicon carbide recuperator for higher temperature and thermal shock capability. This prototype will be able of producing a gross power of 165 Watts at a gross efficiency of 9%. The net power and efficiency will be 130 Watts and 7.1% respectively, after allowing power for the ancillary equipment.

Parameter	Low Air-Preheat Prototype	High Air-Preheat Prototype		
	Gen0-98	Gen1-98	Gen2-99	Gen3-00
Emitter Temperature – K	2000	2035	2070	2100
Air-Preheat Temperature – K	550	1250	1350	1650
Average Convertible Exitance at Emitter – W/cm^2	3.50	4.00	4.50	5.00
Average Exitance View Factor	0.80	0.80	0.80	0.80
Average Convertible Exitance at Array – W/cm^2	2.80	3.20	3.60	4.00
Emitter Area – cm^2	144	144	144	144
PV Array Area – cm^2	144	144	144	144
PV Cell Efficiency	32%	32%	32%	32%
PV Array Uniformity Factor	85%	85%	90%	90%
Gross Power – W	110	125	150	165
Parasitic – W	30	35	35	35
Net Power – W	80	90	115	130
Gross System Efficiency	2.0%	4.0%	7.0%	9.0%
Net System Efficiency	1.5%	2.9%	5.4%	7.1%

TABLE 2. TPV Power Source Specifications

ACKNOWLEDGMENTS

This work has been supported by the U.S. Department of Defense through DARPA and NASA-Lewis. The authors acknowledge the guidance provided by Dr. Bob Rosenfeld of DARPA and Mr. Dave Wilt of NASA-Lewis. We also acknowledge contributions of Mr. Phil Jenkins of Essential Research Inc.

REFERENCES

1. Becker, F.E., Doyle, E.F., Shukla, K., "Development of a Portable Thermophotovoltaic Power Generator," The Third NREL Conference on the Thermophotovoltaic Generation of Electricity, Colorado Springs, 1997.

2. Becker, F.E., Doyle, E.F., Mastronardi, R., Shukla, K., Linder, E.B., and Garverick, L.M., "Development of a 500 Watt Portable Thermophotovoltaic Power Generator," Twenty-Fifth IEEE Photovoltaic Specialists Conference, Washington, D.C., 1996, pp. 1413-1416.

3. Becker, F.E., "Recuperators for Thermophotovoltaic Energy Conversion Systems," Prospector VIII: Thermophotovoltaics – An Update on DOD, Academic, and Commercial Research, sponsored by Space Power Institute. Auburn University, Alabama and Army Research Office, Research Triangle Park, North Carolina, July 14-17, 1996, Edited by M. Frank Rose, pp. 173--194.

4. Doyle, E.F., "System Aspects of TPV Energy Conversion," Prospector VIII: Thermophotovoltaics – An Update on DOD, Academic, and Commercial Research, sponsored by Space Power Institute. Auburn University, Alabama and Army Research Office, Research Triangle park, North Carolina, July 14-17, 1996, Edited by M. Frank Rose, pp. 173-194.

5. Nelson, R.E., "Grid Independent Residential Power Systems," The Second NREL Conference on Thermophotovoltaic Generation of Electricity, Colorado Springs, 1995, pp. 221-237.

6. Nelson, R.E., "Thermophotovoltaic Emitter Development," The First NREL Conference on Thermophotovoltaic Generation of Electricity, Copper Mountain, 1994, pp. 80-95.

Interfacing a Small Thermophotovoltaic Generator to the Grid

W. Durisch, B. Grob, J.-C. Mayor, J.-C. Panitz and A. Rosselet*

Paul Scherrer Institut, PSI, CH-5232 Villigen PSI, Switzerland
* University of Lausanne, CH-1015 Lausanne, Switzerland

Abstract. A prototype thermophotovoltaic generator and grid-interfacing device have been developed to demonstrate the feasibility of grid-connected operation. For this purpose a conventional butane burner (rated power 1.35 kW$_{th}$) was equipped with a ceramic composite emitter made of rare earth oxides. A water layer between emitter and photocells was used to protect the photocells against overheating. It absorbs the non-convertible emitter radiation and is heated up thereby. The hot water so produced in larger units of this type could be used in a primary recirculation loop to transfer heat to a secondary domestic hot water system. For the photovoltaic generator, commercial grade silicon solar cells with 16 % efficiency (under standard test conditions) were used. With the radiation of the emitter, a current of 4.6 A at a maximum power point voltage of 3.3 V was produced, corresponding to a DC output of 15 W and a thermal to DC power conversion efficiency of 1.1 %. A specially developed high efficiency DC/DC converter and a modified, commercially available inverter were used to feed the generated power to the local grid. Under the experimental conditions in question the DC/DC-converter and the grid-inverter had efficiencies of 98 and 91 %, respectively resulting in an overall interface efficiency of 89 %. From modeling of the measured electrical characteristics of the photo cell generator under solar and emitter radiation, it is concluded that the photo current was about three times higher under the filtered emitter radiation. Under these conditions the electrical losses of the photocells were significantly higher than under sunlight.

INTRODUCTION

The total Swiss energy consumption in 1997 amounted to 224 billion kWh [1]. Approximately half of this is used by heating systems. The consumption for domestic space heating and hot water production amounted to 62 billion kWh, of which 46 billion kWh were supplied by fuel oil (35) and fuel gas (11). Taking into account that the average space heating and hot water demand of a typical Swiss household corresponds to about 30,000 kWh$_{th}$ of primary energy per year, then about 1.5 million oil-burning and gas-fired domestic central heating systems could be equipped with grid-connected thermophotovoltaic (TPV) generators. Assuming that a thermal-to-alternating-current conversion efficiency of 5 % can be achieved (the theoretical limit at a combustion temperature of 1700 K is 82 %), the existing household potential of TPV co-generation systems in Switzerland corresponds to

2.3 billion kWh of electricity per year. Furthermore, if domestic space heating and hot water production presently based on electricity were replaced by oil-burning and gas-fired TPV co-generation systems, an additional contribution of about 0.5 billion kWh/yr. could be produced, resulting in an effective TPV potential of 2.8 billion kWh$_{el}$/yr. This corresponds to 6 % of the total annual electricity consumption of Switzerland. This means that decentralized TPV co-generation systems could - at long sight - provide a valuable contribution to the Swiss electricity supply system. From a simple economic consideration (based on typical assumptions of life time, interest rate and operational cost) it is concluded that a characteristic domestic central heating system equipped with a five-percent-efficient grid-connected TPV generator with extra investment cost of 4000 Swiss francs (including grid-inverter) would lead to an electricity price of 21 Swiss Cents per kWh, comparable to the average electricity price for Swiss households of 13 to 28 Swiss Cents per kWh. This, too, demonstrates that the prospect for a widespread application of TPV co-generation systems in Switzerland looks attractive and that further research, development and demonstration efforts are justified. A particular advantage of TPV in Switzerland is the fact that most TPV heat and electricity would be produced simultaneously with the high heating and electricity demand in winter, when conventional electricity might become scarce. Furthermore, heat and by-product electricity generated simultaneously in TPV heating systems fit well into the Swiss household energy demand mix of about 90 % for heat and 10 % for electricity. The same applies to many other European countries. The purpose of the present paper is to report on the progress of PSI's TPV project, which started in January 1997.

EXPERIMENTAL TEST SET-UP

The experimental test assembly consists of a small prototype TPV generator, Figure 1, a DC/DC converter and a grid-inverter. The TPV generator is based on a conventional butane burner with a nominal thermal output power of 1.35 kW. Gaseous butane is provided by evaporation of liquid butane contained in a steel cylinder (vapor pressure at 20 °C: 2.13 atm). The butane vapor flows through a gas valve into the injector needle and from there into the self sucking injector tube where the combustion air is taken in and mixed with the butane. The butane/air mixture then flows to the perforated burner tube which is surrounded by the emitter. The emitter is made of a porous tissue-like ceramic composite structure containing a mixture of several different rare earth oxides. Its spectral emittance is matched to the spectral sensitivity of crystalline silicon photo cells. To protect the silicon photo cells from overheating by infrared radiation with wavelengths longer than approximately 1200 nm, a cylindrical layer of water is arranged around the emitter. It is enclosed between two highly transparent con-

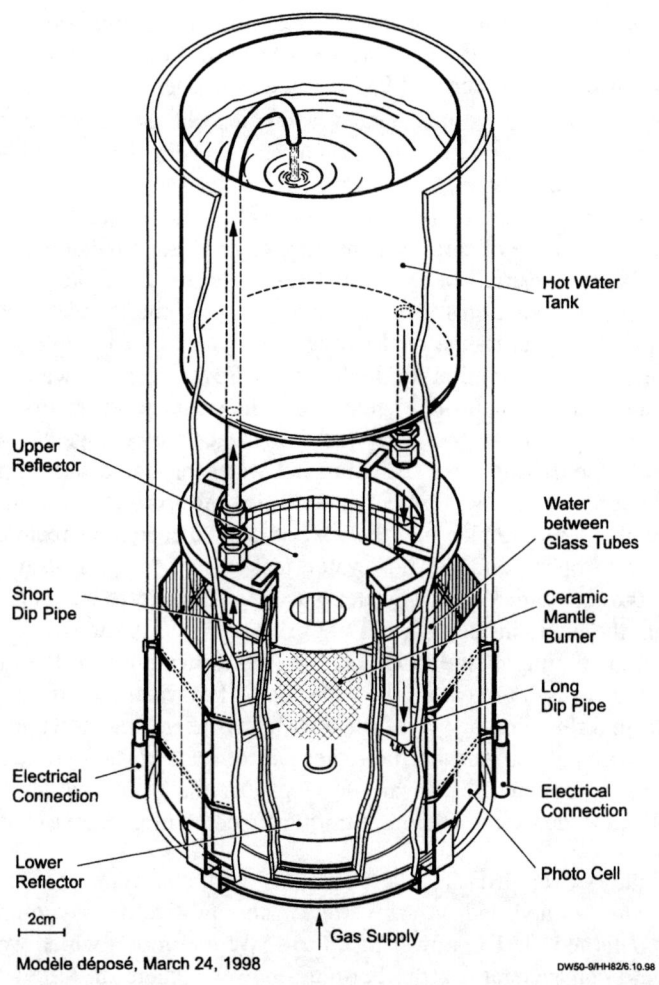

FIGURE 1. Prototype TPV generator with integrated hot water recirculation.

centric glass tubes which are glued to metallic flanges on both sides. It is known that water layers of a certain thickness have an excellent transmittance in the sensitivity range of crystalline silicon photo cells, and that water absorbs infrared radiation very efficiently. These properties make water an ideal medium in silicon based thermophotovoltaics. The inner tube of this selective filter protects the photo cells from direct contact with hot flue gas. For the photo cells, affordable crystalline silicon solar cells with a standard test condition (STC) efficiency of 16 % were used. To reduce electrical losses, standard cells of 10 cm by 10 cm were cut down across the tabbing bands to a size of 5 cm by 10 cm. To achieve an output voltage as high as possible, the cells were series-connected. As supporting structure for the cells, a cylinder of cheap polymethylmethacrylate, PMMA (Plexiglas) was taken and the cells were glued in in such a way that the different thermal expansions of silicon and PMMA at elevated temperatures did not lead to critical stresses. The active cell area was found to be 380 cm^2. Reflectors were used to reduce losses of emitter radiation at both ends of the cylindrical generator configuration. Their convex geometry was determined via a simplified ray tracing method. The upper reflector has a hole in its center to allow the flue gas to escape. The reflectors are made of ceramic substrates coated with a metallic alloy. Besides reflecting a useful fraction of the emitter radiation to the photo cells, they also lower the infrared radiation losses, leading to a welcome increase of the emitter temperature. The upper flange of the water filter is equipped with inlet and outlet tubes for a water recirculation loop. Hot water flows from the water filter into a hot water tank on top of the TPV generator. The steam pressure from the boiling water in the upper part of the water filter ring gap is used together with natural convection as the driving force. It pumps hot water through a short dip pipe up to the top of the hot water tank, while water is flowing from the bottom of the tank down through a long dip pipe into the water filter. To check this recirculation loop for continuous operation, the hot water tank was equipped with a coil heat exchanger through which cooling water was led. Its flow rate was adjusted to such a value that the water in the tank did not exceed 75 °C. A run over several hours proved the functioning of the loop. The flue gas flowing out of the inner tube of the water filter is used to heat up the water in the hot water tank. Furthermore, the thermal lift in a flue pipe on top of the PMMA cylinder (not shown in Fig. 1) was used to suck in ambient air into the lower part of the PMMA cylinder, where the photo cells are mounted. Fresh air flowing over the front and rear side of the silicon cells provided some cooling, limiting the cell temperature to about 55 °C.

A specially-developed DC/DC converter was connected between the cell generator and a modified grid-inverter made by the Swiss company, LEC, Küsnacht. This fed the DC power of up to 15 W and more, which was produced by the photo cell generator at an optimum power voltage of 3.3 V, to the local grid. The modification was carried out in cooperation with LEC on a product developed by that company for solar applications. An overall converter/inverter interface efficiency of 89 % was achieved. The converter was required because no inverters were available on the market working at input voltages lower than 24 V DC. The first grid interfacing experiments in November 1997 with commercially available DC/DC converters led to unsatisfactory results (very low interface efficiency due to low efficiency of a series connection of two industrial DC/DC converters).

For additional details of the TPV generator described above, the reader is referred to [2,3,4].

RESULTS

The mass flow rate of butane was determined by the weight loss of the gas cylinder within a measured period of time. The average of five measurements demonstrated a consumption of (113 ± 3.8) g/h, corresponding to a thermal power of (1435 ± 48) W, according to the lower calorific value of 45.73 MJ/kg. The photo cell generator was characterized using a slightly modified version of PSI's test facility for solar cells and modules [5,6]. A typical test record is shown in Figure 2. The characteristic data in Table 1 were derived from this.

TABLE 1. Characteristic data of the photo cell generator exposed to filtered emitter radiation.

Open circuit voltage, U_{oc}	4.48 V
Short circuit current, I_{sc}	5.01 A
Maximum power point voltage, U_m	3.33 V
Maximum power point current, I_m	4.56 A
Maximum power, P_m	15.2 W
Fill factor, FF	0.68

The short circuit current of 5.01 A is about three times higher than measured under a sunlight intensity of 909 W/m^2. The fill factor is relatively low as compared to the value of 0.75 under sunlight, indicating enhanced electrical losses in cells not designed for concentrated radiation. However, it is still comparable to that of one of the less efficient silicon cells under a one sun insolation.

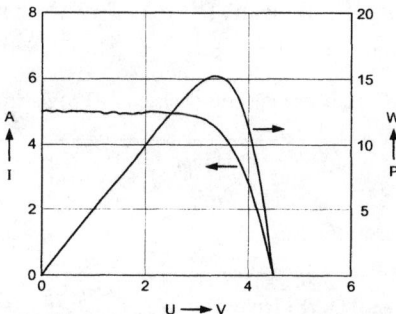

FIGURE 2. Current and power vs. voltage of the photo cell generator shown in Fig. 1 under actual operating conditions. The power characteristic exhibits a maximum power of 15.2 W at 3.3 V DC. The generator consists of eight series connected silicon cells with a combined active cell area of 380 cm^2. The fluctuation of the current in the otherwise constant region comes from the flaring emitter light.

The DC/DC converter and the grid-inverter were characterized using calibrated wideband power analyzers (VALHALLA, 2101 and Voltech PM 3000A) suitable for DC, sinusoidal and non-sinusoidal waveforms. The test set-up is shown in Figure 3.

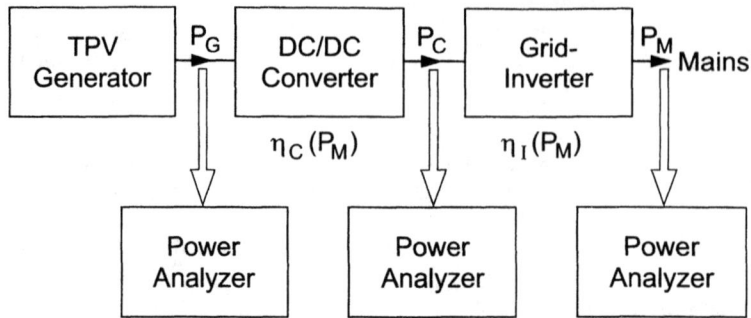

FIGURE 3. Measurement set-up for the electrical characterization of the grid-connected TPV generator shown in Fig. 1.

The fitted measurement results are shown in Figure 4. The efficiency of the DC/DC converter vs. the output power of the grid-inverter rises from 96 % at about 2 W to a maximum value of 98.3 % at 7.5 W and falls to 91.7 % at 50 W. Compared to commercially available DC/DC converters this is a remarkable result. The inverter efficiency reaches a maximum of 92.2 % at an output power of 25 W and sinks to 90.5 % at 50 W. Figure 4 also shows the overall efficiency of the converter and inverter, exhibiting a maximum efficiency of 89.3 % at 18.3 W and going down to 83 % at 50 W. The results in Figure 4 clearly demonstrate the importance of carefully designing and selecting the components to obtain an efficient system. The results of a typical TPV experiment interfaced to the grid are summarized in Table 2.

TABLE 2. System performance of the grid-connected TPV generator shown in Fig. 1.

Thermal (chemical) power of burner, W	1435
DC power of TPV generator, W	15.2
Thermal-to-DC efficiency, %	1.1
Efficiency of DC/DC converter, %	97.9
Output power of DC/DC converter, W	14.9
Efficiency of grid-inverter, %	90.8
Output power of grid-inverter, W	13.5
Overall (system) efficiency, %	0.94

Additional information on inverters useful for stand-alone and grid-connected operation of thermophotovoltaic generators is found in [7,8,9].

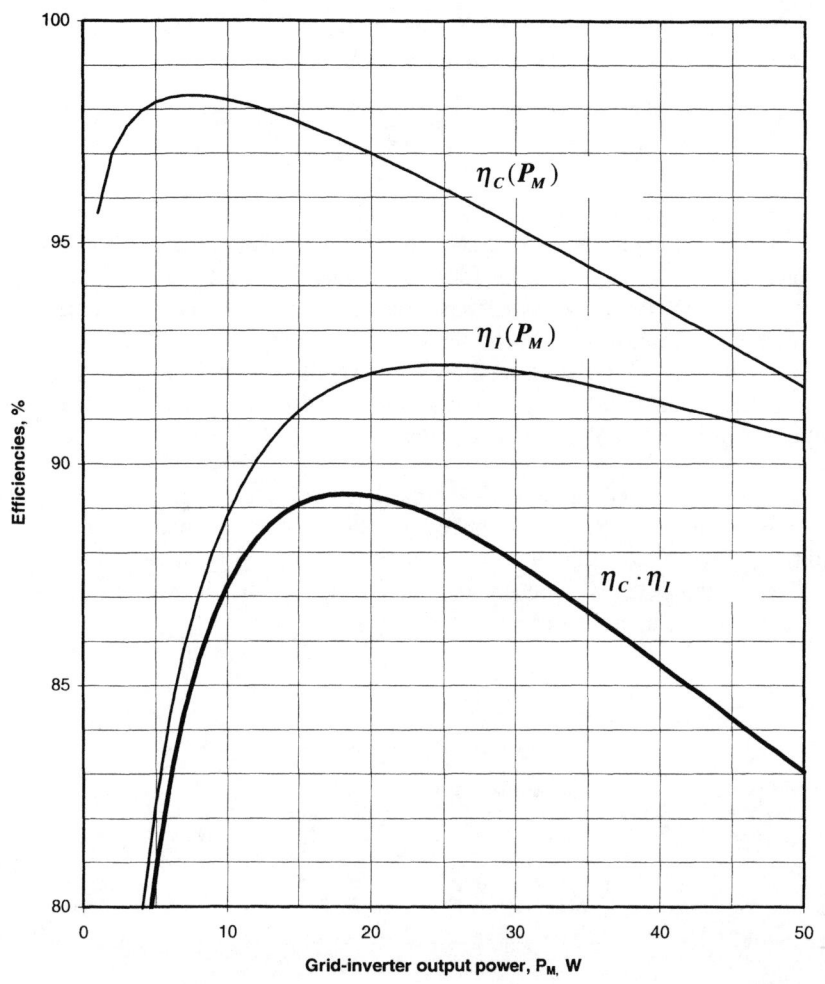

FIGURE 4. Efficiencies vs. grid-inverter output power. Upper curve: DC/DC converter, middle: grid-inverter, lower: overall efficiency of the series connection of the DC/DC converter and grid-inverter. The curves represent fitted measurements.

The output of the grid-inverter was analyzed using a Gould 7100 Power Analyzer. The RMS voltage and the power factor were found to be 232 V and 0.99. Furthermore it was found that the current harmonics meet the European Standards

EN 61,000.3.2 and EN 61,000.3.3 with respect to grid-interference. Additional information regarding these measurements is found in [9].

MODELLING OF PV GENERATOR

The photocells utilized in a TPV generator are exposed to radiation fluxes substantially higher than in the case of solar photovoltaics. Fluxes as high as 5 times, or even more, of the sun intensity for photon energies greater than the band gap energy are expected. At the same time, and in spite of active cell cooling, the working temperatures of the photocells in a TPV generator are also expected to be higher. A reliable extrapolation method for PV generator characteristics measured under solar conditions is very valuable in order to optimize the layout and design of silicon-cell-based TPV generators (band gap energy 1.12 eV).

The classical One-Diode-Model [10] has been extended in order to extrapolate and predict correctly the performance of solar photocells under TPV working conditions. The model accounts for the higher incident flux of radiation and for the higher temperatures reached in a TPV generator. Figure 5 shows the equivalent electrical circuit used:

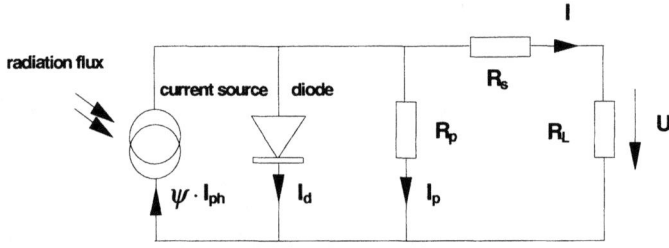

FIGURE 5. Equivalent electrical circuit for photocells.

Applying Kirchhoff's law (current balance) to this circuit leads to the characteristic equation for the current I as a function of the voltage U:

$$(1) \quad I = \psi \cdot I_{ph} - I_s \left[\exp\left(\frac{U + R_s I}{n U_T} \right) - 1 \right] - \frac{U + R_s I}{R_p}$$

where n is an empirical nonideality factor, U_T the thermodynamic voltage ($U_T = kT/e$) and ψ the ratio of the photo currents under the different radiation conditions. The reverse saturation current I_s of the diode (dark current) is a strong function of the temperature [11] and may be written as:

$$(2) \qquad \frac{I_s(T)}{I_s(T_0)} = \left(\frac{T}{T_0}\right)^3 \cdot \exp\left[-\frac{\varepsilon_{G,0}}{k}\left(\frac{1}{T}-\frac{1}{T_0}\right)\right]$$

where k is Boltzmann's constant and $\varepsilon_{G,0}$ the extrapolated band gap energy of the cell at $T = 0$ K ($\varepsilon_{G,0} = 1.17$ eV for c-Si cells). The temperature dependency of the series and parallel resistances R_s and R_p also has to be taken into account:

$$(3) \qquad R_k(T) = R_k(T_0) \cdot \left[1 + \alpha(T - T_0)\right] \quad \text{with } \alpha \approx 0.004$$

The ratio of the photo currents for TPV and solar radiation conditions is defined as:

$$(4) \qquad \psi = \frac{I_{ph,TPV}}{I_{ph,sun}}$$

In order to check the model, I/U measurements of the same photo cell string have been performed, both under solar and TPV radiation conditions. The string, consisting of eight cells connected in series (c-Si, ASE, Heilbronn, Germany), was first measured under the following conditions: $G_n = 909$ W/m² and $T = 37.3$ °C. A least squares fit for the main parameters of Eq. (1) gave the following results for one cell (active area = 47.5 cm², $\psi = 1$), Table 3:

TABLE 3. Parameters of one of the silicon photo cells shown in Fig. 1, under solar illumination

I_{ph}, A	1.605
I_s, nA	44.00
n	1.403
R_s, mΩ	5.65
R_p, Ω	71.1

These values are valid for one single cell at a reference temperature of $T_{ref} = 25\ °C$. However they can be used to calculate the characteristics of any array consisting of p parallel rows, each with s identical cells connected in series. The following relationships hold between one single cell (small letters) and an array consisting of s times p cells (capital letters):

(5) $\quad U = s \cdot u \qquad I = p \cdot i \qquad \dot{E}_{el} = s \cdot p \cdot \dot{e}_{el} \qquad I_s = p \cdot i_s$

(5) $\quad R_s = \dfrac{s}{p} r_s \qquad R_p = \dfrac{s}{p} r_p \qquad N = s \cdot n \qquad I_{ph} = p \cdot i_{ph}$

A second series of I/U measurements were then performed under typical TPV conditions. In this case Eq. (1) was used with the parameters as determined for the solar conditions, Table 3, and only ψ and T were adapted to the new conditions. The values found for these two parameters were: $\psi = 3.11$ and $T = 55\ °C$. Figure 6 shows the comparison between the measured TPV-characteristics (broken line) and the extrapolated solar-characteristics (solid line). The agreement of the model with the measurements is excellent.

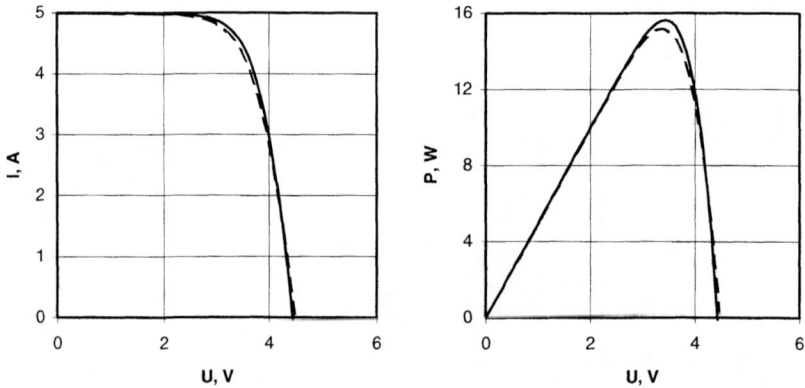

FIGURE 6. Comparison of measured and modeled characteristics of the eight cell string exposed to the emitter radiation of the thermophotovoltaic generator shown in Fig. 1. Left: current vs. voltage, right: power vs. voltage. Broken lines: measured, solid lines: extrapolated model.

In Fig. 7 the influence of cell temperature on the current I and power P of the eight cell string is shown. A temperature reduction of 30 °C would increase the output power from 15.6 to 20.2 W, clearly demonstrating the importance of good cell cooling.

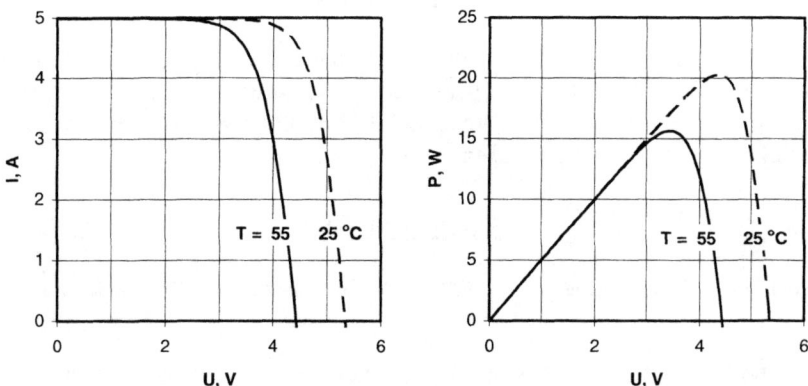

FIGURE 7. Influence of the cell temperature on current I and power P of the eight cell string generator shown in Fig. 1.

CONCLUSIONS AND OUTLOOK

The feasibility of interfacing a TPV generator to the local grid and the effective operation of such a generator under realistic conditions have successfully been demonstrated using a small prototype. Despite its simple construction and the cheap components from which it is made, the thermal-to-DC conversion efficiency of 1.1 % is among the highest achieved for silicon-cell-based small TPV generators without preheating the combustion air. Using PSI's expertise in solar photovoltaics and the know-how of the Swiss company LEC, Küsnacht, it was possible to achieve a grid-interfacing efficiency of 89 %. Even whilst keeping in mind that feeding 15 Watts DC at a voltage of 3.3 V DC to the 230 V AC grid is not within normal operational conditions, the efficiency of 89 % is remarkably high. Due to enhanced radiation intensities in TPV generators, particular attention should be paid to the electrical losses in the photo cell generator. The application of concentrator cells will therefore be taken into consideration. From the experiments performed so far, it is concluded that the conversion of chemically released energy on solid surfaces into useful radiation plays a key role in the TPV. In the future, therefore, more emphasis will be laid on the development of suitable emitters and on the understanding of the conversion mechanism. In the near future, a commercial 20 kW gas-fired household burner/boiler system will be reconstructed to provide a grid-connected TPV generator for co-generation of domestic heat, hot water and electricity. The development of a low band gap cell (silicon/ germanium) was recently started in parallel in PSI's materials department.

ACKNOWLEDGEMENTS

This work has been supported by the Research and Development Foundation of the Swiss Gas Industry (FOGA). Thanks are due to Friederike Geiger for coating ceramic substrates. We also thank the LEC, CH-8700 Küsnacht, Switzerland, for providing a tailored prototype grid-inverter.

REFERENCES

1. Schweizerische Gesamtenergiestatistik 1997. *Sonderdruck aus Bulletin SEV/VSE, Nr. 16/August 1998.*

2. Panitz, J.-Ch., Mayor, J.-C., Brunner, A., Durisch, W. and Schubnell, M. Development of Thermophotovoltaic Energy Converters. *PSI Annual Report 1997/Annex V, 33-34.*

3. Schubnell, M., Durisch, W., Grob, B., Mayor, J.-C. and Panitz, J.-Ch. Koproduktion von Wärme und Strom mit Thermophotovoltaik. *Gwa 78(1998)4, 248-253.*

4. Panitz, J.-Ch., Schubnell, M., Geiger, F. and Durisch, W. Influence of Ytterbium Concentration on the Emissive Properties of Yb:YAG and Yb:Y_2O_3. *Third NREL Conference on Thermophotovoltaic Generation of Electricity, AIP Conference Proceedings 401, 1997, p. 265-276.*

5. Durisch, W., Urban, J. and Smestad, G. Characterization of Solar Cells and Modules under Actual Operating Conditions. *Proceedings of the World Renewable Energy Congress*, June 1996, Denver, Vol. **1**, p. 359, Pergamon, Oxford, 1996.

6. Durisch, W., Tille, D. Wörz, A. and Plapp, W. 1998. Characterization of Photovoltaic Generators. Submitted to *ENERGY SOURCES.*

7. Durisch, W., Leutenegger, S. and Tille, D. Comparison of Small Inverters for Grid-Independent Photovoltaic Systems. *Renewable Energy 15 (1998) 585-589.*

8. Durisch, W. and Tille, D. Testing of Small Sinusoidal Inverters for Photovoltaic Stand-Alone Systems. Submitted to *ENERGY SOURCES.*

9. Durisch, W., Lavric, R. and Struss, O. Drei Netzverbund-Kleinsysteme im Vergleichstest. *Poster, Nationale Photovoltaiktagung, Bern, Switzerland, May 5, 1998*

10. Durisch, W. and Plapp, W. Charakterisierung von Grätzel-Zellen unter realklimatischen Betriebsbedingungen. *PSI-Bericht, August 1997*

11. Chatelain, J. D. and Dessoulavy, R. Traité d'Electricité, Vol.VIII: Electronique. *Presses Polytechniques Romandes, Lausanne. 3ème édition, 1989.*

POSTER SESSIONS

Performance Status of 0.55eV InGaAs Thermophotovoltaic Cells

S. Wojtczuk[a], P. Colter[a], G. Charache[b], and D. DePoy[b]

a) Spire Corporation, One Patriots Park, Bedford, MA 01730-2396
b) Lockheed-Martin Inc., Schenectady, NY 12301

Abstract: Data on ~0.55eV $In_{0.72}Ga_{0.28}As$ cells with an average open-circuit voltage (Voc) of 298mV (standard deviation 7mV) at an average short-circuit current density of 1.16 A/cm² (sdev. 0.1 A/cm²) and an average fill-factor of 61.6% (sdev. 2.8%) is reported. The absorption coefficient of $In_{0.72}Ga_{0.28}As$ was measured by a differential transmission technique. We use a numerical integration of the absorption data to determine the radiative recombination coefficient for $In_{0.72}Ga_{0.28}As$. Using this absorption data and simple one-dimensional analytical formula, the quantum efficiency and dark current/I-V of the above cells are modelled. Models show the cells may be limited more by Auger and radiative recombination rather than Shockley-Read-Hall recombination at dislocation centers caused by the 1.3% lattice mismatch of the cell to the host InP wafer.

Introduction

Low bandgap photovoltaic cells are needed for TPV systems which utilize an appreciable portion of the thermal photons emitted from the heat sources of 1200C or less (Table 1). In this paper, we give material and cell performance data on one important type of TPV cell which uses 0.55eV $In_{0.72}Ga_{0.28}As$, including measured absorption coefficient data.

Table 1 *% of Energy From Blackbody Falling Below Cutoff Wavelength*

Cell Bandgap	Cell Materials	Cell Cutoff	Blackbody Temperature		
			1000C	1100C	1200C
1.1eV	Si	1.1μm	0.9%	1.5%	2.2%
~0.7eV	Ge, GaSb, $In_{0.53}Ga_{0.47}As$	~1.8μm	12%	16%	20%
~0.55eV	$In_{0.72}Ga_{0.28}As$, InGaAsSb	~2.3μm	26%	31%	36%

CP460, *Thermophotovoltaic Generation of Electricity: Fourth NREL Conference*
edited by T. J. Coutts, J. P. Benner, and C. S. Allman
© 1999 The American Institute of Physics 1-56396-828-2/99/$15.00

Absorption Coefficient Data

In order to model quantum efficiency, absorption coefficient data are needed. We measured the transmission versus wavelength through two thicknesses of undoped 0.55eV $In_{0.72}Ga_{0.28}As$, 0.1 and 1μm thick. These epilayers were grown on semi-insulating InP wafers to lower free-carrier absorption (InP transparent at $\lambda > 930$nm). Double-side polished wafers were used to minimize surface scattering.

Transmission data was taken in 20 nm increments from 1200 to 2340nm using light from a grating monochromator at near-normal incidence on the polished InP side of the test sample. This air/InP interface reflects ~27% of the incident light (InP refractive index ~3.15 over 1200-2400nm[1]). The slight non-normal incidence eliminates reflected light. The transmitted light transits through the 600μm InP wafer, impinging on the $In_{0.72}Ga_{0.28}As$ film. The reflection at this middle InP/$In_{0.72}Ga_{0.28}As$ interface is negligible (~0.4%) since the refractive index of $In_{0.72}Ga_{0.28}As$ (estimated as 3.6) is close to InP. A portion of the beam is absorbed as it transits across the InGaAs film. Finally, the light impinges upon the $In_{0.72}Ga_{0.28}As$/air interface, where ~68% of the light exits out the epilayer and is collected into an integrating sphere.

We now discuss multiple reflections in the measurement. About 32% of the light beam is reflected at the final $In_{0.72}Ga_{0.28}As$/air interface, makes a second pass through the InGaAs, passes through the InP, and eventually ~27% is reflected from the initial InP/air interface, which then makes a third pass through the InGaAs before again arriving at the $In_{0.72}Ga_{0.28}As$/air interface. This sequence is repeated infinitely. Ignoring the InP/InGaAs interface due to its small reflection (~0.4%), and calling the air/InP and InGaAs/air reflections equal (R~0.3), the transmission is[2]:

$$T = \frac{(1-R)^2 e^{-\alpha L}}{1 - R^2 e^{-2\alpha L}}$$

To find the InGaAs absorption coefficient α at each wavelength, we ratio the measured transmission intensities T_1 and T_2 for thicknesses L_1 and L_2:

$$\log\frac{T_1}{T_2} = \log\frac{(1-R^2)e^{-\alpha L_1}}{(1-R^2)e^{-\alpha L_2}} + \log\frac{1-R^2 e^{-2\alpha L_2}}{1-R^2 e^{-2\alpha L_1}} \approx \alpha(L_2 - L_1) + O(0.1)$$

Now, exp(-2αL) can at most vary between 0 and 1. Therefore the natural log of the second fraction can at most be 0.105 (i.e., on the order of 0.1 or O(0.1)). Therefore, the data in Figure 1 is calculated from:

$$\alpha = \frac{\log(T_1/T_2)}{L_2 - L_1} \pm \frac{1}{2}\frac{O(0.1)}{L_2 - L_1} = \frac{\log(T_1/T_2)}{L_2 - L_1} \pm 500 cm^{-1}$$

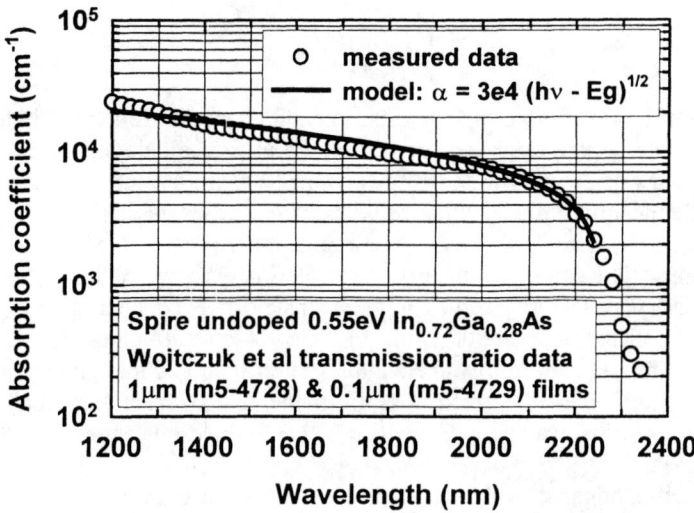

FIGURE 1. Measured 0.55eV $In_{0.72}Ga_{0.28}As$ absorption coefficient data and model fit.

Note that the above absorption data is accurate for a lightly-doped base region of a TPV cell, and for moderately doped P-type regions (before band-tailing due to impurity bands becomes noticeable). However, it is not at all accurate for a heavily-doped (>10^{19} cm^{-3}) N-type emitter region for an N-on-P cell. Charache et al. have shown[3] that heavily-doped N-type InGaAs exhibits a pronounced Moss-Burstein shift (i.e., semiconductors with a small effective mass become optically transparent when the available states at the conduction band edge are filled with equilibrium electrons). Heavily-doped N-type InGaAs is significantly more transparent than would be ordinarily expected. The optical bandgap is about 1 eV for a nominally 0.6eV-electrical-bandgap N-type InGaAs sample doped 10^{19} cm^{-3}.

Lifetime Model

The absorption coefficient data was numerically integrated to obtain the radiative lifetime recombination coefficient[4] B:

$$B = \frac{8\pi(kT)^3}{n_{iMM}^2 c^2 h^3} \sum_\lambda \frac{n_r^2 \alpha(\lambda)\left(\frac{hc}{\lambda kT}\right)^2}{\exp\left(\frac{hc}{\lambda kT}\right)-1}\left(\frac{hc\, d\lambda}{\lambda^2 kT}\right) = 2 \times 10^{-10} \frac{cm^3}{s}$$

where k, T, c, and h have their usual meaning, and the refractive index n_r of the material was assumed constant at 3.5 in this calculation, and the $d\lambda$ increment was 20nm with the range λ from 1200 to 2400nm. The intrinsic carrier concentration was obtained with effective masses that were linearly interpolated for the $In_{0.72}Ga_{0.28}As$ composition from the GaAs and InAs endpoints. The result was the effective electron mass was $0.036 m_o$ and the hole mass was $0.417 m_o$ leading to an intrinsic carrier concentration n_{iMM} of about 1.6×10^{13} cm^{-3} at 300K for $In_{0.72}Ga_{0.28}As$. Although we did not perform the integration to wavelengths shorter than 1200nm, almost all of the contribution to the integral is from the region around the bandgap edge; the omitted wavelengths contribute negligibly to the sum.

For small-bandgap doped semiconductors, Auger recombination is a major limit to the lifetime. Since the Auger recombination coefficient has not been directly measured for $In_{0.72}Ga_{0.28}As$, we will estimate it. For 0.74eV lattice-matched $In_{0.53}Ga_{0.47}As$ the intrinsic Auger recombination coefficient[5,6] C_{LM} is on the order of 10^{-28} cm^6/s. We calculate a similar coefficient C_{MM} for lattice mismatched $In_{0.72}Ga_{0.28}As$ by scaling this value according to the simple theory[7]:

$$C_{MM} \approx C_{LM} \frac{\left(\frac{qEg_{LM}}{kT}\right)^{1.5} \exp\left(\frac{1+2M_{LM}}{1+M_{LM}}\frac{qEg_{LM}}{kT}\right)}{\left(\frac{qEg_{MM}}{kT}\right)^{1.5} \exp\left(\frac{1+2M_{MM}}{1+M_{MM}}\frac{qEg_{MM}}{kT}\right)} \frac{n_{iLM}^2}{n_{iMM}^2} = 5.2 \times 10^{-28} \frac{cm^6}{s}$$

where Eg are the bandgaps, M_{MM} is the mass ratio for lattice mismatched material (m_e/m_h or 0.036/0.417), M_{LM} is lattice-matched (0.044/0.436), and the lattice-matched intrinsic carrier concentration n_{iLM} is 8.7×10^{11} cm^{-3}. The overall lifetime is calculated as:

The overall lifetime is calculated as:

$$\tau(n) = \frac{1}{\dfrac{1}{\tau_{RADIATIVE}} + \dfrac{1}{\tau_{AUGER}} + \dfrac{1}{\tau_{SRH}}} = \frac{1}{Bn + Cn^2 + \dfrac{1}{\tau_{SRH}}}$$

where the above expression is identical for electrons or holes. We have calculated B and C above theoretically. We included the non-radiative Shockley-Read-Hall lifetime τ_{SRH} for completeness; however, the open-circuit voltage of the model described later in this paper agrees with measured cells quite well without the SRH lifetime which is due primarily to recombination at dislocations. The open-circuit voltage is used for this purpose since it is the cell parameter that is most sensitive to dark current and therefore lifetime. **This is a surprising result. The cell is dominated by Auger recombination at doping higher than 4×10^{17} cm^{-3}, and by radiative recombination at lower doping. The SRH dislocation lifetime is large (i.e. negligible) in the present 1.3% mismatch InGaAs.** This data can be used to bound the dislocation lifetime limit as on the order of 100nS or more. Figure 2 shows our lifetime model versus doping.

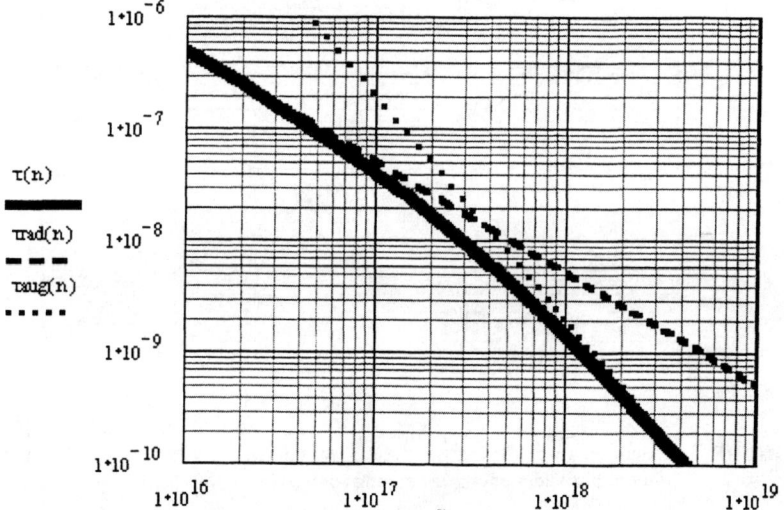

FIGURE 2. Calculated minority carrier (either electron or hole) lifetime vs. doping for 0.55eV $In_{0.72}Ga_{0.28}As$. Thick solid line is overall lifetime; dotted line is Auger recombination, which is dominant at dopings above 4×10^{17} cm^{-3}, and dashed line is the radiative recombination lifetime, which is dominant at lower dopings.

Mobility and Diffusion Length Models

The minority carrier diffusion lengths must be estimated to model a cell. We have measured *majority* carrier electron and hole mobility over a wide range of dopings[8] for 0.55eV InGaAs material. An empirical model fit to the referenced data is given below:

$$\mu e(n) = 1000 + \frac{10000}{1+\sqrt{\frac{n}{4x10^{17}}}} \qquad \mu h(p) = 60 + \frac{120}{1+\sqrt{\frac{p}{4x10^{17}}}} \quad \frac{cm^2}{Vs}$$

We assume that minority carrier electrons in P-material will have a mobility equal to electrons in N-material with the same doping level. This is equivalent to saying that the ionized impurity scattering is proportional to the total number of ionized impurities but is independent of the sign of the impurity charges. Using the above mobility and lifetime models, we estimate the diffusion lengths shown in Figure 3.

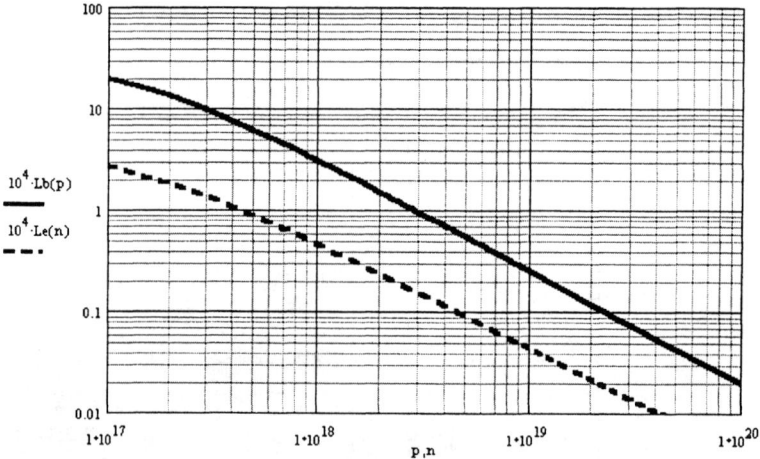

FIGURE 3. Calculated minority carrier base electron (solid line) and emitter hole (dotted) diffusion lengths vs. doping for 0.55eV $In_{0.72}Ga_{0.28}As$.

Quantum Efficiency Models

With the absorption coefficient and diffusion length data given earlier, it is now possible to model the N-on-P ~0.55eV cell described in Table 2.

Table 2 *Structure of N/P $In_{0.72}Ga_{0.28}As$ TPV Cell (Spire 6114-4646-10)*

Layer	Composition	Doping cm^{-3}	Thickness μm
Emitter	$In_{0.72}Ga_{0.28}As$	Se, 4×10^{19}	0.1
Base	$In_{0.72}Ga_{0.28}As$	Zn, 3×10^{17}	3.6
BSF	$InAs_{0.4}P_{0.6}$	Zn, 9×10^{17}	0.05
Grade	$In_{0.72\ to\ 0.53}Ga_{0.28\ to\ 0.22}As$	Zn, 1×10^{19}	4.0
Wafer	InP	Zn, 3×10^{18}	600

Hovel[9] gives a complete model of photocurrent contributions from the emitter, base, and space-charge region. However, the heavily-doped emitter in our N-on-P cell is only 0.1μm, and because of the Moss-Burstein shift, this layer is transparent to wavelengths longer than 1200nm. Since shorter wavelengths are of limited interest for TPV cells, modeling the emitter is not profitable. Similarly, the above dopings give a space-charge region of only 0.05μm. Negligible error is made if we assume all light is absorbed in the base. Thus, we use Hovel's term:

$$\eta = \frac{R\alpha L}{\alpha^2 L^2 - 1}\left(\alpha L - \frac{\frac{SL}{D}\left(\cosh\left(\frac{t}{L}\right) - e^{-\alpha t}\right) + \sinh\left(\frac{t}{L}\right) + \alpha L e^{-\alpha t}}{\frac{SL}{D}\sinh\left(\frac{t}{L}\right) + \cosh\left(\frac{t}{L}\right)}\right)$$

where η is the external quantum efficiency, the term R accounts for the surface reflectivity (~0.7), L is the base electron diffusion length at the doping given in Table 2 (see Fig.3, 10μm), D is the diffusivity at that same doping calculated from the given mobility model (~170 cm^2/s), S is the back surface interface recombination velocity between the $In_{0.72}Ga_{0.28}As$ and $InAs_{0.4}P_{0.6}$ taken as 10^4 cm/s and t is the cell thickness (3.6μm - the sum of the emitter and base epilayers).

Figure 4 shows the fit between the above model and the actual measured data for the cell. The fit is fairly good; the only slight adjustment made to the model described above is that the bandgap was changed from 0.55eV to 0.57eV to match the model cutoff wavelength with the measured cell data. In practice, this is a shift of about 2% indium composition, which is consistent with our long-term run-to-run uniformity in the metal organic chemical vapor deposition (MOCVD) epitaxial growth of these structures.

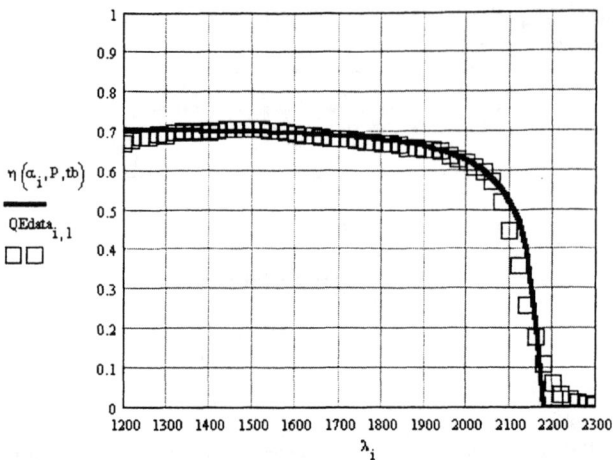

FIGURE 4. Comparison of measured QE data (squares) of cell of Table 2 with model of cell given by the above equation (solid line).

Dark Current Model

Generation-recombination (n=2) dark current is only ~1% of the diffusion dark current for these cells at 0.3V bias. Therefore, as expected, diffusion current dominates these low-bandgap cells, and the space-charge-region generation-recombination term can be ignored. Emitter hole diffusion length L_{EMIT} from Figure 3 is 0.01μm for the structure of Table 2. The emitter is 10 times thicker, and therefore, a simple long-emitter diffusion model is appropriate. **Note that this implies that the cell dark current is quite insensitive to the front surface recombination velocity.**

The base electron diffusion length L_{BASE} is 10μm (same as in QE model) from Figure 3 for the doping in Table 2. The base thickness (3.5μm) is smaller than L_{BASE}; therefore, we expect that a low recombination back surface field $InAs_xP_{1-x}$ layer can help the cell. The dark current[10] model is:

$$JoB = \frac{qD_{BASE}n_i^2}{L_{BASE}P} \left(\frac{\frac{SL_{BASE}}{D_{BASE}}\cosh\left(\frac{t}{L_{BASE}}\right) + \sinh\left(\frac{t}{L_{BASE}}\right)}{\frac{SL_{BASE}}{D_{BASE}}\sinh\left(\frac{t}{L_{BASE}}\right) + \cosh\left(\frac{t}{L_{BASE}}\right)} \right)$$

$$JoB = \frac{qD_{BASE}n_i^2}{L_{BASE}P} 0.385 = 8 \times 10^{-6} \, A/cm^2 \quad JoE = \frac{qD_{EMIT}n_i^2}{L_{EMIT}N} = 1.7 \times 10^{-6} \, A/cm^2$$

All of the terms have the same values and meaning as in the QE model section (e.g. S is 10^4 cm/s) and n_i is the same value (1.6×10^{13} cm^{-3}) as in the radiative lifetime model. Dopings and thicknesses are from Table 2. D_{EMIT} is 1.8 cm^2/s using the mobility formula given earlier. Note that the BSF lowers the base diffusion current factor JoB to only 39% of the value that would occur with a very thick base, adding about kT/q*ln(1/0.39) or ~25mV to the Voc. The dark current from the emitter JoE is not negligible and is roughly 20% of the total diffusion current.

I-V Model

The I-V model is compared using the above dark currents and a cell series resistance[11] Rs (0.025 Ω-cm^2) and short-circuit current density Jsc (1.1 A/cm^2) with an actual I-V (Figure 5) using the model equation:

$$J = (JoE + JoB)\left(\exp\left(\frac{q(V - JRs)}{kT}\right) - 1\right) - Jsc \quad \frac{A}{cm^2}$$

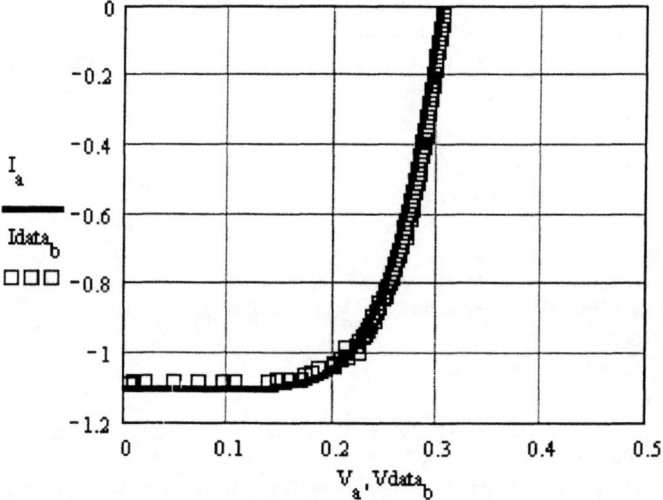

FIGURE 5. Measured IV data (squares) for 1 cm^2 cell of Table 2 vs. model (line).

Table 3 shows the data agreement. As a note, the series resistance Rs is limited mainly by the back alloyed AuZn contact to the InP wafer[10].

Table 3 *Comparison of Measured I-V data vs. Dark-Current I-V Model*

	Cell 6114-4646-10	Wafer Avg. 6114-4646	I-V Model
Voc	305mV	298mV	310mV
Jsc	1.1A/cm^2	1.2A/cm^2	1.1A/cm^2
Fill	64%	62%	65%

Summary

We present a complete quantum efficiency and dark current model of an N/P $In_{0.72}Ga_{0.28}As$ cell. Fits between model and measured data are good. We describe a measurement and give absorption coefficient data for 0.55eV $In_{0.72}Ga_{0.28}As$, which is not widely available. Interesting results are: (1) lifetime in 1.3% lattice-mismatched $In_{0.72}Ga_{0.28}As$ seems limited by Auger and radiative recombination and not dislocation defects; (2) an N/P cell with a heavily doped emitter has a hole diffusion length of ~0.01µm and we predict it is insensitive to front surface recombination since emitters tend to be ~10X thicker than this in order to lower resistance and limit I^2R loss; (3) an effective back surface field can add 25mV to the Voc versus a long base cell with no BSF; and (4) light absorption in N^{++} emitter of an N/P cell is smaller than expected due to the Moss-Burstein shift.

References

[1] Palik, E.D., *Handbook of Optical Constants*, Academic Press, 1985, p. 512.

[2] Pankove, J.I. *Optical Processes in Semiconductors, Dover, 1975, p. 93.*

[3] Charache, G.W., DePoy, D.M., Egley, J.L., Dziendziel, R.J., Freeman, M.J., Baldasaro, P.F., Campbell, B.C., Sharps, P.R., Timmons, M.L., Fahey, R.E., Zhang, K., Borrego, J.M., "Electrical and Optical Properties of Degeneratively-Doped N-type $In_xGa_{1-x}As$", 3rd NREL TPV Conf., AIP Proc. 401, 1997, pp.215-226.

[4] Pankove, op. cit., pp. 108-111.

[5] Landsberg, P.T. *Recombination Process in Semiconductors, Cambridge, 1993.*

[6] Ahrenkiel, R.K., Ellingson, R., Johnston, S. and Wanlass, M., "Recombination lifetime of $In_{0.53}Ga_{0.47}As$ vs. doping density," Appl. Phys. Lett., Vol. 72, 26, 1998, pp. 3470-3472.

[7] Pankove, op. cit., p. 162.

[8] Wojtczuk, S. "Comparison of 0.55eV InGaAs Single-Junction vs. Multi-junction TPV Technology," 3rd NREL TPV Conf., AIP Proc. 401, 1997, pp.205-213.

[9] Hovel, *Solar Cells*, Semiconductors and Semimetals v.11, Academic, 1975, pp.17-20.

[10] Hovel, op. cit., p. 51.

[11] Wojtczuk, S. "$In_xGa_{1-x}As$ TPV Experiment-based Performance Models," 2nd NREL TPV Conf., AIP Proc. 358, 1995, pp. 387-393.

Fabrication and Electrical Characterization of 0.55eV N-on-P InGaAs TPV Devices

W. Nishikawa, D. Joslin, D. Krut, J. Eldredge, A. Narayanan*, M. Takahashi, M. Haddad, M. M. Al-Jassim ** and N.H. Karam

Spectrolab Inc., 12500 Gladstone Ave., Sylmar, CA, 91342
*HRL Laboratories L.L.C., Malibu, CA, 90265
** National Renewable Energy Laboratory, Golden, CO, 80401

Abstract: Results are presented on the characterization and testing of lattice-mismatched 0.55 eV InGaAs/InP thermophotovoltaic (TPV) cells. A robust cell fabrication technique amenable to high throughput production is presented. A versatile light and dark I-V set up capable of fast screening of the TPV cells and an innovative approach for screening high performance cells are presented. We also report on the effect of lattice-matched InAsP and InAlAs back surface field layers on the performance of the TPV cells.

INTRODUCTION

Thermo-photovoltaic (TPV) energy conversion has recently experienced a strong resurgence in research efforts based on the demand for self-contained power sources and on technological developments in InGaAs and InSb tandem solar cells. Current TPV systems have focused on low-temperature blackbody emitters between 1000°C and 1500°C, where, direct low-bandgap PV cells and system hardware can operate practically. Ternary InGaAs and quarternary InGaAsSb material systems are being engineered to match emitter peak wavelengths between 1.9-2.6 μm. As a major provider of compound semiconductor multi-junction space solar power, Spectrolab sees considerable synergy in TPV devices and PV solar cells. This paper summarizes the achievement of the relatively short effort to produce lattice-mismatched 0.55eV InGaAs devices for 1000C radiator applications.

MATERIAL GROWTH AND CHARACTERIZATION

The lattice mismatched InGaAs/InP TPV device structures were grown using a vertical, rotating-disk, low-pressure MOVPE reactor. Films were grown on p-type (3-5 x10^{18} cm^{-3}) InP substrates with direct (100) and 2° off (100) → [110] orientations. A schematic of the TPV device structure is shown in Figure 1.

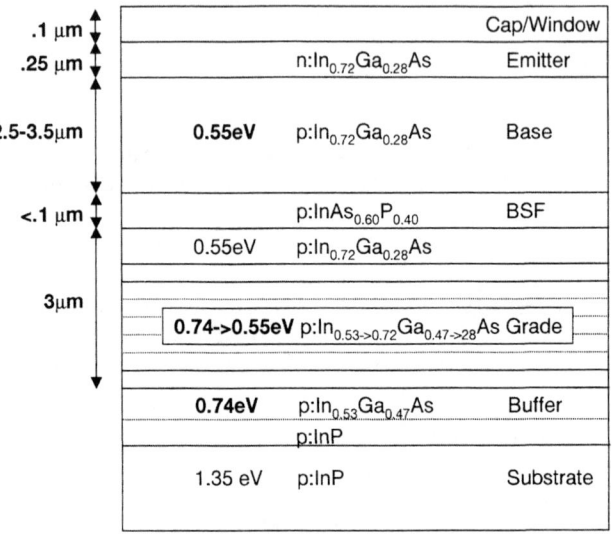

FIGURE 1. Schematic of optimized TPV structure

A step-graded buffer was used to accommodate the lattice-mismatch between the 0.74 eV In$_{0.53}$Ga$_{0.47}$As on InP and the 0.55 eV In$_{0.72}$Ga$_{0.28}$As. The step-graded structure is designed to introduce uniform misfit dislocations at each step and prevent them from threading to the active device region. Figure 2 shows a cross-sectional transmission electron microscopy (XTEM) image of the full device structure, where the dislocations are deflected parallel to the step-graded layers and away from the active region of the device. The surface morphology of good devices had a fine, cross-hatched appearance, as expected from relaxed layers.

The n on p device structures had a base layer thickness in the range of 2.5 -3.5μm and n-emitter thickness ~.25μm. A heavily-doped cap was grown as the last step to form an ohmic contact. The Moss-Burstein shift caused by heavy Sn doping (>3.0x10^{19} cm^{-3}) provided an optical transparency, in wavelength range below the 0.55eV InGaAs bandgap, which simplifies cell processing and improves emitter surface recombination (1).

It is critical to lattice-match the back surface field (BSF) to the 0.55eV InGaAs base layer. Although it is easier to grow $In_{0.72}Al_{0.28}As$ lattice matched to 0.55 eV InGaAs than $InAs_{0.4}P_{0.6}$, the latter is the preferred BSF choice. Oxygen incorporation in Al containing layers is problematic as it increased interfacial defects and back surface recombination. Figure 2b shows a XTEM of a lattice matched $InAs_{0.4}P_{0.6}$ BSF to 0.55 eV InGaAs base.

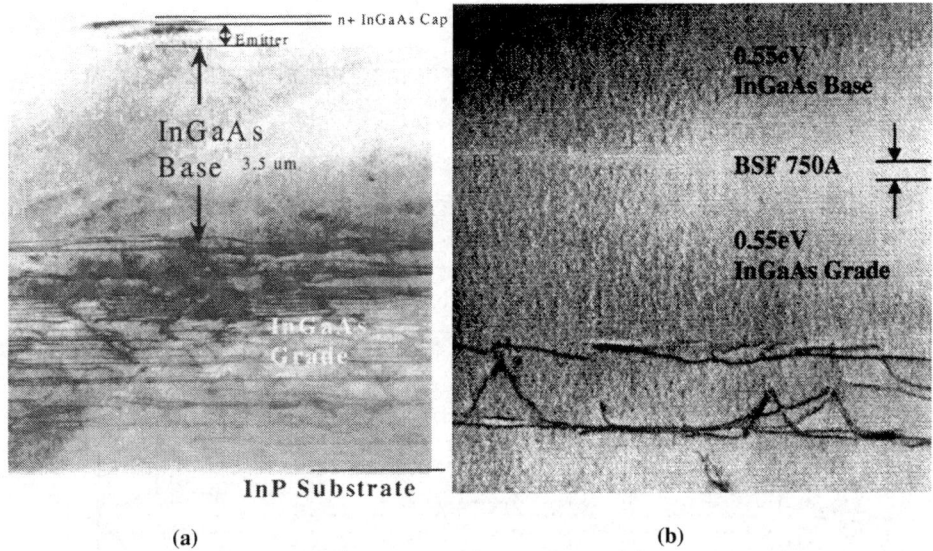

FIGURE 2. XTEM image of a full TPV device structure without a BSF (a), and a higher magnification XTEM illustrating $In_{0.72}Ga_{0.28}As$ Base/$In_{0.40}As_{0.60}P$ BSF interface (b).

TPV DEVICE FABRICATION PROCESS

The conventional device fabrication process for TPV and most other photovoltaic devices involves formation of front and back contacts, deposition of anti-reflection coating (if required), and cell isolation. Typically, front-side grid deposition necessitates a two-step photolithographic process of 1) masking and etching the optically-absorbing cap layer outside of the grid lines, and 2) depositing front metal and forming the grid lines via a lift-off process. On the backside, a low-resistance substrate contact must be formed to mitigate the I^2R power losses, which becomes more severe at high current densities. Forming reliable ohmic contacts to p-type InP is difficult since Zn-doped InP substrates doped greater than $3x10^{18}$ cm^{-3} are not readily available.

We have developed a manufacturable process for TPV cells streamlining both front and back contact steps. Following MOVPE growth, back contacts to the InP substrates (Zn-doped $\sim 3x10^{18}$) were sequentially sputtered Au/Zn/Au, where the composition was determined by layer thickness. The initial Au nucleation layer enhanced the Zn doping at the surface (2). Contact sintering catalyzed Au_2P_3 formation at the semiconductor/metal interface. Specific contact resistances in the range of 1×10^{-4} Ohms-cm^2 were achieved for two temperature conditions at 410°C and 430°C, as shown below in Figure 3. Visually, the back contact appeared matte with a pink color, indicative of the Au_3In phase at the metal surface.

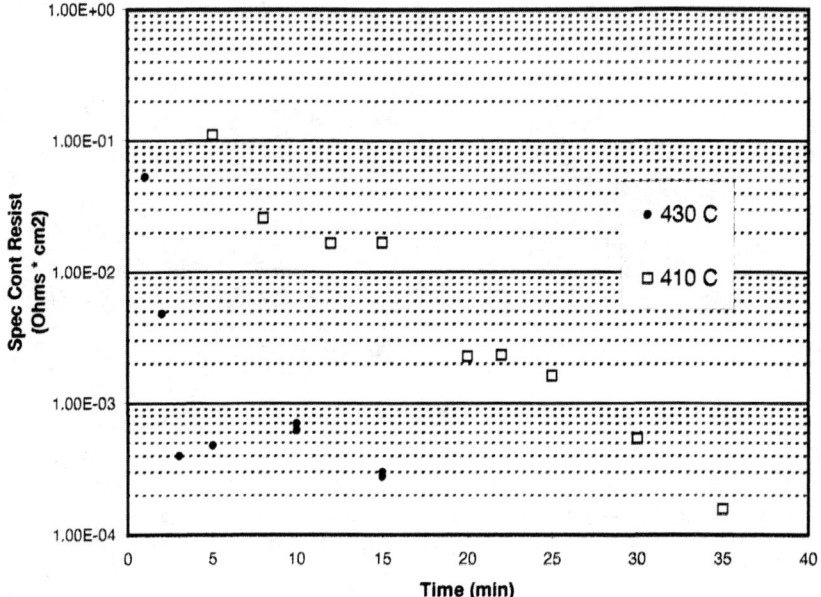

FIGURE 3. Au/Zn/Au Back Contact Optimization to P-Type InP Substrates

Benefiting from the optical properties of the n^{++} cap layer, the front-side processing required only one photomask to define front-metal grids. Narrowly-spaced 10μm grid lines were defined using a two-layer resist process to improve aspect ratios and lift-off profiles.

A non-alloyed Cr/Au/Ag contact is easily achieved for the front contact. Metal adhesion to InGaAs can be improved short sintering. The majority of the metal thickness was contained in the Ag layer (>5 μm). To complete the TPV fabrication sequence, individual cells were saw-diced rather than mesa isolation., thereby eliminating another photolithography and etch process.

From 2-inch InP wafers, twelve 1cm x 1cm cells were fabricated. Cell processing of full devices is routinely accomplished in a single day using this process, compared to the two-day conventional process.

ELECTRICAL TESTING

Full electrical characterization of devices included in-house measurements of light I-V, dark I-V, and cell spectral response(SR) or EQE. Spectral reflectance measurements were also taken on selected samples to convert EQE to internal quantum efficiency (IQE).

An innovative pulsed-light I-V system was developed for fast and easy characterization of the TPV cells. Figure 4 is a schematic of the test station designed to produce necessary current generation in the TPV device under test to simulating in-system performance. The light generation was provided by a tungsten-halogen light source with a gold-coated, non-image forming reflector. A long-pass filter with a cutoff at 1μm was implemented to remove the visible and some near-IR light that is not found in the 1000°C blackbody spectrum.

The custom fixture was designed for a "true" 4-point probe measurement to prevent back contact problems and non-destructive testing of TPV cells. Spectrolab designed an automated cell evaluation (ACE) data acquisition system capable of measuring high current levels up to 5 amps with sweep rates in the 100ms range was used for rapid collection of light and dark IV curves.

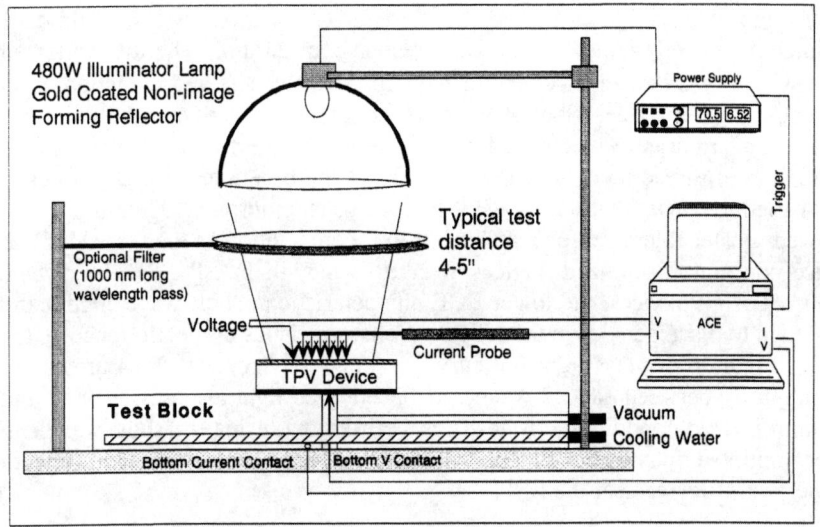

FIGURE 4. TPV Light-IV Test Station with Test Fixture

To avoid thermal heating problems, the device-under-test is exposed to a short (~1sec) illumination and IV response is swept over 100ms. The light source intensity is allowed to reach its maximum where no significant heating of the cell has been observed.

Cells are first tested using in-house SR/EQE system in the range between 900nm and 2300nm at 25nm intervals. EQE measurements are not only valuable tools in measuring and isolating problematic areas in the cell structure, but also provide the integrated Jsc value for devices tested under high concentration conditions. Standard test procedures sets light intensity to generate between 3-3.3A (3.15 nominally) for cells exhibiting near "respectable" QE performance. For cells with below-normal QE, light intensity for the light IV test is adjusted to generate somewhat lower current in the device.

DEVICE PERFORMANCE RESULTS

High performance TPV devices are easily achievable in the lattice mismatched 0.55 eV InGaAs/InP TPV cells by maximizing the diffusion length for a given doping concentration in the base and carefully passivating the front and back side of the cell. The BSF/base interface is critical to efficient carrier collection and high EQE. Figure 5 shows a comparison of the EQE for cells grown with lattice matched InAsP, InAlAs and without a BSF. The 1cm^2 cells of wafer SPL#1 with InAsP BSF produced the highest EQE values peaking at 56.7% at 2.0 µm. These measurements are comparable to the highest reported EQE values reported for 0.55eV InGaAs devices (3). Base and BSF thickness optimizations were guided by the EQE response at 2.0µm. The motivation of a thinner base was the enhancement of the BSF effectiveness and reduction of dark current volumes. As described in the next section, lower dark current levels (I_{dark}) have been empirically correlated to increased V_{oc} performance. To prevent generation of threading dislocation networks at the base interface, BSF thickness was varied to ensure that the InAsP layer was psuedomorphic. Thinning the BSF allowed greater tolerance for matching InAsP composition to the 0.55eV InGaAs lattice constant. Optimized devices were grown with InAsP BSF layers below 1000Å. For all devices, the lower EQE characteristic at short wavelength can be attributed to discrepancies in the reflection characteristics of the reference cell.

Figure 6 displays light I-V curves for TPV devices with V_{oc} outputs up to 301mV at J_{sc} between 3.0-3.7 A/cm^2. Concentrated light testing was done under 1000nm filtered conditions. It has been verified that using the short wavelength filter improved open circuit voltages ~ 5mV due to the increased carrier injection deeper in the device near the BSF.

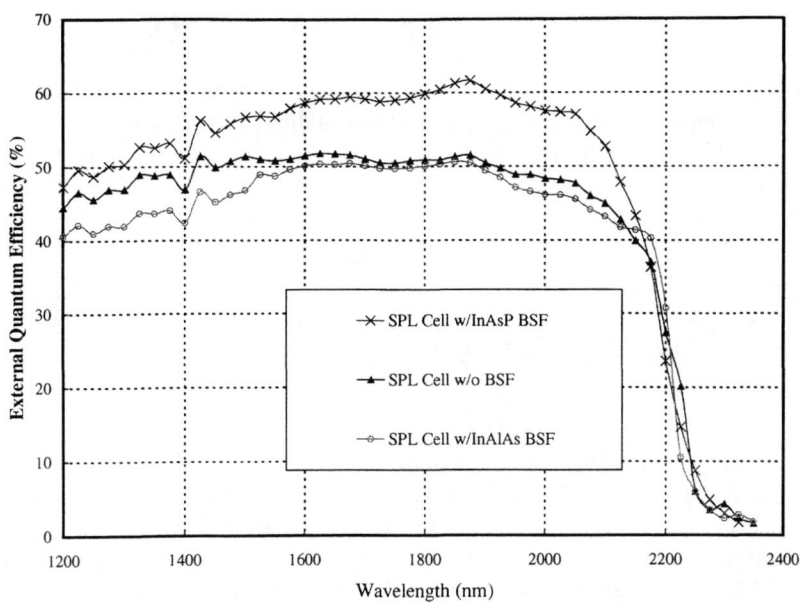

FIGURE 5. External QE Measurements of Spectrolab 0.55eV InGaAs TPV cells.

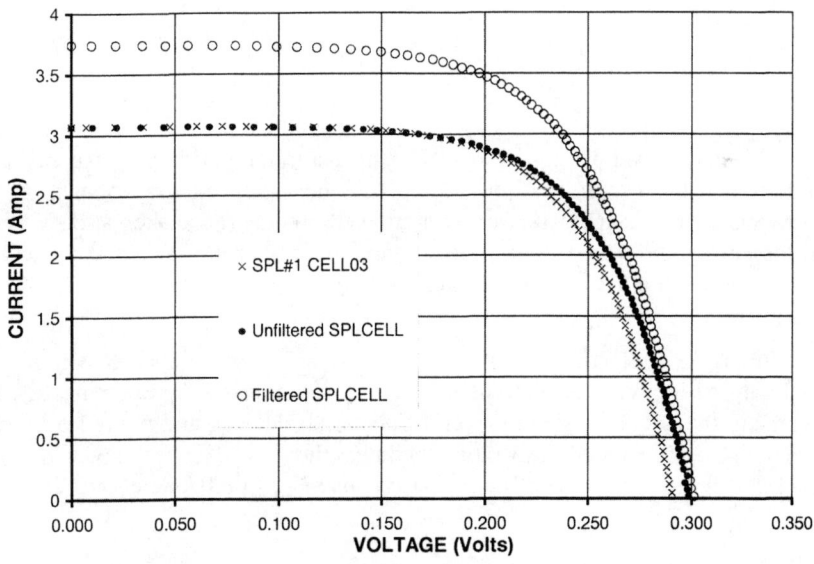

FIGURE 6. Light I-V Measurements of Spectrolab 0.55eV InGaAs TPV cells.

EMPIRICAL DEVICE ANALYSIS

The EQE, light I-V, and dark I-V results provide insight as to what portions of the device are most important to output. Data from the dark I-V characteristic, where the current is measured over several orders up to 3 A/cm^2, can be fit by the simple diode equation. By using a least squares technique to perform this fit, the ideality factor (n), the reverse saturation current (I_o), and the grouped series resistance (R_{series}) terms can be extracted from the simple form of the diode equation (4)

$$I = I_o \left(\exp \frac{q(V + I \cdot R_{series})}{n \cdot k \cdot T} - 1 \right) \tag{1.1}$$

We have calculated the R_{series} to be approximately 3 milliohms (mΩ)-cm^2 for 1x1 cm fully processed cells with Au/Zn/Au back contacts. For purposes of comparison, it can be shown from numerical simulation of the light I-V characteristic that R_{series} value of 10 milliohm-cm^2 will reduce output of the device by 10.5% and must be kept below 5 mΩ-cm^2 to reduce power loss.

When the diode expression is evaluated under illumination at open circuit, it can be shown that

$$\ln(I_o) = \ln(I_{sc}) - \frac{q \cdot V_{oc}}{n \cdot k \cdot T} \tag{1.2}$$

If I_{sc} is held constant in equation (1.2), the natural logarithm of the reverse saturation current I_o can be plotted versus the open circuit voltage. I_o is proportional to the device dark current at a given voltage (I_{dark} taken at 0.2V near load), assuming series and shunt components are negligible. When the ideality factor for various devices is similar, I_{dark} values may be compared and taken as a metric for I_o.

In Figure 7, I_{dark} @ 0.2V is plotted on a log scale against V_{oc} where correlation of high V_{oc} correlate to low I_o can be understood. For purposes of comparison, the data was fit for several Spectrolab devices and plotted over the voltage range that included the V_{oc} of benchmark devices. The best device SPL#1 Cell03, shown in Figure 6 with V_{oc} of 291mV, had I_{dark} of 0.07A/cm^2 at 0.2V.

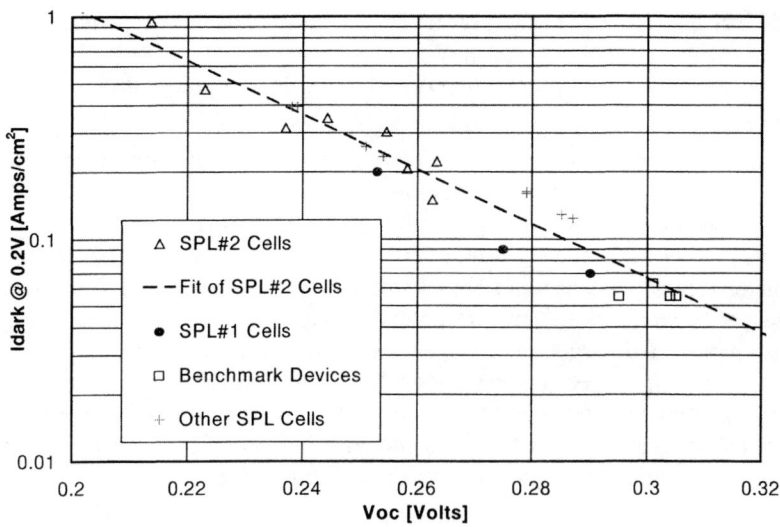

FIGURE 7. Semi-log Plot of I_{dark}@0.2V versus Voc

In a similar fashion, EQE data can be related to the dark current parameters. From one-dimensional transport equations, QE or current generation can be derived as a function of minority carrier diffusion length (L_n) assuming a semi-infinite absorber (base thickness) (5). Since the saturation current I_o is inversely proportional to L_n, the QE function can be approximated as directly proportional to $1/(1+I_{dark})$. In the following Figure 8, the QE results are plotted in this fashion for the same devices in Figure 7.

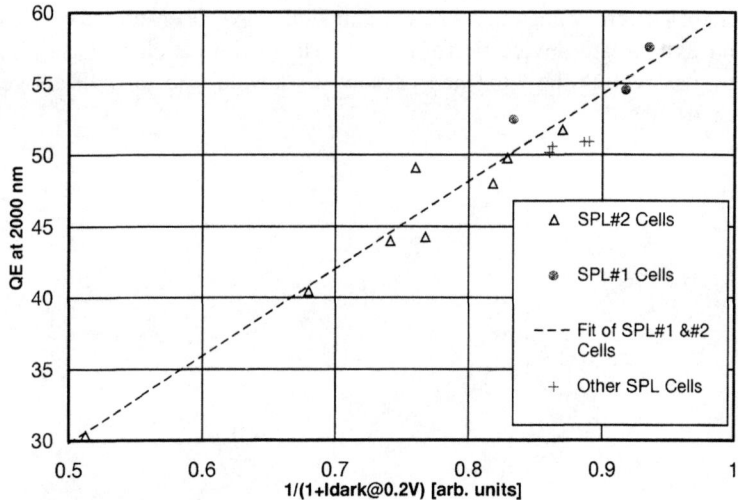

FIGURE 8. Plot of EQE at 2000nm versus 1/(1+Idark) for Spectrolab devices

SPL#1 Cell03 yielded the highest EQE measured at 57.6% corresponding to the highest V_{oc} at 291mV and lowest I_{dark} of that wafer. These diagnostic methods were valuable in anticipating V_{oc} through least-squares fitting of EQE and dark I-V data.

The implication from these analysis methods is that current collection from deeply absorbed photons is improved with low I_o values. These low I_o values are shown to correlate strongly with high V_{oc} performance. When EQE at 2.0μm and V_{oc} is high, the corresponding device efficiency will also be high. This reduces sole dependence on light I-V measurement to assess whether a given device performs well. However, these plotted relationships cannot reveal certain materials characteristics such as diffusion length or back surface recombination. Both EQE and V_{oc} losses are controlled by short diffusion lengths and high interface recombination generated from defected material and high I_o values. For this assessment TEM and material characterization techniques such as time-resolved photo-luminescence gather more direct information.

SUMMARY

Device processing and electrical characterization techniques for 0.55eV InGaAs TPV cells on InP have been investigated. InAsP BSF layers have shown to obviate oxygen gettering and improve interface quality. The device structure, with the heavily-doped cap layer, offered the advantage of 1-mask processing and high throughput manufacturing. Full electrical testing (light I-V, dark I-V, EQE) was taken to understand how sections of the device are operating. The link between reduced dark current levels and improved Voc and EQE was formulated. Plotting dark current values against V_{oc} and EQE have been effectively used to gauge and extrapolate device performance. The optimization of base diffusion length and passivating the front and back of the device lead to EQE above 55% and V_{oc} of 301mV.

REFERENCES

(1) Charache, G., Depoy, D., et. al.,"Electrical and optical properties of degenerately-doped n-type $In_xGa_{1-x}As$," *3rd NREL TPV Conference Proceedings*, CIP 401, 1997, p.215.

(2) Malina, V., Micheli, V., et. al.,"Effect of deposition parameters on the electrical metallurgical properties of Au-Zn contact to p-type InP," *Semicond. Sci. Techno.*, **9**, 1994,pp. 1523-1528.

(3) Wojtczuk S., Colter P., Charache G., et. al. "Production data on 0.55eV InGaAs thermophotovoltaic cells, *Proc. 25th IEEE Photovoltaic Specialist Conference*, 1996, pp. 77-80.

(4) Sze, S.M., *Physics of Semiconductor Devices*, New York: John Wiley & Sons, 1981, 2nd Ed., p. 806.

(5) Fahrenbrush, A. and Bube, Richard, *Fundamentals of Solar Cells*, New York: AP, 1983, p. 85.

Effect of Elevated Temperatures on the Performance of an InP Cell Illuminated by a Selective Emitter

Zheng Chen and Henry W. Brandhorst, Jr.

Space Power Institute, 231 Leach Center
Auburn, AL 36849

ABSTRACT

The thermophotovoltaic (TPV) option was not selected for further deep space mission technology development in NASA for several reasons. Chief among them was the large radiator required to keep the photovoltaic cells at a sufficiently low operating temperature. This led to significant integration problems with the spacecraft and limited sensor view angles. It is clear that the issue of cell temperature is crucial for space applications because of radiator size and system impact. Many efforts have focused on matching cell band gap to appropriate emitters in the 1 to 2 μm range, resulting in band gaps in the 0.5 to 0.8 eV range. However, low band gaps lead to low open circuit voltages (~0.25 to 0.45 V) caused by high intrinsic carrier concentrations (n_i^2). Thus, in order to obtain high performance, Photovoltaic cell temperatures must be kept near room temperature. This leads to the inevitable consequence of very large radiators for space applications. Thus in order to make the TPV system suitable for space, this temperature problem must be resolved. However, by turning the problem around and assuming that the system must have an operating temperature in the 150 to 225 °C range, new opportunities for solution arise. If one selects a cell with a wider band gap, open circuit voltage will be larger and the temperature coefficient of voltage will generally be somewhat lower percentage wise. Thus voltages higher than typical low band gap cells are possible even at these higher temperatures. In this paper, InP is selected to demonstrate this concept because it is a well-researched cell, it has a direct band gap and its various temperature coefficients are known. Response of InP cell at the temperature range of 28 °C to 225 °C is investigated under Yb_2O_3 emitter illumination. The variation of the open circuit voltage and band gap with temperature are also discussed in the paper.

INTRODUCTION

Thermophotovoltaic conversion (TPV) is a direct energy conversion process. As a result, there are several advantages for the TPV system; these are easy coupling to any thermal source, such as combustion or solar, quiet, nonpolluting, and easy maintenance. The efficiency of TPV energy conversion is mainly dependent upon emitter efficiency (η_E), which is defined as the ratio of photon convertible power (P_E) to total photon power (P_T) from the same emitter area,

$$\eta_E = \frac{P_E}{P_T} \qquad (1)$$

and photovoltaic (PV) cell efficiency, η_{PV}, which is the ratio of photon power to output electrical power. Emitter efficiency has been continuously improved recently by using radiation tailored emitter such as selective line emitter (1-5) or selective band emitter (6). Both emitters have large emittance for photon energies above the bandgap energy of PV cells and small emittance for the photon energies less than the bandgap energy.

The PV cell efficiency also has been greatly improved using high quality and new low bandgap energy semiconducting materials, such as, InGaAs and InSb. However, in order to maintain the high efficiency of those low bandgap PV cells, the operating temperature has been kept around 300 K. This requires extra components in the system, such as fans, cooling pumps or radiators, to reject heat from the cells. As a result, not only will extra energy be wasted on operating these components, but also the system will become more complex. Finally, there is only one heat transfer mechanism in space: radiation. This requires large radiators to keep the cell operating at high efficiency. Large radiators are not accepted in most space applications for the reasons outlined above.

In order to keep the TPV system simple and applicable for space applications, the large area radiator needs to be eliminated. This generates a concept where the PV cells are operated at a relative high temperature while maintaining a reasonably high efficiency. In this paper, authors use solid state theories as well as experimental data to demonstrate this concept.

2. BACKGROUND

There are two fundamental questions that need to be understood: what causes material temperature increases and how temperature affects the PV performance.

2.1 Temperature

Temperature of materials is dependent upon the energy of the atoms of which the material is made. The energy required for raising temperature goes into: (1) vibration energy by which atoms vibrate around their lattice positions with an amplitude and frequency that determines the temperature, and (2) raising the energy level of electrons in the structure. The vibration energy, which mainly determines the material temperature, can be also described as phonon energy. The energy required for raising temperature one degree is a material property called heat capacity in terms of cal/mole•K. Therefore the temperature of a material depends upon how much energy being transferred into the materials from external source or environment. In space applications, the energy transfer mechanism is simple: energy can be only transferred by radiation, in terms of photon energy. The temperature of a material in space is determined by a photon to phonon conversion process. In the case of a photovoltaic cell illuminated by a typical emitter, (e.g. blackbody), most of the radiation energy outside the portion of energy that is being converted into electric energy (shown covered by the PV cell quantum efficiency curve figure 1) results in raising the cell temperature by vibrating atoms. Much less energy will be transfered to the atoms if a selective emitter is used to illuminate the cell (figure 1). Therefore, the cell illuminated by a selective emitter will be cooler than the cell illuminated by a blackbody emitter of the same total energy. Thus the cell temperature may be controlled by tailoring the radiation spectra from the selective emitter.

2.2 Effect of temperature on PV cells

In general, the approach to TPV conversion has focused on producing cells with band gaps in the 0.5 to 0.8 eV range (7). This range has been chosen to maximize the cell response to emitters generally in the range of 1 to 2 µm. Modeling studies have indicated that the optimum band gap for TPV applications is near 0.5 eV (8), but those results did not account for issues such as operating temperature, recombination processes, and parasitic series resistance losses that complicate the picture. Most of the effort has been directed toward cells made from various compositions of GaSb, GaInAsSb and InGaAs (9-12). These references are not intended to be a comprehensive review of all the TPV cell work that has been done recently, but do provide specific examples of the thrust of the research.

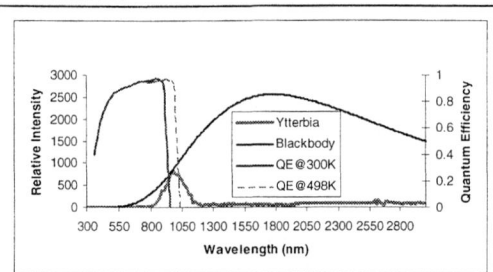

Figure 1 InP cell's quantum efficiency curve superimposed with radiation curves from a black body and an Ytterbia emitter. The dashed quantum efficiency curve of InP at 498K is estimated.

It is clear from the discussion in the INTRODUCTION that the issue of cell temperature is crucial for space applications in order to minimize radiator size and system impact. The efforts cited (7-12) have focused on matching cell band gap to appropriate emitters in the 1 to 2 µm range, resulting in band gaps in the 0.5 to 0.8 eV range. However, low band gaps lead to low open circuit voltages (~0.25 to 0.45 V) caused by high intrinsic carrier concentrations (n_i^2). Thus, in order to obtain high performance, the cell temperatures must be kept near room temperature. This leads to the inevitable consequence noted by Schock (13-14) of very large radiators for space applications or extra cooling components for many terrestrial applications.

Thus in order to make the TPV system suitable for space, this temperature problem must be resolved. However, by turning the problem around and assuming that the system must have an operating temperature in the 150 to 225 °C range, new opportunities for solution arise. If one selects a cell with a wider band gap, open circuit voltage will be larger and the temperature coefficient of voltage will generally be somewhat lower percentage wise. Thus voltages of typical high band gap cells can equal that of low bandgap cells even at these higher temperatures. Increasing temperature however also leads to band gap shrinkage. This shrinkage expands the spectral response to longer wavelength radiation. A estimated spectral response of quantum efficiency curve of InP cell at temperature of

498 K is plotted using dashed line in figure 1 Thus a solution is to walk a very careful line between band gap, temperature and selective emitter wavelength.

InP was selected to demonstrate this concept because it is a well-researched cell, it has a direct band gap and its various temperature coefficients are known. Figure 2 shows the variation of the open circuit voltage (15) and band gap with temperature. The variation of InP band gap with temperature follows the following relationship above room temperature (16) (300 K):

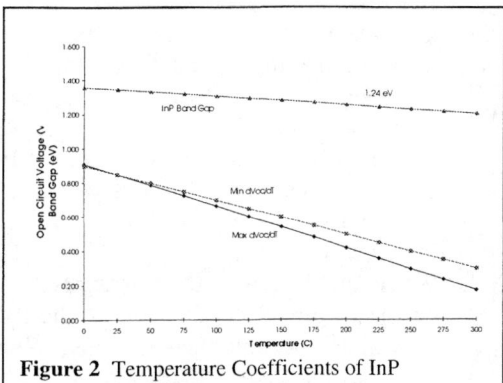

Figure 2 Temperature Coefficients of InP

$$Eg_o(T) = 1.344 - (5.1 \times 10^{-4})(T-300) \qquad (2)$$

Two curves are shown for the variation of cell open circuit voltage indicating the range of values that were determined from the referenced data. It is important to note that those data cover the temperature range from 90 to 400 K. The data above 400 K (127 °C) were extrapolated from the data below 400 K. Depending on the points selected, the temperature coefficients were slightly different as noted in Figure 2, hence two different curves are shown.

Several interesting insights can be drawn from this figure. First, the band gap of the cell at 225 °C is 1.24 eV. This corresponds to light of 1.0 μm wavelength - like the ytterbium selective emitter. Furthermore, InP is a direct band gap cell, so the absorption coefficient at the band edge will be large ($\sim 10^5$). Finally, the open circuit voltage is about 0.4 V at this temperature - equivalent to the room temperature voltage of low band gap cells like GaSb. Furthermore the junction quality of wide band gap cells is generally superior to that of narrow band gap cells. Beyond these points, it is clear that the higher operating temperature will reduce the size of the radiator for space use by a factor of at least five or may be totally eliminated depending upon the heat sources.

Thus there is strong theoretical evidence for a new approach to TPV systems for space use.

3. EXPERIMENTAL PROCEDURES

An InP photovoltaic cell was obtained from SPIRE, Corp. The band gap of the cell at room temperature is 1.34 eV. The cell spectral response was obtained from the literature (17) and is shown in figure 1. The cell was used to determine its response to 1 μm light at elevated temperatures. Two light sources were used in this study: a halogen light and a fibrous Yb_2O_3 emitter. For the halogen light, a 1000 nm narrow bandpass filter was placed between the PV cell and the light. The transmission curve of this filter is shown in figure 3. As shown, the light in the wavelength range of 950 nm to 1050 nm pass through the filter and illuminate the cell. A hot air gun was used to elevate the cell temperature. The cell temperature was measured with a thermocouple directly on the cell surface. The objective for this experimental setup was to confirm the theoretical predictions by measuring the cell's open-circuit voltage and its short circuit current response (2 ohm load) at elevated temperatures. In addition, the cell's current-voltage curves were obtained as a function of temperature.

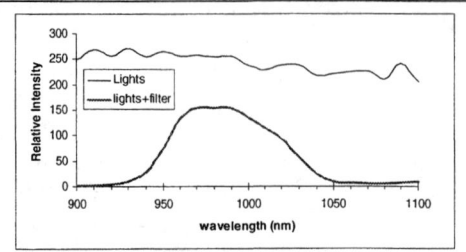

Figure 3 The radiation spectra from the Halogen light and the light passing through the filter.

When the Yb_2O_3 emitter was used, it was placed directly in the front of the PV cell without any intermediate filter. The emitter was heated to 1373 K by a H_2/air torch. The radiation spectrum of the emitter is also shown in figure 1. For this setup, we examined the PV cell response to illumination from the selective emitter for two different cases: one where the cell kept at 300 K by constantly blowing the cooling air to the cell; the other where the cell temperature was raised to 375 K without using any cooling. The temperature increase in the cell resulted from the emitter radiation only. The results from this experiment were used to examine the concept of using a high operating temperature PV cell coupled with a selective emitters in a TPV system so as to eliminate cooling components or extra radiators.

4. RESULTS AND DISCUSION

The variation of the cell open-circuit voltage with operating temperature follows predictions, in that the voltage decreased at about 2.0 mV/°C. The measured open-circuit voltage is plotted in figure 4. At temperature of 200 °C. the open-circuit voltage still is over 440 mV, which is still higher than the open-circuit voltage of 0.75 eV InGaAs cell at room temperature.

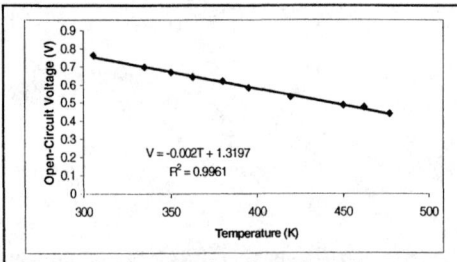

Figure 4 Open-circuit voltage dependence on the operating temperature.

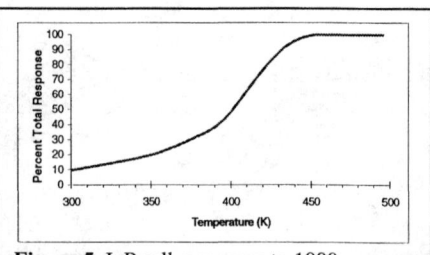

Figure 5 InP cell responses to 1000 nm illumination at various temperatures.

The InP cell response to 1000 nm light from the narrow bandpass filter, should start at about 75 °C based on the information provided by the manufacturer. However, the experimental observation showed the cell response starting at room temperature and then gradually increasing as the temperature increased to about 100 °C. As shown in figure 3, the measured transmission of the filter actually starts at 920 nm. Furthermore, the transmission curve of the filter was also measured at short wavelengths as shown in figure 6. There is a large transmission of short wavelength photons (or higher energy photons) passing through the filter. This explains the cell response seen at room temperature under illumination through the filter. As temperature further increases, the cell shows more rapid response up to about 225 °C. This indicates that the bandgap of the cell has been reduced to around 1.30 eV at 100 °C and 1.24 eV at 225 °C, which moves the cell spectral response into the filter transmission band between 950 nm and 1000 nm.

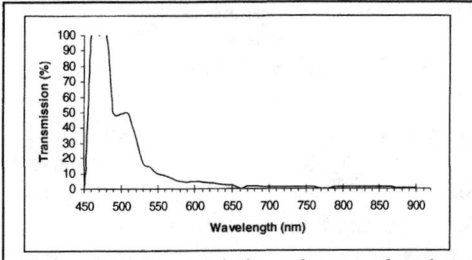

Figure 6 High transmission at low wavelengths from the filter causes in the early response of the cell.

We expected that saturation of the cell response should start above 225 °C. The experimental data (figure 5) show that the saturation actually starts at 175 °C. After carefully examining the cell and the electric contact points in the circuit, we concluded that early saturation is caused by the increase of contact resistance between the cell and conducting leads at elevated temperatures.

For most low bandgap cells, the fill factor calculated from current and voltage curves decreases as operating temperature increases. Solid state theory indicates that high bandgap cell should hold its fill factor relatively constant over an extended range of temperature. The fill factor of the InP cell at elevated temperatures was measured and is shown in figure 7. Decrease of the fill factor with temperatures shows at a rate of 0.09 %/K. The results have confirmed that the high bandgap cell can be operated at a relatively high temperature range without significantly reducing its fill factor.

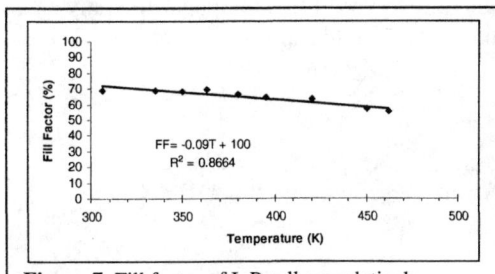

Figure 7 Fill factor of InP cell are relatively insensitive to temperature.

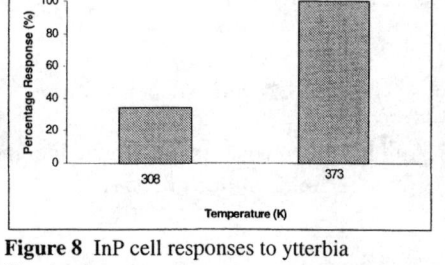

Figure 8 InP cell responses to ytterbia illumination at two temperatures.

These experimental data support the theory that the bandgap of InP decreases with temperature following Equation 2. This has been demonstrated by increase of the cell response to 1000 nm illumination as cell temperature increases. In addition, the InP cell shows that it can be operated at a wide temperature range such as 300 K to 500 K and still have higher fill factor and open-circuit voltage compared to low bandgap cells.

Because the cell shows its high temperature capability, we can use a selective emitter such as Ytterbia (Yb_2O_3) emitter, its peak with selective emission around 960 nm (figure 1) to replace the 1000 nm filtered halogen lights. The low but controlled out-of-band radiation (non-convertible radiation) from the emitter can be used to raise the cell to a desired temperature where the cell can fully respond to the selective emission. The preliminary data shown in figure 8 demonstrate that the InP cell has low response to ytterbia illumination at room temperature while the cell was air cooled. After stopping cool air, the cell temperature was raised to 100 °C by the radiation from the emitter. The cell response to the ytterbia illumination was more than doubled as the cell temperature increased from 33 °C to 100 °C.

5 CONCLUSION

This study has demonstrated that the feasibility of eliminating cooling components and/or reducing radiator size in TPV systems. These data confirm that high bandgap PV cells can be operated at elevated temperatures (up to 500 K) and still have reasonably high open-circuit voltage and fill factor compared to low bandgap cells. The preliminary experiment of coupling an InP cell with a selective emitter indicates that InP has much better response to the ytterbia selective emitter as expected.

InP presents an interesting option, even though its band gap at the temperature range of 175 °C to 225 °C may be a bit large for using ytterbia as the selective emitter. Fortunately, there are other materials systems that will also match this emitter. Several materials systems fit into this range very well, such as GaSb-AlSb and the InP-InAs systems. The InP-InAs system also offers the benefits of excellent radiation resistance. Further investigation is needed to select an optimum couple between the PV cell and emitter in order to the system being operated at high efficiency and high power output without cooling components or large radiators. However, it would appear that any solar cell with bandgap in the appropriate range can be used in this approach.

6 ACKNOWLEDGEMENT

Authors also wish to acknowledge the financial support of the DOE Office under contract DE-AC03-98SF21557.

7 REFERENCES

[1] Guazzoni, G.E., 1972, "High Temperature Spectral emittance of Oxides of Erbium, Samarium, Neodymium, and Ytterbium," Applied Spectroscopy, Vol. 26, [1], pp.60-65.

[2] Chubb, D.L., Lowe, R.A., and Good, B.S., 1994, "Emittance Theory for Thin Film Selective Emitter," Proceedings of the first NREL Conference on the Thermophotovoltaic Generation of Electricity, AIP Conference Proceedings 321, pp.229-244.

[3] Nelson, R.E., 1986,"Rare Earth Oxide TPV Emitters," in Proceedings of the 32nd International Power Sources Symposium, pp.95-110.

[4] Z. Chen, P.L. Adair, and M. F. Rose, "Fabrication of Fibrous Er_2O_3 Composites for Heat to Light Conversion," High Temperature materials, Vol 37 (1997) p77-80.

[5] Z. Chen, M.F. Rose, Eric B. Clark and Donald L. Chubb, "Reinforced Solid Erbium Oxide Emitters for TPV Application," Proceedings of International Mechanical Engineering, Atlanta, Georgia, 1996.

[6] Z. Chen, P.L. Adair, and M. F. Rose, "Multiple-Dopant Selective Emitter," Proceedings of the Third NREL Conference, Colorado Springs, Co 1997, P181-188.

[7] T. J. Coutts, M. W. Wanlass, J. S. Ward and S. Johnson, *"A Review of Recent Advances in Thermophotovoltaics"*, in Conference Record of the Twenty Fifth IEEE Photovoltaic Specialists Conference-1996, IEEE, New York, pp. 25-30, 1996.

[8] M. W. Wanlass, J. S. Ward, K. A. Emery, and T. J. Coutts, "$Ga_xIn_{1-x}As$ ThermophotovoltaicConverters", in Conference Record of the Twenty Fourth IEEE Photovoltaic Specialists Conference-1994, IEEE, New York, pp. 1685-1691, 1994.

[9] S. Wojtczuk, *"$In_xGa_{1-x}As$ TPV Experiment-based Performance Models"*, in Second NREL Conference on Thermophotovoltaic Generation of Electricity (1995), AIP Conference Proceedings 358, pp. 387-393, (1996).

[10] P. R. Sharps and M. L. Timmons, *"Development of p-on-n GaInAs TPV Devices"*, in Second NREL Conference on Thermophotovoltaic Generation of Electricity (1995), AIP Conference Proceedings 358, pp. 458-468, (1996).

[11] Z. A. Schellenbarger, M. G. Mauk, L. C. DiNetta and G. W. Charache, *"Recent Progress in InGaAsSb/GaSb TPV Devices"*, in Conference Record of the Twenty Fifth IEEE Photovoltaic Specialists Conference-1996, IEEE, New York, pp. 81-84, 1996.

[12] L. Fraas, R. Ballantyne, J. Samaras and N. Seal, *"Electric Power Production Using New GaSb Photovoltaic Cells with Extended Infrared Response"*, in First NREL Conference on Thermophotovoltaic Generation of Electricity (1994), AIP Conference Proceedings 321, pp. 44-53, (1995).

[13] A. Schock, M. Mukunda, C. Or and G. Summers, *"Analysis, Optimization, and Assessment of Radioisotope Thermophotovoltaic System Design for an Illustrative Space Mission"*, in First NREL Conference on Thermophotovoltaic Generation of Electricity 1994), AIP Conference Proceedings 321, pp. 331-356, (1995).

[14] A. Schock, C. Or and M. Mukunda, *"Effect of Expanded Integration Limits and of Measured Infrared Filter Improvements on Performance of RTPV System"*, in Second NREL Conference on Thermophotovoltaic Generation of Electricity (1995), AIP Conference Proceedings 358, pp. 55-80, (1996).

[15] C. J. Keaveny, S. M. Vernon, V. E. Haven, M. J. Nowlan, R. J. Walters, R. L. Statler and G. P. Summers, *"Radiation-Hard, High Efficiency InP Solar Cell and Panel Development"*, in Proceedings of the 26th Intersociety Energy Conversion Engineering Conference, IECEC-91, American Nuclear Society, La Grange Park, IL, pp. 321-326, (1991).

[16] D. Krillov and J. L. Merz, J. Appl. Phys. **54**, 4104, (1983).

[17] R.W. Hoffman, Jr., N.S. Fatemi, P.P. Jenkins, V.G. Weizer and M.A. Stan etc. "Improved performance of p/n solar cells," Proc. Of the 26th IEEE PVSC, Anaheim, CA, pp. 816-818 (1997)

Integrated Development and Testing of Multi-Kilowatt TPV Generator Systems

Edward M. West, William R. Connelly

Vehicle Research Institute, Western Washington University
Bellingham, Washington 98225

Abstract. On going research at Western Washington University has brought about a number of important advances in the state of TPV system development. Specific areas of TPV work include new cylindrical PV cell array designs, improved burner systems, enhanced recuperator performance and overall TPV system development. System development includes the implementation of the TPV8 generator in the purpose built Viking 29 TPV series hybrid demonstration vehicle (1). TPV research capability enhancements include the development of a computerized test and data acquisition system, an array characterization system (2), and an array flash testing station for quality assurance prior to final assembly. TPV system achievements include steady-state cell output levels of 2.4 watts/cm^2 on an array test station and 1.5 watts/cm^2 using a combustion source. Successful development of enhanced recuperators with thermal efficiency above 80%. Array output levels of 1 kilowatt have been achieved with array efficiencies in excess of 7.5%. TPV technology development guidelines are discussed with the hope of focusing the community on these important issues.

INTRODUCTION

This paper summarizes the important accomplishments concerning the development of multi-kilowatt TPV generator systems at WWU. First, this paper covers new results in the development of kilowatt sized TPV arrays, and new designs under development. Second, recent developments on the TPV burner systems are presented. Third, an overview of the issues involved in system development and ancillary component selection is discussed. Fourth, recent improvements of the research facilities at WWU are summarized. Finally, a set of TPV technology development guidelines are presented with the hope of focusing the community on these important issues.

CYLINDRICAL ARRAY DEVELOPMENT

During the development of TPV technology at WWU thorough testing of the hardware developed to date has been carried out. Steady state test station array output levels of 1 kilowatt have been achieved with array efficiencies in excess of 7.5%. This corresponds to steady-state cell output levels of 2.4 watts/cm^2 on an array test station and 1.5 watts/cm^2 using a combustion source.

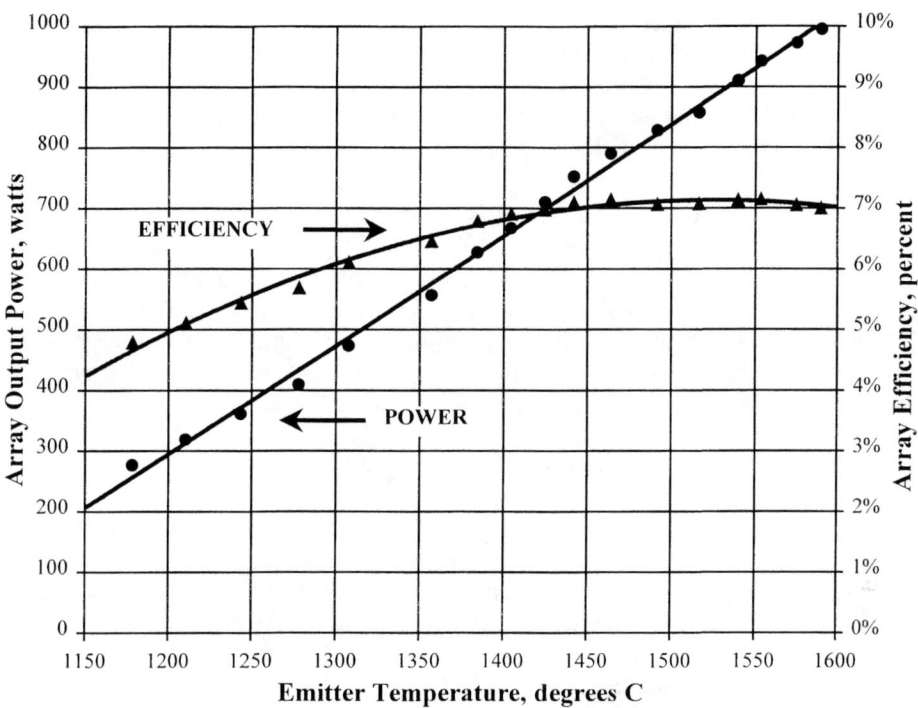

FIGURE 1. Array Output Power and Efficiency versus Emitter Temperature on Array Test Station.

Figure 1 shows the array power output and efficiency as a function of emitter temperature for the array run on the electric test station. Although this was not the best array developed to date, it clearly indicates what these components are capable of.

Tests of Various Array Designs

To evaluate the array cavity flux distribution, an individually pinned-out circuit was developed for measurement of flux distribution. Each one of the individually pinned-out cells was calibrated to compensate for the cell to cell variability. When the in-cavity flux distribution experiments were carried out it was found that the flux was essentially uniform on all of the cells.

A new one-piece machined "T" mirror with an evaporated gold coating has been developed. To date, WWU and JX Crystals have manufactured five sets (100 mirrors) of this type for various contracts. Also, a new gold plated end mirror has been developed which helps close-up the parasitic gaps in the ends of the receiver array. Testing shows that the "T" mirror configuration gives the highest possible array efficiency.

To explore spectral control within the array cavity, a new integral filter receiver array has been completed. The cells in this array have a multi-layer dielectric filter applied directly to the face of the Gallium Antimonide PV cells manufactured by JX Crystals. The expectation was that the integration of the filter with the cell should decrease the transmission losses in the filter system and improve the array efficiency. Testing carried out on this array show no significant difference in performance compared to cylinders with dielectric filters on quartz, glued to the face of the cells.

Development of New Array Designs

Based on the last four years of research on TPV arrays, a new phase of the development is being undertaken at WWU to produce two new TPV array designs. The first of these new arrays will use an all-new 113% larger cell produced by JX Crystals, while the second one will shingle the cells to create the circuit. The next two subsections outline the development of these cylindrical arrays.

Cylinder with Larger Cell

As an outcome of the research during 1997, we concluded that improvements in power and efficiency could be achieved by using a larger size cell in the array. Before JX Crystals could produce such a cell, an analysis tool had to be developed to analyze potential gridline designs. Western Washington University worked closely with JX Crystal in the development of an analysis tool that determines the effects of grid design changes on cell performance. Specifically, the analysis accounts for gridline geometry and spacing, bus bar geometry and PV cell emitter and base characteristics.

Once this tool was developed and checked against JX Crystal's current production cells, it was used in the development of the new cell design. The cells were first laid out on the new 3" wafers, and then the grid geometry was developed. With the die size and grid design fixed, artwork was drawn up at JX Crystals and the production of the 0.65" square cells began.

New array geometry has been developed based on the new 0.65" square cells from JX Crystals. The new array design uses 18 strings of 15 cells each that results in an array of similar geometry to the current array. Based on modeling projections we expect an increase of about 10% in both power and efficiency. This larger cell array will begin the testing phase of development during the beginning of 1999.

Shingled Cell Cylinder

Through cooperation with JX Crystals, the idea of shingling the cells within the TPV array has gained new significance. Although the idea of shingling PV cells to improve the packing factor is not new, this concept had been put aside due to thermal mismatch concerns between the Gallium Antimonide (GaSb) cells and the substrate. JX Crystals has recently located a metal matrix composite product manufactured of Aluminum and Silicon Carbide (AlSiC). This new material has a coefficient of thermal expansion (CTE) which closely matches the CTE of GaSb.

By manufacturing the substrates for the shingled GaSb cells out of the AlSiC material the thermal stresses in the cells should be held to an acceptable level. By shingling the cells, the gaps between the cells will be eliminated and the inactive border area at the edge of the cells is reduced by half. Finally, the interconnects at the edges of the cells will be eliminated and the area currently covered by mirrors will not be required. The resulting increase in active cell area within the cavity should allow for higher power levels at a given emitter temperature, while the reduction of parasitic losses should increase efficiency

Overview of PV Cell Substrate Materials

A variety of materials have been evaluated for use as substrates for the GaSb cells in the PV array. The primary design concern with the cell substrate is the minimization of the thermal resistance to the cooling system. All of the systems assembled to date use an insulated metal circuit board of some type.

Until recently, all WWU arrays used a polyamide film insulated metal circuit board. The board was either of Aluminum or Copper. The thermal resistance of this substrate material is 1.1 K/W and is commercially available as Koolbase.

More recent circuit boards have been fabricated using a ceramic powder filled epoxy layer over the Aluminum or Copper substrate. These materials are commercially available through either Thermagon (3) or Bergquist (4). These circuit boards have a thermal resistance of 0.31 K/W, which represents a 72% reduction in thermal resistance through the substrate compared to the Koolbase material.

Most recently a new material has become commercially available which represents a substantial advancement in high thermal conductivity substrate materials. This material is a metal matrix composite of Aluminum and Silicon Carbide (AlSiC). A Silicon Carbide powder pre-form is infiltrated with molten aluminum creating a substrate material with low thermal resistance and a CTE that closely matches that of the GaSb PV cells. The AlSiC material still requires an insulator to make it a useful substrate. Two types of insulators are available, plasma sprayed alumina or co-infiltrated Aluminum Nitride (AlN). The AlSiC substrate with the AlN insulator has a thermal resistance of 0.15 K/W, which represents a 52% reduction in thermal resistance compared to the Thermagon material. These new substrates, manufactured by Process Systems, Corp. (5), stand out as a new area for enhancing TPV array performance.

BURNER SYSTEMS

To improve system performance, a new three-inch diameter burner has been developed. By increasing the emitter diameter of the previous design from $2\frac{1}{8}$ inches to 3 inches, both the input power and the view factor have been improved. This burner redesign increases both the view factor and the energy input level by 40%. Further, the increased emitter diameter allows for larger flow passages for the combustion gasses which also allows for increased energy input levels.

Recuperator Development

Because of the high gas stream temperature leaving the burner section of the generator, a recuperator is necessary to develop a system with reasonable efficiency. Many different designs have been developed at WWU both for the high temperature and intermediate temperature regions of the recuperator. Recently, new designs have been developed for both the high temperature ceramic recuperator and the intermediate temperature metal recuperator. This section of the paper discusses the specific development of each of these recuperators.

Silicon Carbide Recuperator

To improve the performance of the high temperature section of the recuperator a new convoluted wall ceramic heat exchanger has been developed. The geometric design and manufacturing of the forms was carried out at WWU, while a ceramics vendor carried out the slip casting and firing. To allow for the testing of the new design for the ceramic recuperator, our supplier has manufactured seven prototype units. See figure 2 for a view of the new silicon carbide recuperator. These convoluted tubes were designed with the goal of maximizing the heat exchange area while still maintaining a simple – thermally shock resistant – geometry. Testing of these recuperators will be carried out at the end of 1998.

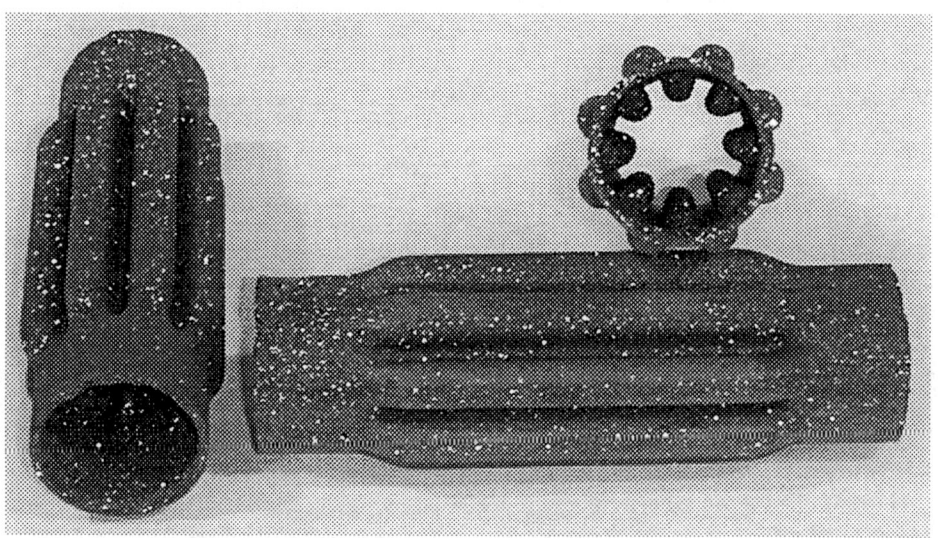

FIGURE 2. Three Convoluted Wall Silicon Carbide Ceramic Recuperators.

FIGURE 3. Two High Efficiency Counter-Flow Welded 310 Stainless Steel Recuperators.

310 Stainless Steel Recuperator

To enhance the TPV system efficiency, a new 310 stainless steel recuperator has been developed. While increasing the exchanger area, this recuperator is more compact than all previous designs. To evaluate the new and previous recuperator performance, a testing process has been developed, including the required analytical procedures. This testing process determines both the recuperator efficiency and the pressure required to drive the gas through the recuperator.

This new recuperator has increased the efficiency from 65% to 82% for the intermediate temperature metal section. Although this only represents a 17% increase in recuperator efficiency, it represents a 50% reduction in the quantity of waste heat compare with the previous designs. Ultimately, it is this type of system enhancement that will push TPV technology into the real world. See figure 3 for a view of the new recuperator.

Burner Control

To allow a TPV system to function in an automatic manner, a burner control system must be developed. A system of this type is currently under development at WWU. The system incorporates a furnace control module that includes a fuel solenoid control, a fan control, an airflow sense circuit, and an automatic flame sensing capability. In addition to the basic on/off capability of the furnace control, a proportion fuel valve is being utilized to allow for the continuous control of the heat-input level and air fuel ratio.

SYSTEM DEVELOPMENT

Although the ancillary components of the TPV system have not received significant scrutiny from the TPV community, their development is paramount to the success of TPV technology. The system components – in order of power usage – that require attention are: the system cooling fan, the combustion air supply, the array coolant pump (if any), the peak power tracker, and the system control hardware. The development of these components is critical because they directly consume the electric power that the generator is producing. As a result, the efficiency of the ancillary components has a direct effect on both the power output and efficiency of the TPV system. To help define the current situation, a general discussion of the aforementioned components follows.

System Cooling Fan

The purpose of the system cooling fan is to circulate ambient air through a heat exchanger to remove the waste heat from the TPV array. Typical low-pressure fans of the size required for a TPV system operate with an efficiency between 10 and 30%. This efficiency is defined by the equation:

$$\eta_{fan} = \frac{\dot{V}(P_2 - P_1)}{P_{elec}} \quad (1)$$

Where P_{elec} is the motor electric input power, $\dot{V}$ is the volume flow of air, and P_1 & P_2 are the pressures before and after the fan. Given the efficiency of a fan – or pump – this equation can be rearranged to give the required input power as follows:

$$P_{elec} = \frac{\dot{V}(P_2 - P_1)}{\eta_{fan}} \quad (2)$$

For the TPV system, the volume flow of cooling air required can be calculated through the use of the equation for sensible heat rise in the cooling air (6). The following equation determines the required volume flow of air $\dot{V}$, given the required heat dissipation $\dot{Q}$ and the inlet and outlet temperatures of the cooling air T_1 & T_2:

$$\dot{V} = \frac{\dot{Q}}{\rho C_p (T_2 - T_1)} \quad (3)$$

Where ρ and C_p are fluid density and specific heat, respectively.

For example, using equation 3, for a system with a 1kW array output, an array efficiency (2) of 7.5%, and an air temperature rise of 20°C, the cooling air requirement is about 1200 cfm. Using equation 2 with a pressure rise of 0.5 in-H₂O and a fan efficiency of 25% the cooling fan power requirement is 275 watts. This represents

28% of the electric power that is produced by the TPV array. Clearly, maximizing the fan efficiency and heat rejection temperature of the TPV system is critical to the overall performance of the system.

Combustion Air Supply

The requirements for the fan used for combustion air supply are similar to the cooling fan. In most TPV systems the combustion fan will need to supply much less air at much higher pressure than the cooling fan. The volume requirement for combustion air is dictated by the necessary input power to the TPV burner, while the sum of the recuperator and burner pressure drop at the desired flow dictates the pressure requirement. To determine the required flow of combustion air the following equation is used:

$$\dot{V} = \frac{\dot{Q}\, AF}{\rho\, H_u} \qquad (4)$$

Where $\dot{Q}$ is the input energy requirement, AF is the air fuel mass ratio, ρ is the density of air, and is H_u the heating value of the fuel.

For a hydrocarbon burner that requires 16 kW of input power, the required volume flow of combustion air is about 11 cfm. For a typical burner system with 20 in-H2O pressure drop and a fan efficiency of 20% the required fan power is about 125 watts. Since this represents 13% of the array output power, maximizing fan efficiency and reducing pressure drop are important design goals for the TPV system.

Array Coolant Pump

If the design of the TPV system uses a water cooling system to remove the heat from the TPV array, then an array coolant pump will be required. The power of the pump can also be predicted using the same type of analysis that was used for the cooling fan. Equation 3 is used to determine the coolant flow, given the maximum acceptable temperature rise in the coolant, the specific heat and density of the coolant, and the heat flow to be removed. Given a maximum temperature rise of 7°C in water and a cooling load of 12 kW the required coolant flow is about 7 gpm. With a known flow and the corresponding pressure drop through the cooling system at that flow rate, the pump power required is calculated using equation 2.

A typical centrifugal water pump has an efficiency between 10 and 15%. For a given system which has a pressure drop of 70 in-H_2O at 7 gpm, and a pump efficiency of 15% the power consumption is about 50 watts. This represents about 5% of the array output, and cuts directly into the power and efficiency of the TPV system. Again, optimizing the pumping efficiency and reducing the system pressure drop will have a very beneficial effect on the overall system performance.

Peak Power Tracker

To maximize the efficiency of most TPV systems, a Peak Power Tracker (PPT) is needed to adjust the load on the TPV array to its optimum operating point. By monitoring the array performance the peak power tracker can vary the gain of a power converter to center the array voltage at the peak power point. Peak power trackers come in two possible configurations, boost converters and buck converters. The boost and buck terminology refers to the converter's influence on the array voltage.

Boost converters use inverters, step-up transformer, rectification, and filtering to increase the output voltage. Buck converters use current choppers and filtering to decrease output voltage. Of the two types, buck converters have higher intrinsic efficiency because they avoid the energy loss through the step-up transformer. Overall efficiencies of a PPT range from a low of 85% to a maximum of about 98% with a typical buck type PPT have an efficiency of about 95%.

For most applications a buck type PPT can be specified which would limit the converter loss to about 5% of the array output. In these applications, it is desirable to have the converter ratio close to one for maximum efficiency, so array voltage and system output voltage should be chosen carefully to maximize system performance.

RESEARCH CAPABILITY ENHANCEMENTS

To insure the successful development of TPV technology at WWU a number of important capability enhancements have been carried out. This section summarizes three important capabilities that have been added to the development arsenal at WWU.

Computerized Data Acquisition and Analysis

To enhance the testing capability of the TPV systems and components, a new computerized data acquisition system has been developed. A new PC was acquired along with a data acquisition card and interface hardware to collect the data into the computer. A new workstation was created to house the computer and all of the interface electronics into a single mobile system. New high accuracy air, fuel, and water flow meters were specified and procured for this system. Finally, we implemented the Labview software to monitor and control testing and running of TPV systems.

Cylindrical Array Characterization Unit (CACU)

A new receiver array characterization unit has been developed which matches the emitter geometry of the three-inch burner. This test station has an input power capability of 15 kW and a maximum temperature capability of 1800 K. A thorough treatment of the development of this array testing system can be found in related WWU papers (2).

Circuit Flash Testing System

A circuit flash testing system has been developed and installed at WWU. This system is similar to flash testing systems developed at JX Crystals and was developed in cooperation with them. The unit uses a large photographic flash to provide an illumination event with a duration of about 1 millisecond. During this illumination event the I-V curve of the circuit being tested is measured. This is accomplished through the use of a computer controlled bipolar amplifier capable of the delivering the necessary current to sink the short circuit current of the circuit being tested.

The measured I-V curve is then analyzed by software developed by JX Crystals. This analysis results in data files that include the raw I-V curve and the following derived results: I_{sc}, V_{oc}, I_{max}, V_{max}, P_{max} and FF. Through the use of this system circuits can be tested and characterized both to determine acceptable function and to allow sorting of circuits into matched sets. It is expected that the use of this system will markedly improve the performance of the arrays produced at WWU.

TPV TECHNOLOGY DEVELOPMENT GUIDELINES

Based on the analysis of the ancillary components, it is clear that half of the output of the PV array can be consumed by these components. This means that the selection of the ancillary components can have a greater effect on the overall system performance than the design of the generator itself. For WWU to accomplish the successful development of a TPV generator system, all aspects of the system have been considered from the initial definition. Figure 4 is an example of a complete prototype TPV generator system developed at WWU.

FIGURE 4. Single Cylinder Demonstration Unit Developed at Western Washington University.

With ancillary losses in mind, it is obvious why serious consideration has been taken when selecting and specifying system components. Given that the community has spent little effort on these aspects of TPV development, it is not surprising that few demonstration systems exist.

There are a number of other aspects of TPV technology that should also be high on the system development priority list:

- Maximize the emitter area – this maximizes power output capability.
- Operate the emitter at the highest temperature the materials will allow.
- Maximize the PV cell area in the cavity – this minimizes emitter temperature.
- Minimize the parasitic losses in the cavity to the greatest possible degree.
- Minimize the thermal resistance between the PV cell and the cooling fluid.
- Maximize the heat rejection surface area to the ambient cooling air.
- Minimize the pressure drops through the burner and the cooling system without compromising their thermal performance.
- Maximize the efficiency of all of the fans and pumps in the system.

These aspects of system design must be given their rightful place in the priorities that define the development of TPV technology. If TPV technology is going to be transformed from a laboratory pursuit into a technological success, greater effort needs to be expended on comprehensive system development.

ACKNOWLEDGMENTS

This material is based upon work supported by the U. S. Army Research Office under contract number 33881-CH-MUR and by the U. S. Department of Energy through the Division of Advanced Energy Projects under contract number DE-FG06-94ER1249. The authors would like to thank Dr. Michael Seal, Orion Morrison, and Eileen Seal at WWU, and Lewis Fraas at JX Crystals, for all of their valuable contributions to this research.

REFERENCES

1. Morrison, O., et al., "Use of a Thermophotovoltaic Generator in a Hybrid Electric Vehicle," presented at The Fourth NREL Conference on the Thermophotovoltaic Generation of Electricity, Denver, Colorado, October 11–14, 1998.
2. Connelly, W. R. and West, E. M., "Cylindrical TPV Array Characterization," presented at The Fourth NREL Conference on the Thermophotovoltaic Generation of Electricity, Denver, Colorado, October 11–14, 1998.
3. Thermagon: http://www.thermagon.com
4. Bergquist: http://www.bergquistcompany.com
5. Process Systems Corp.: http://www.alsic.com
6. Incopera, F. P. and DeWitt D. P., *Fundamentals of Heat and Mass Transfer*, 3rd Edition., New York: John Wiley & Sons, 1990

Theoretical prediction of the plasma frequency and moss-burstein shifts for degenerately doped InAs, $In_{1-x}Ga_x$ As and $InP_{1-y}As_y$

J.E. Raynolds,[*] C.B. Geller,[**] G.W. Charache,[*] T. Holden,[***] F.H. Pollak,[***] W. Maanstadt,[****] R. Asahi,[****] and A. J. Freeman,[****]

[*]Lockheed Martin Corp., P.O. Box 1072, Schenectady, NY; [**]Bettis Atomic Power Laboratory, West Mifflin, PA 15213; [***]Physics Dept. and NYS Center for Advanced Technology in Ultrafast Photonic and Materials Applications, Brooklyn College of CUNY, Brooklyn, NY 11210; [****]Dept. of Physics, Northwestern University, Evanston IL

ABSTRACT

Theoretical predictions for the plasma frequency and Moss-Burstein shift (optical band gap) of degenerately doped ($n > 10^{19} cm^{-3}$) InAs, $In_x Ga_{1-x}$ As, and $InP_{1-y}As_y$ are presented. These systems are of interest because they possess desirable optical properties for thermophotovoltaic (TPV) applications. The studies presented are based on electronic band structures calculated using the Full Potential Linearized Augmented Plane Wave (FLAPW) method which includes non-local screened exchange (sX-LDA) and spin-orbit effects. The plasma frequency and Moss-Burstein shift are calculated vs. doping assuming a "rigid band" approximation (i.e. conduction band filling of the "undoped" bands). The doping dependence of the effective mass (band non-parabolicity) plays an important role at the high dopings considered here. This effect leads to a maximum in the plasma frequency vs. doping ($2 - 3 \times 10^{14}$/s) and a significant departure from the "constant effective mass" prediction for the optical band gap vs. doping. These calculations are in good agreement with measurements.

INTRODUCTION

In order to achieve the highest performance of TPV systems, spectral control filters are used to block radiation in regions of the black-body spectrum which cannot be converted to electricity. The long wavelength portion of the spectrum can be blocked by the use of a "plasma filter" which is a heavily doped semiconductor exhibiting metallic-like reflectivity at long wavelengths. Currently, InAs, InGaAs and InPAs alloys are being investigated experimentally as promising candidate materials for plasma filters.

Theoretical investigations were undertaken to resolve certain puzzling aspects of the measured doping dependence of the plasma frequency (frequency below which reflectivity becomes large) and the optical absorption edge. Classical theory predicts that the plasma frequency should increase linearly with the square root of the doping density, while measurements show that it actually reaches a maximum at high doping densities (on the order of $10^{19} cm^{-3}$). This discrepancy is due to the fact that the effective mass of charge carriers is not constant as assumed by the classical theory, but rather, increases with doping. The Moss-Burstein shift (doping induced shift of the optical absorption edge) is also affected by the fact that the carrier effective mass depends on doping.

CALCULATIONS

A theoretical description of the plasma frequency and Moss-Burstein shift requires an accurate band structure. To this end, we have carried out band structure calculations using the Full Potential Linear Augmented Plane Wave (FLAPW) method[1] which uses a hybrid Screened Exchange/Local Density Approximation (sX-LDA) functional to describe exchange correlation effects. This method has been shown to yield realistic semiconductor band gaps and conduction band dispersions.[2] The Moss-Burstein shift and plasma frequency are calculated from the FLAPW band structures in a rigid band picture. The Moss-Burstein shift is calculated by filling the conduction band (calculated for the undoped system) to a given energy, then determining the corresponding carrier density as

$$\rho = \frac{V_k}{(2\pi)^3}, \quad \text{(EQ 1)}$$

where V_k is the corresponding volume in k space. A simple approximation is employed to determine V_k. The FLAPW bands along high symmetry directions are fit to polynomial expansions. A spherical average is then carried out near each critical point to yield a one dimensional E vs. k relation: $E = E(k)$. The k space volume associated with each critical point is then simply given by the volume of a sphere:

$$V_k = \frac{4\pi k^3}{3}. \quad \text{(EQ 2)}$$

The plasma frequency is determined as the frequency for which the dielectric function:

$$\varepsilon(\omega) = \varepsilon_\infty - \frac{4\pi e^2}{\hbar^2 \omega^2} \int_0^{k_F} \frac{\partial^2 E}{\partial k^2} \frac{2}{(2\pi)^3} d\vec{k}, \qquad (EQ\ 3)$$

vanishes (working in cgs units).[3] In the previous equation e is the electron charge, $\hbar$ is Planck's constant, ε_∞ is the high frequency component of the dielectric function, and ω is the angular frequency of electromagnetic wave. Thus, the plasma frequency ω_p is given by:

$$\omega_p^2 = \frac{4\pi e^2}{\varepsilon_\infty \hbar^2} \int_0^{k_F} \frac{\partial^2 E}{\partial k^2} \frac{2}{(2\pi)^3} d\vec{k}. \qquad (EQ\ 4)$$

This integral is evaluated in a manner similar to that for the Moss-Burstein shift. The bands are spherically averaged and the carrier density is proportional to the k space volume (the volume of a sphere). A doping dependent effective mass, m_s, can be defined in terms of the dielectric function in the standard way:

$$\varepsilon(\omega) = \varepsilon_\infty - \frac{4\pi e^2 N}{\omega^2 m_s m_o}, \qquad (EQ\ 5)$$

where N is the carrier density, and m_o is the electron mass. This effective mass, m_s (referred to as the average effective mass), should not to be confused with the effective mass, m^*, of an electron in a given electronic state. Rather m_s represents the collective behavior of the conduction electrons. At high doping levels, m_s becomes negative. The plasma frequency is undefined when the mass becomes negative because there is no solution to $\varepsilon(\omega) = 0$ for real frequencies. This implies that the plasma excitations are damped at sufficiently high dopings.

RESULTS

The results for the plasma frequency vs. doping are presented in Fig. 1 for InPAs, InGaAs, and InAs along with measurements. These calculations are consistent with the observation that the plasma frequency reaches a maximum. Also shown is the classical prediction for InGaAs showing the failure of the classical prediction at these high doping levels.

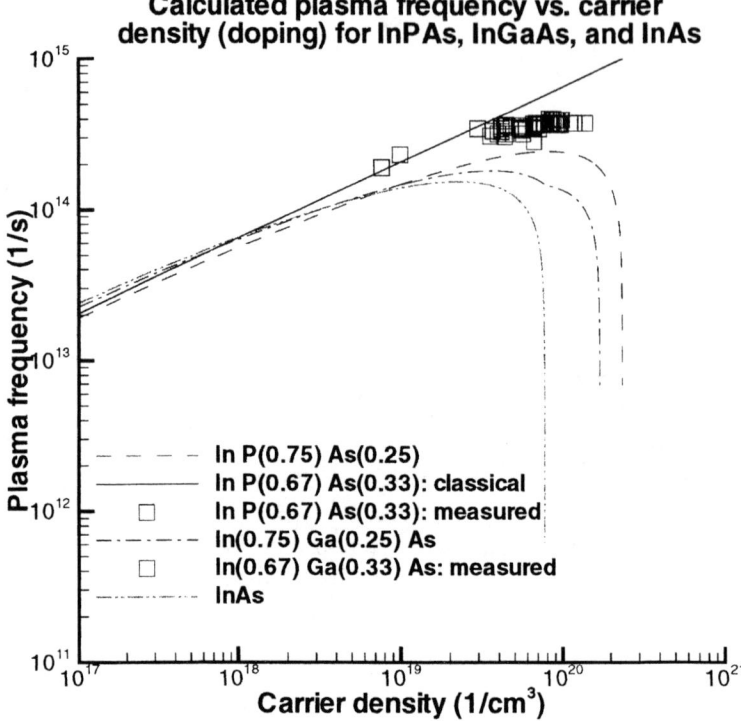

FIGURE 1. Calculated plasma frequency vs. doping density for InPAs, InGaAs, and InAs along with experimental measurements. The calculations for InGaAs and InPAs are based on the idealized compositions: InP(0.75)As(0.25), and In(0.75) Ga(0.25) As. These compositions were chosen for computational convenience as they allow the use of small unit cells (8-atoms). The atomic configurations used in the calculations are the ones giving the minimum energy. These calculations are consistent with the observation that the plasma frequency reaches a maximum.

The predicted decrease in the plasma frequency at high doping is not observed because the doping solubility limit is reached. The decrease in the plasma frequency at high doping results from the fact that the effective mass m^* (mass of an individual state) becomes negative above half filling of the band. The change in sign of the effective mass is due to the change in sign of the curvature of the bands. The states in the upper half of the band are of "anti-bonding" character. Therefore, there may be a connection between the solubility limit and band structure in that dopant atoms may be more likely to go into interstitial sites rather than to fill anti-bonding states by going into substitutional sites.

The calculated results for the doping dependence of the optical absorption edge (band gap) are presented in Fig. 2 along with experimental measurements. Also included in this figure is the prediction obtained assuming a constant effective mass (parabolic approximation for the conduction band dispersion).

The calculated result increases more slowly with doping than the parabolic approximation (constant effective mass) predicts, due to the increase in effective mass with doping. The measured doping dependence of the band gap is also consistent with an effective mass that increases with doping. The calculated band gap is smaller than the measured band gap due to the composition difference between the calculation and experiment (see caption to Fig. 1). A rigid shift of the calculated results to correct for this band gap would improve the agreement with the measurements.

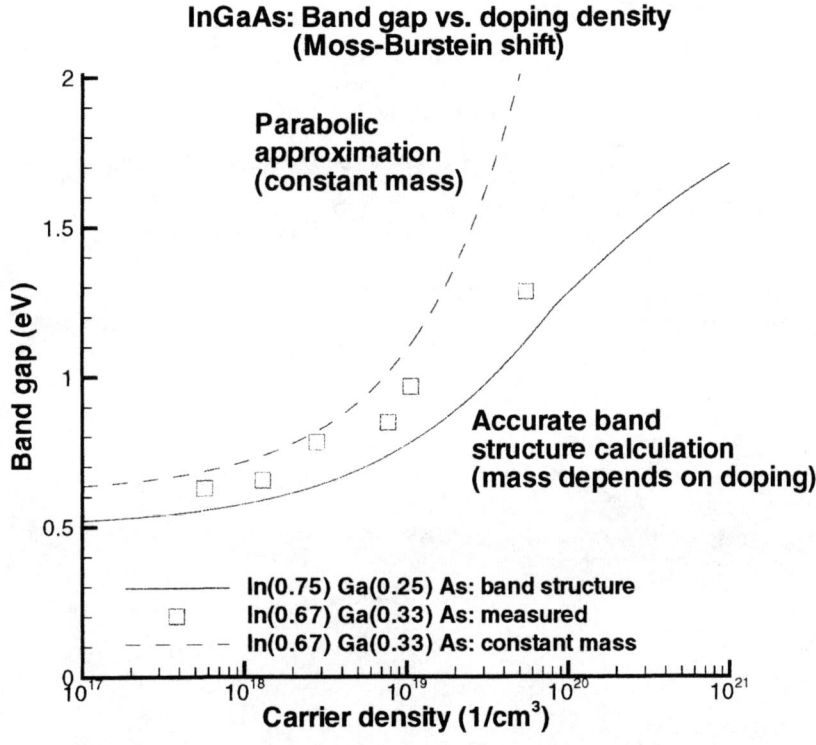

FIGURE 2. Calculated optical band gap vs. doping density for InGaAs along with experimental measurements and parabolic band prediction. The discrepancy between the experiment and the calculation at low doping is largely due to the fact that the composition chosen for the calculation differs from that of the experiment (see Fig. 1 caption). The doping dependent effective mass is evident from the fact that the Fermi energy increases with doping more gradually than predicted by the parabolic band approximation (constant effective mass).

CONCLUSION

The present investigation helped to resolve the interpretation of experimental data that was in conflict with the predictions of simple classical theory. The results of the band structure calculations are in agreement with experiments in which a maximum of the plasma frequency with doping was observed. The classical theory, on the other hand, predicts the plasma frequency to increase as the square root of the doping density. The calculated shift of the optical band gap (to higher frequencies) vs. doping (Moss-Burstein shift) is also in better agreement with the experiments than is the classical prediction. The failure of the classical theory is due to the fact that it does not account for the doping dependence of the effective mass of the charge carriers.

REFERENCES

1. Jansen, H. J. F., and Freeman, A. J., *Phys. Rev.* **B 30**, 561 (1984).

2. Geller, C. B., and Blazeck, T. S., *4th NREL conference on TPV*, session #1, talk #5.

3. Spitzer, W. G., and Fan, H. Y., *Phys. Rev.* **106**, 882 (1957).

Emittance Theory for Cylindrical Fiber Selective Emitter

Donald L. Chubb

National Aeronautics and Space Administration
Lewis Research Center
Cleveland, Ohio 44135

ABSTRACT

A fibrous rare earth selective emitter is approximated as an infinitely long cylinder. The spectral emittance, ε_λ, is obtained by solving the radiative transfer equations with appropriate boundary conditions and uniform temperature. For optical depths, $K_R = \alpha_\lambda R$, where α_λ is the extinction coefficient and R is the cylinder radius, greater than 1 the spectral emittance is nearly at its maximum value. There is an optimum cylinder radius, R_{opt}, for maximum emitter efficiency, η_E. Values for R_{opt} are strongly dependent on the number of emission bands of the material. The optimum radius decreases slowly with increasing emitter temperature, while the maximum efficiency and useful radiated power increase rapidly with increasing temperature.

INTRODUCTION

Fibrous rare earth oxide selective emitters for thermophotovoltaic (TPV) applications have been of research interest for several years. Nelson (1) began working with fibrous emitters in the 1980's. In addition fibrous emitters are being developed at Quantum Group (2) and the Auburn Space Power Institute (3). This paper develops the emittance theory for a fibrous emitter by approximating the emitter as an infinitely long cylinder. Since the fibrous emitters consist of bundles of 1 to 10 μm diameter fibers this theory does not include the effects of the reflectance that occurs when radiation leaves a fiber and enters an adjoining fiber. The whole bundle of fibers is being approximated as a continuous cylinder. If the fibers are closely packed and the reflectance at the interface between a fiber and the medium in the voids between fibers is small, then the error resulting from the approximation should be small.

The spectral emittance of the cylinder is obtained by solving the radiative transfer equation with appropriate boundary conditions. Knowing the spectral emittance allows the emitter efficiency to be calculated. As an example, emitter efficiency is calculated for an erbium-holmium aluminum garnet cylindrical emitter ($Er_{.3}Ho_{2.7}Al_5O_{12}$) which is being studied at NASA Lewis.

TEMPERATURE OF EMITTING CYLINDER

As pointed out earlier (4), temperature drop across a planar or film type emitter causes a major reduction in the spectral emittance in the emission band of a selective emitter. However, in most cases for a cylindrical emitter there will be a negligible temperature drop. This can be seen by considering the steady state energy equation for an infinite cylinder with no internal heat generation and thermal conductivity, k_{th}, and where we assume the temperature, T, and radiation flux, Q, depend only on the radial coordinate, r.

$$r\left[k_{th}\frac{dT}{dr} - Q\right] = \text{constant} \qquad (1)$$

In order to avoid the term in brackets being singular at $r = 0$ it must vanish for all r. Thus at all r the conduction and radiation fluxes balance.

$$k_{th}\frac{dT}{dr} = Q \qquad (2)$$

In other words, for steady state conditions all the thermal energy being conducted into the cylinder at the outer radius, $r = R$, leaves the cylinder as radiation.

The radiation flux, Q, will always be less than the blackbody flux $\sigma_{sb}T^4$, where $\sigma_{sb} = 5.67 \times 10^{-12}$ w/cm²K⁴ is the Stefan Boltzmann constant. Therefore, define the following dimensionless variables.

$$\overline{Q} = \frac{Q}{\sigma_{sb}T_s^4} \qquad \overline{T} = \frac{T}{T_s} \qquad \overline{r} = \frac{r}{R} \qquad (3)$$

Where T_s is the temperature at $r = R$. In this case equation (2) becomes the following.

$$\frac{d\overline{T}}{d\overline{r}} = \gamma \overline{Q} \qquad (4)$$

Where γ is the ratio of the radiation flux to the thermal conduction flux.

$$\gamma \equiv \frac{\sigma_{sb}T_s^4}{k_{th}\frac{T_s}{R}} = \frac{\sigma_{sb}T_s^3 R}{k_{th}} \qquad (5)$$

For the ceramic materials used in most selective emitters, $k_{th} \geq 0.01$ w/cmK. Also, for TPV applications $T_s \leq 2000$K. Therefore, $\gamma \leq 4.5R$(cm). So that if $R < 0.1$ cm it is a reasonable approximation to neglect the right hand side of equation (4) and obtain the result $\bar{T}$ = constant ($T = T_s$). If $\gamma \ll 1$ for a film or planar emitter then a linear temperature variation results (4) rather than a uniform temperature. For the cylinder emittance calculation that follows, a uniform temperature is assumed.

SPECTRAL EMITTANCE OF CYLINDER

The spectral emittance is defined as follows.

$$\varepsilon_\lambda \equiv \frac{q_\lambda(R)}{e_{bs}(\lambda, T_s)} \tag{6}$$

Where $q_\lambda(R)$ is the radiation flux leaving the cylinder at $r = R$ and $e_{bs}(\lambda,T_s)$ is the blackbody emissive power where λ is wavelength and T_s is the cylinder temperature.

$$e_{bs}(\lambda, T_s) = \pi i_{bs}(\lambda, T_s) = \frac{2\pi h c_0^2}{\lambda^5 [\exp(hc_0/\lambda k T_s) - 1]} \tag{7}$$

Appearing in equation (7) is the blackbody intensity $i_{bs}(\lambda,T_s)$, w/cm² nm steradian, Planck's constant, h, Boltzmann's constant, k, and the vacuum speed of light, c_0.

The radiation flux, $q_\lambda(R)$, is obtained by solving the radiation transfer equations for the intensities, $i_\lambda^+(R)$, and $i_\lambda^-(R)$. Where $i_\lambda^+(R)$ is the intensity moving in the +R direction and $i_\lambda^-(R)$ is the intensity moving in the –R direction as shown in figure 1. Assuming the intensities depend only on the radial coordinate leads to transport equations for $i_\lambda^+(R)$ and $i_\lambda^-(R)$ identical to the planar case (5). These equations are written in terms of the optical depth, K, rather than the coordinate, r.

$$K = \alpha_\lambda r \tag{8a}$$

$$K_R = \alpha_\lambda R \tag{8b}$$

Where α_λ is the extinction coefficient, assumed independent of r, and is the sum of the absorption coefficient, a_λ, and the scattering coefficient σ_λ.

$$\alpha_\lambda = a_\lambda + \sigma_\lambda \tag{9}$$

The boundary conditions that must be applied are the following. At $r = K = 0$, from symmetry conditions.

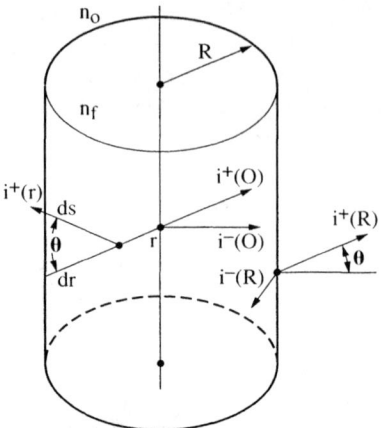

n_f = index of refraction of cylinder
n_o = index of refraction of surroundings
$\cos\theta \, ds = dr$

FIGURE 1. Schematic of emitting cylinder.

$$i_\lambda^+(0) = i_\lambda^-(0) \quad \text{at} \quad K = 0 \tag{10a}$$

At r = R or $K = K_R$ the intensity moving in the $-R$ direction is equal to the reflected intensity.

$$i_\lambda^-(K_R) = \rho_{fo} i_\lambda^+(K_R) \quad \text{at} \quad K = K_R \tag{10b}$$

Where ρ_{fo} is the reflectance at the cylinder outer radius, R. At r = R total reflectance occurs for certain angles of incidence, θ. At an interface between a material with a index of refraction, n_i, and a material with index of refraction, n_j, where $n_i > n_j$, radiation moving from i into j with angles of incidence $\theta > \theta_M$, where θ_M is given by Snell's Law will be totally reflected. Since $n_f > n_o$, for the cylinder-air interface we have the following result for the reflectance, ρ_{fo}.

$$\rho_{fo} = 1 \quad \text{for} \quad \theta \geq \theta_M \quad \text{where} \quad \mu_M^2 \equiv \cos^2\theta_M = 1 - \left(\frac{n_o}{n_f}\right)^2 \tag{11a}$$

For the case where $\theta < \theta_M$ ($\mu > \mu_M$) we approximate ρ_{fo} by the reflectance for normal incidence (6).

$$\rho_{fo} = \left(\frac{n_f - n_o}{n_f + n_o}\right)^2 \quad \theta < \theta_M \ (\mu > \mu_M) \tag{11b}$$

Solution of the radiative transfer equations for $q_\lambda(K_R)$ is presented in (5) for the film or planar case with a linear temperature variation through the film and for no scattering ($\sigma_s = 0$ in eq. (9)). Results for the uniform temperature cylinder can be obtained from these results by setting $\varepsilon_{fs} = 0$ ($\rho_{fs} = 1$) and $\Delta T = 0$ in equations (33) to (36) of reference (5). When this is done the following result is obtained for the spectral emittance.

$$\varepsilon_\lambda = \frac{n_o^2(1-\rho_{fo})\left[1-4E_3^2(K_R)\right]}{1-4E_3(K_R)\left[\rho_{fo}E_3(K_R)+\mu_M^2(1-\rho_{fo})E_3\left(\frac{K_R}{\mu_M}\right)\right]} \tag{12}$$

uniform temperature and no scattering

Appearing in equation (12) is the exponential integral $E_3(x)$.

For single crystal materials, such as rare earth doped yttrium aluminum garnet (YAG), scattering should be negligible. However, for polycrystalline rare earth oxides such as those being considered in references (1) to (3) scattering should be significant. In those cases equation (12) will overestimate the spectral emittance.

In figure 2 the spectral emittance, ε_λ, using equation (12) is shown as a function of optical depth, K_R, for $n_o = 1$ and $n_f = 1, 1.5$ and 2.0. As figure 2 shows ε_λ increases rapidly with K_R and reaches nearly its limiting value for $K_R = 1$. Notice also that for small K_R as n_f increases the spectral emittance rate of increase also increases. For most of the selective emitter materials, $1.5 \leq n_f \leq 2.0$.

Once the extinction coefficient, α_λ, and index of refraction, n_f, are known equation (12) can be used to calculate ε_λ. Figure 3 shows the calculated spectral emittance for an erbium-holmium aluminum garnet cylinder (R = 0.4 cm) with 10% erbium and 90% holmium, $Er_{0.3}Ho_{2.7}Al_5O_{12}$. This single crystal material is being considered for a film type selective emitter (5). The extinction coefficient and index of refraction were obtained from reference (5). Holmium has its main emission band centered at $\lambda \approx 2000$ nm with smaller bands centered at $\lambda \approx 1100, 890$ and 750 nm. Erbium has its main emission band centered at $\lambda \approx 1500$ nm with secondary bands centered at $\lambda \approx 1000, 800$ and 640 nm. All of these bands show up as regions of large emittance in figure 3. In the region $2000 < \lambda < 4500$ nm, $Er_{0.3}Ho_{2.7}Al_5O_{12}$ is nearly transparent ($\alpha_\lambda \to 0$) and thus ε_λ is small. The highly oscillatory result in this region results from numerical error in α_λ (5). For the region $\lambda > \lambda_c = 5000$ nm ε_λ becomes large again. This large ε_λ results from vibrational modes of the crystal lattice and is a characteristic of most rare earth selective emitter materials (1). We call the wavelength, λ_c, the long wavelength cutoff.

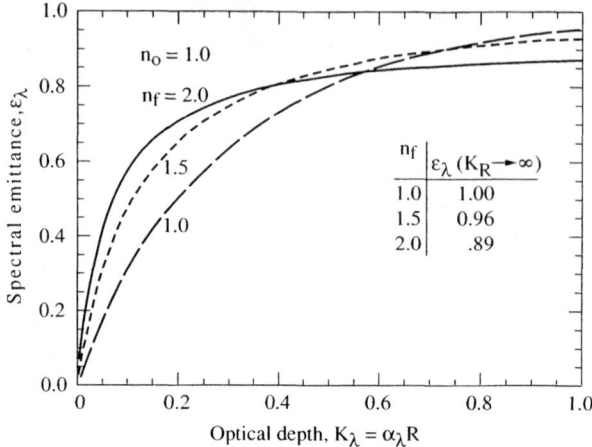

FIGURE 2. Spectral emittance for cylinder of radius, R, at a constant temperature, T_s, as a function of optical depth, $K_\lambda = \alpha_\lambda R$, where α_λ is the extinction coefficient and n_f is the cylinder index of refraction and n_o is the surrounding index of refraction.

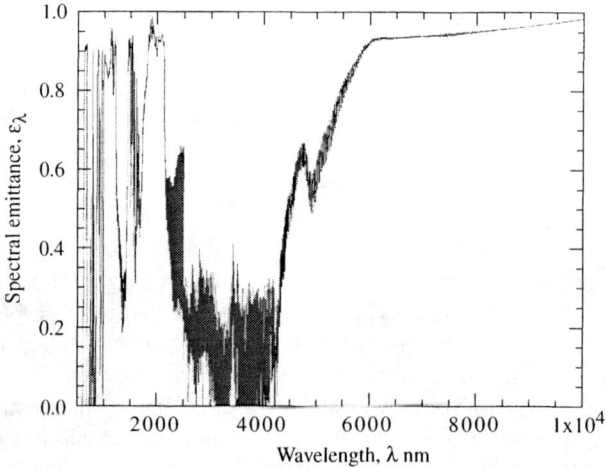

FIGURE 3. Theoretical spectral emittance of $Er_{0.3} Ho_{2.7} Al_5 O_{12}$ cylinder of radius, R = 0.4 cm, calculated using measured extinction coefficient and index of refraction.

EMITTER EFFICIENCY

As a measure of the effectiveness of a selective emitter define the emitter efficiency as follows.

$$\eta_E \equiv \frac{\text{useful radiated power}}{\text{total radiated power}} = \frac{Q_b}{Q_T} = \frac{\int_0^{\lambda_g} q_\lambda(R)d\lambda}{\int_0^\infty q_\lambda(R)d\lambda} = \frac{\int_0^{\lambda_g} \varepsilon_\lambda e_{bs}(\lambda, T_s)d\lambda}{\int_0^\infty \varepsilon_\lambda e_{bs}(\lambda, T_s)d\lambda} \quad (13)$$

The numerator, Q_b, is the power radiated in the wavelength region $0 \leq \lambda \leq \lambda_g$. In a TPV system λ_g corresponds to the bandgap energy, $E_g = hc_o/\lambda_g$, of the PV cell. The denominator is the total radiated power, Q_T.

Consider how η_E will be behave as a function of the cylinder radius, R. As figure 2 shows, ε_λ increases rapidly with optical depth, $K_R = \alpha_\lambda R$. Therefore, for the emission bands (large α_λ and $0 \leq \lambda \leq \lambda_g$) ε_λ and thus Q_b will quickly approach their limiting values as R increases from zero. However, for the regions between the emission bands and the long wavelength cutoff (small α_λ and $\lambda_1 < \lambda < \lambda_c$) ε_λ and thus Q_T will increase more slowly to their limiting values as R increases from zero. As a result, there will be an optimum radius, R_{opt}, that will yield maximum η_{EMAX}. This is illustrated in figure 4 for $Er_{0.3}Ho_{2.7}Al_5O_{12}$ where $\lambda_g = 2200$ nm ($E_g = 0.56$ eV) was chosen for use in equation (13) and $T_s = 1700$ K. As can be seen, η_E rises rapidly to η_{EMAX} and then decreases slowly for $R > R_{opt}$. Also shown in figure 4 is the useful power radiated, Q_b. As can be seen, Q_b rises rapidly and then begins to level off. For $T_s = 1700$K the optimum radius is $R_{opt} = 0.34$ cm. It should be mentioned that for radii of this magnitude the uniform

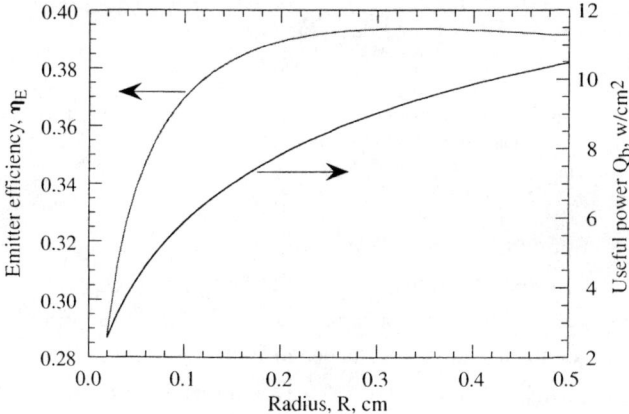

FIGURE 4. Emitter efficiency, η_E, and useful power, Q_b, as a function of cylinder radius, R, at emitter temperature, $T_s = 1700$ K for $Er_{0.3}Ho_{2.7}Al_5O_{12}$.

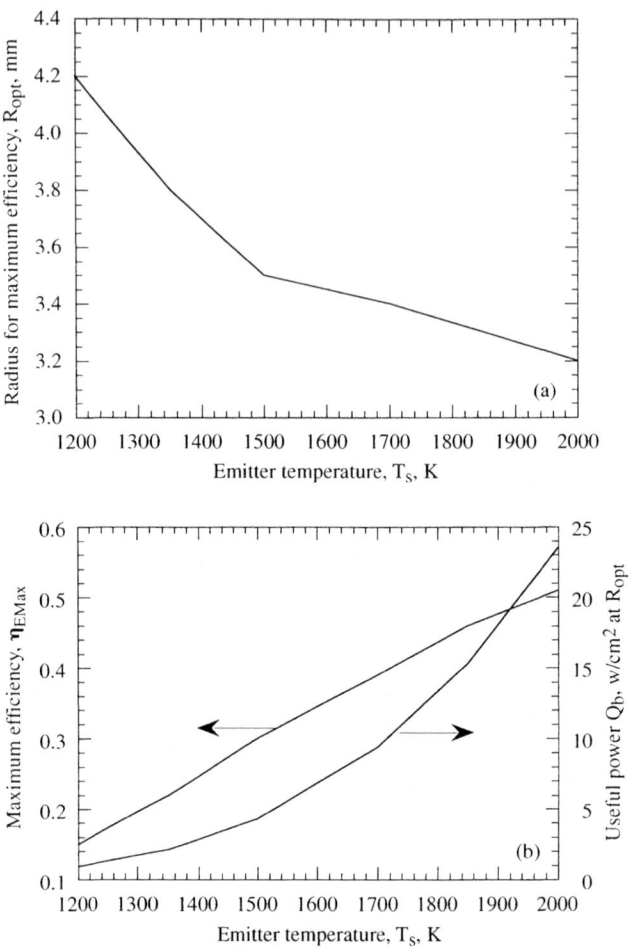

FIGURE 5. Optimum radius, R_{opt}, maximum efficiency, η_{EMax}, and useful power, Q_b, at $R = R_{opt}$ as functions of emitter temperature, T_s, for $Er_{0.3} Ho_{2.7} Al_5 O_{12}$. (a) Optimum radius, R_{opt}. (b) Maximum efficiency and useful power.

temperature assumption becomes questionable. In figures 5(a) and (b) R_{opt}, $\eta_E(R_{opt})$ and $Q_b(R_{opt})$ are shown as functions of T_s. Both $\eta_E(R_{opt})$ and $Q_b(R_{opt})$ increase significantly with temperature while R_{opt} decreases only 25 percent in going from $T_s = 1200$ to 2000K. The large increase in η_E results because the maximum value of the blackbody emissive power, $e_{bs}(\lambda,T_s)$, shifts to shorter wavelengths as T_s increases. Therefore, Q_b increases faster than Q_T as T_s increases. The useful power, Q_b, increases at least as T_s^4.

$Er_{0.3}Ho_{2.7}Al_5O_{12}$ has 8 emission bands and thus a large region of high ε_λ for $0 \leq \lambda \leq \lambda_g$. Other rare earth garnets such as $Tm_3Al_5O_{12}$ with a single major emission band at $\lambda \approx$ 1900 nm have a much smaller region of large ε_λ. Thus, R_{opt} will occur approximately when $K_R = R_{opt} \alpha_\lambda \approx 1$, where α_λ is the extinction coefficient characteristic of the emission band. In the case of $Tm_3Al_5O_{12}$ with $\alpha_\lambda \approx 30$ cm^{-1}, $R_{opt} \approx 0.3$ mm compared to $R_{opt} \approx 3.5$ mm for $Er_{0.3}Ho_{2.7}Al_5O_{12}$.

CONCLUSION

1. For most rare earth selective emitters the temperature is uniform through the cylinder.
2. When the optical depth $K_R = \alpha_\lambda R \geq 1$ the spectral emittance is nearly a maximum.
3. There is an optimum value for the radius, R_{opt}, which yields maximum emitter efficiency.
4. R_{opt} strongly depends on the emitter material. For an emitter with only a single strong emission band R_{opt} is the order of 0.1 mm, whereas for an emitter with many emission bands R_{opt} is the order of 1 mm.
5. R_{opt} decreases slowly with increasing emitter temperature.
6. The maximum efficiency, η_{EMAX}, and useful power, Q_b, increases significantly with temperature.

REFERENCES

1. Nelson, R.E., "Thermophotovoltaic Emitter Development," presented at the First NREL Conference on Thermophotovoltaic Generation of Electricity, AIP Proc. 321, 80.
2. Holmquist, G.A., "TPV Power Source Development for an Unmanned Undersea Vehicle," presented at the First NREL Conference, AIP Proc. 321, 308.
3. Adair, P.L. and Rose, M.F., "Composite Emitters for TPV Systems," presented at the First NREL Conference, AIP Proc. 321, 245.
4. Chubb, D.L., Good, B.S., Clark, E.B. and Chen, Z., "Effect of Temperature Gradient on Thick Film Selective Emitter Emittance," presented at the Third NREL Conference on Thermophotovoltaic Generation of Electricity, AIP Proc. 401, 293.
5. Chubb, D.L., Pal, A.T., Patton, M.O. and Jenkins, P.P., "Rare Earth Doped High Temperature Ceramic Selective Emitters," to be published in the *Journal of the European Ceramic Society*, also NASA/TM—1998-208491.
6. Siegel, R. and Howell, J.R., "Thermal Radiation Heat Transfer," 2nd edition, Washington, D.C., Hemisphere, 1981, Ch. 4.

Fibrous Selective Emitter Structures from Sol-Gel Process

K. C. Chen, Ph.D.

Quantum Group, Inc.
11211 Sorrento Valley Road
San Diego, CA 92121

ABSTRACT

Selective emitters have the potential benefit of high efficiency due to the matching of emission spectra to the response of photovoltaic (PV) cells. Continuous uniform rare-earth oxide selective emitter fibers were successfully fabricated using a viscous solution made from metal organic precursors. Cylindrical- and planar configuration emitter structures were made by direct cross-winding or stacking of precursor fiber layers. The combustion and optical performance of the planar emitter structures were tested. The results indicates that both the designing of the fiber packing density and the thickness is critical for high photon and power output.

INTRODUCTION

Over the past decade, driven by the high-temperature ceramic-matrix-composite development program, several continuous high-performance ceramic fibers are now commercially available in quantity. One thing learned from these efforts is that high-temperature fiber applications require extensive efforts in continuous-fiber manufacturing technology development, as well as in fiber high-temperature property characterization. Successful usage of rare-earth-oxide fibers to provide specific wavelengths at incandescence for thermophotovoltaic (TPV) devices will inevitably follow similar developmental paths. Because the rare-earth-oxide fibers are not regarded as traditional stress-loading structural ceramic fibers, there are only limited studies on fiber manufacturing, grain growth, strength, creep, and corrosion behaviors at high-temperature. Clearly, if the rare-earth oxide selective-emitter fiber and structures are to be improved in mechanical strength and achieve their potential in any application, it is essential to understand key high- temperature properties. This can only be done when emitter fibers of well-defined dimension and microstructure are available.

The current methods of making rare-earth-oxide fibrous ceramics were described by Nelson [1]. These fibrous ceramics consist of discontinuous rare-earth-oxide fibers made by relic processes. The individual fibers in these ceramic articles vary in diameter and microstructure. There are only a few techniques that are capable of producing continuous rare-earth-oxide fibers in quantity at low cost. These include those that are spun from viscous liquids that may contain dispersed oxides in colloid or ultra-fine particulate form, or from polymers of chemical compounds in solution convertible to oxides, or mixtures. The processes are sometimes identified as sol-gel or polymer pyrolysis. The commonality of the processes is that the fibers are first formed, then fired under controlled conditions to a ceramic. Both processes have successfully produced commercial quantities of high-strength alumina, mullite and other oxide fibers. The precursor-method is the better processing technique for rare-earth-oxide fiber based on the high sintering and high melting temperature (2350-2420°C) required for complete densification.

The classic sol-gel approach to produce oxide ceramics is based on hydrolysis and polymerization of metal alkoxide that results in a sol-to-gel transition. During the sol-to gel-transition, fiber is drawn from the liquid of suitable viscosity at a suitable time. Unfortunately, this classic sol-gel approach is not applicable to rare-earth-oxide fiber processing because of rapid hydrolysis and polymerization of rare-earth metal alkoxides. The rapid hydrolysis and polymerization of rare-earth alkoxides causes the solution to precipitate, instead of becoming a viscous liquid. A process was developed to control the rapid hydrolysis by decreasing the number of alkoxy groups, and to ensure the polymerization into a two-dimensional chain-like structure. This approach produces a solution that has a time-invariant rheology allowing spinning of a precursor fiber continuously at speeds sufficient for commercial feasibility, as well as fabricating emitter structures directly by a winding process.

1. SOL-GEL PROCESSING OF RARE-EARTH OXIDE FIBERS

1.1. Precursor Synthesis and Viscosity Control

The rare-earth isopropoxides were synthesized by reaction of high-purity rare-earth metal chips with anhydrous isopropanol, with a small amount of mercuric chloride as catalyst. The solution was refluxed and then vacuum-filtered to a clear solution in a nitrogen glovebox. Ytterbium, erbium and holmium isopropoxide solutions are faint yellow, rose red and dark yellow color, respectively. The concentrations are usually around 0.5M. An organic acid, such as 2-ethylhexanoic acid, was then added slowly into a rare-earth isopropoxide solution at the boiling state with vigorous stirring. The molar ratio of the organic acid to the metal alkoxides was 1.5. The addition of organic acid to rare-earth isopropoxide replaces approximately half of the isopropoxy groups and changes the slow hydrolysis of the rare-earth isopropoxide. Hexane was added to aid homogeneity of the solution during acid addition. After refluxing for 1 hour, water was added to hydrolyze the remaining isopropoxy groups and start the polymerization. The solution became viscous within a few minutes, but was not viscous enough for fiber drawing. Therefore, another approach was developed to precisely control viscosity. The solvent in the solution was first removed by using a rotary evaporator, and subsequently dried in a 60°C oven until a transparent brittle mass was formed. Since the polymeric precursor has the quasi-two-dimensional structure, the dry

Precursor Fiber Spinning Range										
Precursor	Cyclohexane Content in Spinning Liquid (Wt. %)									
Acid/alkoxide	10	20	30	40	50	60	70	80	90	100
1.00	■									
1.50	■■									
1.75	■■■									
2.00					■■					

FIGURE 1. Ytterbium precursors fiber drawing range with cyclohexane as solvent

precursor is expected to have high solubility in organic solvent. This brittle precursor mass was crushed into fine precursor powder to facilitate dissolution and homogeneity. The solubility ranges of the precursor powder vary only with species and amount of the organic acid and the solvent. Figure 1 shows the suitable fiber spinning range for polymeric precursor, using cyclohexane as solvent. The viscosity of the precursor liquid is time-independent and can be controlled either by the amount of the solvent or the temperature.

1.2. Precursor Fiber Spinning and Oxide Fiber Microstructure

A continuous fiber forms when the viscous liquid passes through a 50μm spinneret. Twenty milliliters of viscous liquid can produce close to 3000 meters of precursor fiber of 40μm in diameter at a speed of 35 m/min. (Figure 2). An even faster spinning speed produces fiber of thinner but sacrifices fiber thread-line stability. Depending on the fiber spinning rate and drying rate, the cross-section of the fiber varies from circular to kidney shape (Figure 3). The fiber can be cross-wound without sticking by applying a heat lamp or purposely joining by slower drying. The slow drying allows the in-situ emitter structure to be formed on a mandrel. X-ray diffraction shows the precursor fiber is amorphous. The precursor fiber was then pyrolyzed in a controlled nitrogen atmosphere and heating schedule to convert to oxide fiber. The precursor fiber contains a fugitive 2-ethylhexanoate group. The thermal gravimetric analysis indicates rapid organic weight loss between 230-360°C (Figure 4). Slow removal of the fugitive organic component is crucial so that no pores or cracks are to be generated during the pyrolysis. The fiber shrinks approximately 50% in dimension during the pyrolysis below 500°C. Crystallization occurs at 700°C. Subsequently, the fiber is slowly heat-treated in air to remove any remaining carbon, and sintered at 1500-1600°C. Figure 5 shows the scanning electron micrographs of the erbia fiber. The ytterbia and erbia fibers have fine grain below 1200-1300°C and slow grain growth

FIGURE 2. Continuous ytterbia precursor fiber spinning at a speed of 35 m/min

FIGURE 3. The shape of the fiber cross-section is dictated by diffusion of the solvent from the fiber interior and is affected by spinning speed and drying rate.

below 1500°C. Fiber fracture surface is similar to duPont's zirconia-modified alumina PRD-166 fiber, which has sub-micrometer grain size in this temperature range.

2. EMITTER STRUCTURES FABRICATION AND TESTING

2.1. Emitter Structures Fabrication

While the ultimate goal of this research is to produce high-strength continuous rare-earth oxide fiber, and use fiber architecture to expand the design options for tough and

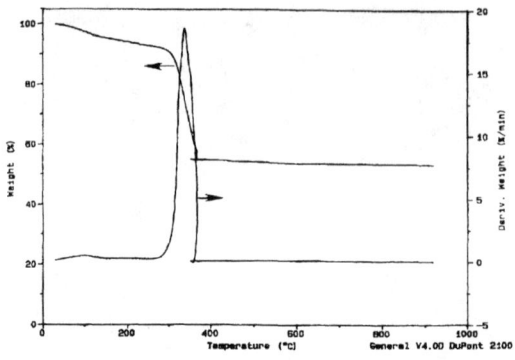

FIGURE 4. Thermal gravimetric analysis of precursor powder in air showing critical pyrolysis region due to rapid weight loss

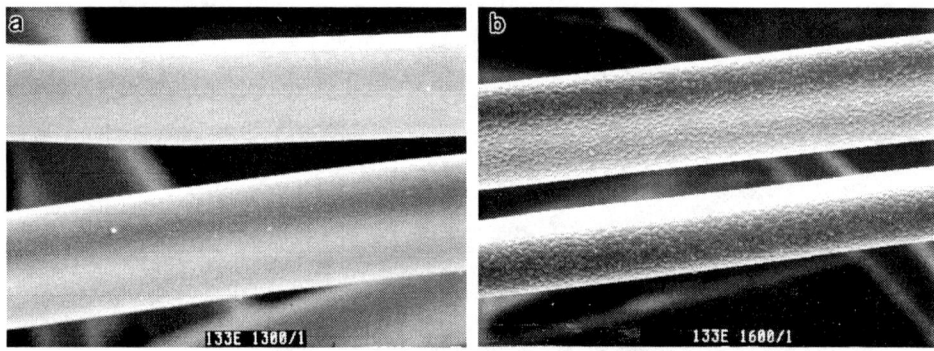

FIGURE 5. The scanning electron micrographs of erbia selective emitter fiber after heat treatment 1 hour at (a) 1300°C and (b) 1600°C

reliable emitter structure, the selective- emitter fiber can not be braided or weaved at the present time. Extensive manufacturing developmental effort will be required to reduce the surface and interior critical flaw size in order to reach the ultimate fiber strength for weaving. To shorten the development time, the precursor fiber is directly cross-wound onto the pickup wheel forming a cross-hatched pattern. The packing density of the fiber layer is determined by the speeds of cross-winding and spinning. The fiber packing density and thickness of the emitter structure affects the combustion characteristics, such as fuel loading rate, back pressure, flashback, temperature gradient, and photon output of the emitter structure. Several emitter-design options were evaluated to determine manufacturing feasibility, and their combustion and power output performance. Several cylindrical- and planar-emitter structures were made. Figure 6 shows two cylindrical and a planar-emitter structures made by sol-gel derived ytterbia fiber. The cylindrical structures are more difficult to manufacture because they require different sizes of support as the structure progressively shrinks during the organic pyrolysis. The planar- emitter structure was thought to be a better emitter-structure design because it has a 1:1 view factor when it is placed in front of a

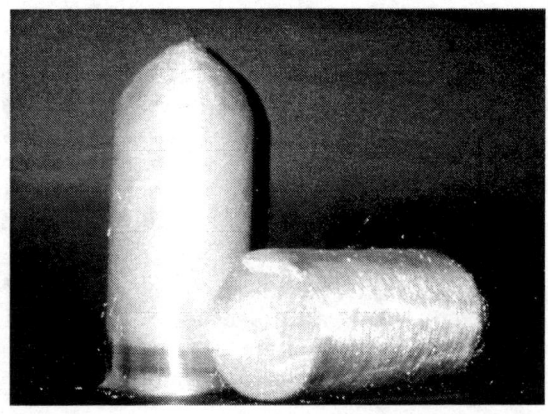

FIGURE 6. Cylindrical emitter structures by direct cross-winding of the precursor fiber

FIGURE 7. Photomicrograph of the flat panel emitter structure (1500°C, 1 hr. 32X)

flat PV array. The planar-emitter structures are made of 10 precursor fiber panels in alternate 0-90° orientation. A total of 30,000 meters of precursor fiber is needed to make a single oxide planar emitter structure of 12 cm by 12 cm in size and 0.7 cm in thickness (Figure 7)!

2.2. Planar Emitter Structures Testing

Before the emitter structure design and manufacturing technology can incorporate all the necessary elements to couple with the combustion portion of the TPV system, a planar-emitter structure using a zirconia block as support was made to test the power output and combustion performance. The emitter structure consists of two fibrous emitter panels attached to a frame made from zircona block (Figure 8). The fuel distributor was also made with perforated zirconia. The inner surface of the zirconia is lined with an ytterbia felt. While the flat emitter was designed for an atmospheric air-breathing TPV application, compressed air was used monitor the fuel-to-air ratio in the performance test. Several fibrous-emitter panels of different fiber packing densities were made.

Tests show the fiber packing density, packing uniformity and the thickness are very critical to control the flame-front location and power output. High fiber-packing density prevents the flame front to move inside the emitter structure, resulting in a low power output in the silicon PV array. Optimal fiber-packing

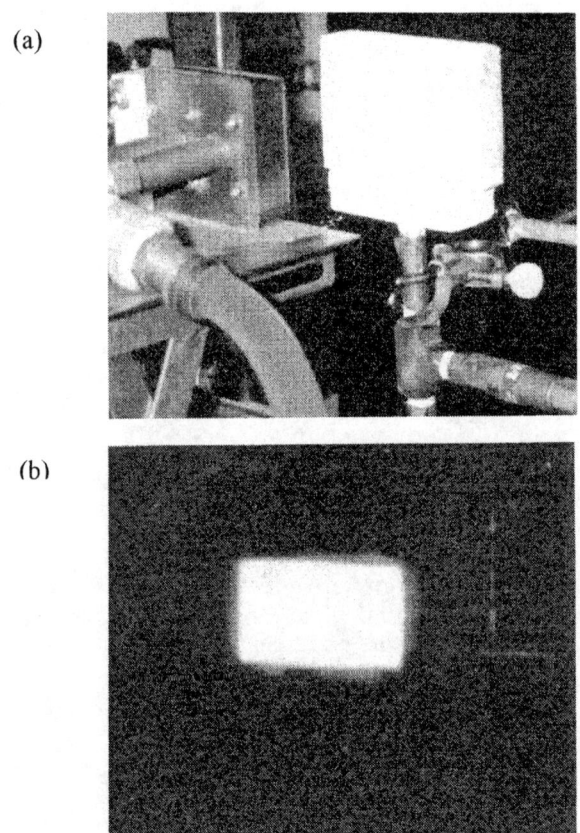

FIGURE 8. The flat emitter structure (a) in combustion testing setup; (b) during testing

density allows the flame to stay in the emitter layer or completely move in between two emitter panels.

Table 1 summarizes the combustion performance and the optical output of the flat-panel emitter structure. Flame was initiated outside, and within seconds moved inside the planar emitters and the zirconia support. The emitter structures have very low back-pressure. The power output of the structure was not high due to a slightly denser and a somewhat variation in fiber packing-density. The zirconia block also removes heat from the flame.

Table 1. The combustion characteristics and power output of the flat-panel Yb_2O_3 emitter structure

Gas Rotometer (USCFH*)	Pressure (psig)	Air Rotometer (USCFH*)	Fuel Input Rate (BTU/hr.) (W)	Back Pressure (iwc)	Volts (V) Voc	Ampere (mA) Isc	Power (Watts)	Fuel Loading (W/cm^2)
Distance from PV = 2.0 inch									
1.90	1.20	52	4000 (1170W)		16	160	2.6	
3.50	1.40	65	7400 (2165W)	0.3	16.1	195	3.1	31.8
Distance from PV = 1.5 inch									
3.5	1.5	75	7450 (2180W)	0.3	16.3	230	3.8	32.1

* UCSCFH: Uncorreted Standard Cubic Foot Per Hour;
Emitter Area = 7.9 cm x 8.6 cm = 68 cm^2

In conclusion, we believe that fabrication of a lower density emitter panel will produce high photon and power output.

REFERENCES

[1] Nelson, R. E., *AIP Conference Proceedings* 358 Eds. By T. J. Coutts and J.P. Benner, New York: American Institute of Physics, 1995, pp.80-98.

Commercial GaSb Cell and Circuit Development for the Midnight Sun® TPV Stove

Lewis Fraas, Russ Ballantyne, She Hui, Shi-Zhong Ye,
Sean Gregory, Jason Keyes, James Avery, David Lamson,
and Bert Daniels

JX Crystals Inc., Issaquah, WA 98027

Abstract. Most residential heating applications for thermophotovoltaic generation are very price-sensitive. Some applications are possible at a few dollars per Watt, but widespread on-grid acceptance will require electric generation pricing no greater than $1 per Watt. Two recent discoveries make such pricing possible using GaSb cells. First, cells have been successfully processed on polycrystalline material, which means that drastic cost reduction is possible in crystal growth. This also means that larger wafers can be fabricated, dropping cell processing costs. And second, it has been demonstrated that circuits can be assembled in a shingled pattern on a thermally-matched substrate. This has the dual benefit of greatly reducing the required labor for circuit assembly and substantially improving system packing factors.

INTRODUCTION

Several recent papers describe thermophotovoltaic (TPV) cells and circuits designed to be used with radioisotope heat sources for military or space applications. Because of the safety issues associated with these sources, emitter temperatures are limited to below 1100° C. This explains the interest in the InGaAs and GaInSbAs materials systems with band gaps in the 0.55 eV range. Unfortunately, these cells are inherently expensive because the process for fabricating them is complex and requires the use of very toxic gases. Cost is not a serious constraint for these applications.

JX Crystals' interest, however, is in commercial applications using hydrocarbon-fueled heat sources. Fortunately, the infrared (IR) emitter in this case can run in the 1200° C to 1400° C temperature range. This higher emitter temperature leads to higher cell power densities and, thus, lower costs. Furthermore, the optimum cell bandgap for this emitter temperature shifts to 0.7 eV and GaSb cells can be used. The GaSb cell fabrication process uses a simple diffusion step for junction formation without the use of toxic gases. For this application, cost is a serious constraint. This paper will describe some innovations in GaSb cell and circuit fabrication techniques designed to lower TPV commercial system costs.

THE MIDNIGHT SUN® STOVE

Provided here first is a description of a prototypical TPV commercial application: JX Crystals' Midnight Sun® gas-fired TPV stove that cogenerates both heat and electricity for off-grid homes. It has previously been noted that TPV generators of heat and electricity are complimentary to solar photovoltaic systems; they can be used in colder climates and during the winter months when the sun is not shining. Customers that live away from the electric power grid who have solar panels and batteries can purchase a TPV heating stove to keep themselves warm and keep their batteries charged in winter months. These customers represent an off-grid market, and JX Crystals has developed the small TPV heating stove shown in figure 1 for this market.

FIGURE 1. The Midnight Sun® Stove, cogenerating 25,000 BTU/hr and 100 Watts.

GaSb infrared sensitive photovoltaic cells responding to wavelengths out to 1.8 microns now make it practical to generate electricity using ceramic infrared emitters operating at moderate temperatures. The Midnight Sun® Stove can be used as a fireplace insert or as a vented stove located in the corner of the family room. It generates approximately 25,000 BTU per hour of heat and 100 Watts of electricity for battery charging, and a warm glow is visible through the front window.

There is also a grid-connected market for the Midnight Sun® TPV Stove. Recall that during the last winter as a result of ice storms, approximately one million people in Canada and another million people in the US were without electricity for several weeks. Because heating furnaces require electric power for air or water circulation, these people were also very cold. Previously, this market has been described as a self-powered furnace market. However, this market can be reformulated in more fundamental terms. The customer base here is grid-connected homeowners who suffer electric power outages during winter storms at which time they loose both heat and electricity. The self-powered furnace idea represents one way of providing for heat during the power outage. Noted here is that the Midnight Sun® Stove for the family room represents an alternative economical and beneficial solution for this market. During a power outage, it would be sufficient to provide heat for the family room rather than the whole house. Meanwhile, the Midnight Sun® Stove also provides enough electricity to operate the TV, a small light, a computer, and a security system as needed. Finally while providing enough heat to keep the family room cozy, it will also provide enough heat to keep the house water pipes from freezing.

The Stove shown in figure 1 has an emitter/filter/cell insert schematically depicted in figure 2. It is equipped with a fan for forced air circulation and a blower for supplying combustion air to the burner. It is also equipped with a spark igniter, flame sense, and thermostat control card for safe automated operation. These auxiliaries consume 12 W DC. It can be configured for 12 or 24 V output and propane or natural gas fuel sources. Dimensions are 56 cm tall × 51 cm wide × 23 cm deep.

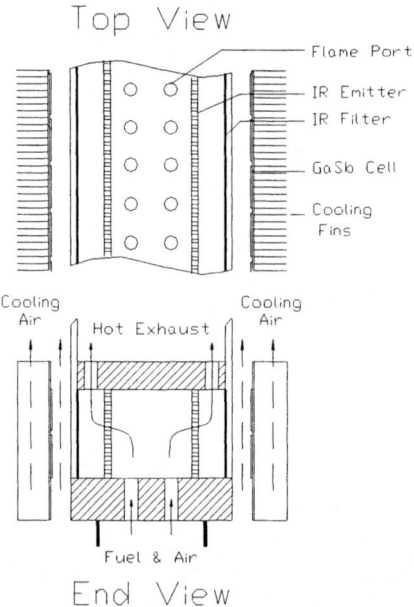

FIGURE 2. Top view and end view of TPV insert for the Midnight Sun® Stove showing flame heated IR emitter, GaSb converter, and IR filters.

Except for the GaSb cells, circuits, and emitter, the Midnight Sun® Stove uses commercially available parts similar to those used in the heating industry. The emitter is a SiC honeycomb emitter similar to those used in radiant heaters. Comparable heating stoves without the TPV benefits sell for approximately $1,000. The near term goal is to fabricate the two 50 W panels in each stove at $250 each. In the following sections, recent cost reducing breakthroughs that have been made in both the cell fabrication and circuit fabrication areas are described (patents pending).

POLYCRYSTALLINE GaSb TPV CELLS

JX Crystals has been fabricating gallium antimonide (GaSb) TPV cells for electric power generation using infrared radiation from hydrocarbon fired radiant heat sources. The market for these cells can be expanded dramatically if the cell fabrication cost can be reduced. Currently, 16 GaSb cells are laid out on 3 inch diameter single crystal GaSb wafers, cut from 3 inch diameter GaSb single crystals grown at JX Crystals. There are two primary contributors to cost: the cost of the single-crystal wafer itself and the cost of the wafer processing steps to obtain the final cells.

In the semiconductor industry, it is known that device cost can be reduced by increasing wafer diameter. This comes about because the wafer processing costs are generally per wafer costs rather than per device costs. It is very desirable to increase the diameter of a GaSb wafer in order to reduce GaSb cell costs. For example, one might obtain 64 cells from a 6 inch wafer at nearly the same cost that one obtains 16 cells on a 3 inch wafer now. However, the largest diameter GaSb crystal grown to date is a 4 inch diameter crystal at JX Crystals. Growth of a larger crystal will require both new equipment and considerable experimentation. It is desirable to obtain a large diameter low cost GaSb wafer that can be processed into high performance GaSb TPV cells.

In the solar photovoltaic field, it is known that silicon photovoltaic cells can be fabricated on course grain polycrystalline wafers. Although there is some degradation in performance, these polycrystalline silicon solar cells are commercially available. In this case, the polycrystalline silicon wafers are obtained from cast polycrystalline silicon ingots. Is it possible that high performance GaSb TPV cells can be fabricated using polycrystalline GaSb wafers?

The answer to this question is not obvious for several reasons. First, GaSb is a different material than silicon with a different crystal structure. Second, there is no theory describing how grain boundaries effect thermophotovoltaic cell performance. And third, a TPV cell operates in a vastly different environment than a solar cell.

Within the TPV field, there is no empirical information relating to polycrystalline material. Starting from the solar cell case, theoretical arguments can be made both for and against polycrystalline materials. The argument against polycrystalline material is that TPV cell materials have lower band-gaps than solar cell materials (0.7 eV instead of 1.1 eV) and therefore, TPV materials generate lower voltages. So, any loss in cell voltage for TPV cells from grain boundary shunts will be more significant and intolerable. The argument in favor of polycrystalline material is that TPV cells operate at much higher current densities and, therefore, shunting currents from grain boundaries will be less significant. The fact is that the answer to the question of whether polycrystalline GaSb wafers can be used to make high performance GaSb TPV cells was not known. It is desirable to answer this question.

JX Crystals is the only company in the world that has GaSb ingot formation and extensive GaSb TPV cell processing capabilities in the same facility. Therefore, it is in a good position to explore the effects of polycrystalline wafers on

GaSb cell performance. To answer the question posed above, the company intentionally processed several polycrystalline GaSb wafers and fabricated several polycrystalline GaSb cells. The GaSb cell size fabricated was 1.7 cm^2 and the average crystal size on the processed wafers was between 0.5 to 1.0 cm^2. Table 1 summarizes the results. This table compares the average performance parameters for 12 single crystal control cells with 12 polycrystalline cells.

TABLE 1. Polycrystalline vs. single crystal GaSb cell performance summary.

	FF	Voc	Isc	Vmax	Pmax
Single	0.756	0.463	6.2	0.387	2.166
Poly	0.726	0.438	6.2	0.363	1.985

The results are that high performance polycrystalline GaSb TPV cells can in fact be made. Under the conditions stated, the polycrystalline cell average performance is down from the single crystal case by 8.7%. However, the potential cost savings are expected to dramatically outweigh this small performance drop.

Figure 3 shows a top view of our polycrystalline GaSb TPV cell and a cross section through this cell. The cell is fabricated on several GaSb crystals of varying orientations joined together at grain boundaries. A cell contains a base region of Tellurium doped N-type GaSb with a thin Zinc doped P-type surface region on its top side, the side facing the infrared source. A metal grid connected to a bus region is located in contact with the P-type region on the top of the cell and a continuous metal layer is in contact with the N-type region at the back of the cell. A multi-layer coating covers the front side of the cell. This multi-layer coating serves as an infrared filter. Its purpose is to transmit convertible infrared energy to the cell and reflect as much non-convertible infrared energy back to the IR source as possible. It is expected that GaSb cells from polycrystalline material can be made in high volume production for $1 per Watt.

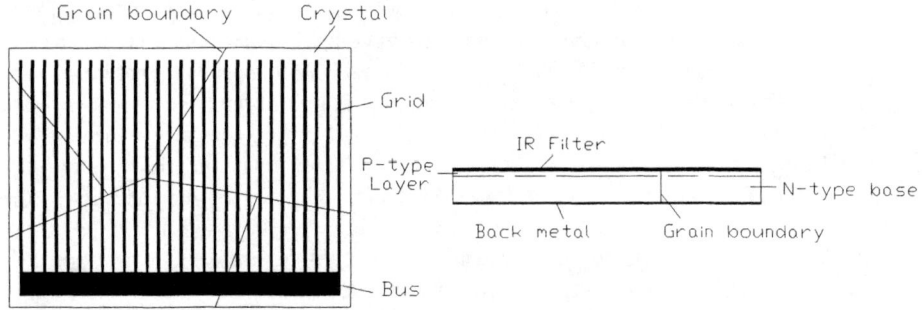

FIGURE 3. Top view and cross section through polycrystalline GaSb TPV cell.

SHINGLED CIRCUITS FOR TPV SYSTEMS

In the current TPV circuit design in production at JX Crystals, GaSb TPV cells are soldered down onto metal pads on a copper substrate where a thin insulating layer isolates the metal pads from the copper substrate. Then, flexible interconnects are bonded from pads on the top of the cell over to pads at the sides of the circuit. Finally, mirrors are attached over the lead bond areas. Since the leads have to carry large currents in TPV circuits, the lead area is fairly large and, consequently, the mirrors can cover typically 40% of the circuit area.

This configuration has two disadvantages. First, while the lead-bonding step can be automated, it is still quite time consuming. It would be desirable to eliminate this step. Second, in a stove panel, the mirror area leads to a loss in system power since the panel area is approximately equal to the emitter area. It is therefore desirable to reduce the mirror area and to increase the fractional active cell area.

There is an alternative circuit concept that was proposed many years ago for solar photovoltaic applications. The idea is to shingle the cells with the top pad attached underneath to the back of the next cell. Unfortunately, this idea has been tried without success in solar panels. The problem is that during thermal cycling, the substrate material expands at a different rate than the cell so that the rigid bond joint eventually is pulled apart, forming an open circuit. The flexible leads in the present circuit design avoid this failure mechanism.

The solar photovoltaic shingle concept did not work because the silicon solar cells are quite large and silicon has a very low thermal expansion coefficient (4.2×10^{-6} / °C) compared to substrate materials. Furthermore, the temperature excursions for space solar panels are very large. However, GaSb TPV cells are smaller and have a higher thermal expansion coefficient of 7×10^{-6} / °C. Also, there have been recent developments in the field of GaAs microwave device packaging. The thermal expansion coefficient for GaAs is 6.5×10^{-6} / °C.

Packaging materials now available for GaAs devices are AlSiC composites and metal laminates. AlSiC composites consist of SiC particles in an aluminum matrix. This material can be cast in various shapes and has a thermal expansion coefficient of 7.5×10^{-6} / °C.

Figure 4 shows a TPV shingle circuit designed for our Midnight Sun® Stove panel. This circuit utilizes a terraced AlSiC substrate. An insulating film is deposited over the terraces and then copper pads are deposited on the terrace top faces. GaSb TPV cells are then bonded to the copper pads and connected in series in a shingle pattern. This shingle circuit concept now has the advantage of a low-cost simple circuit assembly process. Front to back interconnect is now made in a single solder-flow process step.

In order to demonstrate this shingle circuit concept, JX Crystals has now successfully fabricated the TPV shingle circuit shown in figure 5. The particular circuit fabricated has 22 cells in series and a total active area of 30 cm². This circuit has been tested at various IR illumination levels such that the electric power it produced ranged from 10 to 44 Watts. Table 2 summarizes these test results.

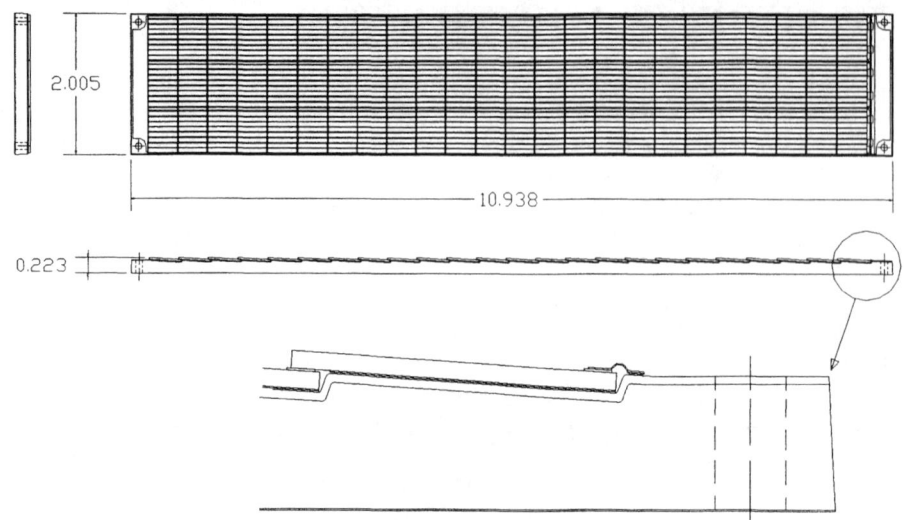

FIGURE 4. One end of a shingled circuit of GaSb cells on an AlSiC substrate.

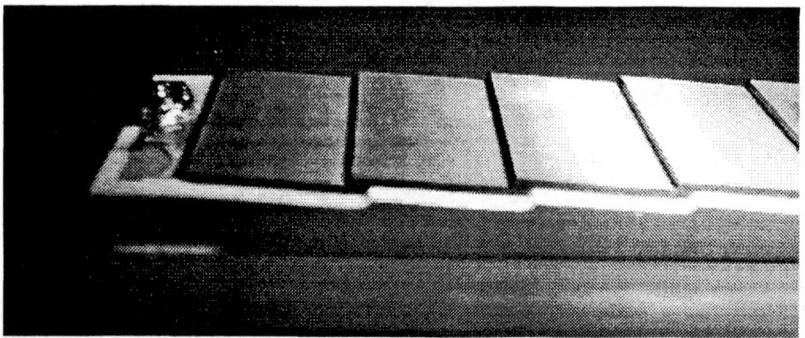

FIGURE 5. One end of a shingled circuit of GaSb cells on an AlSiC substrate.

TABLE 2. Shingled circuit test results, stepping up in IR illumination.

Isc (Amps)	Pmax (Watts)	Fill Factor	Voc (Volts)	Vmax (Volts)	Imax (Amps)
1.50	10.1	0.69	9.7	7.9	1.27
1.83	12.4	0.69	9.8	7.9	1.58
2.28	15.8	0.69	10.0	8.1	1.96
4.60	32.7	0.68	10.4	8.1	4.01
6.25	44.4	0.67	10.6	8.0	5.55

A key requirement for a successful shingle circuit is that it survive multiple thermal cycles. This should be expected if a good coefficient of thermal expansion match between cell and substrate is obtained. For an AlSiC shingle circuit, the match is expected to be within 1 part in 10 million per degree C. We have performed thermal cycle testing and, as shown in table 3, we have not seen any change in circuit performance.

TABLE 3. Thermal cycle test results.

Test cycle	FF	Voc	Isc	Imax	Vmax	Pmax
0	0.694	10.095	2.563	2.18	8.234	17.952
1	0.698	10.065	2.67	2.307	8.126	18.744
3	0.696	10.034	2.607	2.214	8.222	18.201
10	0.698	10.001	2.546	2.204	8.066	17.779
30	0.695	10.038	2.656	2.279	8.126	18.521

CONCLUSIONS

The Midnight Sun® Stove described in this paper provides both heat and electricity for the family room in a home not connected to the electric power grid. This represents a first TPV commercial product. GaSb cells that respond to infrared radiation from a silicon carbide honeycomb emitter make this system possible because these cells allow the emitter to operate at a moderate 1300° C temperature. Manufacturing costs are important for this application, and breakthroughs in the fabricating of both GaSb cells and circuits are described here. These breakthroughs should dramatically reduce future costs.

In the cell area, polycrystalline GaSb cells have been fabricated for the first time. JX Crystals demonstrated that the power produced by these cells is nearly as high as that produced by single crystal cells. This finding should allow the fabrication of larger diameter cast polycrystalline ingots with cheaper equipment in shorter times compared to the present single crystal case.

In the circuit fabrication area, JX Crystals designed and demonstrated TPV high performance shingle circuits for the first time. These shingle circuits have a high active area of 80% compared to previous circuits. Furthermore, they are simple to fabricate, requiring only a single solder reflow step. These circuits are not degraded after multiple thermal cycles. The key to these shingled circuits is the use of a substrate material with a coefficient of thermal expansion (CTE) that is closely matched to the CTE of the GaSb cells.

Use of a Thermophotovoltaic Generator in a Hybrid Electric Vehicle

Orion Morrison, Dr. Michael Seal, Edward West, and William Connelly

Vehicle Research Institute
Western Washington University
Bellingham, Washington 98225

Abstract: Viking 29 is the World's first thermophotovoltaic (TPV) powered automobile. The prototype was funded by the Department of Energy and designed and built by students and faculty at the Vehicle Research Institute (VRI) at Western Washington University. Viking 29 is a series hybrid electric vehicle that utilizes TPV generators to charge its battery pack. Acceleration, speed, and handling compare to modern high performance sports cars, while emissions are cleaner than current internal combustion engine vehicles.

INTRODUCTION

The Vehicle Research Institute (VRI) at Western Washington University was founded in 1974 as a program of research in vehicle design, prototype construction and engineering. Funded by grants from the Department of Energy (DoE) and Department of Defense (DoD), the VRI and its industrial partner, JX Crystals Inc., have been involved with TPV research since 1993. This research has focused on low band gap gallium antimonide (GaSb) cells and broad band emitters. The VRI received a DoE grant to design and build a hybrid electric car that utilizes a TPV generator. This paper focuses on the design, fabrication, systems, and performance of Viking 29, the world's first TPV powered automobile (Figure 1).

FIGURE 1. Viking 29, the world's first TPV powered automobile.

CP460, *Thermophotovoltaic Generation of Electricity: Fourth NREL Conference*
edited by T. J. Coutts, J. P. Benner, and C. S. Allman
© 1999 The American Institute of Physics 1-56396-828-2/99/$15.00

Hybrid Vehicles

The performance limits of pure electric vehicles have curbed the public's acceptance of the electric vehicle as a replacement for internal combustion powered automobiles. Range is most commonly pointed out as the electric vehicle's greatest limitation. While the range of today's electric vehicles would be sufficient for a majority of commuters, families are not willing to give up the convenience of having an automobile that can make extended trips. Electric Vehicle supporters believed major advances in battery technology would extend the range of electric vehicles to be comparable with those of internal combustion powered automobiles. Despite substantial improvements, most notably in nickel cadmium and nickel/metal hydride research, battery technology has fallen short of expectations.

In the interim, industry has turned to hybrid electric vehicles (HEVs). Toyota is in the forefront with the Prius, the world's first production hybrid electric vehicle. The Prius utilizes an internal combustion engine to supplement the electric motor and charge the nickel/metal hydride battery pack. While the public has been very receptive of the Prius, production rates have doubled to meet demand (1), it has not eliminated reliance on the internal combustion engine. The hydrogen fuel cell and Thermophotovoltaic (TPV) generator are two less polluting options being researched for use in HEVs.

VEHICLE DESIGN

The initial proposal was to install the TPV generator into one of the existing Viking vehicles built by Western Washington University (2). Upon further study, it was determined that current Viking cars would not be able to accommodate the TPV generator and support systems necessary to power an automobile. It was decided that a new vehicle would be designed and constructed for the purpose of installing a TPV generator. The generator would charge a battery pack that in turn powers the vehicle's electric motor.

Design requirements were driven by the need for performance comparable to current internal combustion automobiles. The body and chassis of the vehicle was designed to minimize vehicle weight while maintaining sufficient rigidity. A body shape with low aerodynamic drag was important to maximize the range of the automobile. Wind tunnel tests with a 1/10 scale model were used to determine the optimum functional body shape.

Acceleration, speed, handling, and ergonomics were all major considerations. Acceleration needed to be sufficient for safe city driving. A Minimum top speed of 70 miles per hour was necessary to allow freeway driving. Handling requirements included full ackerman steering geometry, zero bump steer, and appropriate camber gain with roll. The ergonomics of Viking 29 would need to be similar to today's automobiles to ease the transition for users.

The TPV generator was to be made modular to allow the installation of other generator configurations in the future. An opening under the automobile would allow access to remove and install modular generators. All connections between the automobile and generator needed to be easily disconnected for quick generator removal and installation. Numerous components were required to support the multiple systems in the vehicle. The main components required for the TPV generator are a fuel source, power supply (12 and 50 volts), radiators, water pump, and peak power trackers.

Body/Chassis

Viking 29 is a unibody design, the body and chassis are a single structure. The monocoque design distributes loads throughout the chassis and body of the automobile. The force distribution allows for a lighter chassis and body design.

A full scale plug in the shape of the car was constructed. Female molds for the different components of the body were created from the plug. Additional molds had to be constructed for the internal components including the seats, battery boxes, and wheel wells. Each component was constructed in its respective mold, using a vacuum to remove excess resin and voids and increase laminar bonding. An E-glass and vinyl-ester resin composite was used to give the body and chassis the optimum combination of impact absorbing and load carrying properties. An additional advantage of using E-glass is the low conductivity of the material will help avoid electrical problems that can potentially occur in carbon fiber chassises. A carbon fiber composite was used for the doors and rear hatch for increased strength to weight ratio. The number of layers and fiber orientation varied depending on the component's function.

Molded components were then assembled using overlapping joints (Figure 2). Honeycomb matrix was used for the firewall bulkhead to provide a rigid structure to attach components and to support the dash and windshield. Square steel tube frames were bonded in the front and rear to support the steering and suspension systems. A finish, matte, and veil is used on the exterior to provide a smooth contour. After applying multiple layers of primer, the vehicle was painted British racing green with acrylic urethane paint.

FIGURE 2. Fiberglass unibody chassis.

Semi-gullwing doors provide easier entry and exit over the battery boxes. A detented strut assembly holds the doors open. The front window is from a Dodge Neon while the other windows and headlight covers are thermoformed 3/16" from polycarbonate.

Windows are bonded in place with a urethane adhesive. The dash is a clear coated carbon fiber layup. Custom leather seats surround the carbon fiber seat frames. The driver's seat is adjustable forward and aft. Electric side mirrors provide for easy adjustment from the driver's seat.

Vehicle Systems

Short and long arm wishbone suspension links are used in the front and rear of the vehicle. A-arms are mounted to the steel frames that are bonded into the chassis. Each corner uses direct acting coil-over shocks. The large 254 mm front and rear disc brakes provide ample braking without requiring power assist. An adjustable brake bias bar allows easy tuning of the front/rear braking ratio. Collapsible tilt steering turns the wheels through a 16:1 ratio rack and pinion. Low front end loading eliminates the need for power steering. EV1 tires mounted on custom machined rims provide a low drag power transfer to the road.

Power Plant

The main power plant is a Unique Mobility 100 hp 3 phase DC brushless motor. See Figure 3 for the location of major vehicle components. The motor's operating efficiency is greater than 90%. A Unique Mobility processor and motor controller serves as the interface between the user, motor, and batteries. The motor and controller are water cooled. Plumbing the water through an off-the-shelf auxiliary heater box provides adequate cooling. Temperature and water flow gauges mounted in the instrument panel alert the driver of cooling problems.

Two hundred seventy-three Saft™ nickel cadmium aircraft batteries occupy the two battery boxes on either side of the passenger compartment combining for a total nominal voltage of 327.6 volts. Each battery is rated at 25 Amp hours per charge giving the battery pack a total power rating of 8.19 Kwhr (Equations 1 and 2). As weight is a major consideration in electric vehicles, specific energy is used to rate

$$P_{(per\ cell)} = I * V = 25 * 1.2 = 30\ Whr. \tag{1}$$

$$P_{(pack)} = P_{(per\ cell)} * \#\ of\ batteries = 30 * 273 = 8.19\ KWhr. \tag{2}$$

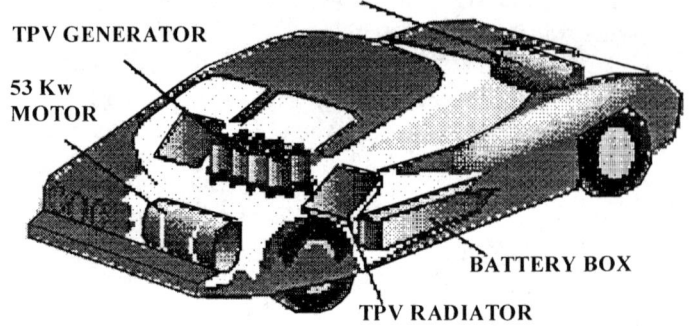

FIGURE 3. Viking 29 major components.

battery packs by the amount of KWhr per Kg. Each battery weighs .915 Kg, giving the battery pack a specific energy of 32.8 Whr/Kg (Equation 3).

$$\text{Specific Energy} = \text{Whr} / \text{Kg} = 30/.915 = 32.8 \text{ Whr/Kg} \qquad (3)$$

A machined adapter allows the motor to be mounted end on to a manual four speed Volkswagon transaxle. The dry plate clutch assembly runs on a ball bearing assembly hub that removes all thrust loading from the electric motor. The hydraulic clutch provides smooth shifting even at the motor's limit of 8000 rpm. Marine shifting cables were used to span the distance between the gearshift and transmission. A bell crank attached to the shifting cables allowed clearance of the generator compartment and reversed direction so a standard shifting pattern could be maintained. Driveshafts with inner and outer continuous velocity joints take the drive to the rear wheels.

The processor adjusts the motor's power usage to maintain a constant torque. A linear potentiometer mounted on the accelerator pedal provides the torque signal to the processor. The motor is also used as a generator that recaptures power from the car's momentum to charge the batteries. A potentiometer mounted on the brake pedal controls the regenerative braking signal.

Array Design

Infrared sensitive gallium antimonide (GaSb) cells manufactured by JX Crystals are used in the TPV arrays. Twenty circuit boards containing 19 cells each surround a black body emitter constructed of silicon carbide (see Figure 4). The emitter is heated by natural gas to a temperature of 1700C. A stainless steel recuperator preheats incoming air. A diffuser mixes the gas and air as they enter the combustion zone (3).

Photovoltaic circuit boards are bonded to aluminum receivers that conduct excess heat into the cooling water through fins. Two quartz crystal shields surrounding the emitter help to reduce convective losses and reduce exposure of cells to excess temperatures. Polished aluminum mirrors mounted in the gaps between cells reflect back photons that would other wise be lost as waste heat. An array output of 1 kW has been achieved on an electric powered test station. Past experiments indicate similar results are obtainable with a combustion powered system.

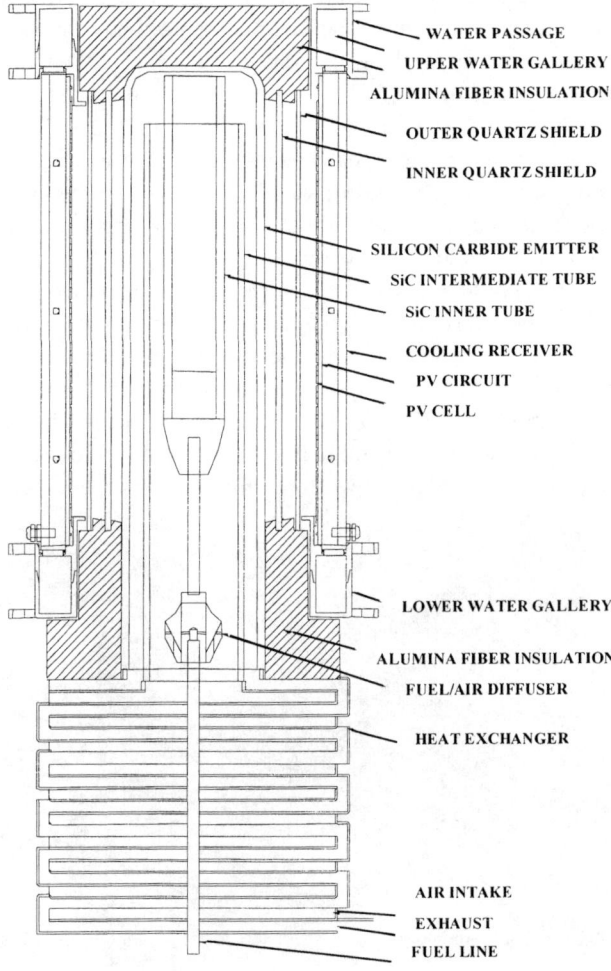

FIGURE 4. Diagram of TPV array components.

Generator

The goal of installing the TPV generator was to increase the range of Viking 29 to 200 miles to be comparable with gasoline powered automobiles (4). The battery pack has 8.19 kWhrs of energy and gives Viking 29 a range of 50 miles. The generator must provide 24.57 kWhrs of energy to make up the difference of 150 miles. With an output of eight kilowatts the generator will need to operate for 3.07 hours to increase Viking 29's range to 200 miles.

The eight arrays are arranged in groups of four (see Figure 5). Each array has a stainless steel recuperator that sits in the generators main frame. Alumina insulation is wrapped around the recuperators to decrease heat loss. Each of the variable speed fans supplies combustion air for two arrays. Exhaust, containing mostly water and carbon dioxide, exits the recuperator under the generator.

A compressed natural gas tank in the front of the vehicle fuels the TPV generators. The carbon fiber wound tank has a maximum pressure of 3500 psi. The tank is filled through a quick connect fitting under a hatch on the hood of the vehicle. A two stage regulator brings the pressure down to 28 psi. Flow control valves on each gas line regulate the amount of fuel entering the individual arrays. A manifold mounted on the generator allows for easy disconnect of the fuel lines.

The efficiency of the photovoltaic cells decreases as their temperature increases. The water cooling system minimizes the operating temperature of the cells while maximizing their efficiency and decreasing their exposure to thermal shock. Each group of four arrays has its own cooling system consisting of a pump, radiator, fan, and fill cap. Large radiators (28" X 18") are used to keep the cooling water temperature as low as possible. The radiators are mounted below slats in the rear hatch. Fans circulate air from under the car, through the radiator, and out the slats.

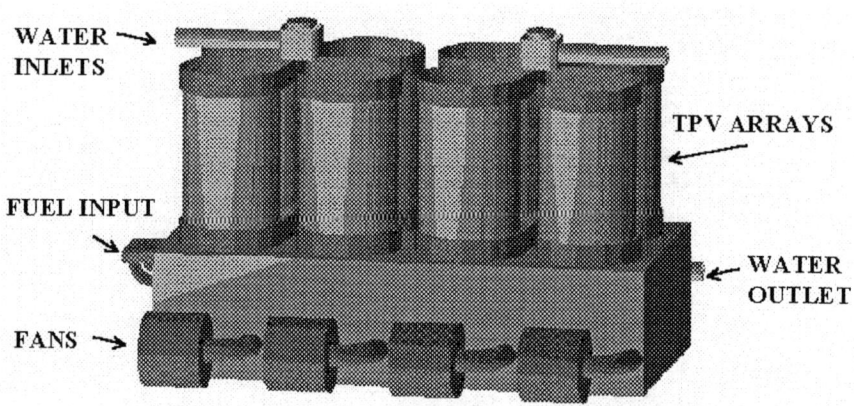

FIGURE 5. The TPV generator is capable of supplying 8kW to Viking 29's battery pack.

The arrays have an average output of 35 volts. Peak Power Trackers (PPT's) adjust the load to maximize array power and output a higher voltage suitable to charge Viking 29's battery pack.

Performance

Viking 29 performs very well as an electric vehicle. With a top speed of more than 100 miles per hour, it can easily maintain freeway speeds. Viking 29's 0-60 time of ten seconds is more than sufficient acceleration for safe city driving. The battery pack alone provides a range of 50 miles, well above the average commute. A coefficient of drag of .335 was calculated from coast down test data. Viking 29's aerodynamic body and weight of 2330 pounds contributes to the low coefficient of drag.

While the generators have been operated in the vehicle, their performance has not yet been tested. Research is currently under way to automate the operation of the generator. Automation requires the air/fuel mixture be ramped up slowly to avoid thermal shock to the components. In addition, the circuit will monitor buss voltage and turn on the appropriate number of arrays when necessary.

CONCLUSION

The goal of demonstrating the use of a TPV generator in an automobile has been accomplished. However, substantial improvements need to be made for the application to be a feasible alternative to other methods of transportation. The two main focuses of improvement are the automation of the generator and increased efficiency of both the arrays and their supporting components, both of which are being researched at Western. The robustness of the ceramic components inside the arrays is another concern being addressed.

With the success of Viking 29 and other hybrid electric vehicles, it appears a feasible alternative to relying on fossil fuels is on the horizon. Continued research will improve current technologies and explore future possibilities.

REFERENCES

1. Associated Press, "Toyota previews gas-electric car." *Bellingham Herald*, November 13, 1997.

2. Seal, M. R., Final Report, DOE Award DE-FG06-94ER12149, Vehicle Research Institute, Bellingham, WA, 1998.

3. Connelly, W. R., West, E. M., "Cylindrical TPV Array Characterization", Vehicle Research Institute, Bellingham, WA.

4. Seal, M. R., and Fraas, L., "A Thermophotovoltaic Generator For Use As An Auxiliary Power Unit In A Hybrid Electric Vehicle", Vehicle Research Institute, Bellingham, WA.

Cylindrical TPV Array Characterization

William R. Connelly, Edward M. West

Vehicle Research Institute, Western Washington University
Bellingham, Washington 98225

Abstract. During the past three years, the research staff at Western Washington University's Vehicle Research Institute has developed a system for characterizing and comparing Thermophotovoltaic (TPV) arrays. The function of this system is to enable the measurement of array power output and efficiency. This makes possible the evaluation and comparison of array geometry and component changes, and assists in the evaluation of different materials used in TPV array construction. Development of two of these test stations is thoroughly described.

INTRODUCTION

As thermophotovoltaic systems progress into the next stages of development, i.e. the construction of practical functional prototypes and demonstration systems, a standardized method of TPV array characterization has become essential for the purposes of performance evaluation and discussion. During the past three years, three distinctly different thermophotovoltaic arrays have been constructed at the Vehicle Research Institute (VRI). Two new arrays are planned for the coming year. A method for comparing TPV array designs has been developed at the VRI and implemented at JX Crystals, Issaquah, WA. This paper will discuss the 6kW and 15kW Cylindrical Array Characterization Unit (CACU) currently in use at the VRI. It will detail the components used in the construction of the unit, the data which is collected by the system, and the methodology and equations used to evaluate array performance.

History

Burner system and TPV array development occurring in parallel during the first year of research at the VRI made the need for a standard method of cylindrical array characterization obvious. The early burners are sensitive to small geometric changes during assembly, which affect both startup and warm up characteristics. Diffuser geometry giving poor fuel mixing affects system performance at startup. The emitter temperature profile is affected by internal geometric changes such as combustion chamber length and position in the annular flow paths. Changes being made to the burner system were affecting emitter temperature distributions and performance.

CP460, *Thermophotovoltaic Generation of Electricity: Fourth NREL Conference*
edited by T. J. Coutts, J. P. Benner, and C. S. Allman
© 1999 The American Institute of Physics 1-56396-828-2/99/$15.00

At the same time, design and development of a 380-cell TPV array prototype for the Department of Energy was underway. Speculation about the affect of mirror geometry on photon recycling needed to be explored. What effect the quartz insulation shield has on array performance was also of interest, as was the reflectivity of the refractory insulation used in the cavity. After just a few unsuccessful attempts to characterize the effect of changing mirror geometry or cavity design with an assembled burner system as the power source, work began on a simple reliable way to provide a stable standard input power to the TPV array cavity.

The 6KW Array Characterization Unit

The array characterization units developed at the VRI use Silicon Carbide (SiC) resistive heating elements. The heating element in the 6kW CACU, manufactured by Cesiwid, Inc., is 2.125 inches in diameter by 18 inches long, with a 12-inch long hot zone and a nominal resistance of 2.75 Ω. This element is chosen because it matches the diameter and emissive properties of the emitters used in the fuel burning units under development at this time.

An Enirex Pro-Troller® silicon controlled rectifier (SCR) is used for precise control of electric power to the element. A step-down isolation transformer is used to drop the input voltage to the element from 208V to 120V. The power level can be set manually at the SCR using a 10kΩ potentiometer or by a proportional integrating differentiating (PID) process controller that receives input from a 'C-Type' thermocouple inside the SiC heating element.

A Flex-Core® current transformer (100:5 ratio) on the input line to the element measures current and is connected to a Flex-Core® watt transducer that also measures the voltage drop across the heating element. The product of these two values is output to a Texmate® digital panel meter. The watt transducer was added to the system to indicate true power by accounting for AC phase shift due to inductive components of the load. These are the essential components used in both the 6kW and 15kW cylindrical array characterization units.

Testing performed during the second year of array development at the VRI was done entirely with the 6kW CACU. The system is limited in terms of input power because of constraints within the building. It became apparent during this period that the available input power was limiting the ability of the system to reach operating temperatures near those that are achieved with gas fired burner systems.

The diameter of the emitter used in the gas fired burner system was enlarged during this period of research to improve the view factor. This improves radiation coupling between the emitter and cell which increases power output for a given emitter temperature. To accurately evaluate the expected performance of the new fuel burning systems, a second CACU was designed and implemented. This unit is referred to as the 15kW Cylindrical Array Characterization Unit.

THE 15KW CYLINDRICAL ARRAY CHARACTERIZATION UNIT

The 15kW cylindrical array characterization unit receives input power from a separate building circuit installed for this application. This new circuit provides 480V input on a 60A circuit. Like the 6kW CACU, this system also uses an SCR to regulate power to the heating element. An Acme buck transformer drops the voltage to the element from 480V to 436V. A 480V to 115V transformer is used to power the onboard PID control, watt transducer, panel meter, and cooling fan. The heating element is a three-inch diameter I-Squared-R, Inc. Starbar™ with an 11Ω nominal resistance. The connection of these components is shown in figure 1.

The Starbar™ is surrounded by Zircar SALI™ alumina fiber insulation at the connection end of the element, which is air-cooled. The arrangement of the heating element, SALI™ insulation, and TPV array is shown in figure 2. A current-transformer on the input line to the heating element measures element input current. The current-transformer is connected to a watt transducer that measures the voltage drop across the heating element, and outputs the product of these two values to a digital panel meter in the same manner as the 6KW unit.

The 15KW system also sends the signal from the transducer to a National Instruments™ data-acquisition system (NIDAQ) to enable real time data analysis and data logging. The NIDAQ uses Lab-View™ data analysis software running on a Pentium-II® 300MHz PC for data collection and analysis purposes.

The NIDAQ can also be used to control the input power to the heating element, via temperature information from three 'R-Type' thermocouples placed inside the heating element. These thermocouples present a temperature profile of the heating element and the middle position thermocouple provides an input to the onboard PID and the data acquisition system for external process control, for example, automatic ramp-up to a preset input power level or temperature setting.

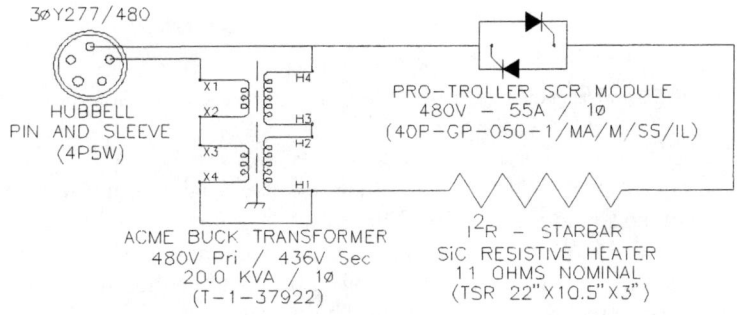

FIGURE 1. Schematic Diagram of Test Station Power Circuit

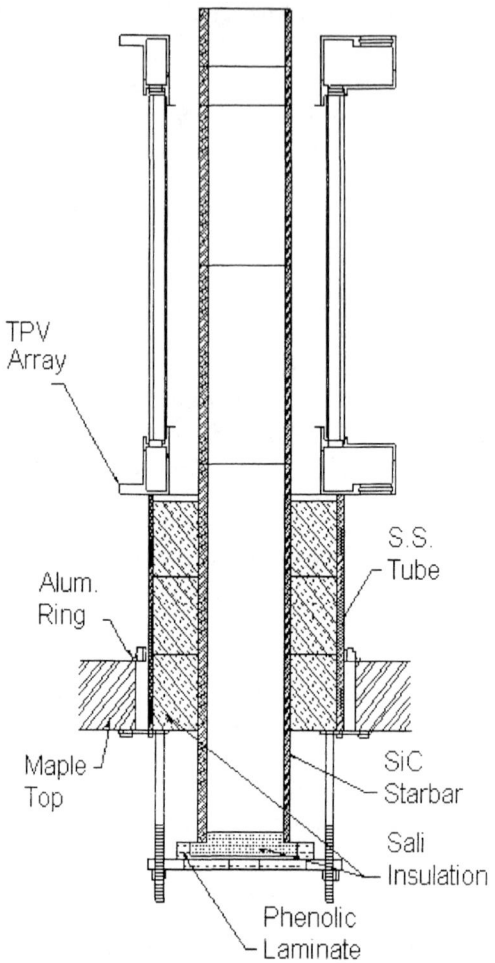

FIGURE 2. Arrangement of Starbar™ Heating Element and Cylindrical Array

The TPV array is connected to an external resistive load through a peak power tracker developed and manufactured by Xantrex of Burnaby, BC Canada. The new CACU is capable of 15kW of input power to the element and achieves temperatures equal to those seen in gas fired burner systems. Data that is collected by the NIDAQ system is processed in real time and stored for later retrieval.

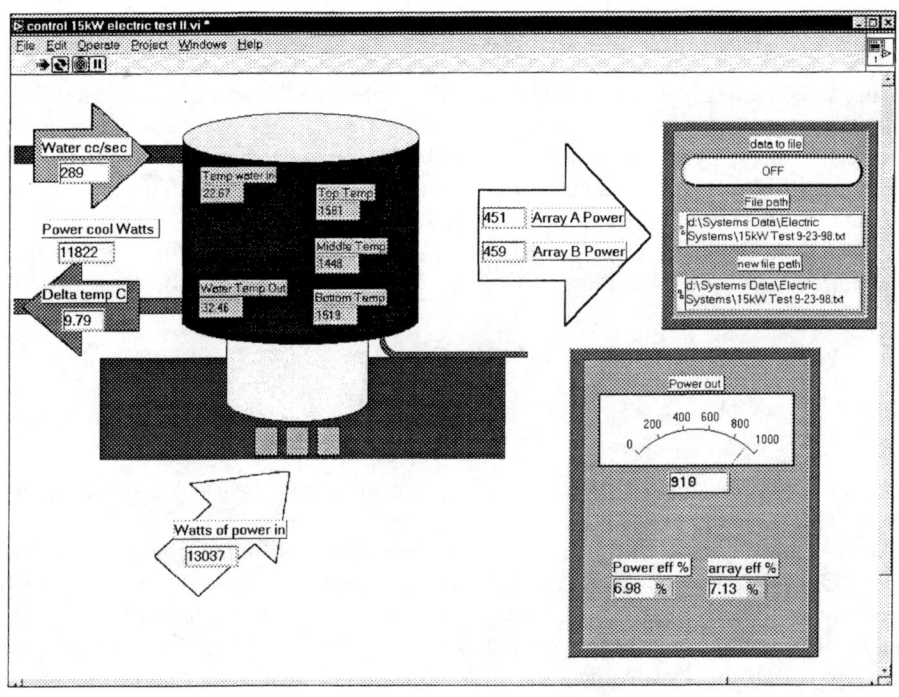

FIGURE 3. Graphical user interface screen with LabView™ virtual instruments.

Data Collection

The National Instruments™ data-acquisition system collects data from several input sources in the CACU system. First, the cooling water circulating in the TPV array is monitored. The flow in cc/second is processed from a Data Industrial™ series 4000 flow meter and the water inlet and outlet temperatures are measured using Omega™ thermistors. Second, temperature inputs from three separate 'R-Type' thermocouples inside the heating element are recorded. Third, the signal from the watt transducer, indicating heating element input power, is recorded. Finally, the power going to the external load through the Xantrex™ peak power tracker is monitored and recorded.

Figure 3 shows the Lab-View software and system information as presented to the user through a graphical interface. The signals sent by the CACU are processed in real time by the Lab-View software using the information collected by the NIDAQ system, and are displayed to the user as a set of virtual instruments. The results are also recorded in a file on the PC for later retrieval. This system saves valuable time compared to hand calculation. Many hours had formerly been spent using the following equations and methods to perform data analysis after each array characterization test.

Methodology and Equations

Once the data has been collected, it is used to calculate the important characteristics of the TPV array. The characteristics compared for each array design or configuration change are: the power efficiency of the cylindrical array characterization unit, the thermophotovoltaic array efficiency, and the array power output. These values are characterized as a function of emitter temperature.

The power efficiency (η_{power}) of the CACU electrical system is defined as the output power produced by the cylindrical TPV array being evaluated (P_{out}), divided by the power input to the CACU (P_{in}).

$$\eta_{power} = \frac{P_{out}}{P_{in}} \qquad (1)$$

This gives an indication of the electrical system efficiency due to the overall power losses in the system itself.

Array efficiency is determined by dividing the electrical output power by the total energy input to the array. Unfortunately, the total energy input to the array cannot be directly measured, but assuming a steady state energy balance, the term ($P_{cool} + P_{out}$) is equal to the energy input to the array through convection, conduction, and radiation. The term P_{cool} is the sensible heat rise of the cooling fluid calculated using equation 2. The array efficiency (η_{array}) is then calculated using equation 3.

$$P_{cool} = \dot{V} \cdot \rho \cdot Cp \cdot \Delta T \qquad (2)$$

$$\eta_{array} = \frac{P_{out}}{(P_{cool} + P_{out})} \qquad (3)$$

The array efficiency is defined as the ability of the array to convert useable photons into electric current. The quantity P_{cool} includes all the energy that is not removed from the system as electricity by the thermophotovoltaic array, but does not include energy lost (P_{loss}) due to the design of the cylindrical array characterization unit. This quantity is calculated as the total input energy P_{in}, minus the sum of P_{cool} and P_{out}.

$$P_{loss} = P_{in} - (P_{cool} + P_{out}) \qquad (4)$$

As every system will have some loss associated with it, this number is only of value when determining what steps may taken to improve the design of the CACU system itself.

Array performance evaluation

The two important factors used to evaluate the performance of cylindrical thermophotovoltaic arrays are the array efficiency and the array power output. A typical performance evaluation from a cylindrical array characterization test is shown in figure 4. Both the array efficiency and power output are graphed relative to the temperature of the emitter.

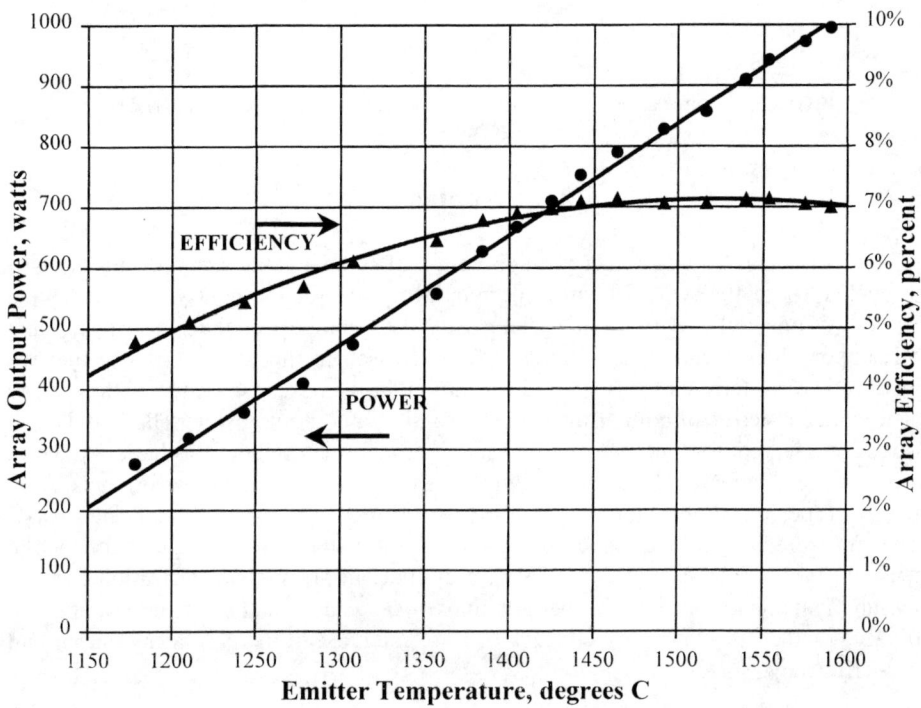

FIGURE 4. Typical performance evaluation curves.

A series of cavity geometry changes, specifically the internal mirror design (see figure 5), were evaluated using the CACU. Design one (V-shaped mirror) is compared as the baseline for evaluation of the other two designs. Relative to design one, design two (U-shaped mirror) produced slightly lower array efficiency. The test performed with design three (T-shaped mirror) resulted in an array efficiency that was notably higher.

A series of tests performed using three different materials to construct the T-mirror, and two different coatings on the same substrate, showed differences in array efficiencies for the same mirror geometry. The cylindrical array characterization unit facilitated the selection of the best material, coating, and geometry for the construction of the array cavity mirror.

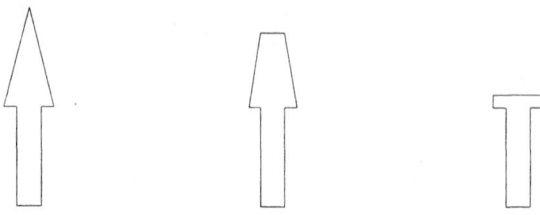

FIGURE 5. Three array mirror geometry designs (1 through 3 from left to right).

Future Applications

The Cylindrical Array Characterization Unit will be a valuable analysis tool to assist the evaluation of the two TPV array designs proposed for the next year of research. The 15KW unit will be used when comparing the geometry of the two designs, and will support the evaluation of different materials used in the construction of the two arrays. The system will assist in determining how these differences affect array performance in terms of both array efficiency and array output power. The CACU will also be employed to tune the mathematical model that predicts radiation transfer within the TPV cavity. The CACU will also be used to evaluate the performance from different types of filter materials being placed on the TPV cells. A capacitance I-V curve measurement system is being added to allow the full I-V data curve to be measured by the NIDAQ system. Using a cylindrical array characterization unit for measurement and comparison of thermophotovoltaic array designs is a time saving and effective method of evaluating changes that are intended to increase array output and array efficiency.

AKNOWLEDGMENTS

The authors would like to thank Dr. Michael Seal, Eileen Seal, the students, and staff of the Vehicle Research Institute. Their countless hours of dedication and contribution to the project have made our work both enjoyable and productive.

High Power Density AEM Combustion for TPV Applications

Aleksandr S. Kushch and Steven M. Skinner

Quantum Group, Inc.
11211 Sorrento Valley Road, San Diego, CA 92121
Telephone: 619-457-3048 Fax: 619-457-3229

ABSTRACT

Various emitter designs and compositions are investigated to improve the performance and reduce the cost of thermophotovoltaic (TPV) devices. In order to maximize the overall system efficiency of combustion-powered TPV devices, it is desirable to design an emitter system that will provide a high and uniform volumetric heat release producing a high intensity and uniform in-band photons flux, thus enabling reduced size and weight of TPV generators.

Quantum Group, Inc. (QGI) has developed a high power density, radiant burner/emitter technology which we have called Advanced Emissive Matrix (AEM). The AEM system is applicable to broad-band and narrow-band TPV approaches. Several AEM combustors - both broad-band and narrow-band - have been built, tested and have consistently demonstrated extremely high power density (up to 30 W/cm^3). Radiant emissions greater than 27 W/cm^2 were measured from an AEM combustor fired with natural gas at 15 % excess air. This level of radiant heat flux enables the utilization of concentrator-type PV cells, which is a method of reducing cost, size and weight of TPV devices.

Additionally, it was found that the AEM structure could be engineered to provide a desired radiant emission profile. The influence of the primary combustion parameters on emitter performance will also be presented. Therefore, integration of the narrow-band emitter materials into an AEM combustor results in the delivery of high intensity, selective and uniform photons to the PV cell face that improves TPV device performance.

INTRODUCTION

TPV system efficiency is the major factor that determines the cost of TPV devices, and restricts the large-scale commercialization of TPV products. TPV system-energy budget analysis shows that the major losses in TPV devices occur due to limited fuel-to-radiant-energy conversion, poor photon collection efficiency, and low photon-to-electron photovoltaic (PV) conversion. Radiant (surface combustion) burners typically convert 15-20% of fuel chemical energy to radiant emission. Mantle-type structures have better efficiency (about 25%), but even in this case 75% of the fuel energy escapes from the combustor with the exhaust. It is possible to return the portion of exhaust energy into the emitter heating cycle by using recuperators, but this approach increases the cost of TPV devices and the complexity of unit design. It is desirable to utilize a combustion/emitter system that will convert the fuel to radiant energy much more effectively than in existing devices.

An alternative way to improve TPV system efficiency is to supply the PV cells with the "clean" photon emission which can be achieved by using optical filters or narrow-band emitters made of rare-earth oxides or a combination of both. It was reported (1, 2) that photon-to-electron PV conversion reaches about 40% when ytterbia emitter was tested with concentrator silicon (Si) PV cells. The graybody emitters coupled with the same PV cells provide for 26% conversion in the best case (3).

Concentrator-type PV cells can be considered as beneficial for TPV applications because they are capable of accepting high radiant flux (about 250 suns equivalent) that reduces the number of PV cells and eventually the cost of TPV devices. Typically, TPV devices feature the emitter exitance in the range from 3.5 to 9.8 W/cm^2 (2,4) which is too low an intensity for an effective concentrator PV cell operation. Realistically, PV cell illumination is even lower due to photon losses in the volume from the emitter to the PV arrays. In order to utilize the advantage of concentrator PV cells, it is necessary to develop the combustion/emitter system that is capable of providing an adequate photon flux on a PV cell face.

At the same time, it is important to maintain a uniform PV array illumination in order to achieve maximum electric energy production. Our experiments with a five-inch mantle showed that the maximum radiant flux (about 5 W/cm^2) measured along a latitude axis at a distance of 5 cm from the emitter occurred at the center of the emitter; the ends produced about 3 W/cm^2, so the TPV device which utilized a mantle emitter, required special PV cell circuiting.

Portability of the power generator could be another important factor in competition among TPV and other electric power generating technologies. Some application designers (for example military) would definitely prefer to use small, lightweight units if the price of such devices will be comparable with the other choices.

In summary, the extremely desirable features for TPV technology are:
- high-energy density
- intense radiant output from combustion/emitter systems
- narrow-band photon emission that fits to the PV absorption spectrum
- uniform PV illumination
- high photon-collection efficiency by PV arrays

ADVANCED EMISSIVE MATRIX (AEM) COMBUSTION TECHNOLOGY

In the past, several R&D projects that focused on development of self-powered appliances and TPV co-generation system have been successfully completed at QGI (5,6). The power-generating sections in these prototypes were comprised of a ceramic fiber burner (CFB) coupled with Si PV arrays. The surface layer of CFB was made of ytterbia so the emission spectrum of these burners was centered at 980 nm which perfectly fits to the Si PV absorption spectrum. CFBs are easy to manufacture. They provide for complete combustion, and require low electric power to run the combustion blower. These burners could be used in systems which do not require high energy density, such as appliances. In order to increase the energy density of the combustion system, more fuel/air mixture has to be introduced into the combustion chamber, and the combustion process has to be completed within this chamber. CFB and other surface-combustion burners are highly effective when they operate in a narrow range of combustion parameters (such as firing rate and stiochiometry). Within the optimum combustion parameters, the flame stabilizes at close proximity to the burner surface, and the heat from the flame effectively transfers to the burner surface. The CFB surface becomes radiant, and photons are collected and converted into electricity by the PV cells.

The major condition for the flame stabilization is that the flame propagation speed (FPS) must be equal to the speed of the combustible mixture (SCM), which is introduced through the porous layer of the burner. If the SCM is higher than the FPS, the flame front moves away from the surface, and the burner loses an active radiant mode; in other words, it could not effectively convert chemical fuel energy into radiant photon emission. At the opposite condition, when the SCM is smaller than the FPS, flame quenches into the porous layer structure or flashback occurs.

Recognizing that any attempt to reach high-energy density combustion based on utilization of surface-combustion burners cannot realistically be achieved, QGI has developed another combustion technology that we refer as an Advanced Emissive Matrix (AEM) combustion.

The major idea of the AEM combustion is to develop the conditions for a volumetric flame stabilization, instead of the surface stabilization which is typical for CFB. In order to achieve such a condition, the combustion chamber volume is filled with the three-dimensional structure which interfaces with the stream of incoming combustible mixture, resulting in the creation of multiple-vortex zones where flame can be stabilized. This structure becomes an emitter in a few seconds after flame has stabilized in some places, and then the flame propagates in the whole volume of the AEM. When the AEM reaches a radiant mode, the stable combustion takes place in the whole volume, because the AEM acts as a source of ignition and flame stabilization simultaneously. The incoming mixture enters into a high turbulence/temperature environment, rapidly preheats due to conduction and radiation heat transfer from the AEM, then ignites and burns in the AEM volume.

In order to achieve high radiant output from the AEM, this structure has to be optically thin; in other words, the majority photons generated by the AEM components should be able to escape from the emitter throughout the multiple openings. Photon-collecting devices (for TPV application PV cells) located outside of the AEM capture the emission and convert it to electricity.

Practically, the AEM structure can be fabricated in different shapes, including cylindrical, conical, spherical, and rectangular-box shaped; however, the AEM structural density and its members dimensions do affect the combustion performance and the radiant output.

COMBUSTION EXPERIMENTS

The first experiments were conducted at QGI with metal-based AEM structures. We made the emitter system of high temperature-resistive alloy (wire screen) with different mesh sizes. Two types of structures were investigated at this time: the set of concentric cylinders and "flat spiral." Both structures are shown in Figure 1. Both emitters have similar overall dimensions: about 13 cm in diameter and 25 cm in height. The combustion chamber was made from a quartz tube, in order to observe the AEM during the test, and to conduct optical and combustion measurements. A natural gas/air mixture was introduced into the combustion chamber throughout the porous distributive layer located at the bottom of the combustion chamber.

FIGURE 1. Metal-Based AEM Structures:
a) Set of Concentric Cylinders: b) "Flat Spiral"

Major objectives of these experiments were: (1) to define the potentials in volumetric heat release, combustion completeness, and radiant output at different firing rates and fuel/air mixtures; and (2) to investigate the capability of controlling uniformity of the radiant flux from the AEM emitter.

The combustion performance was measured by a combustion-efficiency analyzer, Testo 33, manufactured by Testoterm, Inc. This unit is capable of measuring the oxygen, carbon monoxide, carbon dioxide, and nitrogen oxides (separately NO and NO_2) content in the exhaust, as well as excess air and appliance efficiency. The fuel input was measured by a gas flowmeter that allowed us to calculate the thermal energy input for each experiment. The radiant emission was measured by a heat flux-meter, H-201, manufactured by Medtherm Corporation (Huntsville, AL) and the rate or the fuel-to-radiant-energy conversion was evaluated with the water jacket based on the water temperature rise at particulate water flow rate.

Later, the other metal-based AEM structures were fabricated and tested in order to evaluate the performance of the different emitter-system designs, and to define the cost-effective way for the AEM manufacturing process. A variety of metal AEM structures were fabricated and tested, including the set of rods, tubes, coils, and a combination of cylinders and star-type structures (star fin). The results were similar when the major condition (optically thin emitter) was embedded into the combustion chamber.

TEST RESULTS

Energy Density

The AEM combustion devices demonstrated an exceptional energy-density feature. The maximum volumetric heat release of the majority of the units was in order of 3 million Btu/hr-ft^3 (31 W/cm^3). This number is very impressive in comparison with the other combustion systems. The volumetric heat release in the large scale boilers (electric power plants) is typically in order of 50 thousand Btu/hr-ft^3 (0.5 W/cm^3). Industrial boilers usually designed for 80-100 thousand Btu/hr-ft^3 (0.8-1.0 W/cm^3). A well-stirred reactor (WSR) was developed at QGI (1) as a methane/oxygen TPV combustor. It demonstrated multi-kilowatt electric output, but had an energy density of about 0.5 million Btu/hr-ft^3 (5.0 W/cm^3) which is much less than AEM. This outstanding energy density of AEM devices allows for designing extremely portable combustion devices. Our experiments with the appliances (water heaters), which are currently available on a market, showed that by retrofitting them into AEM combustion systems the capacity of the devices can be increased by a factor of ten.

Radiant Emission

Radiant emission from the AEM structure was measured at a distance of 6.67 cm from the outer emitter surface. The quartz tube separated the AEM emitter from the ambient. Figure 2 presents the axial radiant-flux profile from the concentric cylinder AEM structure.

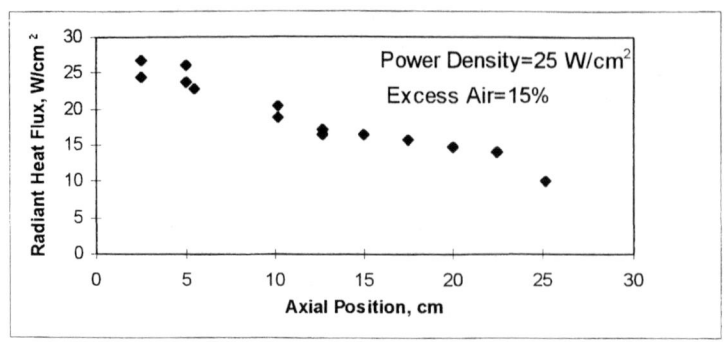

FIGURE 2. Axial Radiant-Flux Profile Measured at 6.67 cm from Concentric Cylinder AEM Structure

This data was obtained at 15% excess air, and a power density of 25 W/cm^3. Maximum radiant flux of 27 W/cm^2 was measured at the bottom section of the AEM. It is important to mention that the measured value of the radiant flux does not reflect the maximum achievable photon density for AEM technology. This experiment was conducted at 15% excess air in order to prevent rapid AEM degradation due to high emitter temperature. By using the other AEM structural material (such as rare earth oxides) the combustion mixture can be tuned close to its stiochiometry, and as the result, the flame and emitter temperatures will be significantly higher, which will result in higher radiant flux. Even assuming that the radiant flux of 25 W/cm^2 will be delivered to the PV surface (which is the limit for concentrator-type Si PV cells) and PV photon-to-electron conversion rate will be between 25 and 40% (depending on emission spectrum), the expected electric power density can be estimated in the range from 6.25 to 10.0 W/cm^2.

The other factor which significantly controls the PV arrays performance is the uniformity of PV illumination. Ideally, the best PV array performance can be achieved when each PV cell that is connected in series is illuminated uniformly. Our experiments showed that by changing the structural density of the AEM, it is possible to engineer the radiant profile according to the PV array configuration. Three different AEM structures were tested in the same combustion chamber with the same energy density and combustion parameters. The first AEM (circle symbols) was similar to that which was described above. The second and third emitters (square and triangle symbols) represent star-fin structures of different configurations. The axial radiant profiles were measured at 6.67 cm from the emitter. The results of these tests are depicted in Figure 3.

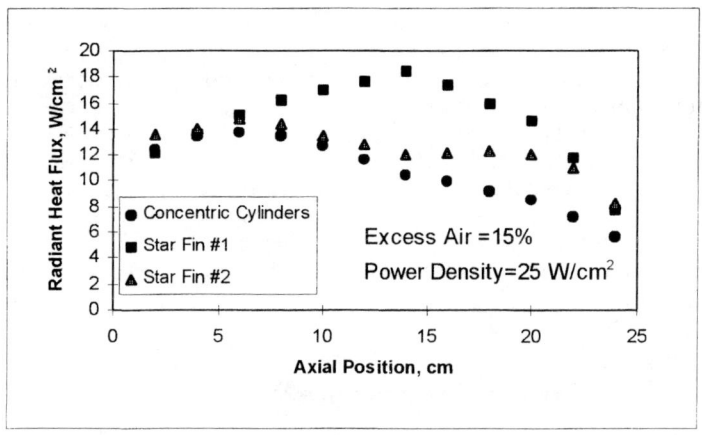

FIGURE 3. Axial Radiant Profiles of the AEM Emitters Measured at 6.67 cm from the AEM

It is clear that the radiant profile can be controlled by AEM configuration. It is possible to concentrate the peak of the emission at some particular areas, or make it almost flat as presented by triangle symbols. Again, the absolute value of the radiant flux is expected to be much higher if the fuel/air ratio will be maintained close to stoichiometry.

Several experiments were conducted in order to estimate the rate of fuel-chemical-energy-to-radiant-emission conversion. The radiant efficiency was estimated based on water temperature rise with the stable water flow through the water jacket which surrounds the AEM emitter. The energy density ranged from 3.9 to 15.8 W/cm^2 during the experiment. Excess air was varied from 40 to 80 % in order to prevent rapid AEM degradation. Test results for this set of experiments are depicted in Figure 4.

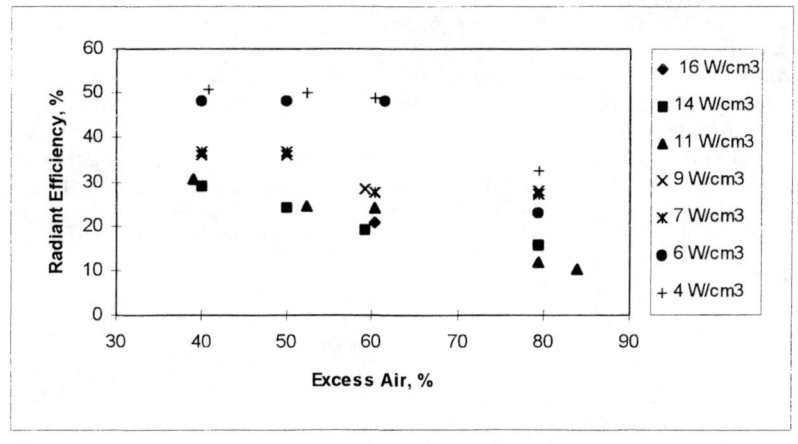

FIGURE 4. AEM Radiant Efficiency

The maximum radiant-conversion efficiency (of about 52%) was achieved at the lowest excess air and fuel input conditions. It is not surprising that at elevated excess air the radiant efficiency drops. It is well known that the flame temperature reaches the maximum close to stoichiometry, so additional air reduces the flame and emitter temperatures, and as a result, decreases the radiant output. At this point it was important to prove that an AEM system can provide for over that 50% radiant efficiency. Note that this data was obtained by testing metal-based AEM structures. The emitter temperature was estimated around 1100 °C. It is expected that the rate of the fuel-to-photon-energy conversion will be significantly higher if the AEM will be fabricated of rare earth oxides which are capable of withstanding the 2000 °C environment.

Combustion Performance

Two major concerns were investigated during the experiments. The first was the combustion completeness in terms of TPV efficiency, and the second addressed the safety issue. The carbon monoxide (CO) level was considered as a criteria for combustion completeness. It was found that the AEM technology allows for very clean combustion. The CO level at excess air of about 7 to 10% was less than 50 ppm, which is much less than required for appliances by existing regulations.

SUPEREMISSIVE AEM

In order to utilize the unique features of AEM combustion for TPV applications, QGI fabricated several superemissive (narrow band) AEM structures. The major objectives of this project were to learn how to build self-supportive emitters, and to integrate them into TPV systems. There were two approaches: 1) to construct the emitter of high-temperature material and apply superemissive coating on its surface, and 2) to fabricate AEM of pure superemitters such as rare earth oxides.

The first approach was realized by fabricating the AEM of alumina as the supportive structure and coating it with the ytterbia. Alumina was selected to be a supportive structure because of its high temperature resistance and low emissivity. It is expected that a blackbody background emission from the alumina structure will be less than from the metal-based AEM. At the same time ytterbia will emit the photons with the wavelength centered at 980 nm. Two AEM emitters were fabricated using this technology. The first was comprised of seven single elements as presented in Figure 5.

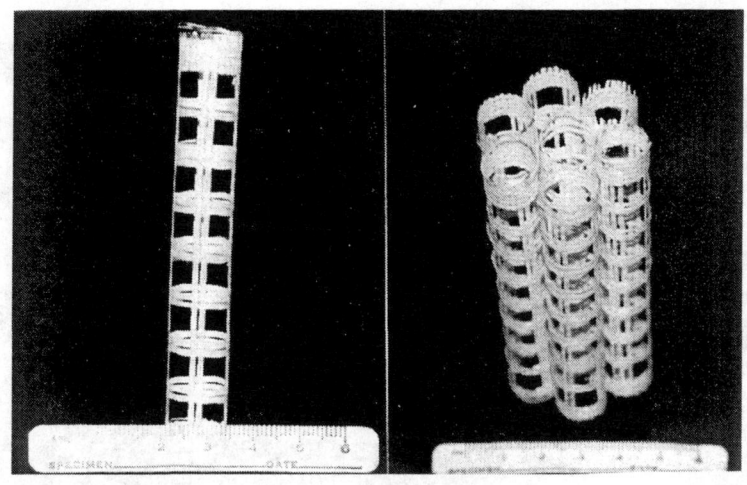

FIGURE 5. Single Component and Assembled Tubular Coated AEM

This emitter was about 4 cm in diameter and 7.6 cm in height. Single AEM components were glued together by special high-temperature-resistive glue also developed at QGI. It was found that the superemissive AEM exhibits the same level of energy density, and high radiant output as well as combustion performance.

The second coated AEM was the mimic of the metal "flat spiral" AEM with the same dimensions. It was also fabricated of alumina and coated with ytterbia. Figure 6 illustrates the superemissive flat spiral AEM, Figure 7 presents this emitter during the test.

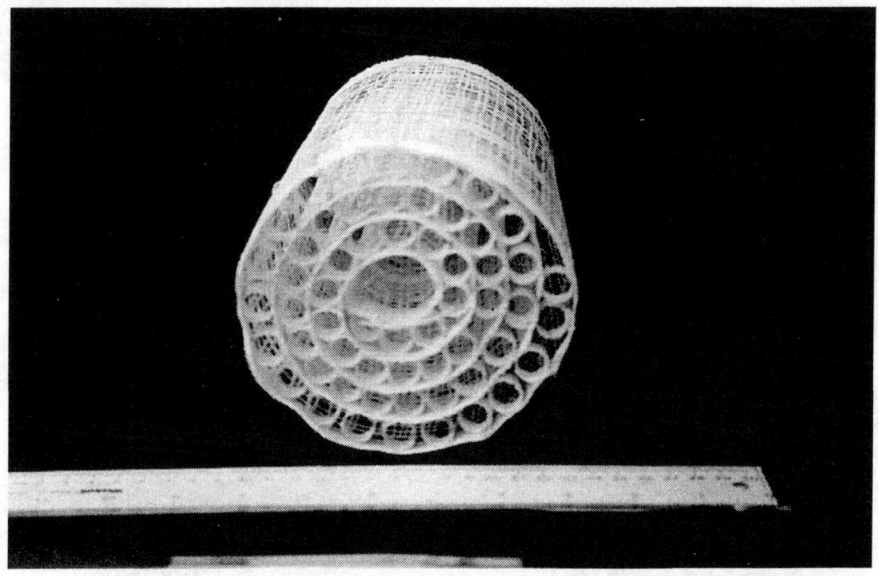

FIGURE 6. Flat Spiral Superemissive AEM

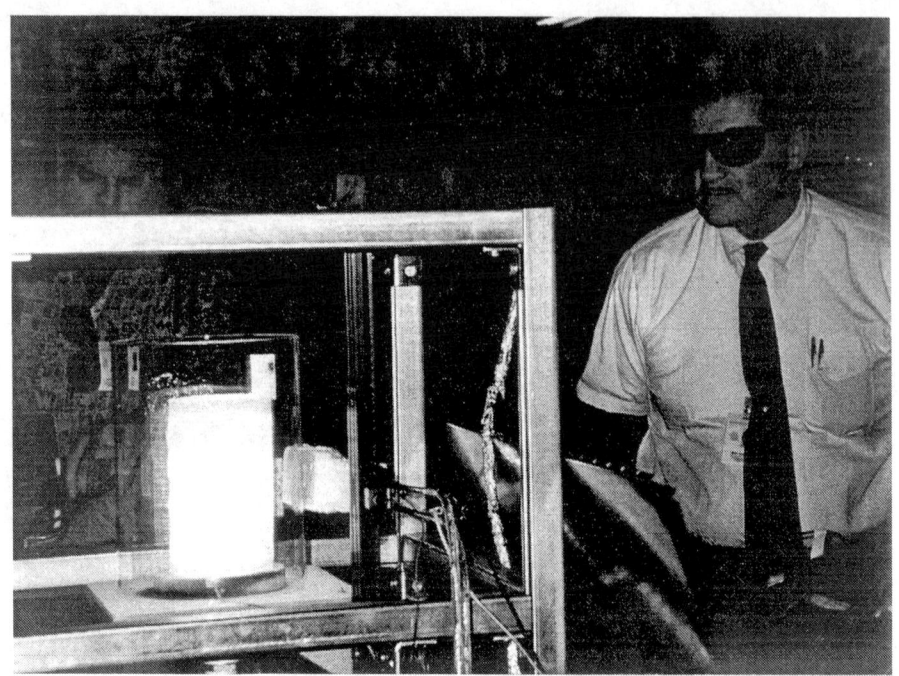

FIGURE 7 Flat Spiral Superemissive AEM During the Test

Another AEM (Log Cabin), 4 by 4 by 4 inch made of pure ytterbia felt, which was sliced and glued together into a three-dimensional matrix as shown in Figure 8. The same high temperature glue was used as a joint compound for this structure. Figure 9 depicts this AEM during the test.

FIGURE 8. Log Cabin Ytterbia AEM

FIGURE 9. Log Cabin AEM During the Test

All these AEM structures demonstrated an outstanding energy density and radiant output features. The emitter temperature was so high during the experiments that the alumina-based AEM was partially melted, and the log cabin structure was damaged due to high temperature creep.
Nevertheless, QGI continue to pursue the AEM approach by running an internal R&D program in order to improve high-temperature stability of superemissive structures.

CONCLUSIONS

Quantum Group Inc. has developed, fabricated and tested the AEM combustion devices, which allow us to reach an exceptional volumetric heat release accompanied with an extremely intense radiant output. QGI has proven that by engineering AEM structural density, the emitter can provide for uniform PV illumination. Several ways to fabricate superemissive AEM were investigated and the results shown in this paper.

REFERENCES

1. ARPA report, "Development of Thermophotovoltaic Generators for (Unmanned Underwater Vehicle) Military Applications," ARPA Contract No. MDA972-93-C-0042.

2. Becker, F. E., Doyle, E. F., and Shukla, K, "Development of a Portable Thermophotovoltaic Generator," presented at the Third NREL Conference Thermophotovoltaic Generation of Electricity, Colorado Springs, CO, May 1997

3. Photovoltaic Insider's Report (1996) Vol. XV No. 5 page 3, May 1996

4. Fraas, L. M., Ferguson, L., McCoy, L. C., and Pernisz, U. C. "SiC IR Emitter Design for Thermophotovoltaic Generators," presented at the Second NREL Conference on Thermophotovoltaics Generation of Electricity, Colorado Springs, CO. July 1995

5. Goldstein, M. K., and Kushch, A. S., "Low NOx Thermophotovoltaic Generator," *IEEE Transactions on Industry Applications, 32, No 1, 41-46 (1996)*

6. Kushch A. S., Skinner S. M., Sarmiento P. A., and Brennan R "Development of a Co-generating Thermophotovoltaic Powered Combination Hot Water Heater/Hydronic Boiler," presented at the Third NREL Conference Thermophotovoltaic Generation of Electricity, Colorado Springs, CO, May 1997

A Study of Contacts and Back-Surface Reflectors for 0.6-eV $Ga_{0.32}In_{0.68}As/InAs_{0.32}P_{0.68}$ Thermophotovoltaic Monolithically Interconnected Modules

X. Wu, A. Duda, J.J. Carapella, J.S. Ward, J.D. Webb, and M.W. Wanlass

National Renewable Energy Laboratory, Golden, Colorado 80401

Abstract

Thermophotovoltaic (TPV) systems have recently rekindled a high level of interest for a number of applications. In order to meet the requirement of low-temperature (~1000°C) TPV systems, 0.6-eV $Ga_{0.32}In_{0.68}As/InAs_{0.32}P_{0.68}$ TPV monolithically interconnected modules (MIMs) have been developed at the National Renewable energy Laboratory (NREL)[1]. The successful fabrication of $Ga_{0.32}In_{0.68}As/InAs_{0.32}P_{0.68}$ MIMs depends on developing and optimizing of several key processes. Some results regarding the chemical vapor deposition (CVD)-SiO_2 insulating layer, selective chemical etch via sidewall profiles, double-layer antireflection coatings, and metallization via interconnects have previously been given elsewhere [2]. In this paper, we report on the study of contacts and back-surface reflectors. In the first part of this paper, Ti/Pd/Ag and Cr/Pd/Ag contact to n-$InAs_{0.32}P_{0.68}$ and p-$Ga_{0.32}In_{0.68}As$ are investigated. The transfer length method (TLM) was used for measuring of specific contact resistance R_c. The dependence of R_c on different doping levels and different pre-treatment of the two semiconductors will be reported. Also, the adhesion and the thermal stability of Ti/Pd/Ag and Cr/Pd/Ag contacts to n-$InAs_{0.32}P_{0.68}$ and p-$Ga_{0.32}In_{0.68}As$ will be presented. In the second part of this paper, we discuss an optimum back-surface reflector (BSR) that has been developed for 0.6-eV $Ga_{0.32}In_{0.68}As/InAs_{0.32}P_{0.68}$ TPV MIM devices. The optimum BSR consists of three layers: ~1300Å MgF_2 (or ~1300Å CVD SiO_2) dielectric layer, ~25Å Ti adhesion layer, and ~1500Å Au reflection layer. This optimum BSR has high reflectance, good adhesion, and excellent thermal stability.

Contacts

0.6-eV TPV MIMs usually operate at high current densities of more than 2 A/cm^2. A suitable contact system for these devices has to satisfy the following three requirements: low specific contact resistance (our modeling work indicated that this value should be $<1 \times 10^{-4}$ Ω cm^2), excellent long-term thermal stability, and good adhesion. Also, low-cost and simple fabrication processing need to be considered for larger-scale use. In this study, two contact systems, including Ti/Pd/Ag and Cr/Pd/Ag, have been investigated. We show that low contact resistance and good adhesion of Ti and Cr are obtained with both n-$InAs_{0.32}P_{0.68}$ and p-$Ga_{0.32}In_{0.68}As$. Ag is used as the conduction layer

CP460, *Thermophotovoltaic Generation of Electricity: Fourth NREL Conference*
edited by T. J. Coutts, J. P. Benner, and C. S. Allman
© 1999 The American Institute of Physics 1-56396-828-2/99/$15.00

because of low cost and high conductivity. In these two contact systems, the Pd film works as a Ag diffusion barrier layer.

Dependence of specific contact resistance on the doping level of n-InAs$_{0.32}$P$_{0.68}$

We investigated the dependence of R_c to both Ti and Cr for different doping levels of n-InAs$_{0.32}$P$_{0.68}$. In the 0.6-eV Ga$_{0.32}$In$_{0.68}$As/InAs$_{0.32}$P$_{0.68}$ TPV converter structure, n-InAs$_{0.32}$P$_{0.68}$ and p-Ga$_{0.32}$In$_{0.68}$As serve as contact layers. The n-InAs$_{0.32}$P$_{0.68}$ layer also serves as an emitter layer of the cell-isolated diode (CID). To minimize the absorption of the n-InAs$_{0.32}$P$_{0.68}$ layer and to optimize the quality of the CID, the n-InAs$_{0.32}$P$_{0.68}$ layer cannot be doped to high levels. Therefore, one must study the dependence of R_c on the doping level of the n-InAs$_{0.32}$P$_{0.68}$ layer. Figure 1 shows the dependence of R_c on different doping levels (N_e) of the n-InAs$_{0.32}$P$_{0.68}$ layer. As shown in Figure 1, R_c increases with decreasing N_e. But, even if N_e is reduced to 1×10^{18} cm^{-3}, R_c can still meet the requirement for these devices.

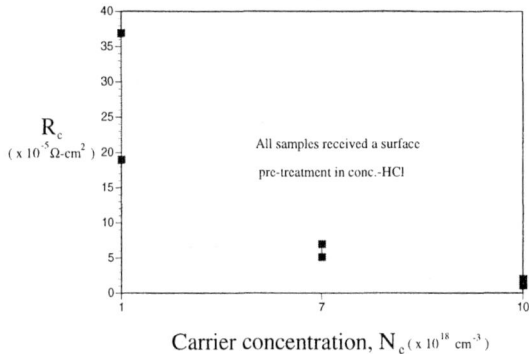

Figure 1. The dependence of R_c on the doping N_e level of n-InAs$_{0.32}$P$_{0.68}$.

Dependence of specific contact resistance on pre-treatment of semiconductors

The dependence of R_c on different pre-treatments of n-InAs$_{0.32}$P$_{0.68}$ and p-Ga$_{0.32}$In$_{0.68}$As was investigated. We found that removing the native oxide from the surface of semiconductors can improve adhesion and reduce specific contact resistance. Figure 2 shows that the n-InAs$_{0.32}$P$_{0.68}$ sample received a dip in concentrated HCl before metallization has the lowest R_c. The results for p-Ga$_{0.32}$In$_{0.68}$As are similar.

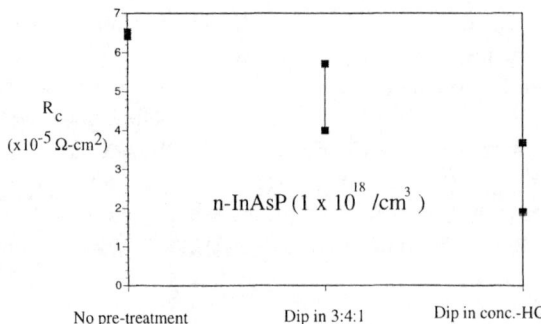

Figure 2. The dependence of R_c on different pre-treatments of n-InAs$_{0.32}$P$_{0.68}$.

Specific contact resistance

Figure 3 illustrates the specific contact resistance of Ti/Pd/Ag and Cr/Pd/Ag contacts to n-InAs$_{0.32}$P$_{0.68}$ and p-Ga$_{0.32}$In$_{0.68}$As. The carrier concentration of n-InAs$_{0.32}$P$_{0.68}$ and p-Ga$_{0.32}$In$_{0.68}$As are 1×10^{18} cm^{-3} and 1×10^{19} cm^{-3}, respectively. It can be seen that all of these results can meet the requirements for the MIM design. R_c of the Cr contact system is slightly higher than that of the Ti contact system.

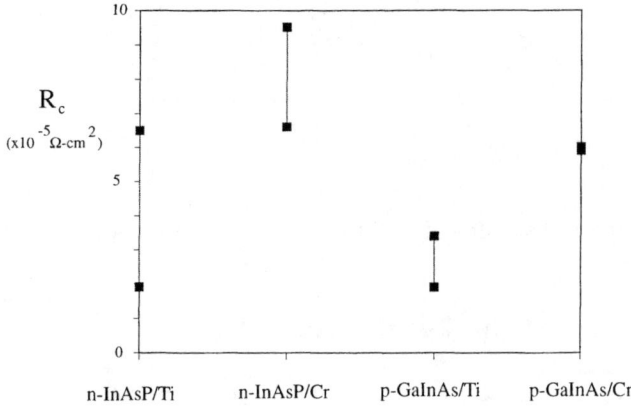

Figure 3. The specific contact resistance of Ti/Pd/Ag and Cr/Pd/Ag contacts to n-InAs$_{0.32}$P$_{0.68}$ and p-Ga$_{0.32}$In$_{0.68}$As.

Adhesion

All samples with Ti/Pd/Ag and Cr/Pd/Ag contacts to n-InAs$_{0.32}$P$_{0.68}$ and p-Ga$_{0.32}$In$_{0.68}$As have passed a Scotch tape test. Cr/Pd/Ag fine contact grids have better adhesion to p-Ga$_{0.32}$In$_{0.68}$As than Ti/Pd/Ag fine grids.

Thermal stability

Finally, the thermal stability of Ti/Pd/Ag contacts to n-InAs$_{0.32}$P$_{0.68}$ and p-Ga$_{0.32}$In$_{0.68}$As has been inspected. Two samples, one of Ti/Pd/Ag contacts to n-InAs$_{0.32}$P$_{0.68}$ (1×10^{18} cm^{-3}) and another of Ti/Pd/Ag contacts to p-Ga$_{0.32}$In$_{0.68}$As (1×10^{19} cm^{-3}), have been exposed to 80°C for 120 hours. Figure 4 shows the changes of specific contact resistance for these two samples before and after thermal stress. It can be seen that R_c is slightly improved for both samples after the thermal treatment.

In summary, experimental results, including specific contact resistance, adhesion, and thermal stability, demonstrate that Ti/Pd/Ag or Cr/Pd/Ag contacts to n-InAs$_{0.32}$P$_{0.68}$ and p-Ga$_{0.32}$In$_{0.68}$As can meet the requirements of the TPV MIM. These two contact systems have already been integrated into NREL's 0.6-eV Ga$_{0.32}$In$_{0.68}$As/InAs$_{0.32}$P$_{0.68}$ TPV MIMs. We have fabricated a 4-cell 0.25 cm^2, 0.6-eV Ga$_{0.32}$In$_{0.68}$As/InAs$_{0.32}$P$_{0.68}$ TPV MIM with a fill factor of 71%.

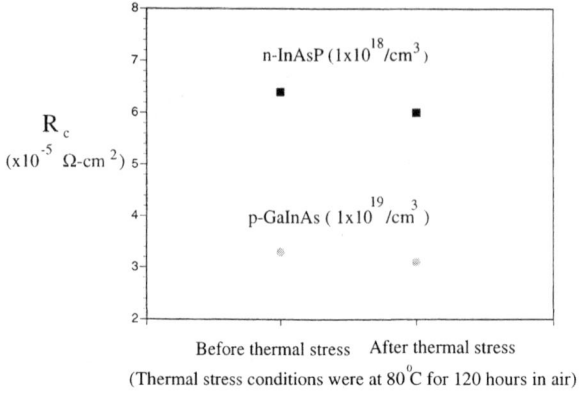

Figure 4. The thermal stability of Ti/Pd/Ag contacts to n-InAs$_{0.32}$P$_{0.68}$ and p-Ga$_{0.32}$In$_{0.68}$As.

Back-surface reflector

The TPV system requires a spectral control component to recoup low-energy photons, which is crucial to the efficiency of the system. Two techniques have been used for realizing the spectra control in TPV systems: selective filters serve as the front-surface spectral control and back-surface reflectors (BSRs) serve as back-surface spectral control. Compared to the selective filters, the back-surface reflector has much less parasitic absorption of the above-bandgap radiation, a wider bandwidth, and a lower reflection of above-bandgap photons. In general, there are a number of criteria that must be simultaneously satisfied for a BSR to succeed. These criteria include high reflectance, low contact resistance, strong adhesion, and thermal stability. Because the BSR can be directly prepared on a semi-insulating substrate, the contact resistance is not an issue in the TPV MIMs. NREL's MIMs use a novel, interdigitated contacting scheme. This new design offers an attractive solution to a number of problems confronting TPV device performance, including problems with the BSR. Our interdigitation design allows us to realize the full performance potential of a well-designed BSR by minimizing parasitic free-carrier absorption. In this section, we summarize the experimental results and present the optimum BSR structure for 0.6-eV Ga$_{0.32}$In$_{0.68}$As/InAs$_{0.32}$P$_{0.68}$ TPV MIMs.

We use a BSR structure with three layers, which includes a dielectric layer, an adhesion layer, and a reflection layer. In general, the metals with best infrared (IR) reflectance are the noble metals. We selected a Au film with a thickness of 1500Å as the reflection layer. In a BSR structure, the dielectric layer serves not only as a diffusion barrier, but also, as an optical mismatch between the metal and semiconductor for improving the reflectance in the IR wavelength range. Both SiO$_2$ and MgF$_2$ films have been used as the dielectric layer in our BSR structure. They both have excellent insulating properties and lower refractive index. SiO$_2$ films were prepared by the CVD technique at about 350°C. MgF$_2$ films were deposited by the thermal evaporation technique at room temperature. Our experimental results indicate that the reflectance of the BSR with a MgF$_2$ dielectric layer is slightly higher than that with a SiO$_2$ layer. This is reasonable because the refractive index of MgF$_2$ (n=1.38) is slightly lower than that of the SiO$_2$ film (n=1.46). In addition, there is an absorption peak at about 7 μm for the BSR with a SiO$_2$ dielectric layer from Fourier transform infrared (FTIR) measurement. Conversely, the BSR with a MgF$_2$ layer does not have this absorption peak (see Figure 5). Based on these results, we prefer to use MgF$_2$ film as a dielectric layer in our TPV devices.

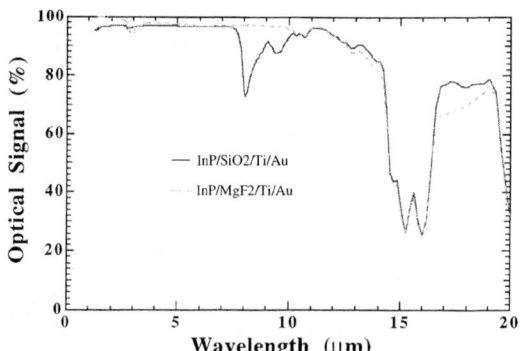

Figure 5. Comparison of the reflectance between two BSRs with a SiO$_2$ and MgF$_2$ dielectric layer in the IR-wavelength range.

Figure 6 shows a comparison of the reflectance of four BSRs with different thicknesses of the SiO$_2$ dielectric layer. As shown in this figure, the BSR with a SiO$_2$ dielectric layer having a thickness of ~1300Å has the highest reflectance. The thin Ti film serves as an adhesion layer in our BSR structure. The improvement of the BSR's adhesion always results in reducing reflectance; therefore, it is necessary to optimize the Ti thickness. We found that the BSR with a 25Å Ti film has the highest reflectance and excellent adhesion.

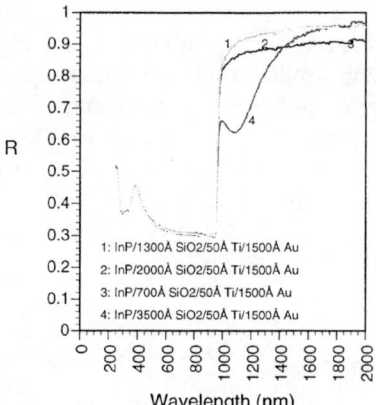

Figure 6. Comparison of the reflectance of four BSRs with different thicknesses of the SiO$_2$ dielectric layer.

We also studied the thermal stability of the BSRs. Two BSR samples, which included SiO_2 and MgF_2 dielectric layers individually, were prepared for this study. Both BSRs received a thermal annealing treatment at 80°C in air for 120 hours. Two techniques, reflectance measurement and secondary-ion mass spectroscopy (SIMS) analysis, have been used to inspect the change of the two BSR samples. Figures 7a and 7b show the change of the reflectance before and after the thermal treatment for two BSRs with SiO_2 and MgF_2 dielectric layers, respectively. As shown in these two figures, there is no change of the reflectance after the thermal treatment. SIMS results also indicate that the Au profile does not change after the thermal treatment. Therefore, the optical measurement and SIMS analysis demonstrate that our BSRs have excellent thermal stability.

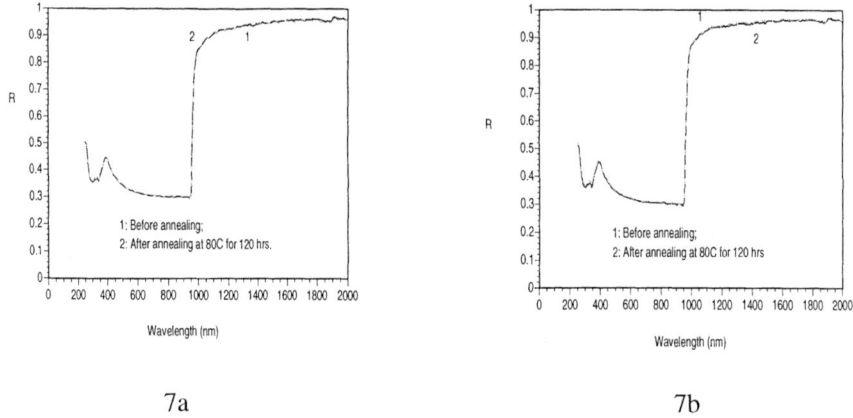

Figure 7. The change of the reflectance after the thermal treatment for two BSRs with a SiO_2 (7a) and a MgF_2 (7b) dielectric layer, respectively.

As a result of this investigation of BSRs, the optimum BSR should consist of a ~1300Å MgF_2 (or SiO_2) dielectric layer, ~25Å Ti adhesion layer, and ~1500Å Au reflection layer. The optimum BSR structure has been used on two samples: one is a semi-insulated (SI) InP wafer, and the other is a 0.6-eV $Ga_{0.32}In_{0.68}As/InAs_{0.32}P_{0.68}$ device on a SI-InP substrate without contact grids and antireflection coating layer. As shown in Figure 8, an optimum BSR on a semi-insulated InP substrate has a very high reflectance of more than 90% in the wavelength range between 2–12 μm. It can also be seen from Figure 9 that the reflectance of an optimum BSR on a 0.6-eV $Ga_{0.32}In_{0.68}As/InAs_{0.32}P_{0.68}$ device is in the range of 80%-92%. The somewhat lower reflectance is due to the free-carrier absorption of the doped layers of the device.

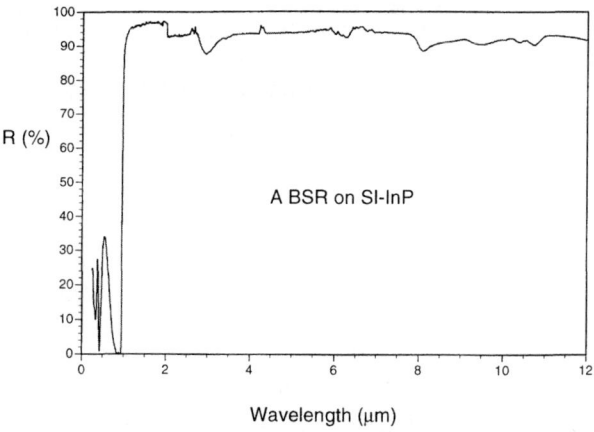

Figure 8. The reflectance of an optimum BSR on a SI-InP substrate with a double antireflection (AR) layer.

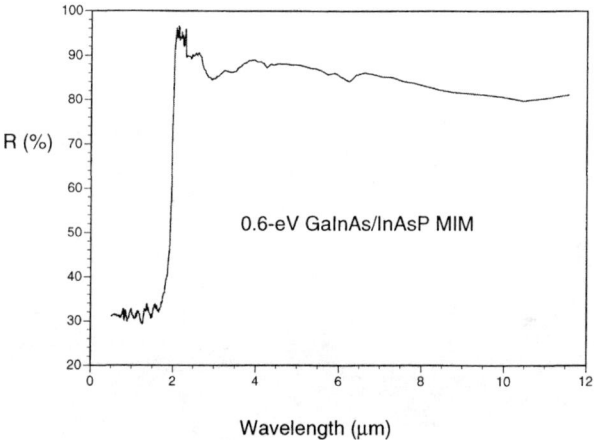

Figure 9. The reflectance of an optimum BSR on a 0.6-eV $Ga_{0.32}In_{0.68}As/InAs_{0.32}P_{0.68}$ device without metal contact grids and AR coating layer.

In summary, an optimum back-surface reflector has been developed for NREL's 0.6-eV $Ga_{0.32}In_{0.68}As/InAs_{0.32}P_{0.68}$ TPV MIMs. The optimum BSR consists of three layers: a ~1300Å MgF_2 (or ~1300Å CVD SiO_2) dielectric layer, a ~25Å Ti adhesion layer, and a ~1500Å Au reflection layer. This optimum BSR has high reflectance, strong adhesion, and excellent thermal stability. A reflectance of 80%-92% in the wavelength range between 2–12 μm has been achieved for a BSR on a 0.6-eV $Ga_{0.32}In_{0.68}As/InAs_{0.32}P_{0.68}$ device.

Acknowledgment

The authors would like to thank L.M. Gedvilas for FTIR measurements.

References

[1] M.W. Wanlass et al., "High-performance, 0.6-eV, $Ga_{0.32}In_{0.68}As/InAs_{0.32}P_{0.68}$ Thermophotovoltaic Converters and Monolithically Interconnected Modules," in this conference.

[2] J.S. Ward et al., "A Novel Design for Monolithically Interconnected Modules (MIMs) for Thermophotovoltaic Power Conversion," Proc. Third NREL Conf. on TPV Gen. of Elect., 1997, AIP 401, pp.227.

p-GaSb/n-GaAs Heterojunctions for Thermophotovoltaic Cells Grown by MOVPE

L.Zheng*, G.M.Sweileh*, S.K.Haywood*, C.G.Scott[†]
M.Lakrimi[‡], N.J.Mason[‡] and P.J.Walker[‡]

*School of Engineering, University of Hull, Cottingham Road, Hull, HU6 7RX, U.K.
[†]Dept of Physics, University of Hull, Cottingham Road, Hull, HU6 7RX, U.K.
[‡]Dept. of Physics, Clarendon Laboratory, University of Oxford, Parks Road, Oxford, OX1 3PU, U.K.

Abstract: The reverse bias dark current of p-GaSb/n-GaAs junctions has been found to be unexpectedly low despite the 7% lattice mismatch between the layers. MOVPE-grown diodes exhibited dark currents up to two orders of magnitude lower than comparable GaSb homojunctions. The initial nucleation of GaSb on GaAs was found to play an important role in determining the properties of the bulk layer. A larger number of small nucleation sites produced diodes with lower reverse bias leakage currents and growth temperature was in turn critical in determining the size of these nucleation islands. Here we present a study of the characteristics of diodes grown at different temperatures and with varying GaSb doping levels. Diodes were grown at temperatures ranging from 500-550°C using TMGa and TMSb. Several orders of magnitude reduction in reverse bias dark current was observed across this narrow temperature range. The 'turn-on' voltage also increases with reduction in growth temperature. Under illumination V_{oc} reached a maximum value of ~0.33V before decreasing slightly with further increase in light intensity. The corresponding fill factor was 0.63. V_{oc} is low compared to GaSb homojunctions (typically 0.45V) but these devices have not been optimised for thickness or doping level. Since most layers are <1μm they show low efficiency (~1%) and heating effects may be significant in reducing V_{oc}.

1. INTRODUCTION

GaSb photovoltaic devices are likely to find application in both the recycling of waste heat to produce electricity [1] and as booster cells for GaAs solar cells used in space. [2]. A potentially significant market for thermophotovoltaics [TPVs] is in association with domestic gas-fired boilers to enable heating systems to function without mains power in remote areas or during a power failure. Hybrid car engines, recycling of waste heat from power stations and 'quiet' military applications are other possibilities.

GaSb homojunction devices have been used with GaAs solar cells in mechanically-stacked tandem systems to produce high efficiency photovoltaic arrays for space applications. The bandgaps of these two materials are close to the optimum combination for a dual cell and efficiencies of ~35% have been achieved compared to ~25% for GaAs alone [2]. Even small improvements in cell efficiency can be beneficial since, used with concentrator technology, this can lead to a critical reduction in the surface area required for power generation on satellites.

A TPV electric generator using GaSb cells with a hydrocarbon burner [3] is already available commercially but as yet the technology is expensive for domestic applications. Efficient cells have also been reported using InGaAs/InP [4] and these two materials are the main competitors for simple cells in this wavelength range. High efficiency devices are not necessarily needed for TPV because of the high radiation levels available close to a burner ($\sim 10W/cm^2$). However, to meet the demands of large-scale TPV, the production technology must be simple, reproducible and cheap. Most GaSb photovoltaics incorporate diffused junctions for economic reasons despite the fact that diffusion tends to encourage grading and it appears to be difficult to form abrupt p-n junctions in GaSb by any method [5,6,7]. More complex structures must offer a significant advantage to trade-off against cost.

Here we investigate p-GaSb/n-GaAs heterojunctions grown by MOVPE on GaAs substrates. If cheap GaAs substrates can be used to replace GaSb substrates in GaSb TPV cells, the cost of MOVPE growth may be compensated. In addition there is the potential for future development of a GaAs/GaSb epitaxial tandem solar cell providing a cheaper version of existing tandem cells, possibly with improved efficiency. GaSb and GaAs are 7% lattice-mismatched and such heterojunctions would not normally be considered for photovoltaics because of the expected high recombination resulting from dislocations. However, we have found that very good I-V characteristics can be obtained in p-GaSb/n-GaAs devices grown by MOVPE deposition of p-GaSb onto an n-GaAs substrate at low temperatures. If suitable growth conditions are chosen dislocations and defects at the interface appear not to be electrically active.

Our devices show $V_{oc} = 0.33V$ with a fill factor of 0.63 despite sub-micron GaSb layers. The best GaSb homojunction devices grown by zinc diffusion into an n-type GaSb wafer, showed $V_{oc} = 0.48V$ with a fill factor of 0.71 under one-sun AM0 illumination[2,8]. However, the heterojunction structures studied here have not been optimised in terms of doping levels, layer thicknesses or metallisation.

2. GROWTH AND STRUCTURAL CHARACTERISATION

2.1 Growth and Processing

GaSb epitaxial layers were grown directly onto n-type (100) GaAs substrates by MOVPE using TMGa and TMSb as the starting materials. The substrates were doped with silicon to a carrier concentration $\sim 4 \times 10^{17} cm^{-3}$. A second piece of semi-insulating GaAs was put into each run to allow Hall measurements to be carried out on the GaSb epilayer. Substrate temperature during growth was between 500 and 550°C. GaSb layers were grown both undoped and with p-type doping using DMZn. It has proved difficult to achieve p-type material with controllable carrier concentrations between $10^{19} cm^{-3}$ and the background level of $10^{16} cm^{-3}$. Control of the pressure in the dopant diffusion line is being improved to address this problem.

Table 1 shows the growth and structural details of the samples processed into mesa diodes for electrical and optical characterisation along with the diode dark current at -0.2V, carrier concentrations and mobilities measured via the Hall effect. The reduction in dark current associated with lower growth temperature suggests that dislocations become less important as conduction pathways. The mobility of the thicker layers approaches the limiting value for unintentionally doped material (800-900 cm^2/Vs) and is indicative of a low level of defects able to cause carrier scattering. The mobility reduces significantly for the sub-micron layers due to the higher dislocation density near the GaSb/GaAs interface. Carrier concentrations of the undoped layers show a corresponding slight increase with reducing growth temperature/thickness. Carrier concentration in the doped layers decreased with increasing growth temperature due to the increased volatility of the zinc. Higher carrier concentration reduces the saturation current and decreases the mobility of the GaSb layers as expected.

All the samples were processed into circular mesa diodes of various diameters using standard photolithography techniques. Some diodes had optical access for spectral response and photocurrent measurements.

Table 1: Sample details: Diode dark current (at -0.2V), 300K carrier concentration and Hall mobility of p-GaSb layers grown on n-GaAs substrates

Sample No:	Growth temp. [°C]	Thickness [μm]	Dark current (A)	Carrier Conc^n. (cm^{-3})	Hall mobility [cm^2/Vs]
2980	550	4.2	2.8×10^{-7}	1.0×10^{16}	714
3114	520	1.2	6.0×10^{-9}	1.4×10^{16}	627
3112	500	0.5	4.1×10^{-10}	1.9×10^{16}	473
3289	520	0.9	5.0×10^{-9}	1.6×10^{19}	170
3288	510	0.5	1.8×10^{-9}	5.4×10^{19}	150
3185	500	0.3	1.6×10^{-10}	6.0×10^{19}	90

2.2 Initial Nucleation

It was known from previous studies that the initial nucleation conditions are of key importance to the properties of strain-relaxed layers [5,9,10]. The mismatch causes the GaSb to grow via the Stranski-Krastanov mode, that is, to initially grow a monolayer or so of pseudomorphic GaSb and then to nucleate islands that grow three-dimensionally. Examination of island growth has previously been carried out [9,11] by atomic force microscopy (AFM). Fig 1 shows the GaAs surface after 3mins growth at temperatures from 560-520°C. Average island size reduces from 350 to 100nm across as the temperature falls over this range with a corresponding increase in nucleation site density. Reduction of the growth temperature to 480°C using TIPGa and t-DMASb has shown even smaller nucleation sites with an average island size of only ~6nm [9]. Doping the GaSb p-type and growth onto a GaAs epitaxial layer rather than directly onto the substrate also influence nucleation and these effects are under investigation.

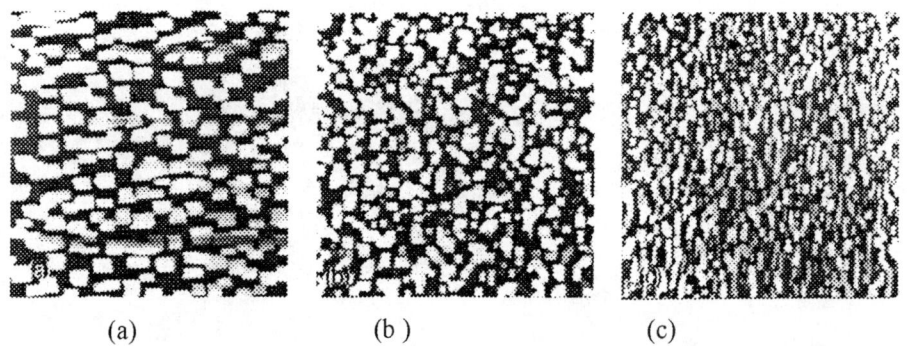

FIGURE 1. AFM images of 20μm square on GaAs substrate surface following 3min GaSb growth. Deposition temperatures and average GaSb island sizes are: (a) 560 °C/~350nm, (b) 540°C/~180nm & (c) 520 °C/~100nm.

3. ELECTRICAL AND OPTICAL RESULTS

3.1 I-V Measurements

I-V measurements were carried out at room temperature on the samples shown in Table 1 and results are shown in Figure 2 for 400μm diameter devices. A reduction in dark current with reducing growth temperature can be seen on going from 550 to 500ºC. Unfortunately the thickness of the layers also falls with reducing growth temperature as the alkyl pyrolysis rate and hence the GaSb deposition rate is falling off rapidly. There is little change in background doping level (Table 1) and hence we would not expect the thickness to have much effect on the diode saturation current.

However, there is an obvious correlation between the size and density of nucleation sites and diode characteristics. While dislocations must form at the junction to relax the strain in the layer, these appear to be less active electrically in the samples grown at lower temperature (cf. dark current in Table 1.) It is possible that more dislocations are created at the interface at lower temperature but that they turn over closer to the interface rather than extending up through the layer. Some evidence for this type of dislocation in GaSb has come from previous microscopy studies. [12]

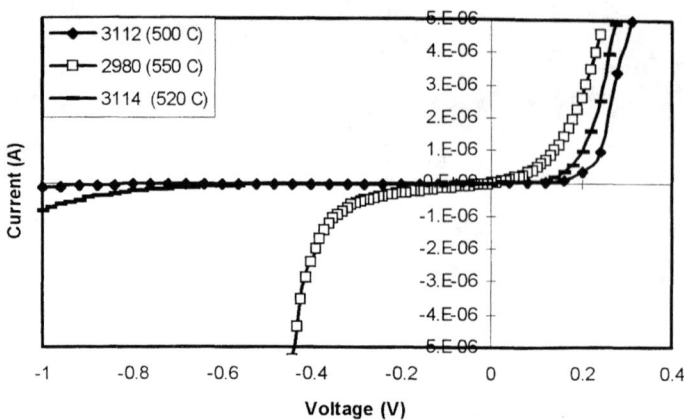

FIGURE 2. I-V characteristics for p-GaSb/n-GaAs heterojunctions grown at temperatures from 500°C-550°C.

3.2 Photocurrent and Spectral Response

I-V curves in the dark and under white light illumination (under 100W halogen lamp) are shown in Fig. 3(a) and 3(b) respectively for samples 3112 (unintentionally doped) and 3288 (doped). It can be seen that the doped sample has a much flatter I-V curve in reverse bias than the undoped sample and hence a higher fill factor (0.62 and 0.54 respectively). The devices illustrated are of similar thickness and grown at similar temperatures but the general shape of the curve was found to be determined by doping for all samples; 3114 showed the sloping curve and low fill factor similar to Fig 3(a) whereas 3185 and 3288 show a flat curve.

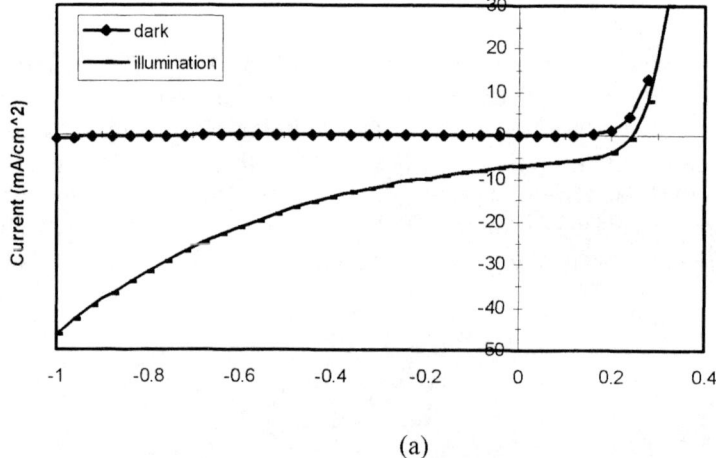

(a)

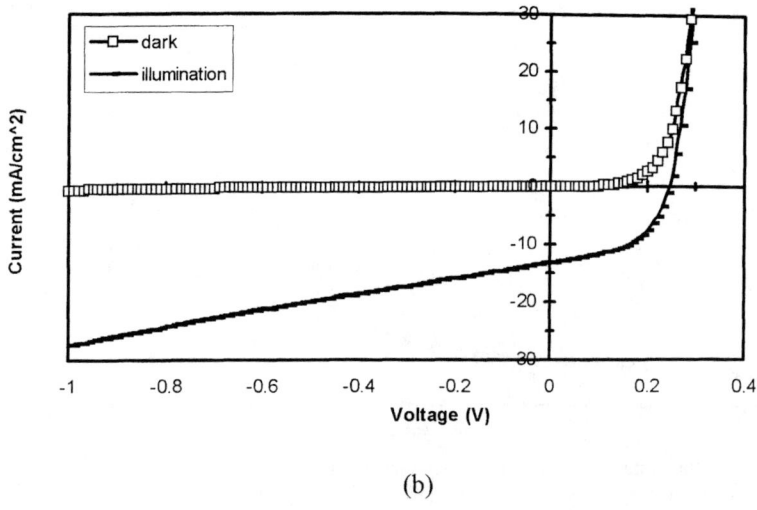

(b)

FIGURE 3. Dark and illuminated I-V curves for devices from (a) sample 3112 and (b) sample 3288.

Both high series resistance and a decrease in shunt resistance on illumination may contribute to the observed reduction in V_{oc} compared to GaSb homojunctions, although this will also be affected by the thin active layers. Apart from the absorption being lower, surface recombination and heating may also contribute proportionately more in thinner layers to reducing Voc. The substantial improvement in performance of the doped GaSb devices suggests that series resistance from contacts to the undoped GaSb layer and/or the layer resistance itself plays a significant role but we cannot rule out a contribution from shunt pathways due to interface defects.

Fig. 4 shows I-V curves as a function of illumination for a second doped sample 3185. Here again we see a flat I-V curve in reverse bias and a fill factor of 0.63, similar to 3288. V_{oc} saturated with illumination intensity at a maximum value of 0.33; thereafter it decreased slightly with further increase in light intensity.

The spectral response for 3112 is shown in Fig.5 normalised to an Ge photodetector. The highest efficiency was achieved from 3112, one of the undoped smples. Higher carrier concentration may reduce the photocurrent through increased recombination at dopant ions and reduced mobility. The high peak just below 870nm is from absorption in the GaAs substrate. This response is much stronger than that from the GaSb layer. The peak at around 1700nm is from the harmonic output of the monochromator and can be removed by choice of a different filter. The response was found to be similar for all the samples studied with the underlying GaAs substrate showing a much stronger response than the GaSb layer.

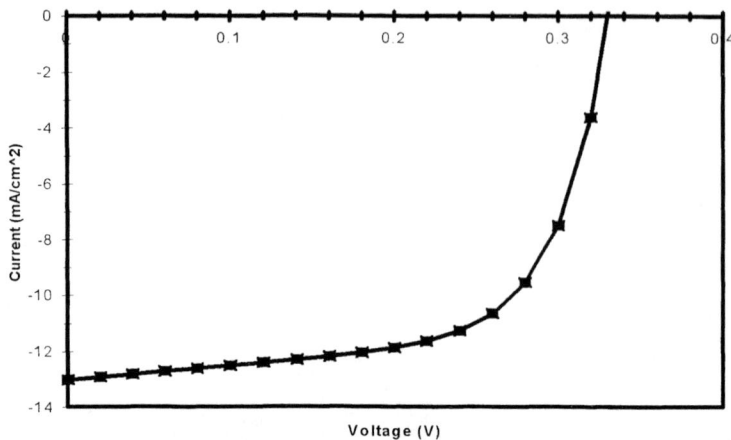

FIGURE 4. I-V curve for sample 3185 under varying degrees of illumination; Voc reaches a maximum value of 0.33V with 0.63 fill factor.

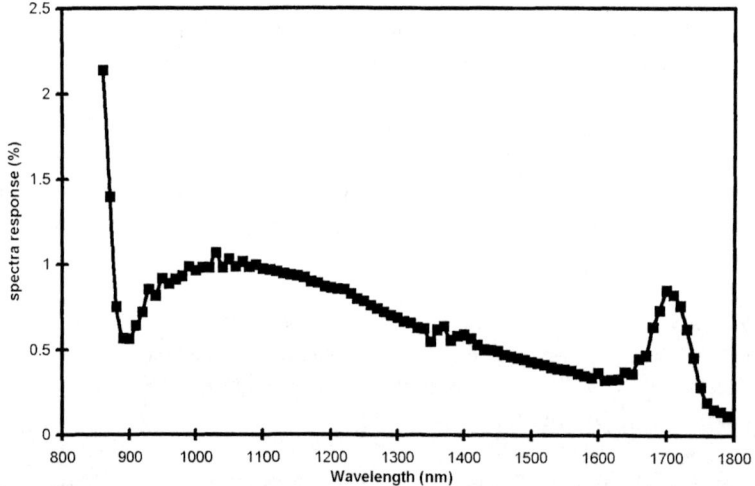

FIGURE 5. Normalised spectral response of 400μm diameter devices from wafer 3112 (500°C growth); Quantum efficiency is ~1% for the GaSb layer.

The efficiency of the GaSb response is not expected to be very high from these thin layers and surface recombination will further degrade the response as there is no window layer or AR coating on the surface. In addition, our modeling suggests an ideal carrier concentration is in the low $10^{17} cm^{-3}$ range rather than the $\sim 10^{16} cm^{-3}$ level of this device.

4. CONCLUSIONS

In conclusion, we have shown that the diode characteristics of MOVPE grown p-GaSb/n-GaAs junctions are strongly influenced by the growth temperature. The reverse bias dark current falls by 3 orders of magnitude (at -0.2V) on going from 550°C to 500°C. The responsivity of diodes under white light illumination was similar for samples grown at 520°C and below. AFM studies suggest that a more rectifying interface is associated with a high density of initial nucleation sites and hence also with immediate formation of dislocations at the interface.

V_{oc} saturated at ~0.33V before decreasing slightly with further increase in light intensity. The corresponding fill factor was 0.63. V_{oc} is low compared to GaSb homojunctions (typically 0.45V) but these devices have not been optimised for thickness or doping level. Since most layers are <1μm they show low efficiency (~1%) and heating effects may be significant in reducing V_{oc}.

Systematic studies of the interface under a wider range of growth conditions and optimisation of the structure are in progress. Further reduction of the growth temperature by use of alkyls which pyrolyse at lower temperature may also be beneficial. Future devices will use intermediate GaSb doping levels and be processed with an antireflection coating and passivation layer. This will establish the relative contributions of design parameters and interface defects to the low value of V_{oc}.

REFERENCES:

1. Fraas, L., Ballantyne, R., Samaras, J. and Seal, M., 'A thermophotovoltaic electric generator using GaSb with a hydrocarbon burner'; *Proc 1st WCPEC Conf.;* Hawaii, pp1713-16, (1994)
2. Fraas, L.M., Avery, J.E., Martin, J., Sundaram, V.S., Girard, G., Dinh, V.T., Davenport, T.M., Yerkes, J.W. and O'Neill, M.J., IEEE Trans. Electron Devices, **37**, 443-449, (1990)
3. Fraas, L., Avery, J.E., Ballantyne, R., Custard, P., Ferguson, L., Huang, H.X., Keyes, J., Mulligan, W., Samaras, J. and Williams, D., ' 2-Amp TPV co-generator using forced-air cooled GaSb cells' *Thermophotovoltaic Generation of Electricity,* Proc. of 3rd NREL Conf., Colorado Springs, 1997, pp369-372
4. Wilt, D.M., Fatemi, N.S., Hoffman, R.W., Jenkins, P.P., Brinker, D.J., Scheiman, D., Lowe, R., Fauer, M. and Jain, R.K., Appl.Phys.Lett. **64**, 2415-2418 (1994)
5. Aardvark, A., Allogho, A.A., Bougnot, G., David, J.P.R., Giani, A., Haywood, S.K., Hill, G., Klipstein, P.C., Mansoor, F., Mason, N.J., Nicholas, R.J., Pascal- Delannoy,F., Pate, M., Ponnampalam,L. and Walker, P.J., Semicon.. Sci. Technol., **89**, S380-S385 (1993).
6. Krier, A., Parry, M.K. and Lanchester, D.S., Semicon.Sci.Tech **6** 1066-71 (1991)
7. Pascal-Delannoy, F., Mason, N.J., Giani, A., Bougnot, J., Walker, P.J., Bougnot, G.and Allogho,G., J.Cryst.Growth **124**, 409-414 (1992)
8. Fraas,L., Girard,G., Avery,J.E., Gee, J.M., Arau,B.A., Sundaram,V.S. and.Thompson, A.G., J.Appl.Phys., **66**, 3866-3870, 1989.
9. Grey, R., Mansoor, F., Haywood, S.K., Mason, N.J., Nicholas, R.J. and Walker, P.J., Optical Materials, **6**, 69-74 (1996)
10. Onozawa,S., Ueda,T., and Akiyama,M., Journal Cryst. Growth, **93** 443-448 (1988)

11. Graham, R.M., Jones, A.C., Mason, N.J., Rushworth, S., Smith. L., and Walker, P.J., Journal Crystal Growth, **145,** 363-370 (1994)
12. Mallard, R.E., Wilshaw, P.R., Mason, N.J. Walker, P.J. and Booker, G.R., Microsc. Semicond.Mater., **100**, 33 (1989)

PHASE STABILITY IN $Ga_xIn_{(1-x)}As_ySb_{(1-y)}$ /GaSb HETEROSTRUCTURES

Y-C.Chen[1], V. Bucklen[1], M.Freeman[2], R.P.Cardines Jr[2],
G. Nichols[2], P. Sanders[2], G. Charache[2], and K. Rajan[1]

[1]Department of Materials Science and Engineering
Rensselaer Polytechnic Institute
Troy, NY 12180-3590
and
[2]Lockheed-Martin, Inc.
Schenectady, NY 12301-1072

Abstract. The crystallographic and microstructural characteristics of liquid phase epitaxy lattice-matched $In_xGa_{(1-x)}As_ySb_{(1-y)}$/GaSb (100) heterostructures is presented. Using both transmission electron microscopy and high resolution X-ray diffraction, a variety of diffusional based phase transformations in the epitaxial films are observed, including: spinodal decomposition, compositional modulations of the order of 30 nm, and weak long range ordering. These results are interpreted in terms of the possible influence of substrate surface structure on the phase stability of epitaxial layers.

INTRODUCTION

The application of III-V compound semiconductor materials in device structures is important for high-speed microelectronics and optoelectronics. These materials have allowed the device engineer to tailor material parameters such as the bandgap and carrier mobility to the need of the device by altering their composition. When using the ternary or quaternary materials, the device designer usually presumes that the compound is completely disordered, without any correlation between the atoms on the sublattices. However, the thermodynamics of the compound often produce material that has some degree of macroscopic or microscopic ordering. Control of such ordering can be used as an additional procedure to tune the optoelectronic properties to specific values for particular devices. Lack of control could result in devices with, for example, emission wavelengths significantly different from the designed values.

Composition modulation is a subset of phase separation (1) and it is often referred to as the spontaneous formation of a phase-separated, self-organized periodic structure. Composition modulation in epitaxial growth has been observed to occur either parallel (2-4) (vertical modulation), perpendicular (5,6) (lateral modulation), or both parallel and perpendicular (4,7) to the growth direction. Recently, there has been a great impetus to apply low dimensional structures to novel electronic and photonic devices (8). For example, lasers with quantum wire (QWR) active regions are predicted to have lower threshold currents, wider modulation bandwidths, and better temperature stability (8) than conventional two-dimensional quantum well devices. Deposition of short period superlattices (SPS) of some devices by molecular beam epitaxy (9,10) has been shown to result in lateral composition modulation, which has been exploited to obtain high densities of nanometer-sized QWR, without the processing limitations of other fabrication methods. Application of this technique to the InAlAs system (8), can lead to novel polarization-sensitive devices. In a laser structure, cladding layers consisting of AlAs/InAs SPS can produce QWR behavior in active layers of InGaAs due to composition-modulation induced strain fields (11).

Compound semiconductor materials have been widely applied to various electronic and optoelectronic devices. Since the microstructure of such materials is an important factor that influences the optoelectronic properies, it is quite important to examine the epilayers microscopically. This paper presents the evolution of weak ordering, composition modulation, microstructure and phase stability of the GaInAsSb system, grown by liquid phase epitaxy (LPE). A variety of experimental techniques are used including: high resolution X-ray diffraction (HRXRD), reciprocal space mapping (RSM), and transmission electron microscopy (TEM). It is expected that similar observations will also be present using other growth techniques such as metalorganic vapor phase epitaxy (MOVPE) or molecular beam epitaxy (MBE).

EXPERIMENTAL DETAILS

The growth parameter space used in this study involved independently examining the effects of epilayer chemistry and the role of substrate misorientation. A series of three different liquid compositions were grown by liquid phase epitaxy. The choice of InGaAsSb composition was dictated by the requirement of achieving a ~0.55 eV bandgap material lattice-matched to GaSb, while remaining in a single-phase region (i.e., outside the spinodal boundary, Figure 1). Hence, three compositions were studied which followed the iso-lattice parameter line associated with lattice matched heterostructures ,Figure 1, and were outside the equilibrium 600°C spinodal boundary (12). The samples were $Ga_{0.82}In_{0.18}As_{0.14}Sb_{0.86}$ (sample A), $Ga_{0.87}In_{0.13}As_{0.12}Sb_{0.88}$ (sample B), and $Ga_{0.92}In_{0.08}As_{0.07}Sb_{0.93}$ (sample C) grown on GaSb (001) substrates. In order to study the influence of substrate misorientation on phase stability, two additional samples with the same composition as A (close to the equilibrium spinodal bundary) were also studied: sample D (2° misoriented toward <110>) and sample E (2° misoriented toward <111$_B$>).

A conventional horizontal source-seed graphite sliding-boat in a Pd-diffused hydrogen atmosphere was used for the LPE process and the epilayers were grown on commercial Te-doped GaSb wafers. After preparation of the melts and substrate cleaning by a HF acid etch, the solution was remelted at 560 °C for 1 hour followed by a slow ramp to the liquidus temperature of 532 °C. The growth temperature for all these films was 530 °C. Details of the growth has been presented previously (14).

Plane-view (PV) and cross-sectional (CS) transmission electron microscopy examinations were carried out in a Philips CM12 operated at 120 kV. PV samples were prepared by first mechanically thinning them from substrate side to less than 100 µm. Next, samples were further thinned to electron transparency by ion milling using a 3 – 5 kV, 0.3 – 0.6 mA, Ar$^+$-ion beam incident at 25°, and lowered to 10° when close to finish. CS samples were prepared in the following steps. First, a diamond saw is used to cut two 2 × 4 mm cross-sectional pieces, in <011> directions, from each epi-layer sample. M-bond adhesive is used to attach the two pieces together, face to face (epi-layers). Next using M-bond adhesive glue, the two-pieces were glued onto a hollow copper grid (sometimes we glued one more piece of GaSb substrate to form a three-pieces sample disc for the reason of better supporting in the next polishing step). The subsequent thinning step is the same as plane-view sample preparation described above. HRXRD experiments were delivered by using BEDE D1 high resolution x-ray diffractometer. RSM was used to study the quality and microstructure of the sample by distinguishing between peak broadening resulting from lattice tilts and variations in lattice parameter.

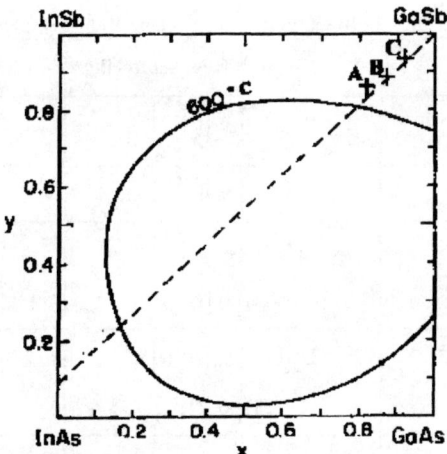

FIGURE 1. Phase diagram shows the relative positions of samples A, B and C. The solid curve is the calculated spinodal isotherm for $Ga_xIn_{1-x}As_ySb_{1-y}$ at temperature 600°C. Dashed line represents compositions for lattice-matched to GaSb (12).

EXPERIMENTAL RESULTS

Table 1 and Figures 2 and 3 summarize the structural changes observed by electron diffraction, associated with the varying compostions of the samples used. What is most significant in these observations is the presence of a {110} type ordering instead of the CuPt type ordering that is often reported. As discussed in our previous study (15), there have been some reports in the literature suggesting the possibility of such transformations but possible mechanisms as described in this paper have not been proposed. The electron diffraction patterns of samples D and E showed more evidence of diffuse streaking compared to those of non-misorientated samples. The diffuse scattering observed in the electron diffraction patterns of sample D and E might indicate the formation of short range ordering (16).

Figure 4(a) is a plane-view dark-field image, with (220) reflection, of sample C. It shows the weak modulation contrast close to the [010] direction. The periodicity of the modulation is about 2.6 nm. The weak modulation contrast is a common feature of plane-view samples B, C and D in the dark-field imaging study, and the results are summarized in Table 2.

Figure 4(b) is a bright-field image of cross-sectional sample E^{cs}, which has the same composition as sample A with a substrate 2° misorientated toward (111_B). It shows the significant effects of substrate misorientation on the phase stability and microstructure of the epi-layer. A fine mottled contrast with a quasi-period of 10-20 nm is distinctly observed along [110] direction, which was not observed on the non-misoriented samples.

TABLE 1. Summary of Electron Diffraction Results of Plane View Samples

Sample	Variants of diffuse scattering	Ordering type reflections
A^{PV}	$[110]$ & $[1\bar{1}0]$	Weak (110) type reflections
B^{PV}	$[110]$ & $[1\bar{1}0]$	None
C^{PV}	$[110]$ & $[1\bar{1}0]$ $[001]$ & $[010]$	Weak (110) reflections
D^{PV} (same composition as A; 2°→(110) on (100))	$[110]$ & $[1\bar{1}0]$ $[001]$ & $[010]$	Weak (110) reflections
E^{PV} (same composition as A; 2°→(111_B) on (100))	$[110]$ & $[1\bar{1}0]$ $[001]$ & $[010]$	Weak (110) reflections

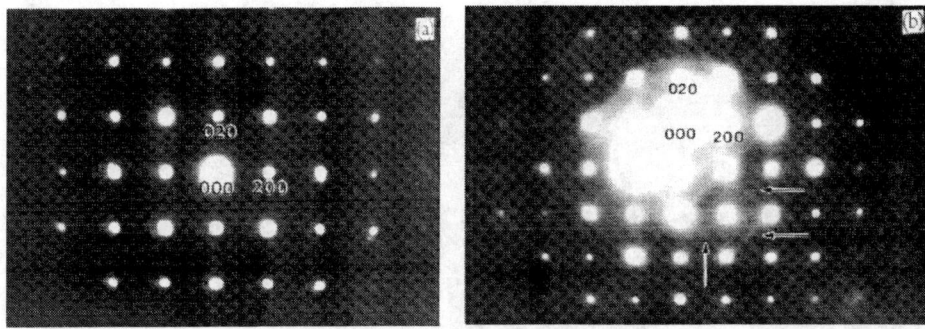

FIGURE 2. (a) Transmission electron microscopy (TEM) diffraction pattern of $Ga_{0.87}In_{0.13}As_{0.12}Sb_{0.88}$, sample B, showing very weak diffuse scattering in [110] and [1$\bar{1}$0] directions. b) TEM diffraction pattern of $Ga_{0.92}In_{0.08}As_{0.07}Sb_{0.93}$, sample C, showing strong diffuse cattering in [110] and [1$\bar{1}$0] directions and in [100] and [010] directions with weak reflection at (110) type positions.

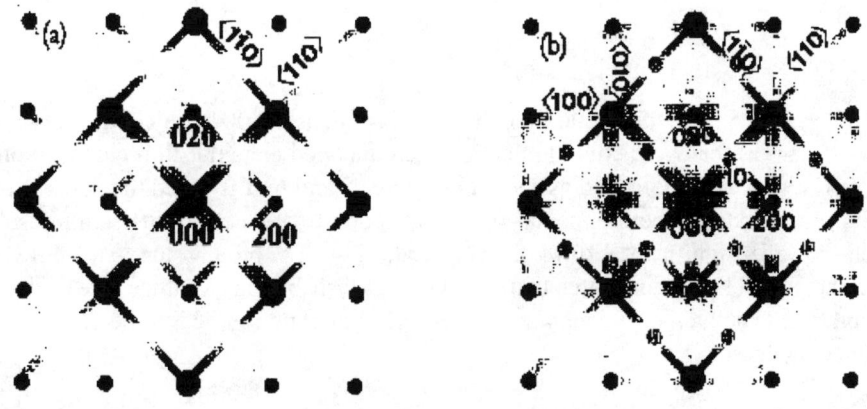

FIGURE 3. Schematic diagrams of the TEM diffraction patterns of (a) $Ga_{0.82}In_{0.18}As_{0.14}Sb_{0.86}$, sample A, and $Ga_{0.87}In_{0.13}As_{0.12}Sb_{0.88}$, sample B, showing weak diffuse scattering in [110] and [1$\bar{1}$0] directions. (b) $Ga_{0.92}In_{0.08}As_{0.07}Sb_{0.93}$, sample C, and $Ga_{0.82}In_{0.18}As_{0.14}Sb_{0.86}$ 2° misorientated toward (110), sample D, showing strong diffuse scattering in [110] and [1$\bar{1}$0] directions and in [100] and [010] directions and the evolution of weak (110) type reflection.

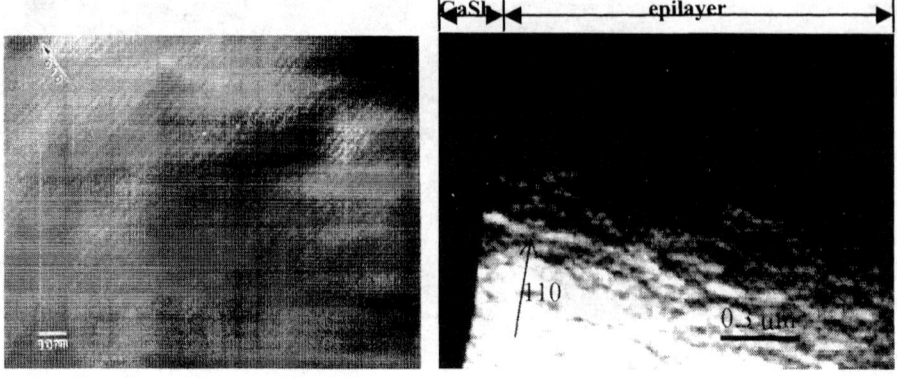

FIGURE 4. (a) Plane-view dark-field image, with (220) reflection, of sample C showing the weak modulation contrast near along the [010] direction. The periodicity of the modulation is about 2.6 nm. (b) Bright-field image of cross-sectional sample E^{cs}, same composition as sample A, substrate 2° misorientated toward (111_B) showing the significant effects of substrate misorientation on the stability and microstructure of the epi-layer. A fine mottled contrast with a quasi-period of 10-20 nm is distinctly observed along the [110] direction.

High resolution x-ray diffraction reciprocal space maps of (004) reflection and triple axis $\theta/2\theta$ curves for two distinct lattice matched undoped epitaxial film compositions, sample C and A were collected as shown in figures 5 and 6. It is noted that these x-ray results corroborate for both epitaxial films and support the electron diffraction results which showed a significant diffuse intensity scattering. The broad returns in the RSMs along the TH/2TH axis indicates that a significant distribution of lattice sizes exist in the epilayer. The lack of broadening in the AXIS2 direction indicates the distribution of lattice tilts is small.

TABLE 2. Summary of Dark-Field Imaging Results

Sample	Reflection plane	Direction of Modulation	Periodicity (nm)
B^{PV}	(220)	~[110]	3.7
C^{PV}	(220)	~[010]	2.6
D^{PV}	(220)	~[110]	3.4

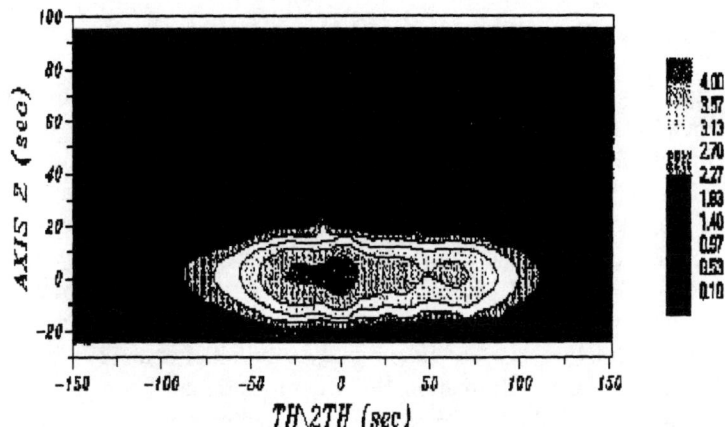

FIGURE 5. HRXRD reciprocal space map of sample C showing significant spread-out lattice parameter TH/2TH) which is consistent with the weak and diffuse scattering observed in the electron diffraction pattern.

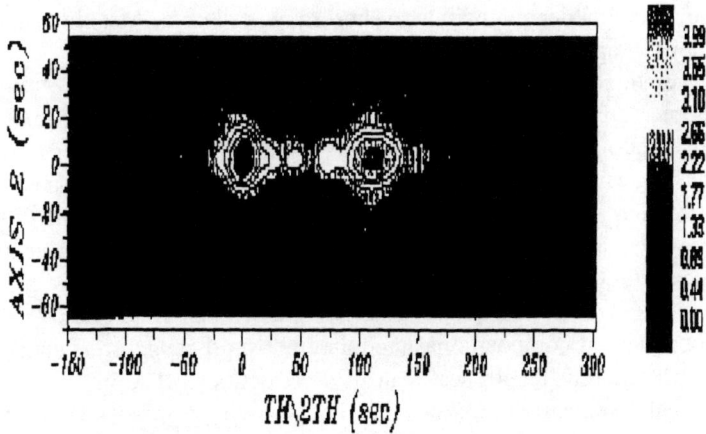

FIGURE 6. HRXRD reciprocal space map of sample A, showing separate peaks of GaSb substrate and epilayer and the indication of lattice parameter variations in the epilayer.

DISCUSSION

The change in crystallography of compositional modulation (Table 2) and microstructures (Fig. 4(b)), with surface tilt implies that surface structure has a significant effect on the phase stability of the epitaxial layers. The study of Henoc et al. (13) on $In_{1-x}Ga_x As_y P_{1-y}$ layers, grown on (001) InP substrates by use of LPE, indicates the presence of two types of contrast modulations in TEM: fine-scale speckle microstructure and coarser scale contrast modulations with wavelengths of 10 and 125 nm, respectively. Figure 4(b), showed the same type of fine scale mottled structure with wavelength, 10-20 nm, is also similar to their results. However there was no coarse-contrast modulation as found in our study. Ahrenkiel et al. (17) found that there was the "self-organized formation of compositionally modulated ZnSeTe superlattices" with a periodicity of 1.8 – 3.2 nm when grown on a >2° tilted substrate along <111> directions. The substrate misorientation condition they used is similar to our sample E, 2° toward (111_B), and the results are consistent to those observed in Figure 4(b). The relationship between this type of composition modulation and substrate misorientation is not well established yet. However, it is believed that it is closely related to the atomic steps associated with the tilted vicinal surface.

Diffuse scattering can arise from short range order in the specimen due to either microdomains of ordered nuclei or local regions of increased order analogous to spinodal decomposition (16). These regions produce diffuse intensity maximum at positions that will eventually correspond to a superlattice spot when the short range order has developed to long range order (18). Combined with our electron diffraction results and fine mottled structure found in TEM bright field image, figure 4 (b), the evolution of long range order may be preceded by different stages of phase stability ranging from spinodal decomposition to short range ordering. Also the tendency of the phase instability appears to increase when epilayer composition approaches the equilibrium spinodal curve.

The high resolution x-ray diffraction results confirm that the diffuse scattering intensity observed in the electron diffraction studies are real effects that are occurring over a much larger volume of material than originally indicated when sampled by the electron diffraction. The distribution of lattice sizes observed in the HRXRD curves is an indication of changes in the local composition (while the overall composition is lattice-matched to GaSb). This indicates that this composition is not randomly distributed. These compositional changes may be manifested in spinodal decomposition, lateral compositional modulations, short range order, and long range order covering different length scales in these structures. The implications of these changes in local composition indicates that anisotropy in properties would be expected and this has been seen in previous studies of ternary III-V compound semiconductors.

CONCLUSIONS

In this study we have shown the phase stability, microstructure and the crystallographic results for LPE grown $In_xGa_{1-x}As_ySb_{1-y}/GaSb$ epitaxial layers. Our conclusion in this study include:

1. Diffuse scattering was found in <100> and <110> directions. Different crystallographic variants in compositional modulation have been observed along with the development of weak (110) ordering.

2. Composition modulation with periodicity of 2.5 – 4.0 nm was found in [110] and [010] directions in the lattice-matched epilayers.

3. Substrate misorientation has significant effect on the growth condition of the vicinal growth surface. Composition modulation due to spinodal decomposition was found in epilayer grown on 2° misoriented toward (111_B) substrate.

4. HRXRD results are consistent with the TEM studies. The spreading of lattice parameter is consistent with the diffuse scattering intensity seen in electron diffraction patterns.

REFERENCES

1. Porter D. A. and Easterling K. E., *Phase Transformation in Metals and Alloys*, London, Chapman & Hall Press, 1992, pp308.
2. Ueda O., Isozumi S. and Komiya S., *Japanese Journal of Applied Physics* **23** L241 (1984)
3. Millunchick J. M., Twesten R. D., Lee S. R., Follstaedt D. M., and E. D. Jones; *Journal of Electronic Materials*, **26** 1048 (1997)
4. Chu S. N. G., Nakahara S., Strege K. E., and Johnston W. D., Jr; *Journal of Applied Physics* **57** 4610 (1985)
5. Mahajan S., Dutt B. V., Temkin H., Cava R. J. and Bonner W. A., *Journal of Crystal Growth* **68** 589 (1984)
6. McDevitt T. L., Mahajan S., and Laughlin D. E.; *Physical Review B* **45** 6614 (1992)
7. Hua G. C., Otsuka N., Grillo D. C., Han J., He L. and Gunshor R. L., *Journal of Crystal Growth* **138** 367 (1994)
8. Abraham P., Garcia Perez M. A., Benyattou T., Guilot G., Sacilotti M., and Letartre X., *Semicond. Sci. Technol.* **10** 1585 (1995)
9. Chou S. T., Cheng K. Y., Chou L. J., and Hseih K. C., *Journal of Applied Physics* **78** 6270 (1995)
10. Chen A. C., Moy A. M., Chou L. J., Hsieh K. C., and Cheng K. Y.; *Applied Physics Letters* **66** 2694 (1995)
11. Peiro F., Cornet A., Ferrer J. C., Merante J. Halkias R., G., and Georgakilas A.; *Materials Research Society Symposium Proceedings*, edited by E. D. Jones, A. Mascarenhas, and P. Petroff (MRS, Pittsburgh, 1996) p265
12. Onabe K., *Japanese Journal of Applied Physics* **21** 964 (1982)
13. Henoc P., . Izrael A, Quillec M. and Launois H., *Appl.Phys Lett.* **40** 963 (1982)
14. DeWinter J. C., Pollack M. A., Srivastava A. K., and Zyskind J. L., *Journal of Electronic Materials*, **14** 729 (1985)
15. Chen Y-C., Bucklen V., Freeman M., Cardines Jr R. P., and . Rajan K; 40th Electronic Materials Conference (1998)

16. Williams D. B. and Carter C. B., *Transmission Electron Microscopy II*, Plenum Press, New York, p259 (1996)
17. Ahrenkiel S. P., Xin S. H., Reimer P. M., Berry J. J., Luo H., Short S., Bode M., Al-Jassim M., Buschert J. R., and . Furdyna J. K, *Physical Review Letters* **75** 1586 (1995)
18. Cowley J. M., *Acta Crystallographica* **A29** 529 (1973)

$Al_xGa_{1-x}Sb$ Window Layers for InGaAsSb / GaSb Thermophotovoltaic Cells

Joseph T. South[1,†], Zane A. Shellenbarger[2], Michael G. Mauk[1,2,‡],
Jeffrey A. Cox[2], Paul E. Sims[2], Robert A. Mueller[3], and John D. Meakin[1]

[1] Materials Science and Engineering
University of Delaware
Newark, Delaware 19716-2000

[2] AstroPower, Inc.
Solar Park
Newark, Delaware 19716-2000

[3] Jet Propulsion Laboratory
Pasadena, California

Abstract

We report results of a comparative study of AlSb-based window layers for InGaAsSb/GaSb TPV cells made by liquid-phase epitaxy (LPE). Previous work has shown that an AlGaAsSb window layer significantly improves the performance of InGaAsSb/GaSb TPV cells. As expected, the window layer enhances short-wavelength spectral response and increases the open-circuit voltage by reducing the reverse-saturation current of the diode. We present results for a simpler alternative window layer based on the ternary AlGaSb alloy. We fabricated, characterized, and compared InGaAsAsSb TPV of three types: 1. with an AlGaAsSb window layer, 2. with an AlAsSb window layer, and 3. with no window layer. Both p-on-n (p-type InGaAsSb emitter; n-type InGaAsSb base) and n-on-p (n-type InGaAsSb emitter; p-type InGaAsSb base) homojunction cell configurations were investigated. The InGaAsSb layers have a ~0.55-ev bandgap and are lattice-matched to a GaSb substrate. As anticipated, both AlGaSb and AlGaAsSb passivated TPV cells were superior to cells without a window layer. The AlGaAsSb / InGaAsSb window / emitter interface is closely lattice matched and would be expected to provide better surface passivation than AlGaSb / InGaAsSb window /emitter interface which has a 0.6% lattice mismatch. On the contrary, our experimental results showed that, based on a comparison of spectral response, AlGaSb window layers were somewhat more effective than AlGaAsSb in passivating the front surface of InGaAsSb diodes. These results demonstrate the viability of a simpler ternary AlGaSb alloy as an alternative to the quaternary AlGaAsSb alloy as a window layer material.

†present address: Materials Science Dept., Virginia Polytechnic Inst. and State Univ., Blacksburg, VA
‡author for correspondence: mauk@astropower.com   tel: 302-366-0400 fax: 302-368-6474

1. INTRODUCTION

Thermophotovoltaic cells based on the InGaAsSb quaternary alloy have demonstrated good performance in the 2300-nm spectral range [1-5]. Both *epitaxial* [1-3] and *diffused* junction [4,5] TPV devices achieve high open-circuit voltages (> 250 mV), excellent long wavelength spectral responses (~0.8 A/W), and good fill factors (60 to 70%). An improved InGaAsSb TPV cell design incorporates a lattice-matched, epitaxial wide-bandgap AlGaAsSb "window" layer over the emitter —primarily to reduce the front surface minority carrier recombination velocity. A reduced front surface recombination velocity yields a better short wavelength response and lower reverse-saturation currents, which in turn translates to higher open-circuit voltages. A secondary benefit of the window layer is a potential reduction in emitter sheet resistance for an increased fill factor. These design approaches are well established for GaAs-based solar cells and LEDs. The utility of a passivating, lattice-matched AlGaAsSb window layer has been demonstrated by WANG et al. [2] for MOCVD-grown TPV cells and SHELLENBARGER et al. [1] for LPE-grown TPV cells. Additionally, the inclusion of a wide bandgap (e.g., AlGaAsSb) back-surface field (BSF) cladding layer between the InGaAsSb base layer and GaSb substrate may also be beneficial, but is not expected to be as significant a factor as the front surface window layer in increasing cell conversion efficiency. The full optimization and potential of an AlGaAsSb /InGaAsSb/AlGaAsSb double heterostructure TPV cell has yet to be realized.

In the interests of various commercial applications, there is much incentive for simplifying the InGaAsSb TPV structure, and for making the epitaxial growth and/or junction formation processes more amenable to low-cost, high-throughput, large-area device production. In this work, we show that a *ternary* AlGaSb window layer provides comparable, if not more effective, front surface passivation than an AlGaAsSb *quaternary* window layer. The $Al_xGa_{1-x}Sb$ alloy has a direct-to-indirect bandgap transition at $x = 0.20$. At $x = 0.5$, the indirect bandgap is 1.3 eV. The room-temperature lattice constants of the two component binaries are $a_{GaSb} = 6.09593$ Å and $a_{AlSb} = 6.1355$ Å. In general ($x > 0$), the lattice constant of the $Al_xGa_{1-x}Sb$ ternary can *not* be exactly lattice-matched to an InGaAsSb epilayer which is lattice-matched to a GaSb substrate. For $x = 0.5$, there is a 0.6 % lattice mismatch between AlGaSb and InGaAsSb lattice matched to GaSb. Accurate data for the degree of lattice mismatch at the growth temperature is not known: the lattice mismatch may be more or less severe at the growth temperature. On this basis alone, it might be expected that the AlGaSb ternary is inferior to the near-exactly lattice-matched AlGaAsSb quaternary window layer since a lattice-mismatched interface has a high density defects that act as minority carrier recombination centers. For the (100) orientation, the minority carrier recombination velocity S dependence on misfit at the window / emitter interface can be roughtly estimated from [6,7]

$$S \propto v_{th} \cdot \frac{\left|(a_W)^2 - (a_E)^2\right|}{(a_W \cdot a_E)^2}$$

where a_w and a_E are the lattice constants of window and emitter layer, and v_{th} is the thermal velocity of minority carriers.

In liquid-phase epitaxy of TPV cells, the use of an AlGaSb ternary window layer instead of an AlGaAsSb quaternary window layer has several distinct advantages. The growth solution for the AlGaAsSb is relatively dilute in arsenic (melt atomic fraction ~0.01%), and depletion of arsenic or its variability due to errors in formulating the melt is problematic. This feature further complicates processes envisioned for growing large-area devices or many layers from the same melt in a semi-continuous batch mode of operation. The ternary avoids problems associated with arsenic depletion. The ternary can also be grown over a wider range of temperatures and is less sensitive to variations or fluctuations in growth temperature and/or melt composition. It is thus easier to match the initial growth temperature of the ternary AlGaSb window with the terminal growth temperature of the InGaAsSb emitter, than is the case with an AlGaAsSb window layer on InGaAsSb. For the AlGaAsSb quaternaries, a ±1 °C change in growth temperature can significantly alter the results with respect to layer composition and uniformity. With regard to process simplification, it should also be noted that a junction diffusion step can be combined with epitaxial growth of the window layer to avoid the need for separately growing an InGaAsSb emitter layer.

For purposes of comparison, both $Al_{1-x}Ga_xAs_{1-y}Sb_y$ and $Al_xGa_{1-x}Sb$ window layers were grown over InGaAsSb epitaxial p-on-n and n-on-p homojunctions. A description of the liquid-phase epitaxial growth processes is given below. We contrast the spectral response of otherwise similar devices with different window layers.

2. LIQUID-PHASE EPITAXY PROCESS AND DEVICE FABRICATION

Liquid-phase epitaxy of the III-V antimonides is well established, and detailed descriptions of the relevant phase equilibria and growth techniques are available in the literature [8-21]. In this work, we use a horizontal slideboat technique. The substrates are chemically polished, 500-μm thick, (100) oriented, n-type or p-type GaSb wafers. The n-type substrate were Te doped to a carrier concentration of $0.7 - 3 \times 10^{17}$ cm^{-3} with an average electron mobility of 3000 cm^2/V·sec at 300 K. The p-type substrates were doped with zinc to a concentration of 10^{18} cm^{-3}. The wafers were diced into 1.6-cm x 1.6-cm substrates. Immediately prior to the LPE step, the substrate was cleaned by five minute immersions in hydrofluoric acid, methanol and isopropanol. Indium, gallium and antimony were added as high purity (99.9999%) shot pieces and aluminum as segments of wire. The metal components of the melt were baked out at 700 °C for fourteen hours prior to growth under a flow of palladium-diffused hydrogen. The arsenic was added (after bake-out) as an undoped InAs polycrystalline "float" wafer of size 0.50 cm x 0.50 cm. The total melt weight was approximately 7 g. This was performed in order to outgas any residual impurities and to de-oxidize the melt constituents. TABLE 1 summarizes the various melt compositions and growth temperatures used here for LPE of InGaAsSb, AlGaAsSb, and AlGaSb.

TABLE 1
Summary for LPE of InGaAsSb, AlGaAsSb, and AlGaSb

	Solid Composition	Liquid Composition					Growth Temp (°C)	Bandgap (eV)
		X^L_{Al}	X^L_{Ga}	X^L_{In}	X^L_{As}	X^L_{Sb}		
A	$In_{0.15}Ga_{0.85}As_{0.17}Sb_{0.83}$	-	0.19	0.59	0.01	0.21	519	0.53
B	"	-	0.169	0.5849	0.001	0.245	518	0.53
C	"	-	0.1252	0.1534	0.001	0.7204	613	0.59
D	$Al_{0.28}Ga_{0.72}As_{0.015}Sb_{0.985}$	0.015	0.957	-	0.001	0.028	515	1.1
E	$Al_{0.67}Ga_{0.33}Sb$	0.053	0.924	-	-	0.023	515	1.3
F	"	0.053	0.924	-	-	0.023	545	1.0

The solid alloy compositions in TABLE 1 were determined by Energy Dispersive Spectroscopy (EDS) or Electron Microprobe Analysis. Compositions A and B were grown from In-rich melts while composition **C** was grown from an Sb-rich melt. The use of an Sb-rich solution has been shown to improve material quality as well as reducing the background carrier concentration [16]. The ramp cooling rate for all layers was 1 °C /min. The growth times for the InGaAsSb base, InGaAsSb emitter, and window layer were 2 minutes, 15 seconds, and 15 seconds, respectively. This resulted in a base thickness of 4 to 6 microns, an emitter thickness of ~0.8 microns, and a window layer thickness of ~0.5 microns.

Both 1-cm x 1-cm TPV cells (FIGURE **1a**) and test diodes structures with 200-μm diameter active areas (FIGURE **1b**) were fabricated. The metallization for the n-type contacts was Au/Sn, while for the p-type contacts was Au/Zn.

 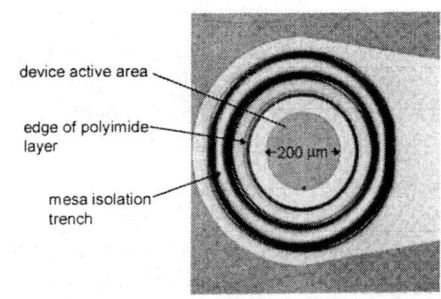

FIGURE 1A: Top-view of 1x 1-cm InGaAsSb-based TPV cell.

FIGURE 1B: Top-view of diode test structure.

3. EVALUATION OF WINDOW LAYERS

The material quality of the AlGaSb layers was evaluated by growing two-layer *p-n* homojunctions on a GaSb substrate. The AlGaSb layers were grown according to specifications of **E** (sample J165-02-A 04) and **F** (sample J165-02-A 09) in TABLE 1. FIGURE 2 shows the normalized spectral response for two *p*-on-*n* AlGaSb diodes.

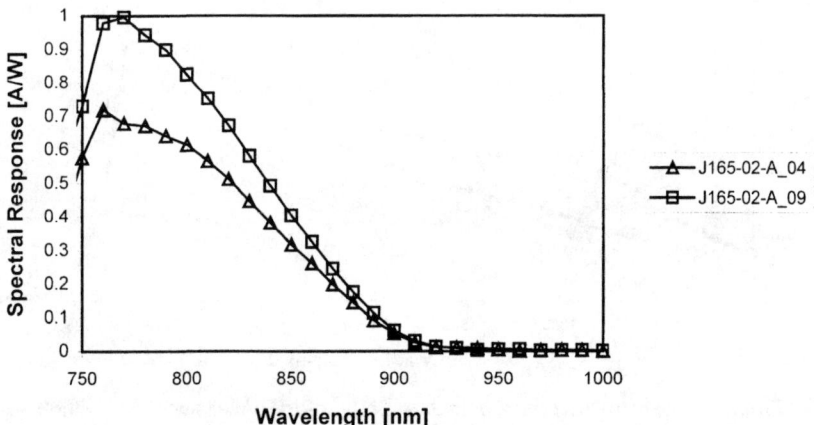

FIGURE 2: Normalized spectral response of AlGaSb *p-n* homojunction diode test structures.

FIGURE 3 compares spectral response for *p*-on-*n* (*p*-type InGaAsSb emitter; *n*-type InGaAsSb base; *n*-type GaSb substrate) InGaAsSb TPV cells of three types: 1. with no window layer; 2. with an AlGaAsSb window layer; and 3. with an AlGaSb window layer. The InGaAsSb emitter and base layers were grown according to the specfications of **A** in TABLE 1. The AlGaAsSb and AlGaSb window layers are grown according to specifications of **D** and **E**, respectively. As anticipated, the spectral response of TPV cells with a wide-bandgap passivating window layer is improved over InGaAsSb TPV cells with no window layer, especially at shorter wavelengths. However, the passivation effected by the AlGaSb window layer was actually somewhat better than the AlGaAsSb window layer–at least for the specific cases investigated.

The spectral response for an *n*-on-*p n* (*n*-type InGaAsSb emitter; *p*-type InGaAsSb base; *p*-type GaSb substrate) InGaAsSb cell with an AlGaSb window is shown in FIGURE 4. Results were similar to *p*-on-*n* devices except that the spectral response was somewhat lower. This is partly attributable to the fact that relatively less effort has been devoted to the optimization of layer thicknesses and doping levels in the *n*-on-*p* cells as compared to *p*-on-*n* devices

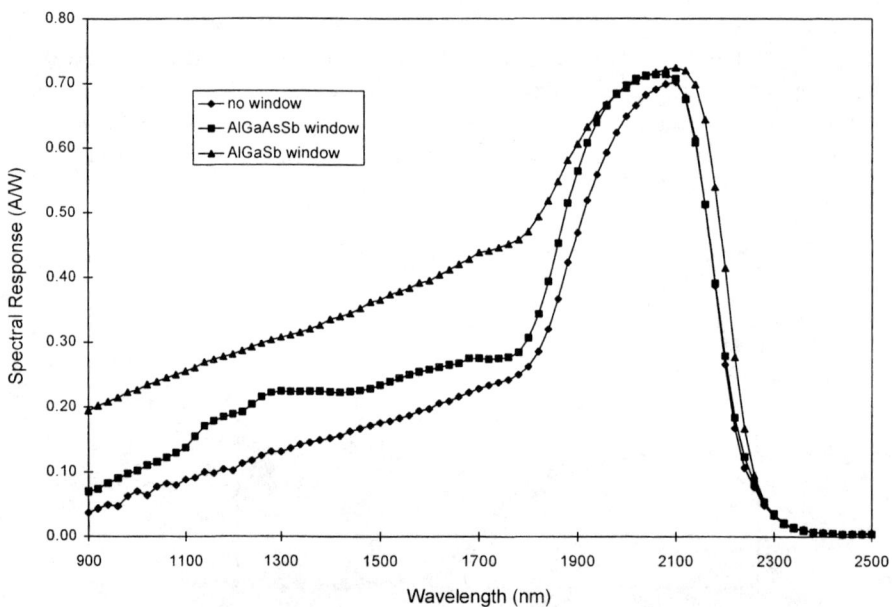

FIGURE 3: Spectral response of InGaAsSb *p*-on-*n* TPV cells with and without AlGaSb or AlGaAsSb window layers.

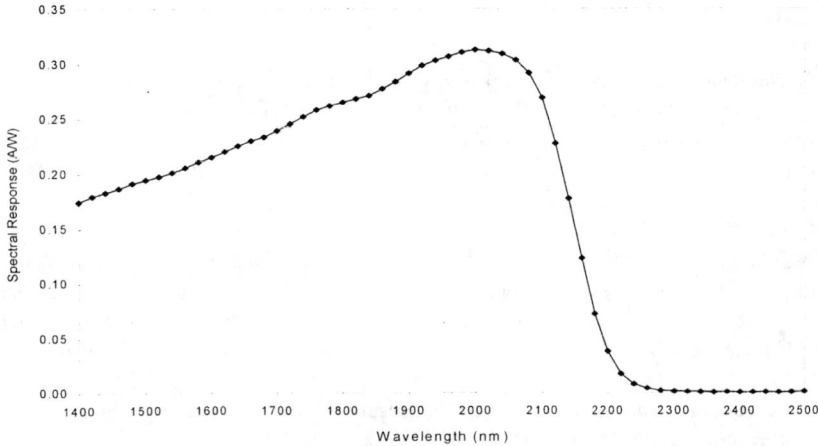

Figure 4: Spectral response of n-on-p InGaAsSb cell with AlGaSb window.

The spectral response for a p-on-n cell with InGaAsSb emitter and base layers grown according to specification **B** (Sb-rich melts) in TABLE 1 is given in FIGURE 5. This device included an AlGaAsSb window layer (as per specification **D** in Table 1). Results were similar to devices using the previous melt compositions.

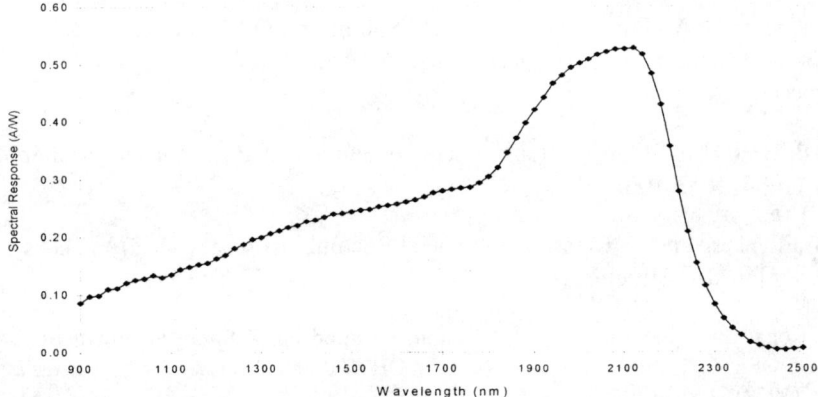

FIGURE 5. Spectral response of InGaAsSb cell using melt composition **B** in TABLE 1.

In conclusion, these results indicate that with regard to front surface passivation, a ternary AlGaSb window layer can perform at least as well as quaternary AlGaAsSb window layers previously used for InGaAsSb-based TPV cells.

Acknowledgment. This work was performed as part of a Department of Energy STTR Phase I Program, Award DE FG02-97ER86057.

References

1. Z.A. Shellenbarger, M.G. Mauk, J.A. Cox, M.I. Gottfried, P.E. Sims, J.D. Lesko, J.B. McNeely, L.C. DiNetta, and R.L. Mueller, "Improvements in GaSb-Based Thermophotovoltaic Cells," *Thermophotovoltaic Generation of Electricity: Third NREL Conference* T.J. Coutts, C.S. Allman, and J.P. Benner, eds. AIP Conf. Proc. **401**, (1997) 117-128.

2. C.A. Wang, H.K. Choi, G.W. Turner, D.L. Spears, and M.J. Manfra, "Lattice-Matched Epitaxial GaInAsSb/GaSb Thermophotovoltaic Devices" *Thermophotovoltaic Generation of Electricity: Third NREL Conference* T.J. Coutts, C.S. Allman, and J.P. Benner, eds. AIP Conf. Proc. **401**, (1997) 75-83.

3. C. Hitchcock, R. Gutmann, J. Borrego, H. Ehsani, I. Bhat, M. Freeman, and G. Charache, "GaInSb and GaInAsSb Thermophotovoltaic Device Fabrication and Characterization" *Thermophotovoltaic Generation of Electricity : Third NREL Conference* T.J. Coutts, C.S. Allman, and J.P. Benner, eds. AIP Conf. Proc. **401**, (1997) 89-103.

4. V.S. Sunaram, S.B. Saban, M.D. Morgan, W.E. Horne, B.D. Evans, J.R. Ketterl, M.B.Z. Morosini, N.B. Patel, and H. Field, "GaSb Based Ternary and Quaternary Diffused Junction Devices for TPV Applications" *Thermophotovoltaic Generation of Electricity: Third NREL Conference* T.J. Coutts, C.S. Allman, and J.P. Benner, eds. AIP Conf. Proc. **401**, (1997)

5. A.W. Bett, B.Y. Ber, M.G. Mauk, J.T. South, and O.V. Sulima, "Pseudo-Closed Box Diffusion of Zn into InGaAsSb and AlGaSb for TPV Devices" (1998) in this proceedings

6. D.B. Holt, "Misfit Dislocations in Semiconductors" *J. Phys. Chemistry of Solids* **27** (1966) 1053-1067.

7. M. Ettenberg and H. Kressel, "Interfacial Recombination at (AlGa)As/GaAs Heterojunction Structures" *J. Applied Physics* **47, 4** (1976) 1538-1544.

8. N. Kobayashi, Y. Horikoshi, and C. Uemura, "Liquid-Phase-Epitaxial Growth of InGaAsSb/GaSb and InGaAsSb/AlGaAsSb DH Wafers," *Japanese J. of Applied Physics* **18** (1979) 2169-2170.

9. M. Astles, H. Hill, A.J. Williams, P.J. Wright, and M.L. Young, "Studies of the $Ga_{1-x}In_xAs_{1-y}Sb_y$ Quaternary Alloy System I. Liquid-Phase Epitaxial Growth and Assessment," *J. Electronic Materials*, **15**, 1 (1986) 41-49.

10. A.S. Jordan and M. Ilegems, "Solid-Liquid Equilibria for Quaternary Solid Solutions Involving Compound Semiconductors in the Regular Solution Approximation," *J. Physical Chemistry of Solids* **36**, (1975) 329.

11. G. Xiuying, Y. Bauhua, M. Yindi, G. Fensheng, Y. Ying, H. Wenjian, L. Xuefeng, X. Jinying, W. Zuahguo, L. and Lanying, "Liquid Phase Epitaxy Growth and Properties of GaInAsSb/AlGaAsSb/GaSb Heterostructures," *Japanese J. of Applied Physics* **30**, 7 (1991) 1343-1347.

12. T.I. Voronina, T.S. Lagunova, M.P. Mikhalova, K.D. Moiseev, Yu.P. Yakovlev, "High Carrier Mobility in *p*-type GaInAsSb/p-InAs Heterostructures," *Semiconductors* **60** (1996) 523-526.

13. J.C. DeWinter, M.A. Pollack, A.K. Srivastava, and J.L. Zyskind, "Liquid Phase Epitaxial the $Ga_{1-x}In_xAs_{1-y}Sb_y$ Lattice-Matched To (100) GaSb Over the 1.71 to 2.33 µm Wavelength Range," *J. Electronic Materials*, **14, 6** (1985) 729-747.

14. Joullié, F. Jia Hua, F. Karouta, H. Mani and C. Alibert, "III-V Alloys Based on GaSb for Optical Communications at 2.0 – 4.5 µm" *Optical Fibers Sources and Detectors* SPIE Proc. **587** (1985) 46-57.

15. E. Tournié, J.-L. Lazzari, H. Mani, F. Pitard, C. Alibert, and A. Joullié, "Growth by Liquid Phase Epitaxy and Characterization of GaInAsSb and InAsSbP Alloys for Mid-Infrared Applications (2-3 µm)" *Physical Concepts of Materials for Novel Optoelectronic Device Applications* SPIE Proc. **1361** (1990) 641-656.

16. J.-M. Wang, Y-M.Sun, and M.-C. Wu, "High-Quality $Ga_{1-x}In_xAs_{1-y}Sb_y$ Quaternary Lasers Grown from Antimonide-Rich Solutions by Liquid-Phase Epitaxy" *J. Crystal Growth* **172** (1997) 514-520.

17. J. Bhan, A. Joullié, H. Mani, A.M. Joullié and C. Alibert, "III-V Heterostructures for Laser Emission in the 2.55 µm Wavelength Region," *Materials and Technologies for Optical Communications* SPIE Proc. **866** (1987) 126-134.

18. A.N. Baranov, S.G. Konnikov, T.B. Popova, V.E. Umansky, and Yu.P. Yakovlev, "Liquid Phase Epitaxy of $Ga_{1-x}Al_xSb_{1-y}As_y$ /GaSb and The Effect of Strain on Phase Equilibria," *J. Crystal Growth* **66** (1984) 547-552.

19. J.R. Pessetto and G.B. Stingfellow, "$Al_xGa_{1-x}As_ySb_{1-y}$ Phase Diagram," *J. Crystal Growth* **62**, (1983) pp. 1-6.

20. K.Y. Cheng and G.L. Pearson, "The Al-Ga-Sb Ternary Phase Diagram and Its Application to Liquid Phase Epitaxial Growth," *J. Electrochemical Society* **124**, No. 5, (1977) 753-757.

21. K. Osamura, K. Nakajima, and Y. Murakami Y. "Experiments and Calculation of the Al-Ga-Sb Ternary Phase Diagram," *J. Electrochemical Society* **126, 11** (1979) 1992-1997.

22. A. Joullié and P. Gautier, "The Al-Ga-Sb Ternary Phase Diagram and Its Application To Solution Growth," *J. Crystal Growth*, **47** (1979) 100-108.

AUTHOR INDEX

A

Ahrenkiel, R. K., 282
Al-Jassim, M. M., 427
Andreev, V. M., 384
Asahi, R., 457
Avery, J., 480

B

Ballantyne, R., 480
Ballard, I., 83
Ballinger, C. T., 161, 335
Barnham, K. W. J., 83
Becker, F. E., 351, 394
Benner, J. P., 4
Ber, B. Y., 237
Bett, A. W., 237
Blazeck, T. S., 74
Boerner, V., 191
Borrego, J. M., 289
Brandhorst, Jr., H. W., 438
Brinker, D., 121
Broman, L., 115
Bucklen, V., 535
Button, C., 83

C

Carapella, J. J., 132, 282, 517
Cardines, Jr., R. P., 535
Charache, G. W., 10, 39, 161, 256, 289, 317, 327, 417, 457, 535
Chen, K. C., 472
Chen, Y-C., 535
Chen, Z., 438
Choi, H. K., 256
Chubb, D. L., 214, 463
Clark, J., 83
Clevenger, M. B., 142, 327
Cody, G. D., 58
Colter, P., 417
Connelly, W. R., 446, 488, 497
Connolly, J. C., 247
Connolly, J. P., 83

Coutts, T. J., 3
Cox, J. A., 545
Crowley, C. J., 197
Custard, P., 312

D

Daniels, B., 480
Danielson, L. R., 317
DeBellis, C. L., 364
Depoy, D. M., 317, 327, 335, 417
Doyle, E. F., 351, 394
Duda, A., 132, 269, 517
Durisch, W., 403
Dutta, P. S., 227, 289

E

Eldredge, J., 427
El-Husseini, A. M., 68, 93
Elkouh, N. A., 197
Ellingson, R., 282
Emery, K., 132, 301

F

Fatemi, N. S., 121, 152
Fraas, L. M., 312, 351, 364, 371, 480
Freeman, A. J., 457
Freeman, M., 535
Freeouf, J. L., 39

G

Garbuzov, D. Z., 247
Gedvilas, L. M., 132, 269
Geller, C. B., 74, 457
Gethers, C. K., 335
Gombert, A., 191
Good, B. S., 214
Gray, J. L., 68, 93
Gregory, S., 480
Griffin, P. R., 83

Grob, B., 403
Groeneveld, M., 312
Guazzoni, G., 30, 177
Gutmann, R. J., 227, 289

H

Haddad, M., 427
Haywood, S. K., 525
Heinzel, A., 103, 191
Hoffman, Jr., R. W., 121
Holden, T., 39, 457
Huang, H.-X., 371
Hui, S., 480

J

Jarefors, K., 115
Jenkins, P. P., 121
Johnston, S., 282
Jones, K. M., 269
Joslin, D., 427

K

Karam, N. H., 427
Keyes, J., 480
Khalfin, V. B., 247
Khan, O. S., 121
Khvostikov, V. P., 384
Krommenhoek, D., 14
Kruger, J. S., 30
Krut, D., 427
Kushch, A. S., 505

L

Lakrimi, M., 525
Lamson, D., 480
Lee, H., 247
Lindberg, E., 115
Luther, J., 12, 103, 191

M

Maanstadt, W., 457
Magari, P. J., 197
Magendanz, G., 312
Mannstadt, W., 74
Marks, J., 115
Martinelli, R. U., 247
Mason, N. J., 525
Mauk, M. G., 237, 545
Mayor, J.-C., 403
Mazer, J. A., 9
Meakin, J. D., 545
Moriarty, T., 132, 301
Morris, N., 247
Morrison, O., 488
Mueller, R. A., 545
Murray, C. S., 121, 142, 152, 161, 327

N

Narayanan, A., 427
Nawrocki, S. J., 30
Nelson, J., 83
Nichols, G. J., 317, 535
Nishikawa, W., 427

O

Oakley, D. C., 256
Olson, M. R., 269
Ostrogorsky, A. G., 227, 289

P

Pal, A. M., 214
Panitz, J.-C., 403
Parrington, J. R., 317
Pierce, D. E., 177
Pollak, F. H., 39, 457

Q

Qin, L., 142, 327

R

Rajan, K., 535
Raynolds, J. E., 39, 49, 457
Riley, D. R., 6
Ringel, S. A., 142, 327
Rohr, C., 83
Rosselet, A., 403
Rumyantsev, V. C., 384

S

Sacks, R. N., 142, 327
Samaras, J., 364, 371
Sanders, P., 535
Saroop, S., 289
Scheiman, D., 121
Scoles, S. W., 364
Scott, C. G., 525
Scotto, M. V., 364
Seal, M., 371, 488
Shellenbarger, Z. A., 545
Shukla, K. C., 351, 394
Sims, P. E., 545
Skinner, S. M., 505
Sorokina, S. V., 384
South, J. T., 237, 545
Stan, M. A., 121
Stollwerck, G., 103
Sulima, O. V., 237
Sweileh, G. M., 525

T

Takahashi, M., 427
Taylor, G. C., 247

V

Vasil'ev, V. I., 384

W

Walker, P. J., 525
Wang, C. A., 256
Wanlass, M. W., 132, 269, 282, 517
Ward, J. S., 132, 517
Watson, R. C., 364
Webb, J. D., 132, 269, 282, 517
Weizer, V. G., 121, 152
West, E. M., 371, 446, 488, 497
Wilt, D. M., 121, 152
Wittwer, V., 191
Wojtczuk, S., 417
Wolf, W., 74
Wu, X., 132, 269, 517

Y

Yamaguchi, H., 17
Yamaguchi, M., 17
Ye, S.-Z., 480

Z

Zenker, M., 103
Zheng, L., 525